# Methods in Enzymology

### Volume 279
### VITAMINS AND COENZYMES
### Part I

# METHODS IN ENZYMOLOGY

### EDITORS-IN-CHIEF

## John N. Abelson    Melvin I. Simon

DIVISION OF BIOLOGY
CALIFORNIA INSTITUTE OF TECHNOLOGY
PASADENA, CALIFORNIA

### FOUNDING EDITORS

## Sidney P. Colowick and Nathan O. Kaplan

*Methods in Enzymology*

*Volume 279*

# Vitamins and Coenzymes

*Part I*

EDITED BY

*Donald B. McCormick*

DEPARTMENT OF BIOCHEMISTRY
EMORY UNIVERSITY
ATLANTA, GEORGIA

*John W. Suttie*

DEPARTMENT OF BIOCHEMISTRY
UNIVERSITY OF WISCONSIN
MADISON, WISCONSIN

*Conrad Wagner*

DEPARTMENT OF BIOCHEMISTRY
VANDERBILT UNIVERSITY AND VA MEDICAL CENTER
NASHVILLE, TENNESSEE

ACADEMIC PRESS

San Diego   London   Boston   New York   Sydney   Tokyo   Toronto

Academic Press
15 East 26th Street, 15th Floor, New York, New York 10010
http://www.apnet.com

Academic Press Limited
24-28 Oval Road, London NW1 7DX, UK
http://www.hbuk.co.uk/ap/

International Standard Book Number: 0-12-182180-3

PRINTED IN THE UNITED STATES OF AMERICA
97  98  99  00  01  02  EB  9  8  7  6  5  4  3  2  1

# Table of Contents

## Section I. Ascorbic Acid

## Section II. Thiamin: Phosphates and Analogs

v

## Section III. Lipoic (Thioctic) Acid and Derivatives

# Section IV. Pantothenic Acid, Coenzyme A, and Derivatives

# Section V. Biotin and Derivatives

# Contributors to Volume 279

Article numbers are in parentheses following the names of contributors.
Affiliations listed are current.

ANDREAS ABEND (25), *Department of Biochemistry, Richard-Willstètter-Allee, University of Karlsruhe, 76128 Karlsruhe, Germany*

ELENA AINBINDER (47), *Department of Biological Regulation, The Weizmann Institute of Science, Rehovot, 76100, Israel*

CLAUDE ALBAN (35), *Unité Mixte CNRS/ Rhône-Poulenc (UM 41), Rhône-Poulenc Agrochimie, 69263 Lyon Cedex 09, France*

RONDA M. ALLEN (37), *Central Research and Development, Experimental Station, E. I. du Pont de Nemours and Co., Wilmington, Delaware 19880*

LEONIDAS G. BACHAS (29), *Department of Chemistry, University of Kentucky, Lexington, Kentucky 40506*

HASAN BAGCI (47), *Middle East Technical University, Faculty of Arts and Sciences, Inönü Bulvari, 06531, Ankara, Turkey*

PIERRE BALDET (35), *Unité Mixte CNRS/ Rhône-Poulenc (UM 41), Rhône-Poulenc Agrochimie, 69263 Lyon Cedex 09, France*

RUMA BANERJEE (24), *Department of Biochemistry, University of Nebraska, Lincoln, Nebraska 68588*

KIYOSHI BANNO (23), *Analytical Chemistry Research Laboratory, Tanabe Seiyaku Co., Ltd., Osaka 532, Japan*

GEOFF BARNARD (47), *Department of Chemical Pathology, Regional Endocrine Unit, Southampton General Hospital, Southampton SO16 6YD, United Kingdom*

KIM BARTLETT (26), *Department of Child Health, Sir James Spence Institute of Child Health, University of Newcastle, Royal Victoria Infirmary, Newcastle-upon-Tyne NE4 4LP, United Kingdom*

EDWARD A. BAYER (47), *Department of Membranes and Biophysics, The Weizmann Institute of Science, Rehovot, 76100, Israel*

DOROTHY BECKETT (39, 43), *Department of Chemistry and Biochemistry, University of Maryland Baltimore County, Baltimore, Maryland 21228*

HELGE BELL (8), *Department of Medicine, Aker University Hospital, 0514 Oslo, Norway*

THOMAS BØHMER (8), *Department of Medicine, Aker University Hospital, 0514 Oslo, Norway*

J. W. I. BRUNNEKREEFT (9), *Department of Clinical Chemistry, Rode Kruis Hospital, 2566 MJ Den Haag, The Netherlands*

B. TSE SUM BUI (38), *Laboratoire de Chimie Organique Biologique, URA CNRS 493, Université Paris VI, 75252 Paris Cedex 05, France*

ALESSANDRO F. CASINI (4), *Dipartimento di Biomedicina, Sezione di Patologia Generale, Università di Pisa, I-56100 Pisa, Italy*

RE-JIIN CHANG (34), *Department of Biology and Institute of Life Science, National Sun Yat-Sen University, Kaohsiung, Taiwan 804, Republic of China*

YUO-SHENG CHANG (34), *Department of Biology and Institute of Life Science, National Sun Yat-Sen University, Kaohsiung, Taiwan 804, Republic of China*

MARIO COMPORTI (4), *Istituto di Patologia Generale, Università di Siena, I-53100 Siena, Italy*

JOHN E. CRONAN, JR. (19), *Department of Microbiology, University of Illinois, Urbana-Champaign, Urbana, Illinois 61801*

SYLVIA DAUNERT (31), *Department of Chemistry, University of Kentucky, Lexington, Kentucky 40506*

BARBARA DEL BELLO (4), *Istituto di Patologia Generale, Università di Siena, I-53100 Siena, Italy*

JOHN C. DEUTSCH (2), *Department of Medicine, Divisions of Gastroenterology and Hematology, University of Colorado Health Sciences Center, and Denver VA Hospital, Denver, Colorado 80220*

ROLAND DOUCE (35), *Unité Mixte CNRS/Rhône-Poulenc (UM 41), Rhône-Poulenc Agrochimie, 69263 Lyon Cedex 09, France*

DAVID L. DYER (42), *Medical Research Service, Long Beach VA Medical Center, Long Beach, California, 90822, and Departments of Medicine, Pediatrics, and Physiology/Biophysics, University of California School of Medicine, Irvine, California 92717*

H. EIDHOF (9), *Department of Clinical Chemistry, Twenteborg Hospital, 7609 PP Almelo, The Netherlands*

BETTY A. EIPPER (5), *Department of Neuroscience, The Johns Hopkins University School of Medicine, Baltimore, Maryland 21205*

FUMIO ENJO (13), *Department of Biochemistry, Kyoto Prefectural University of Medicine, Kyoto, Japan*

GREGORY P. EVANGELATOS (46), *Institute of Radioisotopes and Radiodiagnostic Products, Radioimmunochemistry Laboratory, N.C.S.R. Demokritos, 153 10 Aghia Paraskevi Attikis, Greece*

STAVROS A. EVANGELATOS (46), *Pharmaceutical Technology Laboratory, School of Pharmacy, University of Patras, 261 10 Patras, Greece*

VÉRONIQUE FAYOL (7), *Laboratoire des Vitamines, Foundation Marcel Mérieux, 69365 Lyon Cedex 07, France*

AGATHA FELTUS (31), *Department of Chemistry, University of Kentucky, Lexington, Kentucky 40506*

KRISTIN FLEISCHHAUER (44), *Department of Human Genetics, Medical College of Virginia, Virginia Commonwealth University, Richmond, Virginia 23298*

DENNIS H. FLINT (37), *Central Research and Development, Experimental Station, E. I.*

*du Pont de Nemours and Co., Wilmington, Delaware 19880*

KAZUKO FUJIWARA (20), *Institute for Enzyme Research, University of Tokushima, Tokushima 770, Japan*

BATYA GAYER (47), *Department of Biological Regulation, The Weizmann Institute of Science, Rehovot, 76100, Israel*

J. GERRITS (9), *Department of Clinical Chemistry, Twenteborg Hospital, 7609 PP Almelo, The Netherlands*

KAZUHISA HATAKEYAMA (36), *Tsukuba Research Center, Mitsubishi Chemical Company, Ltd., Inashiki, Ibaraki, 300-03, Japan*

KOU HAYAKAWA (22, 45), *Division of Metabolism, National Children's Medical Research Center, Tokyo 154, Japan*

NATHANIEL G. HENTZ (29), *Department of Chemistry, University of Kentucky, Lexington, Kentucky 40506*

J. HESSELS (9), *Department of Clinical Chemistry, Twenteborg Hospital, 7609 PP Almelo, The Netherlands*

RICHARD E. HILLMAN (11), *Department of Child Health, University of Missouri, Columbia, Missouri 65201*

NORITOSHI HIRABAYASHI (18), *Faculty of Pharmaceutical Sciences, Okayama University, Tsushima, Okayama 700, Japan*

ERIC Z. HUANG (32), *Molecular Tool, L.L.C., Hopkins Bayview Research Campus, Alpha Center, Baltimore, Maryland 21224*

GERHARD HÜBNER (15), *Fachbereich Biochemie/Biotechnologie, Institute of Biochemistry, The University of Halle-Wittenberg, 06120 Halle, Germany*

JEANNE HYMES (44), *Department of Human Genetics, Medical College of Virginia, Virginia Commonwealth University, Richmond, Virginia 23298*

DIONYSSIS S. ITHAKISSIOS (46), *Pharmaceutical Technology Laboratory, School of Pharmacy, University of Patras, 261 10 Patras, Greece*

AKIO IWASHIMA (13), *Kyoto Prefectural Institute of Hygienic and Environmental Sciences, Kyoto, Japan*

SEAN W. JORDAN (19), *Department of Microbiology, University of Illinois, Urbana-Champaign, Urbana, Illinois 61801*

SOTIRIS E. KAKABAKOS (46), *Institute of Radioisotopes and Radiodiagnostic Products, Radioimmunochemistry Laboratory, N.C.S.R. Demokritos, 153 10 Aghia Paraskevi Attikis, Greece*

JAN KARLSEN (8), *Department of Pharmaceutics, University of Oslo, Blindern, 0373 Oslo, Norway*

HIROYUKI KATAOKA (18), *Faculty of Pharmaceutical Sciences, Okayama University, Tsushima, Okayama 700, Japan*

CHRISTIAN KIKUTA (10), *Pharm-Analyt Lab. GmbH, Ferdinand Pichler Gasse 2, A-2500 Baden, Austria*

EIJI KIMOTO (1), *Department of Chemistry, Faculty of Science, Fukuoka University, Fukuoka 814-80, Japan*

MIKI KOBAYASHI (36), *Tsukuba Research Center, Mitsubishi Chemical Company, Ltd., Inashiki, Ibaraki, 300-03, Japan*

FORTÜNE KOHEN (47), *Department of Biological Regulation, The Weizmann Institute of Science, Rehovot, 76100, Israel*

APARNA S. KOLHEKAR (5), *Department of Neuroscience, The Johns Hopkins University School of Medicine, Baltimore, Maryland 21205*

HIROSHI KUMAOKA (12), *School of Pharmaceutical Sciences, Mukogawa Women's University, Nishinomiya, Hyogo 663, Japan*

UMBERTO LAFORENZA (14), *Institute of Human Physiology, University of Pavia, 27100 Pavia, Italy*

RALPH H. LAMBALOT (27), *Department of Biological Chemistry and Molecular Pharmacology, Harvard Medical School, Boston, Massachusetts 02115*

MARK LEVINE (6), *Molecular and Clinical Nutrition Section, National Institute of Diabetes and Digestive and Kidney Diseases, National Institutes of Health, Bethesda, Maryland 20892*

YLVA LINDQVIST (40), *Division of Molecular Structural Biology, Department of Medical Biochemistry and Biophysics, Karolinska Institute, S-171 77 Stockholm, Sweden*

EVANGELIA LIVANIOU (46), *Institute of Radioisotopes and Radiodiagnostic Products, Radioimmunochemistry Laboratory, N.C.S.R. Demokritos, 153 10 Aghia Paraskevi Attikis, Greece*

SERGIO LIZANO (31), *Department of Chemistry, University of Kentucky, Lexington, Kentucky 40506*

EMILIA MAELLARO (4), *Istituto di Patologia Generale, Università di Siena, I-53100 Siena, Italy*

RICHARD E. MAINS (5), *Department of Neuroscience, The Johns Hopkins University School of Medicine, Baltimore, Maryland 21205*

MASAMI MAKITA (18), *Faculty of Pharmaceutical Sciences, Okayama University, Tsushima, Okayama 700, Japan*

A. MARQUET (38), *Laboratoire de Chimie Organique Biologique, URA CNRS 493, Université Paris VI, 75252 Paris Cedex 05, France*

HERMANN J. MASCHER (10), *Pharm-Analyt Lab. GmbH, A-2500 Baden, Austria*

BRIAN W. MATTHEWS (39), *Institute of Molecular Biology, Howard Hughes Medical Institute, and Department of Physics, University of Oregon, Eugene, Oregon 97403*

DONALD M. MOCK (28), *Department of Pediatrics, Arkansas Children's Hospital, Little Rock, Arkansas 72202*

YUTARO MOTOKAWA (20), *Institute for Enzyme Research, University of Tokushima, Tokushima 770, Japan*

KUNIAKI NARISAWA (41), *Department of Biochemical Genetics, Tohoku University School of Medicine, Sendai 980-77, Japan*

HOLGER NEEF (15), *Fachbereich Biochemie/Biotechnologie, Institute of Biochemistry, The University of Halle-Wittenberg, 06120 Halle, Germany*

MORIMITSU NISHIKIMI (3), *Department of Biochemistry, Wakayama Medical College, Wakayama 640, Japan*

HIROSHI NISHIMURA (13), *Department of Biochemistry, Kyoto Prefectural University of Medicine, Kyoto, Japan*

KAZUTO NOSAKA (13), *Department of Biochemistry, Kyoto Prefectural University of Medicine, Kyoto, Japan*

JUN OIZUMI (22, 45), *Division of Metabolism, National Children's Medical Research Center, Tokyo 154, Japan*

KAZUKO OKAMURA-IKEDA (20), *Institute for Enzyme Research, University of Tokushima, Tokushima 770, Japan*

RAGHAVAKAIMAL PADMAKUMAR (24), *Department of Biochemistry, University of Nebraska, Lincoln, Nebraska 68588*

RUGMINI PADMAKUMAR (24), *Department of Biochemistry, University of Nebraska, Lincoln, Nebraska 68588*

MORTEZA POURFARZAM (26), *Department of Child Health, Sir James Spence Institute of Child Health, University of Newcastle, Royal Victoria Infirmary, Newcastle-upon-Tyne NE4 4LP, United Kingdom*

RAINER PREISS (17), *Institute of Clinical Pharmacology, University of Leipzig, D-04107 Leipzig, Germany*

JANET QUINN (21), *Departments of Biochemistry, Genetics and Medicine, University of Newcastle upon Tyne, The Medical School, Framlington Place, Newcastle upon Tyne, Tyne & Wear NE2 4HH, United Kingdom*

SRIDHAR RAMANATHAN (31), *Department of Chemistry, University of Kentucky, Lexington, Kentucky 40506*

GERTRUD I. REHNER (30), *Institute of Nutritional Sciences, Justus Liebig University Giessen, D-35392 Giessen, Germany*

JÁNOS RÉTEY (25), *Department of Biochemistry, Richard-Willstätter-Alle, University of Karlsruhe, 76128 Karlsruhe, Germany*

GIANGUIDO RINDI (14), *Institute of Human Physiology, University of Pavia, 27100 Pavia, Italy*

YU-HUI ROGERS (32), *Molecular Tool, L.L.C., Hopkins Bayview Research Campus, Alpha Center, Baltimore, Maryland 21224*

STEVEN RUMSEY (6), *Molecular and Clinical Nutrition Section, National Institute of Diabetes and Digestive and Kidney Diseases, National Institutes of Health, Bethesda, Maryland 20892*

HAMID M. SAID (42), *Medical Research Service, Long Beach VA Medical Center, Long Beach, California, 90822, and Departments of Medicine, Pediatrics, and Physiology/Biophysics, University of California School of Medicine, Irvine, California 92717*

ALFRED SCHELLENBERGER (15), *Fachbereich Biochemie/Biotechnologie, Institute of Biochemistry, The University of Halle-Wittenberg, 06120 Halle, Germany*

DANIEL G. SCHINDLER (47), *Department of Immunology, The Weizmann Institute of Science, Rehovot, 76100, Israel*

GUNTER SCHNEIDER (40), *Division of Molecular Structural Biology, Department of Medical Biochemistry and Biophysics, Karolinska Institute, S-171 77 Stockholm, Sweden*

DAVID SHIUAN (34), *Department of Biology and Institute of Life Science, National Sun Yat-Sen University, Kaohsiung, Taiwan 804, Republic of China*

JÜRGEN STEIN (30), *Institute of Nutritional Sciences, Justus Liebig University Giessen, D-35392 Giessen, Germany*

LIDIA SUGHERINI (4), *Istituto di Patologia Generale, Università di Siena, I-53100 Siena, Italy*

YOICHI SUZUKI (41), *Department of Biochemical Genetics, Tohoku University School of Medicine, Sendai 980-77, Japan*

YUKIO SUZUKI (16), *Research Institute for Bioresources, Okayama University, Kurashiki 710, Japan*

CHANTAL M. E. TALLAKSEN (8), *Department of Neurology, The National Hospital, Oslo, 0027 Oslo, Norway*

KEIKO TAZUYA (12), *School of Pharmaceutical Sciences, Mukogawa Women's University, Nishinomiya, Hyogo 663, Japan*

JENS TEICHERT (17), *Institute of Clinical Pharmacology, University of Leipzig, D-04107 Leipzig, Germany*

SHIGEYUKI TERADA (1), *Department of Chemistry, Faculty of Science, Fukuoka University, Fukuoka 814-80, Japan*

KEI UCHIDA (16), *Research Institute for Bioresources, Okayama University, Kurashiki 710, Japan*

CHRISTOPHER T. WALSH (27), *Department of Biological Chemistry and Molecular Pharmacology, Harvard Medical School, Boston, Massachusetts 02115*

YAOHUI WANG (6), *Molecular and Clinical Nutrition Section, National Institute of Diabetes and Digestive and Kidney Diseases, National Institutes of Health, Bethesda, Maryland 20892*

HAROLD B. WHITE III (48), *Department of Chemistry and Biochemistry, University of Delaware, Newark, Delaware 19716*

MEIR WILCHEK (47), *Department of Membranes and Biophysics, The Weizmann Institute of Science, Rehovot, 76100, Israel*

ALLAN WITKOWSKI (31), *Department of Chemistry, University of Kentucky, Lexington, Kentucky 40506*

BARRY WOLF (44), *Departments of Human Genetics and Pediatrics, Medical College of Virginia, Virginia Commonwealth University, Richmond, Virginia 23298*

CHWEN-HUEY WU (34), *Department of Biology and Institute of Life Science, National Sun Yat-Sen University, Kaohsiung, Taiwan 804, Republic of China*

DAVID T. WYATT (11), *Department of Pediatrics, Medical College of Wisconsin, Milwaukee, Wisconsin 53226*

YAN XU (43), *Department of Chemistry and Biochemistry, University of Maryland, Baltimore County, Baltimore, Maryland 21045*

KUNIO YAGI (3), *Institute of Applied Biochemistry, Yagi Memorial Park, Mitake, Gifu 505-01, Japan*

KAZUKO YAMADA (12), *School of Pharmaceutical Sciences, Mukogawa Women's University, Nishinomiya, Hyogo 663, Japan*

TAKEO YAMAGUCHI (1), *Department of Chemistry, Faculty of Science, Fukuoka University, Fukuoka 814-80, Japan*

KUNIO YAMAUCHI (45), *Department of Food Technology, College of Agriculture and Veterinary Medicine, Nihon University, Tokyo 154, Japan*

KAZUYUKI YOSHIKAWA (45), *Department of Food Technology, College of Agriculture and Veterinary Medicine, Nihon University, Tokyo 154, Japan*

HIDEAKI YUKAWA (36), *Tsukuba Research Center, Mitsubishi Chemical Company, Ltd., Inashiki, Ibaraki, 300-03, Japan*

# Preface

From 1970 through 1986, eight "Vitamins and Coenzymes" volumes were published in the *Methods in Enzymology* series. Volumes XVIII A, B, and C appeared in 1970–1971 and Volumes 62 (D), 66 (E), and 67 (F) in 1979–1980. These volumes were edited by D. B. McCormick and L. D. Wright. Volumes 122 (G) and 123 (H), published in 1986, were edited by F. Chytil and D. B. McCormick. In the decade that has elapsed since the last volume was published, considerable progress has been made, so it was reasonable to update the subject of "Vitamins and Coenzymes."

In this current set of volumes (279, 280, 281, and 282) we have attempted to collect and collate many of the newer techniques and methodologies attendant to assays, isolations, and characterizations of vitamins, coenzymes, and those systems responsible for their biosynthesis, transport, and metabolism. There are examples of procedures that are modifications of earlier ones as well as of those that have newly evolved. As before, there has been an attempt to allow such overlap as would offer flexibility in the choice of methods, rather than presume any one is best for all laboratories. Where there is no inclusion of a particular subject covered in earlier volumes, we feel the subject was adequately treated and the reader should refer to those volumes.

The information provided reflects the efforts of our numerous contributors to whom we express our gratitude. We are also grateful to our secretaries at our academic home bases and to Shirley Light and the staff of Academic Press. Finally, one of us (D. B. M.) recalls fondly the encouragement proffered years ago by Drs. Nathan Kaplan and Sidney Colowick who saw the need for "Vitamins and Coenzymes" within the *Methods in Enzymology* series which they initiated.

DONALD B. MCCORMICK
JOHN W. SUTTIE
CONRAD WAGNER

# METHODS IN ENZYMOLOGY

VOLUME 73. Immunochemical Techniques (Part B)
*Edited by* JOHN J. LANGONE AND HELEN VAN VUNAKIS

VOLUME 74. Immunochemical Techniques (Part C)
*Edited by* JOHN J. LANGONE AND HELEN VAN VUNAKIS

VOLUME 75. Cumulative Subject Index Volumes XXXI, XXXII, XXXIV–LX
*Edited by* EDWARD A. DENNIS AND MARTHA G. DENNIS

VOLUME 76. Hemoglobins
*Edited by* ERALDO ANTONINI, LUIGI ROSSI-BERNARDI, AND EMILIA CHIANCONE

VOLUME 77. Detoxication and Drug Metabolism
*Edited by* WILLIAM B. JAKOBY

VOLUME 78. Interferons (Part A)
*Edited by* SIDNEY PESTKA

VOLUME 79. Interferons (Part B)
*Edited by* SIDNEY PESTKA

VOLUME 80. Proteolytic Enzymes (Part C)
*Edited by* LASZLO LORAND

VOLUME 81. Biomembranes (Part H: Visual Pigments and Purple Membranes, I)
*Edited by* LESTER PACKER

VOLUME 82. Structural and Contractile Proteins (Part A: Extracellular Matrix)
*Edited by* LEON W. CUNNINGHAM AND DIXIE W. FREDERIKSEN

VOLUME 83. Complex Carbohydrates (Part D)
*Edited by* VICTOR GINSBURG

VOLUME 84. Immunochemical Techniques (Part D: Selected Immunoassays)
*Edited by* JOHN J. LANGONE AND HELEN VAN VUNAKIS

VOLUME 85. Structural and Contractile Proteins (Part B: The Contractile Apparatus and the Cytoskeleton)
*Edited by* DIXIE W. FREDERIKSEN AND LEON W. CUNNINGHAM

VOLUME 86. Prostaglandins and Arachidonate Metabolites
*Edited by* WILLIAM E. M. LANDS AND WILLIAM L. SMITH

VOLUME 87. Enzyme Kinetics and Mechanism (Part C: Intermediates, Stereochemistry, and Rate Studies)
*Edited by* DANIEL L. PURICH

VOLUME 88. Biomembranes (Part I: Visual Pigments and Purple Membranes, II)
*Edited by* LESTER PACKER

VOLUME 89. Carbohydrate Metabolism (Part D)
*Edited by* WILLIS A. WOOD

VOLUME 90. Carbohydrate Metabolism (Part E)
*Edited by* WILLIS A. WOOD

# Section I

# Ascorbic Acid

## [1] Analysis of Ascorbic Acid, Dehydroascorbic Acid, and Transformation Products by Ion-Pairing High-Performance Liquid Chromatography with Multiwavelength Ultraviolet and Electrochemical Detection

*By* Eiji Kimoto, Shigeyuki Terada, and Takeo Yamaguchi

### Introduction

The primary basis for the physiological role of L-ascorbic acid (AsA) is generally believed to be the oxidation–reduction system between AsA and its oxidation form, dehydro-L-ascorbic acid (DHA). Early workers[1] reported that, in living tissues, DHA is reduced enzymatically or chemically back to AsA and that therefore AsA is the predominant form. However, more recent studies have indicated that a considerable concentration of DHA is present in a variety of biological tissues or fluids, especially in disease conditions[2-4] or senescent processes,[5] and in food products.[6-8] Numerous assay methods have been devised to determine AsA and DHA simultaneously. However, they suffer from the "four-S syndrome," as mentioned by Washko *et al.*[9]: inattention to stability, sensitivity, specificity, and substance interference. The quantitative determination of DHA is still difficult. Dehydro-L-ascorbic acid is unstable in aqueous solution and is easily delactonized to 2,3-diketo-L-gulonic acid (DKG), which is further transformed and degraded, leading to uncontrolled variables.[10,11]

Polarographic studies of neutral aqueous solutions of DHA[12] have shown complicated oxidation waves rather than reduction waves, suggesting the occurrence of several easily oxidizable reductants. The autoxidation of

[1] D. Horning, *Ann. N.Y. Acad. Sci.* **258**, 103 (1975).

[2] I. B. Chatterjee and A. Banerjee, *Anal. Biochem.* **98**, 368 (1979).

[3] M. S. Yew, *Horm. Metab. Res.* **15**, 158 (1983).

[4] S. Karp, C. Ciambra, and S. Miklean, *J. Chromatogr.* **504**, 434 (1990).

[5] K. G. Bensch, J. E. Fleming, and W. Lohmann, *Proc. Natl. Acad. Sci. U.S.A.* **82**, 7193 (1985).

[6] R. C. Rose and D. L. Nahrwold, *Anal. Biochem.* **114**, 140 (1981).

[7] S. J. Ziegler, B. Meier, and O. Sticher, *J. Chromatogr.* **391**, 419 (1987).

[8] W. D. Graham and D. Annett, *J. Chromatogr.* **594**, 187 (1992).

[9] P. W. Washko, R. W. Welch, K. R. Dhariwal, Y. Wang, and M. Levine, *Anal. Biochem.* **204**, 1 (1992).

[10] W. H. Kalus, W. G. Filby, and R. Munzner, *Z. Naturforsch.* **37c**, 40 (1982).

[11] S. O. Kang, H. Sapper, and W. Lohmann, *Z. Naturforsch.* **37c**, 1064 (1982).

[12] S. Ono, M. Takagi, and T. Wasa, *J. Am. Chem. Soc.* **75**, 4369 (1958).

AsA may constitute a coordinated chain of redox reactions in which AsA is oxidized and subsequently converted, at least in part, to reducing compounds. These transformation products may exist in association with DHA.

Ion-pairing high-performance liquid chromatography (HPLC) with electrochemical detection (ECD) has been widely used for the determination of reducing compounds in complex biological samples as a sensitive and selective assay.[13] The transformation products of DHA, possessing different ultraviolet (UV) absorption spectra, may be monitored by a multiwavelength detector[14] on a three-dimensional (retention time : wavelength : absorbance) chromatographic plot.

### Preparation of Oxidation Products of L-Ascorbic Acid

The ethanolic DHA solution is prepared by the method of Doner and Hicks.[15] L-Ascorbic acid (10 g) is suspended in 500 ml of ethanol containing 12 g of activated charcoal (Norit SX Plus). Oxygen gas is continuously bubbled through the solution at room temperature for about 24 hr until the disappearance of the 245-nm absorption peak of acidic AsA, and the charcoal is then removed by filtration.

For the preparation of the sodium salt, sodium hydroxide is added to an ice-cold ethanolic DHA solution to obtain a yellow precipitate. It is washed twice with ice-cold ethanol and dried under vacuum. In a neutral aqueous solution, it has absorption maxima at 265 and 345 nm. For the preparation of aqueous DHA solutions, the ethanolic DHA solution is evaporated thoroughly to dryness under vacuum and dissolved in the appropriate solutions.

### Analysis by Electrochemical Detection with High-Performance Liquid Chromatography

High-performance liquid chromatography[16] is carried out with a Jasco 880-PU liquid chromatography apparatus. Separation is achieved on a Radial PAK cartridge Resolve $C_{18}$ column (0.8 × 10 cm; Waters, Milford, CT). The mobile phase is a 40 m$M$ potassium phosphate buffer at pH 6.0, containing 1 m$M$ EDTA, 2.5 m$M$ tetrabutylammonium hydrogen sulfate, and 3% (v/v) methanol. The eluent is pumped at a flow rate of 1 ml min$^{-1}$. Three-dimensional (retention time : wavelength : absorbance) chromato-

[13] P. T. Kissinger, *J. Chem. Educ.* **60,** 308 (1983).
[14] T. Alfredson and T. Sheehan, *J. Chromatogr. Sci.* **24,** 473 (1986).
[15] L. W. Doner and K. B. Hicks, *Anal. Biochem.* **115,** 225 (1981).
[16] E. Kimoto, H. Tanaka, T. Ohmoto, and M. Choami, *Anal. Biochem.* **214,** 38 (1993).

graphic data are obtained as contour maps with a Jasco 330 multichannel UV detector, which covers a wavelength range of 200 to 350 nm. The low UV (below 220 nm), however, is the region where the data are often obscured by the mobile phase and also by unidentified degradation products. The reducing compounds in the eluate are monitored with the aid of an amperometric electrochemical detector (model ICA-3060; Toa Electronics, Ltd.) set at 600 mV vs Ag|AgCl.

Isolation and Characterization

Preparative separation of transformation products is carried out by DEAE-Sepharose column chromatography. DEAE-Sepharose Fast Flow (Pharmacia, Piscataway, NJ) is packed into a column (2.5 × 33 cm). Elution is carried out with 10 mM sodium phosphate buffer, pH 6.0. The eluates are examined by absorption spectrum and ECD–HPLC analysis. If necessary, further purification and desalting can be achieved by gel filtration on a Sephadex G-10 or Cellulofine GCL-25-m column.

Chemical characterization and identification are carried out by use of a [1]H and [13]C NMR spectrometer [JEOL (Tokyo, Japan) JNM-GSX 400 spectrometer] and a fast atom bombardment-mass spectroscopy (FAB-MS) spectrometer (JEOL JMS-AX 505 W spectrometer).

Electrochemical Detection with High-Performance Liquid
  Chromatography of Sodium Salt of Oxidized L-Ascorbic Acid

Figure 1 indicates three electrochemically active spots, AsA, *erythro*-L-ascorbic acid (EAsA), and the reductant R-345, possessing an absorption maximum at 345 nm. During the preparation of the sodium salt of oxidized AsA, DHA is partially reduced to AsA, and the other part of DHA is delactonized to DKG, decarboxylated at the C-1 position, and converted to EAsA, which possesses the 2,3-enediol γ-lactone and contains only one carbon moiety at the side chain (Fig. 4). *erythro*-L-Ascorbic acid was synthesized originally by Reichstein[17] as the lactone of 2-ketopentonic acid.

In such a neutralized ethanolic solution, a greater proportion of DKG undergoes ring closure without decarboxylation to give R-345, which is the 2,3-enediol form of 2,3-diketogulono-δ-lactone.[18] R-345 was originally found in human placenta.[19] A term placenta undergoes progressive deterio-

[17] T. Reichstein, *Helv. Chim. Acta* **17,** 1003 (1934).
[18] H. Tanaka and E. Kimoto, *Bull. Chem. Soc. Jpn.* **63,** 2569 (1990).
[19] N. Toh, T. Inoue, M. Kuraya, H. Tanaka, and E. Kimoto, *Int. J. Biol. Res. Pregn. Perinat.* **8,** 47 (1987).

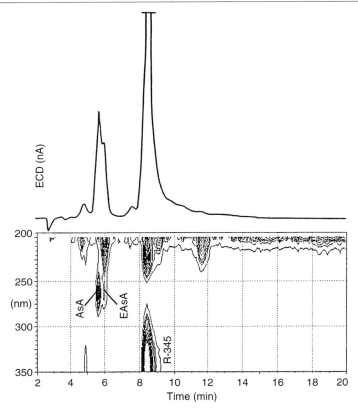

Fig. 1. ECD–HPLC of sodium salt of oxidized AsA.

ration, which may be the reason why so much AsA is oxidized and converted to R-345. R-345 is also found in the 70% (v/v) methanol extract of cucumber or pumpkin, using ECD–HPLC.

Immediately after dissolving DHA in a neutral aqueous solution, [13]C NMR reveals the presence of both DHA and DKG. This shows there is an immediate delactonization of part of DHA in aqueous solution. ECD–HPLC shows only one electrochemically active AsA spot, indicating that a small proportion (less than 1%) of DHA is quickly reduced back to AsA.

After a neutralized aqueous DHA solution is incubated at 37° for 24 hr, it develops a deep brown color and a high absorption peak around 275 nm. The [13]C NMR spectrum reveals mainly the DKG signals. During this incubation time, DHA is almost completely delactonized to give DKG,

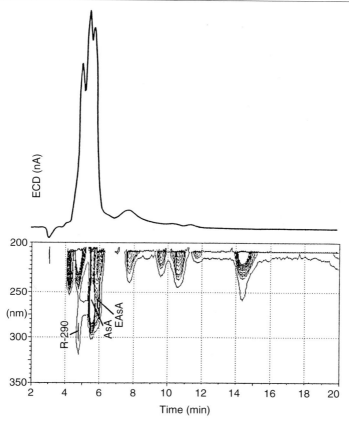

FIG. 2. ECD–HPLC of neutralized aqueous DHA solution after incubation at 37° for 24 hr.

which undergoes further transformation. Analysis by ECD–HPLC (Fig. 2) reveals three main ECD-active spots: AsA, EAsA, and a reductant possessing an absorption maximum at 290 nm. The latter reductant has a retention time a little less than that of AsA, and it is eluted in the void volume on a DEAE-Sepharose column, showing a less negative charge. Its absorption maximum at 290 nm in neutral aqueous solution shifts to 335 nm above pH 9.0. It may possess more consecutive conjugated multiple bonds than those of AsA. The [13]C NMR spectrum shows more than 10 signals, indicating the polymerization of transformation products of DHA or intermediates of the browning reaction of DHA. Its faintly yellow color turns to red or brown on standing at room temperature within several hours.

Borsook *et al.*[20] reported that DHA solutions at pH 9.0 quickly reduced redox dyes, although the compounds responsible for the reduction were not characterized. We have shown that under these conditions, absorption maxima develop at 265 and 330 nm; the latter shifts to 345 nm at pH 6.0. Analysis by ECD–HPLC reveals three electrochemically active spots, AsA, EAsA, and R-345, similar to those in Fig. 1.

Electrochemical Detection with High-Performance Liquid
    Chromatography of Pyrolysis Products of Acidic Dehydro-L-
    ascorbic Acid Solution

When DHA is heated in 1 $N$ sulfuric acid at 90° for 1 hr, two absorption maxima appear at 250 and 295 nm. Analysis by ECD–HPLC (Fig. 3) reveals four main spots, consisting of the electrochemically active AsA, 5-methyl-3,4-dihydroxytetrone (MDT), 3-hydroxy-2-pyrone (3OH2P), and the electrochemically inactive 2-furoic acid (2FA).

5-Methyl-3,4-dihydroxytetrone was first reported by Hasselquist[21] as an *aci*-reductone III derived from the degradation of DHA. It possesses an absorption maximum at 265 nm at pH 7.0 and at 245 nm at pH 3.0, indicating the same chromophore as that of AsA. The side chain of AsA is replaced by the methyl group in the MDT molecule.

3-Hydroxy-2-pyrone possesses an absorption maximum at 295 nm at pH 7.0, which shifts to 315 nm at pH 9.0. 2-Furoic acid possesses an absorption maximum at 245 nm at pH 7.0, which shifts bathochromically to 255 nm in acidic solution. On DEAE-Sepharose column chromatography at pH 6.0, 3OH2P in an undissociated form is eluted in the void volume while 2FA is strongly adsorbed on the column.

Since the late eighteenth century, 2FA and 3OH2P have been known as pyromucic acid and isopyromucic acid, respectively, and as the acid-catalyzed transformation products of mucic acid.[22] Both 2FA and 3OH2P were reported by Velisek *et al.*[23] to be the main volatile degradation products of DHA. These compounds are demonstrable by ECD–HPLC in the smoke condensate of burned AsA or DHA, together with furfural (FF), at a retention time of 16 min. 3-Hydroxy-2-pyrone is found as a storage product of dehydrated (instant) orange juice.[24] Canned single-strength or-

[20] H. Borsook, H. W. Davenport, C. E. P. Jeffreys, and R. C. Warner, *J. Biol. Chem.* **117**, 237 (1937).

[21] H. Hasselquist, *Arkiv. Kemi.* **8**, 381 (1955).

[22] R. H. Wiley and C. H. Jarboe, *J. Am. Chem. Soc.* **78**, 2398 (1956).

[23] J. Velisek, J. Davidek, V. Kubelka, Z. Zelinkova, and J. Pokorny, *Z. Lebensm. Unters.-Forsch.* **162**, 285 (1976).

[24] P. E. Shaw, J. H. Tatum, T. J. Kew, C. J. Wagner, and R. E. Berry, *J. Agric. Food Chem.* **18**, 343 (1970).

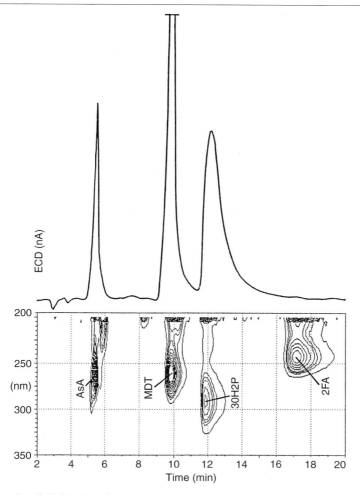

FIG. 3. ECD–HPLC of DHA heated in 1 *N* sulfuric acid at 90° for 1 hr.

ange juice, when stored at warm temperature, develops detrimental off-flavors and contains a variety of degradation products of AsA, including 3OH2P.[25]

When DHA is heated in 0.4 *N* sulfuric acid in a sealed glass tube (under high pressure) at 150° for 2 hr, it develops a high absorption maximum at 245 nm and a plateau absorption around 300 nm. Analysis by ECD–HPLC

[25] J. H. Tatum, S. Nagy, and R. E. Berry, *J. Food Sci.* **40,** 707 (1975).

reveals three spots consisting of electrochemically active 3OH2P and of inactive 2FA and FF. Nomura and Uehara[26] and Kamiya[27] reported that the decomposition of AsA with acid at 160–170° under high pressure gave reductic acid (RA), although only in a minute quantity. However, the RA spot (6.5-min retention time) is barely detectible in the pyrolyzate of AsA or DHA. Under the same pyrolysis conditions, D-galacturonic[28] acid produces abundant RA eluting after EAsA, as determined by ECD–HPLC.

### Possible Reaction Pathway to Produce Transformation Products of Dehydro-L-ascorbic Acid

Under different incubating conditions, DHA converts to a variety of transformation products possessing absorption maxima between 240 and 350 nm. Most of them, as demonstrated in the present ECD–HPLC analysis, are electrochemically active. Many examples of decarboxylation and/or oxidation converting nonreducing compounds into reducing ones are well known: Aldehydic compounds are produced from $\alpha$-keto acids by decarboxylation or from alcohols by oxidation. L-Ascorbic acid itself is produced from L-gulonolactone by its oxidase. Uric acid is the oxidation product of xanthine. It is not surprising that the oxidation of AsA or the transformation of DHA produces the electrochemically active reducing compounds. Most of them may be produced nonenzymatically via DKG, which can undergo decarboxylation and/or intramolecular rearrangement. This may be referred to as the ascorbate cascade (Fig. 4).

In neutralized ethanolic solution, DKG undergoes rapid ring closure and enolic isomerization to form the six-membered ($\delta$-lactone) R-345. At the same time, a small proportion of DKG undergoes decarboxylation, enolization, and ring closure to form the five-membered ($\gamma$-lactone) EAsA. 2,3-Diketo-L-gulonic acid undergoes similar transformation in alkaline aqueous solution.

2,3-Diketo-L-gulonic acid in acidic solution primarily undergoes decarboxylation to form L-xylosone (XLS).[29] Subsequent transformation produces various five-carbon compounds.

L-Xylosone is converted to compound **I** (see Fig. 4), the enolic form of 2-keto-3-deoxypentonic acid, by an intramolecular redox reaction. Compound **I** is dehydrated to form compound **II** (Fig. 4), the unsaturated $\alpha$-keto

[26] D. Nomura and Y. Uehara, *Hakko Kagaku Zasshi* **36,** 290 (1958).

[27] S. Kamiya, *Nippon Nogei Kagaku Kaishi* **33,** 398 (1959).

[28] T. Reichstein and R. Oppenauer, *Helv. Chim. Acta* **16,** 988 (1933).

[29] G. C. Whiting and R. A. Coggins, *Nature* (*London*) **185,** 843 (1960).

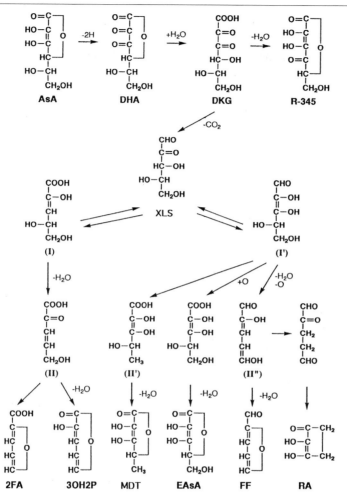

Fig. 4. Possible reaction pathway for the transformation products of DHA. AsA, L-Ascorbic acid; DHA, dehydro-L-ascorbic acid; DKG, 2,3-diketo-L-gulonic acid; EAsA, *erythro*-L-ascorbic acid; 2FA, 2-furoic acid; FF, furfural; MDT, 5-methyl-3,4-dihydroxytetrone; 3OH2P, 3-hydroxy-2-pyrone; RA, reductic acid; R-345, reductant possessing an absorption maximum at 345 nm; XLS, L-xylosone.

acid, which is further dehydrated to produce 2FA, the furan form and/or 3OH2P, the 2-pyrone form. Coggiola[30] reported the production of 2,5-dihydro-2-furoic acid when AsA was heated at 100° under an atmosphere of carbon dioxide for 10 days. The $^1$H NMR spectrum of 2FA indicates three $=CH-$ protons, which differ from $-CH_2-$ protons of AsA. As an alternative pathway, XLS may be converted to the enediol form, compound **I'**. Its aldehyde group may be oxidized to a carboxyl group, which is lactonized to form EAsA. An intramolecular redox of compound **I'** leads to rearrangement from the aldehydic to the carboxylic group and from the hydroxymethyl at the C-5 position to the methyl group, thus producing compound **II'** (Fig. 4). It undergoes ring closure to form MDT. Otherwise, the aldehydic group of compound **I'** can remain intact. It may be dehydrated and reduced (loss of oxygen) to form compound **II''** (Fig. 4) with the diene structure. Compound **II''** may then be dehydrated to form the furan, FF, or rearranged to form the bialdehydic structure as an unstable intermediate that undergoes ring closure easily to form RA. However, RA is produced from AsA or DHA only in minute amounts at the temperatures used in this study. Two types of reactions appear to be responsible for the production of compounds via XLS: oxidation/reduction and intramolecular rearrangements.

Postharvest storage of fruits or vegetables reduces AsA content and gives rise to the formation of oxidation products to a degree relative to the time and temperature of storage. Processing, which causes cellular disorganization, will result in a greatly increased rate of oxidation of AsA, and possibly produce a variety of transformation products of DHA. When formed in living animal tissues, they may be further metabolized. These transformation products may constitute intermediates in the oxidation of AsA.

[30] I. M. Coggiola, *Nature* (*London*) **200,** 954 (1963).

# [2] Gas Chromatographic/Mass Spectrometric Measurement of Ascorbic Acid and Analysis of Ascorbic Acid Degradation in Solution

*By* JOHN C. DEUTSCH

## Introduction

L-Ascorbic acid (AA) is a hexuronic acid, which in biological solutions at neutral pH exists primarily as the ascorbate anion. L-Ascorbic acid and L-dehydroascorbic acid (DHA) are the principal natural compounds with antiscorbutic activity (vitamin C).[1]

L-Ascorbic acid is known to take part in a variety of important biochemical reactions, including the formation of collagen and neurotransmitters as well as the degradation of tyrosine.[2] L-Ascorbic acid is also well known as an antioxidant and can be used as a reducing agent in a variety of reactions *in vitro*.

L-Ascorbic acid is reversibly oxidized through a free-radical intermediate (semidehydroascorbate) to form DHA. Besides this reversible reaction, more than 50 other products may be formed after AA is decarboxylated and further degraded.[3] The first reportedly irreversible step (based on loss of antiscorbutic activity) in AA degradation involves the hydrolysis of DHA to 2,3-diketogulonic acid. The structures of AA, DHA, and diketogulonic acid are shown in Fig. 1. It should be noted that some older literature suggests that this "irreversible step" could be reversed with HI.[4]

In addition, we have described several six-carbon products derived from AA during oxidation that do not arise through the previously described pathways of AA degradation.[5] It appears that the primary species formed depends somewhat on the method of oxidation being employed. Taken together, these data suggest that AA oxidative degradation is a complicated process, and that there is still much to be learned.

---

[1] J. W. Jaffe, *in* "Handbook of Vitamins" (L. J. Machlin, ed.), p. 199. Marcel Dekker, New York, 1984.

[2] M. Levine, *N. Engl. J. Med.* **314,** 892 (1986).

[3] K. Niemela, *J. Chromatogr.* **399,** 235 (1987).

[4] J. R. Penny and S. S. Zilva, *Biochem. J.* **39,** 1 (1945).

[5] J. Deutsch, C. R. Santhosh-Kumar, K. L. Hassell, and J. F. Kolhouse, *Anal. Chem.* **66,** 345 (1994).

 0076-6879/97 $25

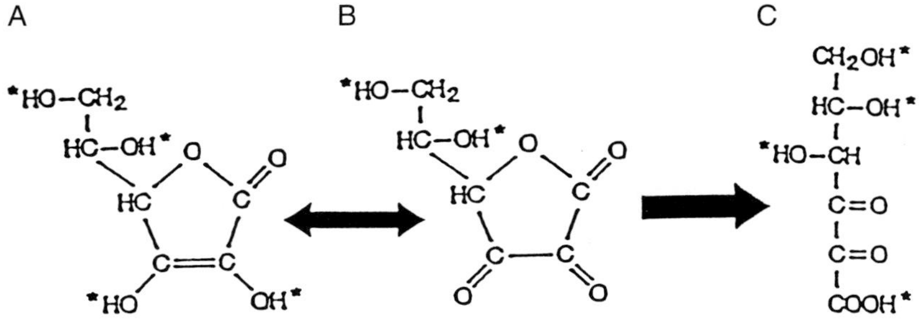

A                              B                              C

FIG. 1. The structures of (A) AA, (B) DHA, and (C) 2,3-diketogulonic acid. The asterisks mark sites that will be derivatized with *t*-Butyldimethylsilyltrifluoroacetamide. [Modified with permission from J. Deutsch *et al.*, *Anal. Chem.* **66**, 345 (1994). Copyright 1994 American Chemical Society.]

## Principle

The gas chromatography/mass spectroscopy (GC/MS) analysis outlined as follows involves sample purification, derivatization, application to a GC column, and analysis by electron impact mass spectrometry. Isotope dilution assays involve adding known quantities of stable isotope-labeled internal standards to the sample prior to purification, and quantifying the endogenous material by measuring the ratio of known stable isotope compound to the unknown endogenous compound. We use $[^{13}C_6]$AA and $[6,6^2$-$H_2]$AA both as internal standards, and as starting materials to make internal standards of other AA-related products. Our analyses are performed on a Hewlett Packard (Palo Alto, CA) 5890 gas chromatograph and 5971A mass detector. Some of the details of instrumentation outlined in this chapter may be specific to this equipment.

## Materials

L-Ascorbic acid can be purchased from a variety of chemical suppliers including Sigma Chemicals (St. Louis, MO) or Fluka Chemicals (Ronkonkoma, NY). The free acid is preferred to the sodium salt. Dehydroascorbic acid monomer can be purchased from ICN Chemicals (Costa Mesa, CA).

The stable isotope-labeled AAs had been available in the past from MSD Isotopes (Montreal, Quebec, Canada). At the current time, however, one must contract with a manufacturer for custom synthesis. *tert*-Butyldimethylsilyltrifluoroacetamide can be purchased from Regis Chemicals (Morton Grove, IL) or Pierce Chemicals (Rockford, IL).

Sample Purification

The stability of AA in solution is highly dependent on the water supply, with degradation probably due to contamination with metals.[6] Stability is also dependent on dissolved oxygen, light, and temperature. Relatively dilute (6 m$M$), unfrozen solutions can be kept for weeks without appreciable loss if the water is highly purified, degassed, and kept at 4° in the dark. These same solutions are reasonably stable when frozen at −80° for years. Heparinized plasma samples frozen at −80° are also stable, but AA is unstable in EDTA-chelated plasma stored at −20°.[7]

The internal standard should be added to the sample as early as possible during processing so that any sample loss will result in proportionate internal standard loss. We add approximately 4- to 10-fold as much [$^{13}C_6$]AA to a sample as the expected AA content. For example, to determine the AA content in plasma (where 5–15 $\mu$g/ml is usual), we add about 5 $\mu$g of [$^{13}C_6$]AA to 100-$\mu$l plasma aliquots prior to precipitation or any extraction. We add approximately 500 ng of [$^{13}C_6$]AA to 5- to 10-$\mu$l urine and plasma samples prior to drying.

Samples from aqueous solutions are dried under centrifugation at room temperature, using a Savant (Holbrook, NY) drying centrifuge system. We have also found that 5- to 10-$\mu$l aliquots of urine or plasma can be dried and analyses can be performed without any purification. The yields are higher if the samples are acidified to pH 1–2 by adding 1 $\mu$l of 50% (w/v) trichloroacetic acid prior to drying. The majority of the analyses on protein-containing samples are performed on samples of 25–100 $\mu$l, which are deproteinized and acidified using a final concentration (weight to volume) of 10% trichloroacetic acid and then diethyl ether extracted to remove the bulk of the trichloroacetic acid.

Derivatization

L-Ascorbic acid and the related products are not volatile, and they decompose with heating. To allow GC analysis, these compounds must be derivatized. Although trimethylsilyl derivatives can be used, we prefer to use *tert*-butyldimethylsilyl derivatives, because the stability is much greater. Obviously, samples must be thoroughly dried prior to derivatization. Our standard procedure is to add 15 $\mu$l of *tert*-butyldimethylsilyltrifluoroacetamide to the dried sample, followed by 35 $\mu$l of acetonitrile, and allow the substances to react for 60 min at 50–60°. Derivatization is probably not

[6] G. R. Buettner, *J. Biochem. Biophys. Methods* **16,** 27 (1988).
[7] J. Deutsch and J. F. Kolhouse, *Anal. Chem.* **65,** 321 (1993).

complete at this point, because the ion abundance tends to go up if the samples are allowed to sit an additional 12 hr at room temperature. Sometimes runover from sample to sample (as much as 5%) occurs, also suggesting incomplete derivatization. Runover is detected by running a *tert*-butyldimethylsilylacetonitrile blank and determining the AA content in that blank. To prevent analytical inaccuracies, a *tert*-butyldimethylsilylace-tonitrile blank should be run before each critical sample.

When we derivatize unpurified samples (e.g., a dried 5-$\mu$l aliquot of urine), we centrifuge the samples after the derivatization reaction for 10 min at 14,000 $g$ at 20° and collect only the supernatant fluid for analysis. We replace the injection port liner at frequent intervals when performing these types of experiments to prevent carbon buildup.

L-Ascorbic acid has four derivatizable sites, giving it an $M^+$ of 632, while DHA has only two derivatizable sites, giving it an $M^+$ of 402. 2,3-Diketogulonic acid (hydrolyzed DHA) has four derivatizable sites and an $M^+$ of 648.

## Gas Chromatography

We have had excellent results using temperature ramps on 10-m polydimethylsiloxane columns having an internal diameter of 0.25 mm and a film thickness of 0.25 $\mu$m (Supelco, Bellefonte, PA). It is important to note, however, that these columns slowly deteriorate, depending on the use and type of sample applied. Column deterioration is evident by shifting retention times and a broadening of peaks. We generally change columns at 6-month intervals, which covers the analysis of approximately 3000 samples.

Samples are applied to a 250° injection port in line with an 80° column, using helium as a carrier gas with a head pressure of 50 kPa and a total flow rate of 10 cm$^3$/min into the system. A 30°C/min temperature ramp (to a maximum temperature of 300°) is applied to the column. This allows AA ($M^+$ of 632) to elute at approximately 6 min (Fig. 2). The retention time of the other substances is approximately related to their derivatized molecular weight, so that DHA ($M^+$ of 402) elutes at 4 min and DHA dimer ($M^+$ of 804) elutes at 8 min.[8] Most of the compounds of interest, including 5- and 4-carbon degradative products, have three or four derivatizable sites so that even though the molecular weight of the 4- or 5-carbon parent is less than that of DHA, the derivatized mass is greater. These compounds usually elute between DHA and AA.

[8] J. C. Deutsch and C. R. Santhosh-Kumar, *J. Chromatogr. A* **724,** 271 (1996).

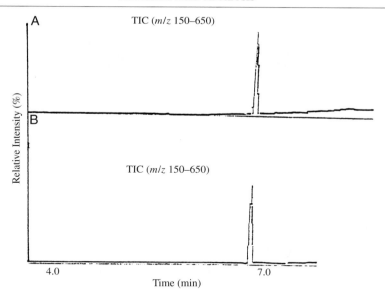

FIG. 2. The total ion chromatogram derived from *tert*-butyldimethylsilyl-derivatized (A) AA and (B) [$^{13}C_6$]AA. [Reprinted with permission from J. Deutsch and J. F. Kolhouse, *Anal. Chem.* **65**, 321 (1993). Copyright 1993 American Chemical Society.]

## Mass Spectrometry

The power of GC/MS is due to the detection system. Ion fragmentography tends to be more specific than ultraviolet (UV) or electrochemical (EC) detection, and structural information can be gathered on unknown compounds. The specificity allows one to electronically increase signal-to-noise ratios, and gives sensitivity that appears to be much greater than that obtained with UV or EC detection. We have been able to detect 9 pg (50 fmol) of AA with a signal-to-noise (*S/N*) ratio of 3:1.[7]

When samples containing 10–50 ng of AA and related products are analyzed, we keep the filament voltage at autotune values (approximately 1500 V). However, for samples from which less is expected, or when complex biological samples are being analyzed, we increase the filament voltage to increase the abundances of the ions of interest to $0.1–10 \times 10^6$ integrated mass units (imu). This often takes 2400 to 3000 V. The ion abundance can also be increased by increasing the repeller from the autotune values of 15 to 19.6 V.

Mass scans are employed when looking for unknown compounds and when trying to determine structures. Selected ion monitoring is used for isotope dilution assays and in following reactions involving known substances.

The ion fragments that are derived during electron impact are dependent both on the structure of the parent compound, and the derivatizing agent. The ions generated are positively charged and are identified on the basis of the mass lost from the parent compound. Butyldimethylsilyl derivatives tend to form abundant ions of $m/z$ $[M - 57]^+$, as well as minor ions of $m/z$ $[M - 15]^+$. Furthermore, tert-butyldimethylsilyl-derivatized carbohydrates often form abundant ions of $m/z$ $[M - 189]^+$ without loss of carbon from the parent molecule (all mass losses of carbon are from the derivatizing agent[7,9]). Figure 3 shows the respective spectra of AA and $[^{13}C_6]AA$, while Fig. 4 shows the spectra of DHA and $[^{13}C_6]DHA$.

## Quantitation by Isotope Dilution Analysis

L-Ascorbic acid is relatively easy to quantitate using isotope dilution techniques. The most accurate, reproducible results occur when we use a predicted 4- to 10-fold excess of $[^{13}C_6]AA$ as an internal standard. Because $[^{13}C_6]AA$ is expensive, we store solutions at $-80°$ and back-quantitate the amount of $[^{13}C_6]AA$, which we add by making a separate calibration curve vs AA standard. Calibration curves using the respective $m/z$ $[M - 57]^+$ ions of AA vs $[^{13}C_6]AA$ and DHA vs $[^2H_2]DHA$ are shown in Fig. 5.

To prepare a plasma sample, for example, 100 $\mu$l of fresh heparinized plasma is placed into 12 × 75 mm borosilicate tubes, to each of which is added 10 $\mu$l from a carefully quantitated solution of $[^{13}C_6]AA$ (containing 400–1000 $\mu$g/ml). Samples are mixed well and 15 $\mu$l of 100% (w/v) trichloroacetic acid is added. Samples are vortexed vigorously, and 875 $\mu$l of water is added. The samples are mixed again, centrifuged at 4° and 2000 $g$ for 10 min, and 750 $\mu$l of supernatant placed in new 12 × 75 mm borosilicate tubes. The supernatant is extracted three times with wet diethyl ether, the organic phase is discarded, and the aqueous phase is dried under vacuum centrifugation in the dark at room temperature. When dry, 15 $\mu$l of tert-butyldimethylsilyltrifluoroacetamide and 35 $\mu$l of acetonitrile are added, the tubes are sealed with Parafilm, and the samples are incubated at 50° for 1 hr. The liquid phase is removed from the precipitate and placed into 1.1-ml crimp-seal glass vials. These samples are then applied to the GC column for analysis.

L-Dehydroascorbic acid is much more difficult to measure, mainly because it undergoes spontaneous hydrolysis on solubilization, making it diffi-

[9] J. Deutsch, R. V. Kolli, C. R. Santhosh-Kumar, and J. F. Kolhouse, *Am. J. Clin. Pathol.* **102,** 595 (1994).

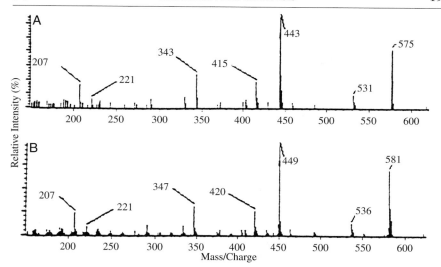

FIG. 3. The spectra derived from *tert*-butyldimethylsilyl-derivatized (A) AA and (B) [$^{13}$C$_6$]AA. [Reprinted with permission from J. Deutsch and J. F. Kolhouse, *Anal. Chem.* **65,** 321 (1993). Copyright 1993 American Chemical Society.]

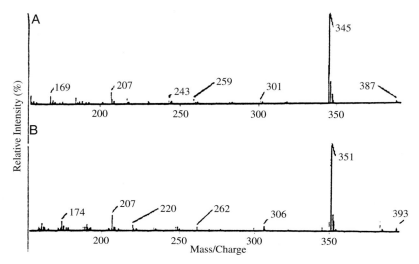

FIG. 4. The spectra derived from *tert*-butyldimethylsilyl-derivatized (A) DHA and (B) [$^{13}$C$_6$]DHA. [Reprinted with permission from J. Deutsch and J. F. Kolhouse, *Anal. Chem.* **65,** 321 (1993). Copyright 1993 American Chemical Society.]

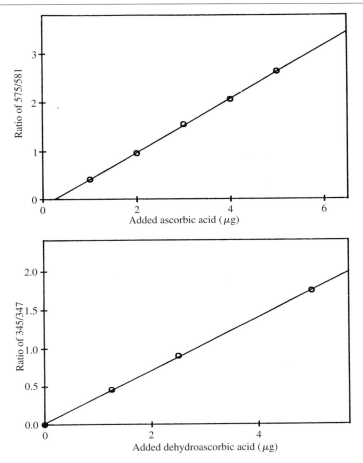

FIG. 5. Calibration curves derived by adding (*top*) increasing amounts of AA to fixed amounts of [$^{13}C_6$]AA or (*bottom*) increasing amounts of DHA to fixed amounts of [$^2H_2$]DHA. [Reprinted with permission from J. Deutsch and J. F. Kolhouse, *Anal. Chem.* **65,** 321 (1993). Copyright 1993 American Chemical Society.]

cult to have a known standard solution from which comparisons can be made.[8] Linear calibration curves with DHA compared to stable isotope standards can be constructed (Fig. 5), but an accurate quantitation is difficult. Figure 6 shows the total ion chromatogram that arises when DHA is solubilized in water. A 10 m$M$ solution made by weighing out DHA may actually contain 3 m$M$ DHA monomer, 3 m$M$ DHA dimer, 3 m$M$ hydrolysis products, and 1 m$M$ contaminants. For our purposes, we make measure-

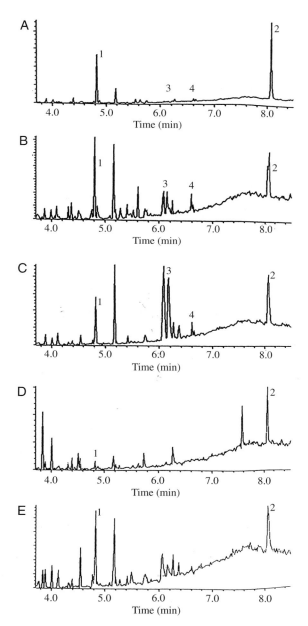

FIG. 6. The total ion chromatograph derived from DHA prior to and during solubilization in water. (A) DHA powder (ICN Biochemicals). (B) DHA powder in water for 30 min. (C) DHA powder in water for 2 hr. (D) DHA dimer powder (Fluka Biochemika). (E) DHA dimer in water for 2 hr. Peak 1 is DHA monomer, peak 2 is DHA dimer, peak 3 is DHA hydrolysis product, and peak 4 is AA. (Reprinted from *J. Chromatogr. A,* in press, J. Deutsch and C. R. Santhosh-Kumar. Copyright 1996 with kind permission from Elsevier Science–NL, Sara Burgerhartstraat 25, 1055 KV Amsterdam, The Netherlands.)

ments relative to an accurate weighing of DHA powder, but this will overestimate the actual quantity of DHA if it is assumed that what is being determined is the compound labeled as DHA in Fig. 1. However, we have shown that the DHA hydrolysis product can be reversed back to AA by using sulfhydryls,[8] which may mean that the hydrolysis product we observe by GC/MS does not lead to the irreversible loss of AA activity. Furthermore, Fig. 6 shows that the DHA dimer is an important component in DHA solutions that have been dried. However, we have shown that the DHA dimer does not appear to be stable in solution (Table I). Therefore, we believe that the isotope dilution assays comparing weighed quantities of DHA to unknown quantities in biological solutions serve as a reasonable estimate of the oxidized AA present. We believe that GC/MS is preferable to high-performance liquid chromatography–electrochemical (HPLC–EC) detection methods for DHA determination because GC/MS measures the abundance of a substance, whereas HPLC–EC detection determinations indirectly estimate DHA content by subtracting AA values in a sample from AA values in that same sample after reduction. Negative values for DHA content (a physical impossibility) have been published by investigators using these HPLC–EC detection subtraction techniques. The most accurate method to determine DHA content by isotope dilution methods involves determination of the exact quantity of $[^{13}C_6]$DHA monomer added to a sample. Currently this can only be estimated by comparing the total ion chromatogram of an internal standard solution to a solution of "pure" DHA.

## Analysis of Biological Solutions for Ascorbic Acid and Ascorbic Acid Degradative Products

L-Ascorbic acid and related products are relatively unstable in solution, and these substances can change dramatically depending on the methods

TABLE I

FORMATION OF L-DEHYDROASCORBIC ACID DIMER BASED ON RELATIVE ABUNDANCE OF $[M - 189]^+$ IONS[a,b]

| Reaction | $m/z$ 615 | $m/z$ 617 | $m/z$ 619 |
|---|---|---|---|
| A × 2 hr, then add B × 2 hr | 1.02 ± 0.04 | 2.13 ± 0.40 | 1.20 ± 0.41 |
| B × 2 hr, then add A × 2 hr | 1.39 ± 0.10 | 2.36 ± 0.42 | 0.84 ± 0.13 |
| A + B × 4 hr | 0.91 ± 0.18 | 1.66 ± 0.26 | 0.74 ± 0.16 |

[a] Reprinted from *J. Chromatogr. A.* **724,** J. C. Deutsch and C. R. Santhosh-Kumar, 271. Copyright 1996 with kind permission from Elsevier Science–NL, Sara Burgerhartstraat 25, 1055 KV Amsterdam, The Netherlands.

[b] From AA and $[^2H_2]$AA during 4-hr incubations with cupric sulfate. A, AA; B, $[^2H_2]$AA.

used for sample processing. For example, we have been able to form 2,3-diketo-4,5,5,6-tetrahydroxyhexanoic acid quantitatively from AA during the use of anion-exchange methods. Stable-isotope dilution methods have a theoretical advantage over other techniques in that the internal standard degrades at the same rate as the endogenous substance, so the ratio remains fixed.

To determine the amount of AA in a sample involves two critical steps. The volume of the sample must be accurately determined and the amount of added internal standard must be accurately measured. Afterward, the ion abundance of the endogenous AA (based on $m/z$ 575) is compared to the ion abundance of the internal standard $[^{13}C_6]AA$ ($m/z$ 581). For example, if 5.0 $\mu$g of $[^{13}C_6]AA$ is added to 100 $\mu$l of plasma and following analysis the integrated ion abundance of $m/z$ 575 is one-fourth the integrated abundance of $m/z$ 581, the AA content in the sample is (5.0 $\mu$g/ 0.100 ml) (1/4) or 12.5 $\mu$g/ml. However, we have also observed that stable isotopes enter into an equilibrium with the endogenous compounds, so that stable-isotope standards that contain proportionate amounts of oxidized AA products must be added carefully when measuring those oxidized compounds. To minimize changes associated with sample purification, we have begun performing analyses of AA and AA-related products on small aliquots of plasma or urine directly. These samples have measurable AA, DHA, and other related organic acids that arise during AA degradation.

L-Ascorbic Acid

Using GC/MS analysis, we have studied the *in vitro* oxidation and reduction reactions that AA and DHA undergo. As mentioned previously for DHA analysis, the situation is complicated. It appears that several competing reactions may be simultaneously occurring. Analysis by GC/MS has the advantage in that we can be sure a product is derived from AA, because we can perform parallel reactions with $[6,6-^2H_2]AA$ and $[^{13}C_6]AA$. These parallel reactions provide structural information (e.g., carbon loss is easily detected by comparing products derived from AA compared to $[^{13}C_6]AA$). We have observed that different AA oxidation products arise depending on the methods used to oxidize the system.[5] This may have profound biological implications. For example, $H_2O_2$ oxidation primarily creates 2,3-diketo-4,5,5,6-tetrahydroxyhexanoic acid. However, cupric sulfate oxidation of AA leads to the formation of a *threo*-hexa-2,4-dienoic acid lactone. Both these species are unstable and would be expected to interact differently with biomolecules. These variable reaction routes may partially explain observations by others in which AA was better than DHA in protecting low-density lipoprotin from oxidation by peroxides, whereas DHA was better than AA in protecting low-density lipoprotein from oxidation by cupric ion.[10]

## Summary

L-Ascorbic acid, DHA, and the oxidized products derived from AA can be accurately measured using GC/MS. Owing to the complex nature of the reactions through which AA proceeds, we believe that GC/MS is currently the procedure of choice in making AA-related measurements. The methods described are useful in defining reactions involving AA. The methods may indicate *in vivo* oxidative injury and may allow the use of AA-derived products to determine if antioxidant modulations are effective.

[10] K. L. Retsky, M. W. Freeman, and B. Frei, *J. Biol. Chem.* **268**, 1304 (1993).

# [3] Expression of Recombinant L-Gulono-γ-lactone Oxidase

*By* KUNIO YAGI and MORIMITSU NISHIKIMI

L-Gulono-γ-lactone oxidase (L-gulono-γ-lactone : oxygen 2-oxido-reductase, EC 1.1.3.8; GLO), which possesses a covalently bound flavin adenine dinucleotide (FAD), catalyzes the oxidation of L-gulono-γ-lactone in the final step of L-ascorbic acid biosynthesis. Most species of higher animals possess the enzyme; however, among mammalian species, humans, other primates, and guinea pigs do not and, as a result, cannot synthesize this vitamin. This is the reason why they are subject to scurvy, a vitamin C deficiency disease, if the vitamin is not supplied sufficiently from their diet. Enzymological and molecular biological studies have demonstrated that their inability to synthesize L-ascorbic acid is due to the lack of GLO stemming from a genetic defect.[1]

To clarify the genetic basis for this enzyme deficiency, we analyzed the sequence of cDNA encoding rat liver GLO[2] and found that both humans

[1] M. Nishikimi and K. Yagi, in "Subcellular Biochemistry" (J. R. Harris, ed.), Vol. 25, p. 17. Plenum, Berlin, 1996.
[2] T. Koshizaka, M. Nishikimi, T. Ozawa, and K. Yagi, *J. Biol. Chem.* **263**, 1619 (1988).

and guinea pigs possess nonfunctional GLO genes.[3] The genes of both species were found to be highly altered in both structure and sequence.[4,5] In contrast, only one point mutation was found to be the cause of the GLO deficiency in the mutant osteogenic disorder Shionogi (ODS) rat, which suffers from osteogenic disorder, a symptom of juvenile scurvy, when fed a vitamin C-deficient diet.[6]

We further confirmed the reality of the etiology of the GLO deficiency of the ODS rat by constructing minigenes for wild-type and mutant GLOs and introducing them into COS-1 cells.[6,7] The enzyme was found to be expressed in the microsomal fraction of the cell, as is the case for rat liver cells. Because the efficiency of expression was not high, owing to low efficiency of transfection, we also utilized a baculovirus vector for the expression.[8] The specific activity of GLO expressed in silkworm cells was comparable to the high level found in rat liver microsomes. Using the baculovirus expression system, we were able to produce the apoenzyme of GLO by culturing virus-infected cells in a riboflavin-deficient medium.

In this chapter, methods utilized for the expression of the recombinant GLO and for detection of the expressed enzyme are described.

## Expression of L-Gulono-γ-lactone Oxidase in COS-1 Cells

The 1.6-kilobase pair (kbp) BclI–BamHI portion of rat GLO cDNA was inserted into pSVL (Pharmacia LKB Biotechnology, Tokyo, Japan) such that it could be expressed under the control of the simian virus 40 (SV40) late promoter. Details of the construction of this GLO minigene (pSVL-GLO) are described in Ref. 7. This minigene was expressed in COS-1 cells, and its expression was assessed either by measurement of GLO activity or by immunohistochemical staining of GLO protein.

Expression of this minigene is carried out as follows: COS-1 cells are cultured in Dullbecco's modified Eagle's medium (DMEM) supplemented with 10% (v/v) fetal bovine serum (FBS) at 37° overnight in humidified 5% $CO_2$–95% air. The cells ($1–2 \times 10^6$) are transfected with pSVL-GLO (10–20 $\mu$g) by the calcium phosphate coprecipitation method,[9] and incubated in

[3] M. Nishikimi, T. Koshizaka, T. Ozawa, and K. Yagi, *Arch. Biochem. Biophys.* **267,** 842 (1988).

[4] M. Nishikimi, T. Kawai, and K. Yagi, *J. Biol. Chem.* **267,** 21967 (1992).

[5] M. Nishikimi, R. Fukuyama, S. Minoshima, N. Shimizu, and K. Yagi, *J. Biol. Chem.* **269,** 13685 (1994).

[6] T. Kawai, M. Nishikimi, T. Ozawa, and K. Yagi, *J. Biol. Chem.* **267,** 21973 (1992).

[7] K. Yagi, T. Koshizaka, M. Kito, T. Ozawa, and M. Nishikimi, *Biochem. Biophys. Res. Commun.* **177,** 659 (1991).

[8] M. Nishikimi, J. Kobayashi, and K. Yagi, *Biochem. Mol. Biol. Int.* **33,** 313 (1994).

[9] F. L. Graham and A. J. C. Van der Eb, *Virology* **52,** 456 (1973).

the same medium at 37° for 4 hr. The medium is then replaced with serum-free DMEM, after which the cells are subjected to glycerol shock and then incubated in DMEM supplemented with 10% (v/v) FBS for 72 hr. The expression of GLO reaches its maximum by the end of this incubation.

For the GLO assay, the transfected cells are disrupted in phosphate-buffered saline (PBS) containing 0.5% (w/v) sodium deoxycholate, and the resulting cell extract is incubated at 37° for 15 min with 2.5 m$M$ L-gulono-$\gamma$-lactone in 50 m$M$ sodium phosphate buffer (pH 7.0) containing 50 m$M$ sodium citrate and 1 m$M$ dithiothreitol (DTT). The L-ascorbic acid formed is measured by high-performance liquid chromatography (HPLC) as detailed in Ref. 10.

For immunohistochemical staining, the cells are cultured on coverglasses in wells of a six-well plate, transfected with pSVL-GLO, cultured as described previously, and fixed for 2 min in methanol–acetone (1:1, v/v). The cells are then washed with PBS and incubated at 4° overnight with 1000-fold diluted anti-rat GLO rabbit antiserum in PBS containing 3% (w/v) skim milk. The coverglasses are washed with three changes of PBS, covered with 2000-fold diluted anti-rabbit IgG–horseradish peroxidase conjugate (Bio-Rad Laboratories, Richmond, CA) in PBS containing 3% (w/v) skim milk, and allowed to stand at room temperature for 1 hr. After the coverglasses have been washed again with three changes of PBS, the color is developed with 3,3′-diaminobenzidine and $H_2O_2$.

The cells used in this experiment were derived from the GLO-deficient kidney of the African green monkey; therefore, positively stained cells are clearly discriminated from cells that have not been transfected. After a 72-hr incubation, 5–10% of the cells are positive and the GLO activity in the cell extract reaches a maximum (0.25 nmol/min/mg protein). In a cell fractionation experiment, the highest activity is found in the microsomal fraction (100,000 $g$ pellet).

## Expression of L-Gulono-$\gamma$-lactone Oxidase with Baculovirus Vector

High-level expression of rat GLO was achieved with a baculovirus expression system using silkworm cells.[8] The 1.6-kbp BclI–BamHI portion of rat GLO cDNA was inserted into the transfer vector pBM030 as specified in Ref. 8, and the resulting expression plasmid was cotransfected with wild-type Bombyx mori nuclear polyhedrosis virus DNA into cells of the silkworm cell line BmN4 by the calcium phosphate coprecipitation method. Recombinant viruses were cloned by visual screening for the inclusion-

[10] M. Kito, N. Ohishi, and K. Yagi, Biochem. Int. 24, 131 (1991).

negative phenotype. The details of these procedures are described by Maeda.[11]

Expression of recombinant GLO is carried out as follows: Monolayers of BmN4 cells $(1 \times 10^6)$ are infected with the recombinant virus at a multiplicity of infection of 10 plaque-forming units per cell, cultured at 27° in MGM-448 medium supplemented with 10% (v/v) FBS, and harvested 5 days later. The cells are sonicated in PBS, and the resulting cell extract is assayed for GLO as described in Ref. 10, except for omission of dithiothreitol from the incubation mixture. The specific activity of GLO in the extract is routinely 16–20 nmol/min/mg protein, which is more than that observed with rat liver microsomes.

The expression of recombinant GLO can also be assessed by Western blot analysis. The cell extracts are electrophoresed on a 7% (w/v) polyacrylamide gel (1-mm thick slab) in the presence of sodium dodecyl sulfate (SDS) by the method of Kadenbach et al.[12] The protein in the gel is electroblotted to a nitrocellulose membrane essentially as described by Towbin et al.,[13] and the GLO on the membrane is visualized by the peroxidase–antiperoxidase (PAP) method[14] as follows: After having been soaked for 1 hr in a solution containing 0.5% (w/v) skim milk, Tris-HCl (pH 8.0), and 0.1 $M$ NaCl, the membrane is incubated overnight with 1000-fold diluted anti-GLO rabbit antiserum with 0.5% (w/v) skim milk solution containing 0.01% (w/v) sodium azide, washed with 0.5% (w/v) skim milk solution twice each time for 10 min, and then incubated for 1 hr with 500-fold diluted anti-rabbit IgG goat antiserum (Medical Biological Laboratories, Tokyo, Japan). The membrane is washed as described previously and then incubated for 1 hr with 1000-fold diluted rabbit PAP (Cappel Laboratories, Cochranville, PA). After having been washed as described, the membrane is soaked for 10 min in a color-developing solution containing 0.5% (w/v) 4-chloro-1-naphthol, 0.015% (v/v) $H_2O_2$, 16.7% (v/v) methanol, 17 m$M$ Tris-HCl (pH 7.5), and 0.4 $M$ NaCl, and is then transferred to water for termination of the reaction. In this immunostaining procedure, the dilution of the reagents is done with 0.5% (w/v) skim milk solution unless otherwise specified, and the incubations are carried out at room temperature. The analysis indicates that GLO expressed in the silkworm cells has the same molecular weight ($\sim$51,000) as that of rat liver GLO. FAD is found to be covalently bound to the recombinant GLO protein by the observation

[11] S. Maeda, in "Invertebrate Cell System Applications" (J. Mitsuhashi, ed.), p. 167. CRC Press, Boca Raton, Florida, 1989.
[12] B. Kadenbach, J. Jarausch, R. Hartman, and P. Merle, Anal. Biochem. **129,** 517 (1983).
[13] H. Towbin, T. Staehelin, and J. Gordon, Proc. Natl. Acad. Sci. U.S.A. **76,** 4350 (1979).
[14] W. F. Glass II, R. C. Briggs, and L. S. Hnilica, Science **211,** 70 (1981).

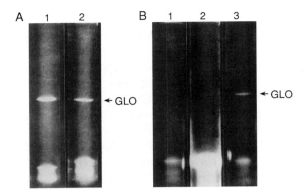

FIG. 1. Visualization of recombinant GLO by its flavin fluorescence. (A) Covalent binding of flavin in recombinant GLO. Recombinant GLO was expressed in BmN4 cells under normal conditions. Lane 1, cell extract (80 μg of protein); lane 2, rat liver microsomes (80 μg of protein). They were subjected to SDS–polyacrylamide gel electrophoresis on disk gels. After the gels were soaked in 10% (v/v) acetic acid for 10 min, GLO was visualized by illumination with UV light of 365-nm wavelength. (B) Test for covalent binding of FAD to the apoprotein of GLO. The apoprotein of GLO was expressed in BmN4 cells under the strictly riboflavin-deficient conditions. Lane 1, cell extract (147 μg of protein) alone; lane 2, cell extract (147 μg of protein in 50 μl of PBS) that had been incubated with FAD (0.2 m*M* final concentration) at 37° for 5 min; lane 3, a reference for such amount of the holoenzyme of recombinant GLO (38 μg of protein as cell extract) that had the same activity as that of the apoprotein-rich cell extract measured in the presence of excess FAD. (From Ref. 8 with permission.)

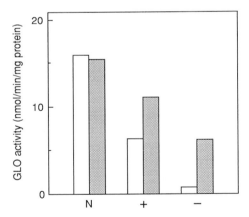

FIG. 2. Expression of GLO in BmN4 cells infected with the recombinant virus under riboflavin-deficient conditions. Cells were cultured for 3 weeks in MGM-448 medium minus riboflavin but supplemented with 10% (v/v) FBS, and then infected with the recombinant virus and cultured for 5 days in the same medium (+) or in MGM-448 medium minus riboflavin without supplementation of FBS (−). The cells cultured under normal conditions (N) were obtained on the fifth day after the start of virus infection. The GLO activity in the cell extract prepared from each culture was measured in the absence (open bars) or presence (solid bars) of FAD. (From Ref. 15 with permission.)

that yellowish-green fluorescence appears when an SDS gel containing the enzyme is soaked in 10% (v/v) acetic acid (see Fig. 1A).

The baculovirus expression system is used to produce the apoprotein of GLO by culturing cells in a medium that is deficient in riboflavin, the precursor of the FAD that is covalently bound to the apoprotein in the holoenzyme of GLO. The experimental procedures are as follows: BmN4 cells are cultured for 3 weeks in MGM-448 medium minus riboflavin but supplemented with 10% (v/v) FBS and then infected with the recombinant virus as described previously. The infected cells are cultured for 5 days in MGM-448 medium minus riboflavin without supplementation of FBS. During this time the GLO activity in the extract prepared from these cells decreases to a great degree, and it is increased about fivefold by the addition of FAD to the GLO assay mixture (Fig. 2).[15] When the recombinant virus-infected cells are cultured in MGM-448 medium minus riboflavin but supplemented with 10% FBS, the effect of FAD addition is less than that attained when they are in the medium without FBS. In contrast, the addition of FAD to an assay mixture containing the extract of recombinant virus-infected cells cultured under normal conditions has no effect on the GLO activity. These results indicate that the cell extract obtained from cells cultured in the riboflavin-deficient medium contains a substantial amount of the apoprotein. In our study the question of whether FAD could bind covalently to the apoprotein is addressed by fluorometry of FAD on SDS–polyacrylamide gels. The extract from recombinant virus-infected cells cultured under the riboflavin-deficient conditions is incubated with FAD (0.2 m$M$) at 37° for 5 min and then subjected to SDS–polyacrylamide gel electrophoresis. The intensity of fluorescence at the position of GLO on the gel soaked in 10% (v/v) acetic acid is not increased compared with that for the cell extract alone, as shown in Fig. 1B, indicating that covalent binding of FAD to the apoprotein does not occur. Thus, it is clear that noncovalent interaction between FAD and the apoprotein can elicit catalytic activity of the enzyme. This finding is in contrast with the observation that the formation of a covalent bond between FAD and the apoprotein parallels the appearance of enzymatic activity in the case of 6-hydroxy-D-nicotine oxidase.[16]

[15] M. Nishikimi, K. Yagi, and J. Kobayashi, in "Flavins and Flavoproteins 1993" (K. Yagi, ed.), p. 791. Walter de Gruyter, Berlin, 1994.
[16] R. Brandsch and V. Bichler, J. Biol. Chem. **266**, 19056 (1991).

## [4] Purification and Characterization of Glutathione-Dependent Dehydroascorbate Reductase from Rat Liver

*By* Emilia Maellaro, Barbara Del Bello, Lidia Sugherini, Mario Comporti, and Alessandro F. Casini

### Introduction

Ascorbic acid (AA) plays a major role as a water-soluble antioxidant, acting by both scavenging aqueous free radicals[1-3] and reducing tocopheroxyl radicals back to $\alpha$-tocopherol.[4,5] Furthermore, it is involved in several metabolic pathways as a reductive agent.[6] In all these instances, AA undergoes a one-electron oxidation to semidehydroascorbate (ascorbyl radical). Dismutation or further oxidation converts semidehydroascorbate to dehydroascorbate (DHA). Dehydroascorbate is metabolized by the cell to diketogulonic acid, which is further catabolized and lost in the urine.

The antioxidant potential of AA is restored by cellular systems capable of reducing its oxidized forms, thus maintaining an effective steady state concentration of the antioxidant.

As far as the reducing process of the two-electron oxidized form (DHA) is concerned, the role of glutathione (GSH) as hydrogen donor (reaction below) has long been recognized.[7-10]

$$DHA + 2GSH \rightarrow AA + GSSG$$

After three decades of conflicting results on the existence of an enzyme-driven reaction,[7-10] Wells *et al.*[11] reported that purified glutaredoxin and protein disulfide-isomerase (PDI) can catalyze the above reaction. A novel cytosolic GSH-dependent DHA reductase has been discovered in our labo-

[1] M. Nishikimi, *Biochem. Biophys. Res. Commun.* **63**, 463 (1975).
[2] B. Halliwell, M. Wasil, and M. Grootveld, *FEBS Lett.* **213**, 15 (1978).
[3] B. Frei, L. England, and B. N. Ames, *Proc. Natl. Acad. Sci. U.S.A.* **86**, 6377 (1989).
[4] J. E. Packer, T. F. Slater, and R. L. Willson, *Nature (London)* **278**, 737 (1979).
[5] M. Scarpa, A. Rigo, M. Maiorino, F. Ursini, and C. Gregolin, *Biochim. Biophys. Acta* **801**, 215 (1984).
[6] H. Padh, *Nutrition Rev.* **49**, 65 (1991).
[7] R. E. Hughes, *Nature (London)* **203**, 1068 (1964).
[8] R. Bigley, M. Riddle, D. Layman, and L. Stankova, *Biochim. Biophys. Acta* **659**, 15 (1981).
[9] S. Basu, S. Som, S. Deb, D. Mukherijee, and I. B. Chatterjee, *Biochem. Biophys. Res. Commun.* **90**, 1335 (1979).
[10] R. L. Stahl, L. F. Liebes, and R. Silber, *Biochim. Biophys. Acta* **839**, 119 (1985).
[11] W. W. Wells, D. P. Xu, Y. Yang, and P. A. Rocque, *J. Biol. Chem.* **265**, 15361 (1990).

ratory.[12] This chapter describes the procedures for the purification of such an enzyme.

## Dehydroascorbate Reductase Assay

*Principle*

Dehydroascorbate reductase catalyzes the reduction of DHA, using GSH as hydrogen donor. The activity is determined by the direct spectrophotometric method essentially according to Stahl *et al.*,[10] i.e., monitoring the change in $A_{265}$ associated with the formation of AA.

*Procedure*

*Reagents*

Buffer A: 100 m$M$ potassium phosphate buffer, pH 7.2
GSH (Boehringer Mannheim, Indianapolis, IN) 20 m$M$ in buffer A
Freshly prepared DHA (Aldrich, Milwaukee, WI), 5 m$M$ in $H_2O$. To avoid the troublesome storage of the easily oxidizable DHA powder, DHA solution can alternatively be prepared by dissolving ascorbic acid (Aldrich) to 5 m$M$ in $H_2O$ and then by exposing it to liquid bromine; excess bromine is removed by bubbling the solution with $N_2$.[13] In both cases DHA solutions must be freshly prepared and kept ice-cold

*Assay.* Two quartz microcuvettes with a 1-cm light path are equilibrated at 30°. In a final volume of 1 ml of buffer A, 50 $\mu$l of GSH stock solution (1 m$M$, final concentration) and 50–100 $\mu$l of the enzyme preparation are added. The reaction is initiated by adding 50 $\mu$l of DHA stock solution (0.25 m$M$, final concentration) and the absorbance at 265 nm is recorded for 2 min; within this time the reaction is linear. The results are calculated as $\Delta A_{265}$ min$^{-1}$ by using a molar absorption coefficient of 14,800. The assay shows a linear dependence on enzyme preparation up to $\Delta A_{265}$ values of 0.120 min$^{-1}$. If required the enzyme fractions are diluted to achieve this activity. The value of nonenzymatic DHA reduction, measured in the absence of enzyme preparation, is subtracted.

One unit of DHA reductase is defined as the net formation of 1 $\mu$mol of AA/min at 30°.

---

[12] E. Maellaro, B. Del Bello, L. Sugherini, A. Santucci, M. Comporti, and A. F. Casini, *Biochem. J.* **301,** 471 (1994).
[13] R. C. Rose, *Am. J. Physiol.* **256,** F52 (1989).

Purification Procedures

Unless otherwise stated all steps are performed at 0–4°.

*Cytosol Preparation*

Male Sprague-Dawley rats (200–250 g) are used. Following ether anesthesia of the animals (10–12 for each preparation), the livers are perfused through the portal vein with saline solution. Liver tissue is homogenized with 2 vol of ice-cold 100 m$M$ potassium phosphate buffer, pH 7.2 (buffer A), using a Potter–Elvehjem homogenizer. The homogenate is centrifuged at 20,000 $g$ for 15 min, and the supernatant is further centrifuged at 100,000 $g$ for 60 min. The cytosolic fraction is diluted 1.5-fold with buffer A and used for the following step.

*Ammonium Sulfate Fractionation*

Powdered $(NH_4)_2SO_4$ (326 g/liter) is slowly added at 0° to cytosol to give 55% saturation. Sufficient 1 $N$ NH$_4$OH is added to keep the pH constant. After being stirred for 30 min, the mixture is centrifuged at 18,000 $g$ for 30 min at 4°. To the supernatant, powdered ammonium sulfate (161 g/liter) is added to give 75% saturation and the mixture is stirred and centrifuged again. The pellet is redissolved in 30 ml of 10 m$M$ potassium phosphate buffer, pH 7.8 (buffer B), and dialyzed overnight at 4° (Visking tubing, molecular weight cutoff 10,000–12,000; Serva, Heidelberg, Germany) against two changes of a total of 4 liters of the same buffer. The dialyzed protein solution is then centrifuged at 1500 $g$ for 10 min at 4° and the resulting supernatant is used for the following step.

*DEAE-Sepharose Chromatography*

The protein solution is applied to a DEAE-Sepharose CL-6B (Sigma, St. Louis, MO) column (2.5 × 9 cm) previously equilibrated with buffer B. After rinsing with 250 ml of the same buffer, the column is eluted with an NaCl linear gradient set up with 300 ml of buffer B and 300 ml of buffer B added with 150 m$M$ NaCl. The mixture is delivered to the column at a flow rate of 90 ml/hr. Eluate fractions of 10 ml are collected and assayed for DHA reductase activity. Fractions with higher specific activity are pooled and concentrated to 1.5 ml by ultrafiltration (Centriprep 30; Amicon, Danvers, MA).

*Gel-Filtration Chromatography*

The sample obtained is applied to a Sephadex G-100 SF (Sigma) column (2.5 × 50 cm) equilibrated with buffer A. Proteins were eluted with the same buffer at a flow rate of 0.25 ml/min. After discarding 100 ml, eluate fractions of 2 ml are collected. Fractions showing higher DHA reductase activity are pooled, concentrated to 0.5 ml by ultrafiltration (Centricon 10; Amicon), and used for the next step.

## Reactive Red 120 Chromatography

The sample obtained is applied to a Reactive Red 120 3000-CL (Sigma) column (1.5 × 2 cm) equilibrated with buffer A containing 1.5 m$M$ $MgCl_2$. Dehydroascorbate reductase activity is recovered by eluting the column with 5 ml of the same buffer at gravity flow. Eluate fractions of 0.5 ml are collected. Fractions containing DHA reductase activity are pooled and stored at 0–4°. This protein solution is used as the final enzyme preparation.

## Yield and Purity

A representative purification is summarized in Table I. The final yield is around 5%. The final enzyme purification is apparently pure on sodium dodecyl sulfate–polyacrylamide gel electrophoresis (SDS–PAGE), whereby a single protein band is observed.[12]

## Physical Properties

### Molecular Weight

Gel filtration carried out either by use of a Sephadex G-100 SF column (same conditions as in Purification Procedures) or by high-performance

TABLE I
PURIFICATION OF GSH-DEPENDENT DEHYDROASCORBATE REDUCTASE FROM
RAT LIVER CYTOSOL[a]

| Purification step | Total protein (mg) | Total activity[b] | Specific activity[c] | Purification (-fold) | Yield (%) |
|---|---|---|---|---|---|
| Cytosol | 9,699.0 | 12,879 | 1.3 | 1.0 | 100 |
| (NH4)2SO4, 55–75% | 2,173.3 | 7,636 | 3.5 | 2.7 | 59 |
| DEAE-Sepharose CL-6B | 20.3 | 3,256 | 160.8 | 123.7 | 25 |
| Sephadex G-100 SF | 2.4 | 1,256 | 519.2 | 399.4 | 10 |
| Reactive Red 120 | 0.85 | 649 | 764.5 | 588.1 | 5 |

[a] Reproduced with minor modifications from Maellaro et al.[12] (with permission from *Biochemical Journal* and The Biochemical Society, London).
[b] Determined as ascorbic acid (AA) formed [(milliunits/ml) × total volume].
[c] Determined as ascorbic acid (AA) formed (milliunits/mg protein).

liquid chromatography (HPLC)[12] is employed to determine the molecular weight of the native enzyme, giving a value of 48,900 and 48,500, respectively. The molecular weight of denatured enzyme is 31,000 as measured by SDS–PAGE. A single protein band is observed both in the presence and absence of 2-mercaptoethanol, which indicates the absence of interchain disulfide bridges. A similar value (31,300) is obtained when molecular weight is determined by HPLC under denaturing conditions.[12] Such a discrepancy between molecular weight values of native and denatured forms could be ascribed to interactions between charged protein groups and gel matrix; the possibility that the native enzyme is a homodimer without interchain covalent bonds must also be considered.

## pH Dependence and Heat Stability

The optimal pH range is 7.5–8.0. Exposure of purified enzyme to 50 or 75° for 5 min decreases its activity by 50 and 90%, respectively.

## Amino Acid Sequence

The N terminus of the protein appears to be blocked. N-terminal primary sequences of five peptides are obtained after fragmentation of the protein with CNBr and $N$-chlorosuccinimide. No sequence similarity is found with any of the known primary structures contained in the protein Identification Researce Databank.[12]

## Catalytic Properties

## Kinetics

The $K_m$ for GSH, calculated in the presence of 1.5 m$M$ DHA (at pH 7.2), is 2.8 m$M$ ± 0.6 with a $V_{max}$ of 4.5 ± 1.3 units/mg protein. The apparent $K_m$ for DHAA, calculated in the presence of 3 m$M$ GSH, is 245 $\mu M$ ± 62 with a $V_{max}$ of 1.9 ± 0.1 units/mg protein.

## Substrate Specificity

The enzyme is capable of reducing isodehydroascorbate as effectively as DHA. Cysteine and acetylcysteine have proven to be poor substrates (8 and 4% of the GSH-driven activity, respectively); no enzymatic DHA reduction is detected with lipoate, coenzyme A, dithiothreitol, or 2-mercaptoethanol.

## Acknowledgments

This work was supported by the Italian Research Council (CNR), Strategy Project ACRO. Additional funds were derived from the Association for International Cancer Research (AICR), United Kingdom, and from the Italian Ministry of University and Scientific Research (40%, Free Radical Pathology).

# [5] Peptidylglycine $\alpha$-Amidating Monooxygenase: An Ascorbate-Requiring Enzyme

*By* Aparna S. Kolhekar, Richard E. Mains, and Betty A. Eipper

Peptidylglycine $\alpha$-amidating monooxygenase (PAM) (also referred to as amidating enzyme) is a bifunctional enzyme that catalyzes the carboxy-terminal amidation of glycine-extended peptides.[1,2] Amidation is a two-step reaction (Fig. 1), with the first step being hydroxylation at the $\alpha$-carbon of the carboxy-terminal glycine, catalyzed by peptidylglycine $\alpha$-hydroxylating monooxygenase (PHM) (EC 1.14.17.3), a copper, ascorbate, and molecular oxygen-dependent enzyme. The second step of the reaction is dealkylation of the peptidyl $\alpha$-hydroxyglycine intermediate, catalyzed by peptidyl $\alpha$-hydroxyglycine $\alpha$-amidating lyase (PAL) (also called peptidyl-amidoglycolate lyase) (EC 4.3.2.5), a divalent metal ion-dependent enzyme. Peptidylglycine $\alpha$-amidating monooxygenase is the only enzyme known to catalyze the $\alpha$-amidation of peptides. Although PAM is encoded by a single gene, soluble and membrane-bound monofunctional and bifunctional forms are generated by tissue-specific alternative splicing and endoproteolytic cleavage.

The PHM-catalyzed reaction requires a reducing cofactor and ascorbate is the most likely physiological reductant,[3-6] although the cofactor requirement *in vivo* can be cell-type specific.[7] Overall, the enzyme converts 2 mol of reduced ascorbate into 2 mol of semidehydroascorbate, consuming

[1] B. A. Eipper, S. L. Milgram, E. J. Husten, H-Y. Yun, and R. E. Mains, *Protein Sci.* **2,** 489 (1993).

[2] B. A. Eipper, D. A. Stoffers, and R. E. Mains, *Annu. Rev. Neurosci.* **15,** 57 (1992).

[3] B. A. Eipper, R. E. Mains, and C. C. Glembotski, *Proc. Natl. Acad. Sci. U.S.A.* **80,** 5144 (1983).

[4] B. A. Eipper and R. E. Mains, *Am. J. Clin. Nutr.* **54,** 1153S (1991).

[5] J. S. Kizer, R. C. Bateman, C. R. Miller, J. Humm, W. H. Busby, and W. W. Youngblood, *Endocrinology* **118,** 2262 (1986).

[6] C. C. Glembotski, B. A. Eipper, and R. E. Mains, *J. Biol. Chem.* **259,** 6385 (1984).

[7] V. May, R. E. Mains, and B. A. Eipper, *Hormone Res.* **32,** 18 (1989).

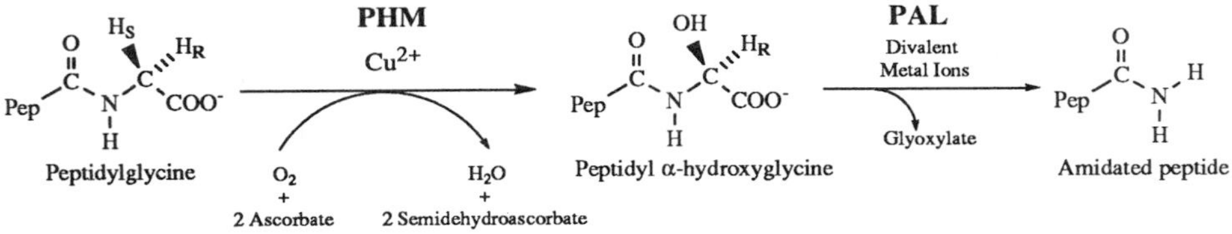

FIG. 1. The two-step peptide $\alpha$-amidation reaction.

1 mol of ascorbate for each mole of $\alpha$-amidated product peptide.[8,9] The optimal ascorbate concentration is typically about 1 m$M$.[3,5,10] Dopamine $\beta$-monooxygenase performs a similar reaction, using the same cofactors.[1,10,11]

## Assay Principle

An N-acetylated tripeptide substrate, $\alpha$-$N$-acetyl-Tyr-Val-Gly or $\alpha$-$N$-acetyl-Tyr-Val-$\alpha$-hydroxyglycine, is used as the PHM or PAL substrate, respectively. Reaction progress is followed by including a trace amount of $^{125}$I-labeled substrate ([$^{125}$I]Ac-Tyr-Val-Gly or [$^{125}$I]Ac-Tyr-Val-$\alpha$-hydroxyglycine) in the reaction. In the presence of $Cu^{2+}$, ascorbate, and molecular oxygen, PHM converts its substrate into Ac-Tyr-Val-$\alpha$-hydroxyglycine, the substrate for PAL. The presence of reduced ascorbate results in the need for catalase to prevent enzyme inactivation.[3,6] At basic pH or in the presence of PAL, this intermediate is dealkylated to form the amidated dipeptide, Ac-Tyr-Val-NH$_2$, and glyoxylate. At pH 7, the PHM substrate and the PAL substrate are both negatively charged while the amidated end product of the reaction is uncharged, allowing separation of substrate and product on the basis of their differential solubility in ethyl acetate. This assay[10] is based on the method developed by Mizuno et al. (1986).[12]

## Extraction of Tissues and Cell Lines

### Reagents for Extraction of Soluble and Membrane PAM

Extraction buffer: 20 m$M$ NaTES [$N$-tris(hydroxymethyl)methyl-2-aminoethanesulfonic acid, sodium salt] and 10 m$M$ mannitol, pH 7.4

[8] A. S. N. Murthy, H. T. Keutmann, and B. A. Eipper, Mol. Endocrinol. 1, 290 (1987).

[9] D. J. Merkler, R. Kulathila, A. P. Consalvo, S. D. Young, and D. E. Ash, Biochemistry 31, 7282 (1992).

[10] B. A. Eipper, A. S. W. Quon, R. E. Mains, J. S. Boswell, and N. J. Blackburn, Biochemistry 34, 2857 (1995).

[11] K. Wimalasema and D. S. Wimalasema, Anal. Biochem. 197, 353 (1991).

[12] K. Mizuno, J. Sakata, M. Kojima, K. Kangawa, and H. Matsuo, Biochem. Biophys. Res. Commun. 137, 984 (1986).

Extraction buffer with detergent: Preceding buffer with 1% (v/v) Triton X-100 [purified Surfact-Amps X-100 from Pierce Chemical Co (Rockford, IL); the quality of the detergent is critical]

Phenylmethylsulfonyl fluoride (PMSF) (GIBCO-BRL, Gaithersburg, MD): Freshly prepared 30-mg/ml solution in ethanol

Protease inhibitor mix, 100× stock: Leupeptin (0.2 mg/ml), pepstatin (0.2 mg/ml), lima bean trypsin inhibitor (5.0 mg/ml; Sigma Chemical Co., St. Louis, MO), and benzamidine (1.6 mg/ml) dissolved in water and stored at −20°.

## Extraction of Total or Soluble and Crude Particulate PAM

For measurement of total PAM activity, tissue or cells are homogenized in 10 vol of extraction buffer with detergent; PMSF and protease inhibitor mix are added before homogenization using a Potter–Elvehjem homogenizer. Tissue culture cell pellets can be effectively extracted by three cycles of freezing (on dry ice) and thawing; tissue requires homogenization. Samples are centrifuged at 1000 $g$ for 5 min at 4° to remove debris and supernatants are stored frozen or kept on ice.

For the separation of soluble from crude particulate PAM, samples are homogenized in extraction buffer without detergent and centrifuged at 1000 $g$ for 5 min at 4° to remove debris. The supernatant is centrifuged at $6 \times 10^6 \, g \cdot \text{min}$ at 4°. On centrifugation, the soluble and crude particulate PAM separate in the supernatant and pellet, respectively. The crude particulate fraction is resuspended in extraction buffer with detergent. Extracts can be stored at −80° for months without loss of activity. Alternatively, subcellular fractionation using differential centrifugation followed by sucrose gradient centrifugation can be carried out before assaying samples for PAM activity.

## Assay Method

### Reagents for Preparation of $[^{125}I]$Ac-Tyr-Val-Gly, $[^{125}I]$Ac-Tyr-Val-α-Hydroxyglycine, and Ac-Tyr-Val-α-Hydroxyglycine

Ac-Tyr-Val-Gly (Peninsula Laboratories, Belmont, CA); stock is 1 mg/ml in water (measured spectrophotometrically using an $\varepsilon_{274}$ of 1340 in 0.1 $M$ HCl)

Sodium phosphate (100 m$M$), pH 6.5

Iodobead (Pierce)

Na$^{125}$I, 0.5 mCi (Amersham, Arlington Heights, IL), 100 mCi/ml, specific activity 2200 Ci/mmol

2-Mercaptoethanol

Sep-Pak $C_{18}$ cartridge (Waters, Milford, CT)

Methanol (2%, v/v) in 5 m$M$ sodium phosphate, pH 5.0

Methanol (80%, v/v) in 5 m$M$ sodium phosphate, pH 5.0

Tris-HCl (1 $M$), pH 7.0, containing bromphenol blue (0.5 g/liter)

Freshly prepared ethyl acetate saturated with water (ethyl acetate must always be kept in polypropylene containers)

Recombinant PHM[10]

Catalase, 20 mg/ml (Boehringer Mannheim, Indianapolis, IN)

Trifluoroacetic acid (TFA) (Pierce) and acetonitrile for reversed-phase high-performance liquid chromatography (RP-HPLC) buffers

### Preparation of [$^{125}$I]Ac-Tyr-Val-Gly, PHM Substrate

Preequilibrate a Sep-Pak cartridge by washing it with several column volumes of 80% (v/v) methanol–phosphate buffer and then 2% (v/v) methanol–phosphate buffer. To a microcentrifuge tube containing 1 $\mu$g of Ac-Tyr-Val-Gly in 70 $\mu$l of sodium phosphate buffer, pH 6.5, add 5 $\mu$l (0.5 mCi) of Na$^{125}$I followed by one Iodobead. All handling of $^{125}$I is carried out in a fume hood approved for usage of this isotope. After 1 min at room temperature, stop the reaction by adding 10 $\mu$l of 2-mercaptoethanol. Add 0.24 ml of 1 $M$ Tris-HCl, pH 7.0, and then 0.64 ml of water-saturated ethyl acetate; shake to mix and allow the phases to separate. Discard the upper phase and use 3.0 ml of 2% (v/v) methanol–phosphate buffer, pH 5.0, to transfer the lower phase into a 10-cm$^3$ syringe attached to the preequilibrated Sep-Pak. After applying the sample, rinse the Sep-Pak with 10 ml of 2% (v/v) methanol–phosphate buffer. Elute the substrate from the Sep-Pak in six fractions of 0.5 ml of 80% (v/v) methanol–phosphate buffer; count an aliquot of each fraction and pool the appropriate fractions for use. The [$^{125}$I]Ac-Tyr-Val-Gly and bromphenol blue coelute from the cartridge. Typically, 30% of the $^{125}$I is incorporated into peptide. The entire process should be monitored with a hand-held Geiger counter and all radioactive waste must be discarded properly. Substrate is stable for at least 2 months when stored at 4°.

### Preparation of [$^{125}$I]Ac-Tyr-Val-α-Hydroxyglycine and Ac-Tyr-Val-α-Hydroxyglycine, PAL Substrate

Purified recombinant PHM[10,13] is used to prepare [$^{125}$I]Ac-Tyr-Val-α-hydroxyglycine from [$^{125}$I]Ac-Tyr-Val-Gly. To 1.65 ml of 150 m$M$ NaMES [2-($N$-morpholino)ethanesulfonic acid], pH 4.5, add 9 $\mu$l of 100 $\mu M$ CuSO$_4$, 9 $\mu$l of catalase (20 mg/ml), 100 $\mu$l of [$^{125}$I]Ac-Tyr-Val-Gly (at least 2 ×

[13] E. J. Husten, F. A. Tausk, H. T. Keutmann, and B. A. Eipper, *J. Biol. Chem.* **268**, 9709 (1993).

$10^7$ cpm), 0.9 $\mu$l of 100 $\mu M$ Ac-Tyr-Val-Gly, 18 $\mu$l of 50 m$M$ ascorbate, and 50 ng of recombinant PHM. After 2 hr at 37°, add 1.0 ml of 1 $M$ Tris, pH 7; with the sample in a polypropylene tube, add 4 ml of water-saturated ethyl acetate and shake. When the phases have separated, discard the upper phase and apply the lower phase to a Sep-Pak cartridge as described previously. The quality of the substrate can be tested by diluting $2 \times 10^4$ cpm into 40 $\mu$l of NaMES buffer, pH 5.0, and converting it to amidated product by adding 15 $\mu$l of 1 $N$ NaOH and allowing it to sit at room temperature for 5 min. Radiolabeled PAL substrate is best used within 2 weeks of preparation, after which the assay background rises.

Ac-Tyr-Val-α-hydroxyglycine, the unlabeled PAL substrate, is synthesized using recombinant PHM. A 1-ml reaction containing 100 $\mu M$ Ac-Tyr-Val-Gly, catalase (0.18 mg/ml), 0.5 $\mu M$ CuSO$_4$, 1.0 m$M$ ascorbate, 140 m$M$ NaMES (pH 4.5), and 50 ng of recombinant PHM is incubated at 37° for 2 hr. The reaction solution is centrifuged to remove any particulate material and the supernatant is filtered through a Centricon 10 microconcentrator (Amicon, Danvers, MA). The filtrate is acidified with TFA (final concentration, 0.1%) and applied to a Waters $C_{18}$ $\mu$Bondapak column equilibrated with 0.1% (v/v) TFA–4% (v/v) acetonitrile. The peptides are eluted with a linear gradient to 0.1% (v/v) TFA–20% (v/v) acetonitrile over 60 min. Ac-Tyr-Val-Gly, Ac-Tyr-Val-α-hydroxyglycine, and Ac-Tyr-Val-amide elute at 8.6, 7.5, and 8.3% (v/v) acetonitrile, respectively. The fractions containing Ac-Tyr-Val-α-hydroxyglycine are pooled and concentrated by vacuum centrifugation. The peptide is redissolved in water and the concentration determined spectrophotometrically ($\varepsilon_{274}$ = 1340 in 0.1 $M$ HCl). Ac-Tyr-Val-α-hydroxyglycine is stable when stored in aliquots at −80°.

## Purification of Recombinant PHM

Recombinant PHM is purified from the spent medium of stably transfected chinese hamster ovary (CHO) or hEK-293 cell lines in milligram quantities.[10,13] The proteins in the medium are precipitated using 70% (w/v) ammonium sulfate. The precipitate is dissolved in 0.5 $M$ (NH$_4$)$_2$SO$_4$–20 m$M$ NaTES, pH 7.4, applied to a Pharmacia (Piscataway, NJ) phenyl-Superose hydrophobic interaction column (HR10/10), and eluted with 50 m$M$ NaTES, pH 7.4. The eluate is brought to 50 m$M$ Tris-HCl, pH 8.0, by repeated concentration with a Centricon 10 and dilution with the desired buffer, and then applied to a Pharmacia Mono Q anion-exchange column (HR5/5). A gradient of NaCl elutes the recombinant PHM at approximately 0.1 $M$ NaCl. Initially, the resuspended ammonium sulfate pellet can be applied to a Sephadex G-75 gel filtration column [in 0.5 $M$ (NH$_4$)$_2$SO$_4$–20 m$M$ NaTES, pH 7.4 buffer] prior to phenyl-Superose chromatography.

*Stock Solutions for PHM, PAL, and PAM Assays*

> Catalase, 20 mg/ml (Boehringer Mannheim)
> Enzyme dilution buffer: 20 m$M$ NaTES, 10 m$M$ mannitol, bovine serum albumin (BSA; 1 mg/ml) (Sigma), 1% (v/v) Triton X-100 (purified Surfact-Amps X-100; Pierce)
> 150 m$M$ NaMES (150 m$M$) at pH 4.5, 5.0, and 5.5
> $CuSO_4$, 100 $\mu M$ in water
> NaOH, 1 $N$
> Ascorbic acid (Sigma), 50 m$M$ in water; prepare fresh daily
> Ac-Tyr-Val-$\alpha$-hydroxyglycine, 100 $\mu M$ in water
> Thesit, 10% (w/v) (Boehringer Mannheim)

*PHM or PAM Assay Protocol*

A standard 40-$\mu$l PHM assay mixture contains 0.5 $\mu M$ $CuSO_4$, 0.5 m$M$ ascorbate, catalase, (0.1 mg/ml), 0.5 $\mu M$ Ac-Tyr-Val-Gly substrate, and 25,000 cpm of [$^{125}$I]Ac-Tyr-Val-Gly substrate diluted in the NaMES buffer and the enzyme sample (8 $\mu$l or less, diluted in the enzyme dilution buffer). Before starting the assay, determine the amount of assay mixture required by adding up the number of enzyme samples to be assayed in duplicate, two blanks containing dilution buffer instead of enzyme, plus four tubes for total counts. The assay mixture is prepared and aliquoted to microcentrifuge tubes; aliquots are also pipetted into $\gamma$ counter tubes to determine the total radioactivity in each assay tube. Enzyme samples are added to begin the reaction. The reactions are carried out at 37° for 30–60 min, although longer times can be used for samples with low activity; samples are centrifuged briefly to remove condensate from the lid. For PHM assays, Ac-Tyr-Val-$\alpha$-hydroxyglycine is converted into Ac-Tyr-Val-$NH_2$ by adding 15 $\mu$l of 1 $N$ NaOH and allowing the sample to sit at room temperature for 5 min. The pH is then lowered by adding 240 $\mu$l of 1.0 $M$ Tris-HCl–bromphenol blue. For PAM assays, base is not added and samples are simply diluted with Tris-HCl–bromphenol blue. Amidated product is extracted by addition of 640 $\mu$l of water-saturated ethyl acetate and vigorous mixing; 320 $\mu$l of the clear upper phase is carefully removed and placed into a polypropylene tube compatible with the $\gamma$ counter to be used.

*PAL Assay Protocol*

A standard 40-$\mu$l assay mixture contains 0.5 $\mu M$ Ac-Tyr-Val-$\alpha$-hydroxyglycine, 10,000 cpm of [$^{125}$I]Ac-Tyr-Val-$\alpha$-hydroxyglycine, and 0.02% (w/v) Thesit diluted into 150 m$M$ NaMES buffer, pH 5.5, and the enzyme sample (8 $\mu$l or less, diluted in the enzyme dilution buffer). Assays are stopped as described except that NaOH is not added at the end of the reaction. The maximum potential conversion of substrate to product is determined in a pair of separate tubes to which NaOH is added at the end of the PHM assay.

## Sample Calculation

Enzyme activity can be calculated in picomoles per microgram per hour, using any standard computer spreadsheet:

$$\text{activity} = \left[\frac{(S \times 2) - (B \times 2)}{T}\right]\left[\frac{[Sub] \cdot [Vol]}{t}\right]\left[\frac{1}{v}\right]\left[\frac{1}{p}\right]$$

where S is $cpm_{sample}$, B is $cpm_{blank}$, T is average $cpm_{total}$, [Sub] is substrate ($\mu M$), [Vol] is assay volume ($\mu l$), t is assay time (hr), v is sample volume ($\mu l$), and p is protein concentration ($\mu g/ml$).

## Optimization of Assay Conditions

The PHM and PAL assays described are performed with substrate concentrations far below the $K_m$ of each enzyme. The substrate concentration can be varied to determine $K_m$ and $V_{max}$ values. The $K_m$ and $V_{max}$ values for PHM when it is part of bifunctional PAM-3 are 14 $\mu M$ and 7 sec$^{-1}$, respectively; the corresponding values for PAL[13] when part of PAM-3 are 37 $\mu M$ and 92 sec$^{-1}$. The copper concentration is critical in a PAM or PHM assay and the optimal level should be determined experimentally. When assaying samples containing more than 1 $\mu g$ of protein, the copper concentration may need to be increased above 0.5 $\mu M$; in our experience, the $Cu^{2+}$ concentration is the most crucial variable for the success of assays using crude tissue homogenates. For samples with high levels of protein we also recommend dilution of the sample to be assayed. The assay is linear up to 20% conversion of substrate to product.

## Comments on Other Assay Methods

We have compared our PHM and PAL assays using [$^{125}$I]Ac-peptide with other assays in the literature in terms of sensitivity, flexibility for

TABLE I
COMPARISON OF PHM AND PAL ASSAYS

| Substrate | Sensitivity[a] (pmol) | Equipment and time required | Ref. |
|---|---|---|---|
| **PHM Assays** | | | |
| [$^{125}$I]Ac-Tyr-Val-Gly | 0.15 | $\gamma$ Counter | 12, 13 |
| D-[$^{125}$I]Tyr-Val-Gly | 0.15 | $\gamma$ Counter | 14, 15 |
| Trinitrophenyl-D-Tyr-Val-Gly (TNP-D-Tyr-Val-Gly) | 400 | HPLC, UV detector, 8 min/sample | 16 |
| N-Dansyl-Tyr-Val-Gly | 25 | HPLC, fluorescence detector, 5 min/sample | 17 |
| Substance P-Gly | 0.005 | RIA, $\gamma$ counter | 18 |
| Dabsyl-Gly-Phe-Gly | 1 | HPLC, UV detector, 5 min/sample | 19 |
| N,N-Dimethyl-1,4-phenylenediamine (DMPD) | 500 | Spectrophotometer, continuous assay | 11, 20 |
| D-Tyr-Val-[$^{14}$C]Gly → [$^{14}$C]glyoxylate | 0.5 | Scintillation counter | 21 |
| 4-Nitrohippuric acid | 600 | HPLC, UV detector, 8 min/sample | 22 |
| Oxygen | 2500 | Oxygen electrode | 23 |
| **PAL Assays** | | | |
| [$^{125}$I]Ac-Tyr-Val-hydroxy-Gly | 0.15 | $\gamma$ Counter | 10, 13 |
| $\alpha$-OH-hippuric acid | 600 | HPLC, UV detector, 8 min/sample | 24 |

[a] Sensitivity has been defined as three times over background when a number was not provided in the cited reference. RIA, Radioimmunoassay.

various peptide substrates, and the equipment required (Table I[10–24]). The $^{125}$I-labeled peptide assays and the radioimmunoassays are the most sensitive; however, the radioimmunoassay method also necessitates raising the appropriately specific antiserum. The assay described here was developed

[14] A. S. N. Murthy, R. E. Mains, and B. A. Eipper, *J. Biol. Chem.* **261,** 1815 (1986).

[15] A. F. Bradbury, M. D. A. Finnie, and D. G. Smyth, *Nature* (*London*) **298,** 686 (1982).

[16] A. G. Katopodis and S. W. May, *Biochem. Biophys. Res. Commun.* **151,** 499 (1988).

[17] B. N. Jones, P. P. Tamburini, A. P. Consalvo, S. D. Young, S. J. Lovato, J. P. Gilligan, A. Y. Jeng, and L. P. Wennogle, *Anal. Biochem.* **168,** 272 (1988).

[18] A. Y. Jeng, M. Wong, S. J. Lovato, M. D. Erion, and J. P. Gilligan, *Anal. Biochem.* **185,** 213 (1990).

[19] T. Chikuma, K. Hanaoka, Y. P. Loh, T. Kato, and Y. Ishii, *Anal. Biochem.* **198,** 263 (1991).

[20] C. Li, C. D. Oldham, and S. W. May, *Biochem. J.* **300,** 31 (1994).

[21] S. E. Ramer, H. Cheng, M. M. Palcic, and J. C. Vederas, *J. Am. Chem. Soc.* **110,** 8526 (1988).

[22] A. G. Katapodis and S. W. May, *Biochemistry* **29,** 4541 (1990).

[23] S. W. May, R. S. Phillips, P. W. Mueller, and H. H. Herman, *J. Biol. Chem.* **256,** 2258 (1981).

[24] A. G. Katapodis, D. Ping, and S. W. May, *Biochemistry* **29,** 6115 (1990).

to detect low levels of endogenous PHM and PAL in tissue extracts and cell lines.

This assay provides substrate flexibility with any uncharged amino acid in the penultimate position of a glycine-extended peptide. The oxygen electrode and the $N,N$-dimethyl-1,4-phenylenediamine (DMPD; ascorbate substitute) assays are also flexible with the peptide substrate, but they are much less sensitive. The trinitrophenyl (TNP)-peptide, dansyl-peptide, dabsyl-peptide, and hippuric acid methods require HPLC separation of substrate and product and can be time consuming without a multiinjector. The oxygen electrode and DMPD (ascorbate substitute) assays are continuous and nondestructive of the peptide product, which can offer major advantages, and may be particularly useful in preparative-scale applications.

### Acknowledgment

Support for this work was provided by the National Institutes of Health Grant DK-32949.

## [6] Principles Involved in Formulating Recommendations for Vitamin C Intake: A Paradigm for Water-Soluble Vitamins

*By* MARK LEVINE, STEVEN RUMSEY, and YAOHUI WANG

### Introduction

Recommendations for vitamin intake can be based on the biochemical and clinical criteria of *in situ* kinetics (Table I).[1-5] This chapter focuses on those clinical criteria that delineate vitamin absorption and distribution. These criteria describe vitamin pharmacokinetics. Pharmacokinetics for

[1] M. Levine, S. Rumsey, Y. Wang, J. B. Park, O. Kwon, W. Xu, and N. Amano, *Methods Enzymol.* **281** (1997).

[2] Food and Nutrition Board, "How Should the Recommended Dietary Allowances Be Revised?" National Academy Press, Washington, DC, 1994.

[3] M. Levine, K. R. Dhariwal, R. W. Welch, Y. Wang, and J. B. Park, *Am. J. Clin. Nutr.* **62**(Suppl.), 1347S (1995).

[4] M. Levine, C. Conry-Cantilena, Y. Wang, R. W. Welch, P. W. Washko, K. R. Dhariwal, J. B. Park, A. Lazarev, J. F. Graumlich, J. King, and L. R. Cantilena, *Proc. Natl. Acad. Sci. U.S.A.* **93**, 3704 (1996).

[5] M. Levine, S. Rumsey, Y. Wang, *et al.*, Vitamin C. *In* "Present Knowledge in Nutrition" (L. J. Filer and E. E. Ziegler, eds.), 7th Ed., pp. 146–159. International Life Sciences Institute, Washington, DC, 1996; 146–159.

TABLE I

*In Situ* KINETICS

Biochemical component: Vitamin biochemical and molecular function in relation to vita-
    min concentration
    Assay for vitamin
    Ability to have different vitamin concentrations *in situ:* transport, depletion, repletion
    Distribution of vitamin *in situ*
    Assay for vitamin function
    Determination of vitamin function *in vitro* and *in situ*
    Localization of vitamin function *in situ* and relationship to vitamin distribution
    Specificity of vitamin function *in vitro* and *in situ*
    Vitamin function in relation to different vitamin concentrations *in situ*
Clinical component: Achieving effective vitamin concentrations in people
    Availability of vitamin in diet
    Steady state vitamin concentration in plasma as function of dose
    Steady state vitamin concentration in tissues as function of dose
    Vitamin bioavailability
    Vitamin excretion
    Vitamin safety and adverse effects
    Beneficial effects in relation to dose: direct effects and epidemiologic observations

vitamins describe steady-state concentration in plasma as a function of
dose, and steady-state concentration in tissues as a function of dose, bioavailability, and excretion. Data for vitamin C are used to illustrate practical
examples of pharmacokinetic principles and determinations.[4]

In discussing how to obtain these data, we assume that a reliable assay
is available to measure the vitamin of interest. The assay must be sensitive,
specific, reproducible, and free from interference by compounds in biological samples. Vitamin stability must be accounted for during sample procure-
ment from subjects, sample handling, sample storage, and the time of the
actual assay. Issues concerning vitamin C assays are discussed in this volume
and elsewhere.[6–8]

Vitamin Intake

A prerequisite for vitamin pharmacokinetic studies is that vitamin intake
from all sources must be certain. Intake can be from foods and/or vitamin
supplements. Vitamin intake from foods may vary, even if food types are
kept constant. For example, the vitamin C content of foods depends on
food source, shipping practices, transit time to market, shelf time, and

[6] E. Kimoto, S. Terada, and T. Yamaguchi, *Methods Enzymol.* **279** [1], 1997 (this volume).
[7] J. C. Deutsch, *Methods Enzymol.* **279** [2], 1997 (this volume).
[8] P. Washko and M. Levine, Inhibition of ascorbic acid transport in human neutrophils by
glucose. *J. Biol. Chem.* **267,** 23568 (1992).

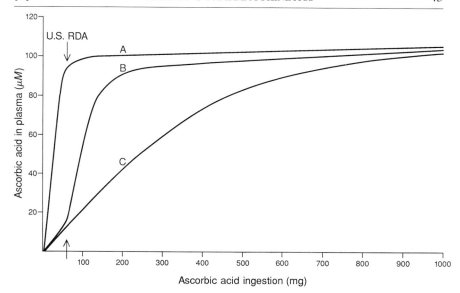

FIG. 1. Theoretical curves of ascorbic acid concentration in plasma in relation to intake.[10] Curves A, B, and C represent three different possibilities of the effects of intake on vitamin concentration. Curves A and B show that small changes in intake over specific dose ranges produce large changes in vitamin concentration. Curve C shows that small differences in intake have little effect on resulting concentration over the entire dose range. The relationships between each curve and the present recommended dietary allowance for vitamin C are also shown. See text and Ref. 10 for details. (Reprinted from Ref. 10, with permission of Karger, Basel.)

cooking methods. The consequence of these factors is that vitamin C content of the same cooked food can be quite different.[9]

It is possible that small differences in intake can have different effects on vitamin concentration in plasma or tissues. Three possibilities are shown in Fig. 1.[10] For curve A (Fig. 1), very little intake produces saturation. At low doses small differences in intake will result in large differences in resulting concentration. For curve B (Fig. 1), there is a sigmoidal relationship between dose and concentration. For the dose range at the sigmoid part of the curve, small differences in intake will result in large differences in resulting concentration. For curve C (Fig. 1), there is only gradual saturation as a function of dose, and small differences in intake will not greatly affect resulting concentrations. The relationship between each curve and the recommended dietary allowance for vitamin C are also shown in Fig. 1.

[9] J. T. Vanderslice and D. J. Higgs, *Am. J. Clin. Nutr.* **54,** 1323S (1991).
[10] M. Levine, C. Conry-Cantilena, and K. R. Dhariwal, *World Rev. Nutr. Diet.* **72,** 114 (1993).

Obviously the pharmacokinetic profile for any vitamin is not known in advance. Because it is possible that small differences in intake can have profound effects on the vitamin concentration achieved, intake should be tightly controlled. An excellent means of controlling vitamin intake for pharmacokinetic studies is to limit intake from foods of the vitamin under investigation. This can be accomplished either with an outpatient or an inpatient diet.

The success of outpatient diets is dependent on food choices, self-reporting, and compliance of subjects. Control of these and other factors is difficult for outpatient diets.[11,12] If small differences in vitamin intake affect resulting concentrations, outpatient diets may not be suitable for pharmacokinetic studies.

We believe that the best means to control vitamin intake is to use inpatient diets that limit ingestion of the vitamin under study. As an example, an inpatient diet was designed to limit vitamin C ingestion to less than 5 mg daily.[12] Because subjects were inpatients, food intake could be controlled carefully and accurate vitamin intake from foods determined. The vitamin under study was provided as a simple solution in the fasted state, as discussed in the section Bioavailability. In this way, accurate vitamin intake was clear.

The inpatient diet developed to restrict vitamin C intake was a selective-menu diet, based on vitamin C content of foods. Deficiencies of other vitamins and minerals were calculated, and deficits were corrected by vitamin and mineral supplements. The principles used to develop this diet can be used to develop diets to restrict intake of other vitamins for pharmacokinetic studies.[12]

By using an inpatient diet deficient in one vitamin, its concentration can be depleted to low values in experimental subjects. Subjects can then be repleted with escalating doses of the vitamin, with pharmacokinetic studies performed at each dose. Steady state can be achieved for each dose, and then the dose increased. To undertake these studies, the inpatient diet is the best way to control intake.

### Steady-State Concentrations as Function of Dose

A constant dose of vitamin over time will result in a plateau concentration in plasma and tissues. This plateau concentration is a steady-state concentration, which describes the concentration resulting from a particular

[11] D. M. Hegsted, *J. Am. Coll. Nutr.* **11**, 241 (1992).
[12] J. King, Y. Wang, R. W. Welch, K. R. Dhariwal, C. Conry-Cantilena, and M. Levine, *Am. J. Clin. Nutr.* in press (1997).

vitamin dose.[13–15] Steady-state concentrations for different vitamin doses define a dose–concentration curve for the vitamin.

Data for vitamin concentrations as a function of dose are essential for making sound recommendations about vitamin ingestion (see Fig. 1).[2–4,16] As examples, if vitamin concentrations saturate at a certain dose, higher doses confer no additional benefit and could be harmful. However, for some vitamins small changes in dose might produce large variations in vitamin concentration. If there is no toxicity over this dose range, this part of the ingestion curve should be avoided for recommended doses.[2,5]

The time to achieve a steady-state concentration for a dose can be estimated as approximately four elimination half-times.[13] This estimate is based on pharmacokinetic assumptions of a constant volume of distribution, constant clearance, and a single-compartment model. Absorption and elimination are assumed to follow first-order kinetics. These assumptions probably are too simple for vitamin pharmacokinetics.[4,17,18] Nevertheless, by using these assumptions the time needed to reach a steady-state plateau can be estimated.

For water-soluble vitamins the time to achieve steady state for any dose may be as short as a few days or as long as a few weeks. These times are short enough so that pharmacokinetic clinical studies with inpatients should be feasible.

For fat-soluble vitamins the volume of distribution may be much larger than for water-soluble vitamins. There may be multiple compartments, and vitamin distribution may be more complex. Repletion time for a single dose may be weeks to months. For similar reasons, depletion time may take many months. Because of these problems, it may be difficult to conduct clinical pharmacokinetic studies for some fat-soluble vitamins. Simpler approaches have been used to determine dose–concentration data for the fat-soluble vitamin $\alpha$-tocopherol,[19] but additional strategies would be useful.

Steady-state plasma concentration can be estimated for a dose of a water-soluble vitamin. These estimates can be verified operationally. For

[13] L. Z. Benet, D. L. Kroetz, and L. B. Sheiner, Pharmacokinetics: The dynamics of drug absorption, distribution, and elimination. *In* "Goodman and Gilman's The Pharmacologic Basis of Therapeutics" (J. G. Hardman, L. E. Limbird, P. B. Molinoff, R. W. Ruddon, and A. G. Gilman, eds.), pp. 3–27. McGraw-Hill, New York; 1996.

[14] M. Gibaldi and D. Perrier, *Pharmacokinetics,* pp. 294–297. Marcel Dekker, New York, 1982.

[15] M. Rowland and T. N. Tozer, *Clinical Pharmacokinetics: Concepts and Applications.* Lea & Febiger, Philadelphia; 1989.

[16] Food and Nutrition Board, "Recommended Dietary Allowances," 10th Ed., pp. 115–124. National Academy Press, Washington, DC, 1989.

[17] A. Kallner, D. Hartmann, and D. Hornig, *Am. J. Clin. Nutr.* **32,** 530 (1979).

[18] J. Graumlich, L. R. Cantilena, T. Ludden, R. Welch, Y. Wang, K. Dhariwal, J. Park, J. King, and M. Levine, *Clin. Pharmacol. Therapeut.* **55,** 200 (1994). [Abstract]

[19] I. Jialal, C. J. Fuller, and B. A. Huet, *Arterioscler. Throm. Vasc. Biol.* **15,** 190 (1995).

example, the steady-state plateau concentration of vitamin C for any dose was determined by using at least five different plasma samples taken over at least 1 week with $<10\%$ SD of the mean.[4] More than 85% of plateau values were based on six or more plasma samples. The first sample included in the plateau calculation was $>90\%$ of the final mean. Using these criteria it was clear when steady state was achieved for a particular dose.

For a water-soluble vitamin, it is reasonable to assume that measurements of steady state in plasma reflect steady state in tissues. For tissues in which vitamin distributes by diffusion from the extracellular milieu, steady state for plasma and tissues should be similar. For tissues that accumulate vitamin by active transport, transporters will achieve equilibrium as a function of external vitamin concentration. Because of active transport, tissues should reach steady state prior to plasma. In addition, tissues with active vitamin transporters should achieve their maximal concentrations at lower doses compared to plasma. Thus, for water-soluble vitamins steady state for plasma should also define steady state for tissues.

Using a diet limiting vitamin intake, steady-state concentrations in plasma and tissues for different vitamin doses can be determined with inpatient subjects. Once the restricted diet is begun, subjects are depleted of the vitamin under study. Samples should be obtained every 2–4 days for vitamin measurements. Subjects are checked frequently, and frank deficiency symptoms are to be avoided. Subjects undergo vitamin depletion so that pharmacokinetics can be studied for low as well as high doses. At a safe nadir of depletion, subjects begin repletion with a low dose of the vitamin. Steady state for this dose is determined, the dose is escalated, and a new steady state is determined for the next dose. In this way a dose–concentration curve can be determined for plasma and tissues. Plasma samples are obtained by simple phlebotomy. Tissues may be obtained using blood samples of circulating cells, or by skin or tissue biopsy.

Using these principles, pharmacokinetics of vitamin C were determined in plasma and tissues.[4] As described, hospitalized inpatient subjects consumed a vitamin C-restricted diet. Steady-state concentrations of vitamin C in plasma were measured for each repletion dose. The data showed that there was a sigmoidal relationship between dose and plateau steady-state concentration of vitamin C (Fig. 2A), similar to curve B in Fig. 1. The first dose beyond the steep portion of the curve was 200 mg daily. There are several explanations for the sigmoid curve, including multicompartment distribution of vitamin C, altered bioavailability at different doses, dose-dependent renal excretion, variations in the volume of distribution at different doses, and nonlinear clearance.[4,17,18] When steady-state plasma concentrations were reached for each dose, vitamin C concentrations were measured in neutrophils, monocytes, and lymphocytes. These cells accumulate vitamin C by active transport.[20–22] As predicted, cells achieved maximal internal concentrations at a lower dose compared to plasma (Fig. 2A and B). Taken together, the data suggest that pharmacokinetics can be determined for vitamin C. These principles should be applicable to other water-soluble vitamins.

### Bioavailability

Bioavailability describes absorption and is measured either as indirect bioavailability or as true bioavailability.[13–15] Indirect bioavailability can be determined in several ways. For example, absorption characteristics of one form of vitamin can be compared to another form. This type of indirect bioavailability is called relative bioavailability. For vitamin C, relative bioavailability has been measured for vitamin C in foods compared to supplements, for different supplement preparations compared to each other, and for different foods compared to each other.[23–25] Indirect bioavailability was also measured when oral absorption was compared to urine excretion.[24,26–28]

Indirect bioavailability does not provide information about absolute absorption: this information is provided by true bioavailability. True bioavailability compares the amount of substance absorbed orally to the same amount administered directly into the blood.[14,15] To determine true bioavailability, subjects must be at steady state for the dose under investigation. Data for two curves are obtained. For the first curve of oral bioavailability, the steady-state baseline is determined, an oral dose is administered, and measurements over minute to hours are taken of the amount of the substance in blood until there is a return to baseline. These data describe a curve that gradually rises and then returns to baseline. For the second curve of intravenous bioavailability, subject plasma values must have returned to baseline from the oral dose. The same dose is given intravenously, and the amount of the substance in blood is measured until there is a return to baseline. When the data from intravenous administration are displayed, plasma values usually rise much faster because the gastrointestinal tract

[20] P. Washko, D. Rotrosen, and M. Levine, *J. Biol. Chem.* **264,** 18996 (1989).
[21] P. Bergsten, G. Amitai, J. Kehrl, K. R. Dhariwal, H. G. Klein, and M. Levine, *J. Biol. Chem.* **265,** 2584 (1990).
[22] P. Bergsten, R. Yu, J. Kehrl, and M. Levine, *Arch. Biochem. Biophys.* **317,** 208 (1995).
[23] E. R. Hartzler. *J. Nutr.* **30,** 355 (1945).
[24] S. Yung, M. Mayersohn, and J. B. Robinson, *J. Pharm. Sci.* **71,** 282 (1982).
[25] A. R. Mangels, G. Block, C. M. Frey, B. H. Patterson, P. R. Taylor, R. P. Norkus, and O. A. Levander, *J. Nutr.* **123,** 1054 (1993).
[26] A. Kallner, D. Hartmann, and D. Hornig, *Int. J. Vitam. Nutr. Res.* **47,** 383 (1977).
[27] M. Mayersohn, *Eur. J. Pharmacol.* **19,** 140 (1972).
[28] S. L. Melethil, W. E. Mason, and C. Chiang, *Int. J. Pharm.* **31,** 83 (1986).

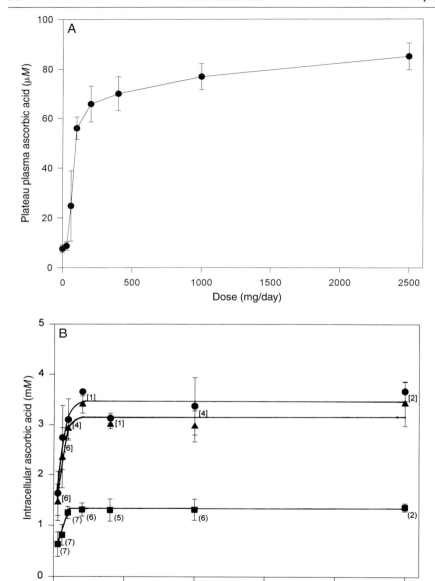

and liver are bypassed. The area under each of the two curves and above baseline is called the AUC, and represents the increment provided by the test dose. True bioavailability is defined as the increment in oral (*per os*, po) administration at steady state divided by the increment in intravenous (iv) administration at steady state, or as $AUC_{po}/AUC_{iv}$. True bioavailability is often displayed as a percentage or as $F$, with complete absorption represented by bioavailability ($F$) of 100%.

True bioavailability is often difficult to determine experimentally because of clinical considerations. Subjects must be at steady state for the tested dose. To obtain enough data for $AUC_{po}$ and $AUC_{iv}$, frequent sampling must be performed after oral and intravenous administration. A sterile intravenous preparation of the test substance must be available, and the oral and intravenous forms should be similar or identical. True bioavailability studies should be performed for several different doses. A hospital setting may be necessary.

True bioavailability studies for a water-soluble vitamin can be performed as part of an in-hospital depletion–repletion study, as has been described for vitamin C.[4] Dietary intake of the vitamin was controlled. Steady-state conditions were achieved as part of pharmacokinetic studies. A hospital setting was necessary for diet control and for the frequent sampling required for bioavailability determinations. Bioavailability was 100% for 200 mg of vitamin C daily, but less than 50% at 1250 mg daily.[4]

For true bioavailability studies, the vitamin should be administered in its simplest form so that absorption is not influenced by other substances. For vitamin C, the intravenous vitamin was administered as such in water. To provide the best comparison, the same preparation was used for oral administration.[4] The vitamin was given in the fasted state to avoid any interference from foods. Bioavailability of the vitamin alone should be determined first. Only after these data are available is it reasonable to determine vitamin bioavailability in the presence of foods. In this way, bioavailability of the vitamin in foods can be compared to bioavailability of the vitamin alone, and effect of foods on bioavailability understood.

---

FIG. 2. (A) Steady-state plateau ascorbic acid concentrations in plasma ($\mu M$) as a function of dose (mg/day) in healthy volunteers. Steady-state plateau concentration was defined as the mean of five or more samples taken over at least 7 days with $\leq$10% SD. The first sample included in all plateau calculations was $\geq$90% of the final plateau mean. (B) Intracellular ascorbic acid concentration at steady state (m$M$) in circulating immune cells as a function of dose (mg/day). Neutrophils (■), monocytes (▲), and lymphocytes (●) were isolated after plasma steady-state plateau was achieved for each dose. Numbers in parentheses at each dose indicate the number of volunteers from whom neutrophils were obtained. Numbers in brackets indicate the number of volunteers from whom lymphocytes and monocytes were obtained by apheresis. For further details see Ref. 4. [Reprinted from M. Levine, *et al.*, *Proc. Natl. Acad. Sci. U.S.A.* **93**, 3704 (1996). Copyright 1996 National Academy of Sciences, U.S.A.]

Bioavailability calculations using AUC values are based on pharmacokinetic assumptions of constant volume of distribution and clearance. The bioavailability data for vitamin C indicated that these assumptions were not valid at doses <200 mg, perhaps because of multiple-compartment distribution.[4,17,18] A model for vitamin C is being developed to account for nonlinearity of clearance and volume of distribution. This modeling is quite complex. Once a new model becomes available, true bioavailability studies can be undertaken in the presence of foods that could affect vitamin C absorption, such as glucose.[8,29]

Urine Excretion

Excretion of a water-soluble vitamin as a function of dose might provide information about saturation of plasma and/or tissues. It is possible that the threshold of urine excretion indicates saturation. Indeed, the threshold for urine excretion is used as one basis for the recommended dietary allowance for vitamin C.[16,30,31] It is also possible that threshold and saturation are not tightly coupled, and that saturation occurs at higher doses compared to threshold. Urine excretion is a function of plasma concentration, glomerular filtration, tubule reabsorption, tubule secretion, and perhaps protein binding in plasma. If there is existing information about these parameters, data concerning urine excretion can be useful. With regard to vitamin C, the vitamin is free in the circulation and is not bound to proteins, undergoes glomerular filtration, and is reabsorbed actively in renal tubules.[32-35] It is not clear whether there is active tubule secretion.

The relationship between saturation and excretion can be investigated as part of pharmacokinetic studies. These studies must be conducted at steady state. The best time to perform them is during bioavailability sampling. A vitamin dose is given orally and urine is collected for 24 hr. The same dose is given intravenously and urine is again collected. Measurements should also be taken for creatinine clearance, possible metabolites, and compounds whose excretion could be affected by the vitamin.

[29] R. W. Welch, Y. Wang, A. Crossman, Jr., J. B. Park, K. L. Kirk, and M. Levine, *J. Biol. Chem.* **270,** 12584 (1995).
[30] E. M. Baker, R. E. Hodges, J. Hood, H. E. Sauberlich, and S. C. March, *Am. J. Clin. Nutr.* **22,** 549 (1968).
[31] E. M. Baker, R. E. Hodges, J. Hood, H. E. Sauberlich, S. C. March, and J. E. Canham, *Am. J. Clin. Nutr.* **24,** 444 (1971).
[32] G. J. Friedman, S. Sherry, and E. Ralli, *J. Clin. Invest.* **19,** 685 (1940).
[33] J. F. Sullivan and A. B. Eisenstein, *Am. J. Clin. Nutr.* **23,** 1339 (1970).
[34] K. R. Dhariwal, W. O. Hartzell, and M. Levine, *Am. J. Clin. Nutr.* **54,** 712 (1991).
[35] D. G. Oreopoulos, R. D. Lindeman, D. J. VanderJagt, A. H. Tzamaloukas, H. N. Bhagavan, and P. J. Garry, *J. Am. Coll. Nutr.* **12,** 537 (1993).

Data were obtained for vitamin C excretion in urine at steady state for each dose.[4] The data indicated that the threshold of urine excretion occurred at the 100-mg dose, which was the saturating dose for cells but not for plasma. At doses of 500 and 1250 mg, virtually all of the absorbed vitamin was excreted in urine. When considered with the dose–concentration curve (Fig. 2), urine excretion data indicate the saturating steady-state dose of vitamin C is 400–1000 mg.

As part of urine excretion studies, samples were taken for measurement of uric acid and oxalic acid. The latter is a metabolite of vitamin C, and excretion of the former may be affected by vitamin C.[36–39] Both substances could be involved in kidney stone formation. The data indicated that excretion of both compounds was increased at the 1000-mg dose of vitamin C. At this dose, cells and plasma are already completely saturated. It is unknown whether enhanced excretion of uric acid and/or oxalate induced by vitamin C could be harmful. However, because vitamin C saturation occurs at lower doses, and because there is potential for harm from uric acid and/or oxalate excretion, vitamin C doses above 500 mg daily should be avoided in healthy people.

Conclusion

Pharmacokinetic principles can be used to characterize absorption and distribution of water-soluble vitamins in healthy humans. These principles were used to obtain vitamin C pharmacokinetic data. New information was obtained on the relationship between dose and steady-state plasma concentrations, dose and steady-state tissue concentrations, bioavailability, and urine excretion. These data are important for making rational recommendations about vitamin C dietary allowances. For example, when pharmacokinetic data are considered with other findings, the evidence indicates that a suitable recommended dietary allowance for vitamin C is 200 mg, rather than the current recommendation of 60 mg.[2,4,16,19,20,22,38–44] Using the

[36] H. B. Stein, A. Hasan, and I. H. Fox, *Ann. Intern. Med.* **84,** 385 (1976).
[37] W. E. Mitch, M. W. Johnson, J. M. Kirshenbaum, and R. E. Lopez, *Clin. Pharmacol. Ther.* **29,** 318 (1981).
[38] M. Urivetzky, D. Kessaris, and A. D. Smith, *J. Urol.* **147,** 1215 (1992).
[39] T. R. Wandzilak, S. D. D'Andre, P. A. Davis, and H. E. Williams, *J. Urol.* **151,** 834 (1994).
[40] P. Lachance and L. Langseth, *Nutr. Rev.* **52,** 266 (1994).
[41] R. W. Welch, P. Bergsten, J. D. Butler, and M. Levine, *Biochem. J.* **294,** 505 (1993).
[42] M. A. Helser, J. H. Hotchkiss, and D. A. Roe, *Carcinogenesis* **13,** 2277 (1992).
[43] T. Byers and N. Guerrero, *Am. J. Clin. Nutr.* **62,** 1385S (1995).
[44] K. F. Gey, U. K. Moser, P. Jordan, H. B. Stahelin, M. Eichholzer, and E. Ludin, *Am. J. Clin. Nutr.* **57,** 787S (1993).

principles described here, it should be possible to obtain similar absorption and distribution information for other water-soluble vitamins. These data may become an essential part of the database for recommended dietary allowances.

# Section II

# Thiamin: Phosphates and Analogs

# [7] High-Performance Liquid Chromatography Determination of Total Thiamin in Biological and Food Products

### By VÉRONIQUE FAYOL

## Introduction

The usual causes of thiamin deficiency are either a primary dietary deficiency as occurs in developing countries or secondary to alcoholism as in developed countries. The rapid and specific analysis of vitamin $B_1$ in blood (to assess vitamin status) and in food is of importance. Several methods for the measurement of thiamin have been described; these include the fluorimetric assay, in which oxidation of thiamin into thiochrome is measured fluorimetrically, and the microbiological method, which is a time-consuming technique. High-performance liquid chromatography (HPLC) methods offer an attractive alternative to the more time-consuming chemical and microbiological assays because of their increased specificity, sensitivity, and reduced analysis time. My purpose is to review HPLC methods for the determination of thiamin in food and blood.

## Sample Extraction

In food and biological material, vitamin $B_1$ is present mainly as the diphosphate ester; the other forms are free thiamin, thiamin monophosphate, and thiamin triphosphate.[1] Blood samples are taken using EDTA or heparin anticoagulant and are hemolysed, generally by freezing at $-20°$, before extraction of thiamin. Thiamin extraction for food and for blood requires an acidic hydrolysis step to break the bonds with proteins, and an enzymatic hydrolysis to convert thiamin phosphates to free thiamin. Acidic hydrolysis is performed using trichloroacetic, perchloric, or sulfuric acid and enzymatic hydrolysis with Clara-Diastase, phosphatase, or Taka-Diastase at alkaline pH. Purification of sample solutions for HPLC requires filtration with membrane filters.

In blood, thiamin is present mainly as thiamin diphosphate, which is the coenzymatically active form. Human erythrocytes contain approximately 80% of total blood thiamin. Blood levels of thiamin in individuals reflect their thiamin status; total thiamin can be assayed in serum but its

[1] W. N. Pearson, *in* "The Vitamins" (P. Gyorgy, ed.), p. 53. Pearson, New York, 1967.

METHODS IN ENZYMOLOGY, VOL. 279

low level has long been a limit to its detection. Under routine conditions, total blood is the more convenient source for the evaluation of vitamin status. Thiamin or its esters gradually decompose during storage of whole blood at 4°; Wielders and Hink[2] reported a loss of 12–24% in total thiamin content and no loss in frozen samples after 11 days of storage. Whole blood can be stored at −20° until assay; this increases its handling convenience and reduces the chance for sampling errors by sample inhomogeneity.[2]

Two sample preparation procedures[2] for whole blood were tested: (1) acid hydrolysis with trichloroacetic acid and enzymatic hydrolysis with Taka-Diastase and (2) acid hydrolysis with perchloroacetic acid and enzymatic hydrolysis with acid phosphatase.

The best result was obtained with the use of perchloracetic acid and phosphatase because the Taka-Diastase contained a small amount of thiamin, whereas no thiamin was detected in the acid phosphatase.

In food, AOAC (Association of Official Analytical Chemists) methods propose the use of Taka-Diastase, clarase, and mylase P, whose activity is due only to the presence of impurities. Hasselmann et al.[3] reported the results of interlaboratory analysis in food, using Taka-Diastase or β-amylase, with large variation in the levels of thiamin. Complete dephosphorylation was obtained with the use of β-amylase and Taka-Diastase, and they suggested that in food containing high amounts of natural vitamins, the quantity of the enzyme used in AOAC methods might be insufficient to release all thiamin from its phosphates esters; so it would be better to use an acid phosphatase in the enzymatic hydrolysis.

## Chromatography

The choice of analytical system depends on the molecular weight, structure, and solubility of thiamin. The HPLC methods used for thiamin are generally based on a reversed-phase column using either ultraviolet (UV) or fluorimetric detection. An external standard is prepared from a thiamin diphosphate working solution. Chromatographic peak areas are calculated (by the use of an electronic integrator) and quantitation is carried out by plotting reported values against a standard calibration curve.

### Techniques Using Ultraviolet Detection

In food a few techniques have been reported for detecting thiamin simultaneously with other vitamins (of the B group) using ion pairing, the

[2] J. P. Wielders and C. J. K. Hink, *J. Chromatogr. A* **277**, 145 (1983).
[3] C. Hasselmann, D. Franck, and P. Grimm, *J. Micronutr. Anal.* **5**, 269 (1989).

effect of which is to select the specific retention time for each vitamin (Table I). Good results were reported: these techniques were reproducible[4–6] and the recovery was greater than 99%.[5,6] There was good agreement between these techniques and the accepted AOAC method[5–7] and the microbiological method using *Lactobacillus viridescens* ATCC 12706.[6]

Sample preparation is simpler than in methods using fluorescence detection and avoids probable interfering compounds in the oxidation reaction leading to thiochrome formation. Nicolas and Pfender[6] used a procedure similar to that of Ayi *et al.*,[5] based on a two-step pH adjustment to deproteinize the sample, followed by gravity filtration. The disadvantage of this simplified procedure was its inadequacy in quantitating the phosphorylated derivatives because there is no dephosphorylation step. It was suggested that this procedure be used only in products containing no appreciable level of phosphate esters. The other disadvantage of UV detection is the low sensibility, making it inadequate for samples with low thiamin content as in blood or nonfortified food.

### Techniques Using Fluorimetric Detection

In techniques involving fluorimetric detection, thiamin and its phosphate esters are quantitatively converted to thiochrome and thiochrome phosphate esters. This reaction can be applied to an HPLC system by using two different approaches. In the first approach, the appropriate reagent is added to the sample containing thiamin compounds and the thiochrome derivatives are chromatographed (precolumn derivatization). In the second approach, the reaction is performed after the chromatography (postcolumn derivatization). These techniques are summarized in Tables II and III. The thiochrome must be extracted into isobutanol before measuring its fluorescence. The excitation wavelength of the fluorimeter is set near to 375 nm and the emission is set in the range of 425–450 nm.

The thiochrome reaction is rapid as performed in highly alkaline solution, where thiochrome and its phosphate esters are stable. Reagents usually used for the alkaline oxidation are potassium hexacyanoferrate (ferricyanide), mercuric chloride, and cyanogen bromide. Cyanogen bromide is an extremely hazardous and rather unstable reagent. Mercuric chloride is stable in solution for 7 days at 4°[8] and the intensity of fluorescence with

[4] J. F. Kamman, J. P. Labuza, and J. J. Warthesen, *J. Food Sci.* **45,** 1497 (1980).

[5] B. K. Ayi, D. A. Yuhas, and K. S. Moffett, *J. Assoc. Off. Anal. Chem.* **68,** 1087 (1985).

[6] E. C. Nicolas and K. A. Egender, *J. Assoc. Off. Anal. Chem.* **73,** 792 (1990).

[7] G. W. Chase, W. O. Londen, and Etenniller, *J. Assoc. Off. Anal. Chem.* **75,** 561 (1992).

[8] W. Weber and H. Kewitz, *Eur. J. Clin. Pharmacol.* **28,** 213 (1985).

TABLE I

HPLC Systems for Total Thiamin with UV Detection

| Compound determined | Sample | Extraction | Column | Mobile phase | UV detection | Ref. |
|---|---|---|---|---|---|---|
| Thiamin and its degradation products | Biological products | | OH-Shodex Pak 250 × 0.6 mm i.d. | Phosphate buffer, 0.05 $M$ pH 3.75 Flow: 0.5 ml/min | 275 nm (thiamin) 250 nm (degradation products) | Panijpan et al.[a] |
| Thiamin Riboflavin Pyridoxine | Blood Sera | Thichloroacetic acid, 1.2 $M$ Clara-Diastase, 50 mg/ml | $C_{18}$ Novapak, 5 $\mu$m 125 × 4.6 mm i.d. | $CH_3OH/H_2O$ (20/80, v/v) + 3 m$M$ hexanesulfonic acid Flow: 1 ml/min | 245 nm ($B_1$, $B_2$) Fluorimetric detection for $B_6$ | Botticher and Botticher[b] |
| Thiamin | Baby flour meal Milk Food | HCl, 6 $N$ | $C_{18}$ $\mu$Bondapak, 10 $\mu$m 300 × 4.6 mm i.d. | $CH_3OH/C_2H_5CO_2H/H_2O$ (0.400/ 0.015/1.585, v/v) + 2 g of EDTA + 3 g of hexanesulfonic acid Flow: 2.5 ml/min | 248 nm | Nicolas and Pfender[c] |
| Thiamin Riboflavin Pyridoxine | Baby flour meal | Perchloric acid, 70% (v/v) | $C_{18}$ Novapak, 5 $\mu$m 150 × 3.9 mm i.d. | Acetonitrile/$H_2O$/$NH_4OH$ (95/ 904.5/0.5, v/v) + 0.95 g of hexanesulfonic acid Flow: 1 ml/min | 254 nm Fluorimetric detection for $B_6$ | Chase et al.[d] |

| | | | | | | |
|---|---|---|---|---|---|---|
| Thiamin | Infant formula products | HCl, 1 $N$ | CN-Zorbax, 6 $\mu$m 250 $\times$ 4.6 mm i.d. | Triethylamine/CH$_3$OH (5.5/90, v/v), pH 7.7 Flow: 1.5 ml/min | 245 nm | Ayi et al.[e] |
| Thiamin Riboflavin | Cereal | HCl, 0.1 $N$ Clarase, 5 g/100 ml | C$_{18}$ $\mu$Bondapak, 10 $\mu$m 300 $\times$ 4.6 mm i.d. | Acetonitrile/phosphate buffer (12.5/87.5, v/v), pH 7.0, + sodium heptanesulfonic acid, 0.005 $M$ Flow: 2.5 ml/min | 254 nm | Kamman et al.[f] |
| Thiamin Riboflavin Niacin | Rice Rice products | H$_2$SO$_4$, 0.1 $N$ Taka-Diastase, 1 $M$ Papain, 10% (w/v) | C$_{18}$ $\mu$Bondapak 10 $\mu$m 100 $\times$ 18 mm i.d. | CH$_3$OH/C$_2$H$_5$CO$_2$H/H$_2$O (390/10/600, v/v) + PIC 5 + PIC 7 Flow: 1 ml/min | 254 nm | Toma and Tabekhia[g] |

[a] B. Panijpan, M. Kimura, and Tokaway, *J. Chromatogr.* **245**, 144 (1982).

[b] B. Botticher and D. Botticher, *Int. J. Vitam. Nutr. Res.* **57**, 273 (1987).

[c] E. C. Nicolas and K. A. Pfender, *J. Assoc. Off. Anal. Chem.* **73**, 792 (1990).

[d] G. W. Chase, W. O. Londen, and J. R. Etenmiller, *J. Assoc. Off. Anal. Chem.* **75**, 561 (1992).

[e] B. K. Ayi, D. A. Yuhas, and K. S. Moffett, *J. Assoc. Off. Anal. Chem.* **68**, 1087 (1985).

[f] J. F. Kamman, T. P. Labuza, and J. J. Warthesen, *J. Food Sci.* **45**, 1497 (1980).

[g] R. B. Toma and M. M. Tabekhia, *J. Food Sci.* **44**, 263 (1979).

TABLE II

HPLC Systems for Total Thiamin Using Fluorimetric Detection and Precolumn Derivatization

| Compound determined | Sample | Extraction | Column | Mobile phase | Fluorimetric detection | Ref. |
|---|---|---|---|---|---|---|
| Thiamin | Food Biological products | $H_2SO_4$ + $C_2H_5COOH$/ TCA Clara-Diastase Oxidation: $K_3Fe(CN)_6$ | $NH_2$-Nucleosil 250 × 4.6 mm | Phosphate buffer/acetonitrile, (25/ 75, v/v) pH 4.4 Flow: 1.8 ml/min | Ex: 370 nm Em: 425 nm | Botticher and Botticher[a] |
| Thiamin | Blood | HCl Taka-Diastase + papain pH 4.5 | Silica Lichrosorb, 5 $\mu$m 250 × 4 mm | $CHCl_3/CH_3OH$ (80/20, v/v) Flow: 2 ml/min | Ex: 375 nm Em: 430 nm | Bailey and Finglas[b] |
| Thiamin | Biological products Food | 0.2 M HCl, 70% (v/v) $HClO_4$ Taka-Diastase: 50 mg/ml pH 4.5 Oxidation: $HgCl_2/NaOH$ (50/50, v/v) | $NH_2$-Lichrosorb, 5 $\mu$m 250 × 4.6 mm | Dichloromethane/$CH_3OH$ (10/90, v/v) Flow: 1 ml/min | Ex: 365 nm Em: 440 nm | Laschi-Loquerie et al.[c] |
| Thiamin phos-phate esters | Blood Sera | TCA, 30% (w/v) Oxidation: $K_3Fe(CN)_6$/ NaOH | PRP-1 Hamilton, 10 $\mu$m 250 × 4, 1 mm | $CH_3OH$/phosphate buffer (10/90, v/v) for esters THF/phosphate buffer (10/90, v/v) for thiamin pH 8.5 Flow: 0.5 ml/min | Ex: 365 nm Em: 433 nm | Bettendorff et al.[d] |
| Thiamin Riboflavin | Food | HCl, 6 N Taka-Diastase (6%, w/v), TCA (50%, w/v) Oxidation: $K_3Fe(CN)_6$/ $NaOH/H_2O$ (1/15/84, v/v) | $C_8$ Radial Pak, 10 $\mu$m 100 × 18 mm i.d. | $CH_3OH$/phosphate buffer (37/63, v/v) Flow: 1.5 ml/min | Ex: 360 nm Em: 415 | Fellman et al.[e] |

| Analyte | Matrix | Extraction/Oxidation | Column | Mobile phase/Flow | Detection | Reference |
|---|---|---|---|---|---|---|
| Thiamin, Thiamin phosphates | Tissues | TCA, 5% (w/v); Oxidation: $K_3FE(CN)_6$/NaOH/$H_2O$ (0.050/0.375/2.125, v/v) | ODS-Ultrasphere, 5 µm, 250 × 4.6 mm i.d. | $CH_3OH$/phosphate buffer, Flow: 1 ml/min | Ex: 390 nm, Em: 475 nm | Bontemps et al.[f] |
| Thiamin | Plasma | $HClO_4$, 70% (v/v), 0.2 N HCl, Phosphatase, Oxidation: $HgCl_2$/NaOH (75/75, v/v) | $NH_2$-Lichrosorb, 5 µm, 120 × 4.6 mm | $CH_3OH$/ether (25/75, v/v), Flow: 2 ml/min | Ex: 365 nm, Em: 440 nm | Weber and Kervitz[g] |
| Thiamin, Riboflavin | Dietetic foods | 0.1 M HCl, β-amylase (50 mg), Taka-Diastase, 500 mg, Oxidation: $K_3Fe(CN)_6$/NaOH (1/24, v/v) | $C_{18}$ µBondapak, 10 µm, 300 × 4.6 mm i.d. | $CH_3OH$/sodium acetate (60/4, v/v), pH: 4.5, Flow: 1 ml/min | Ex: 366 nm, Em: 450 nm | Hasselmann et al.[h] |
| Thiamin and phosphate esters | Blood Sera | 2.44 M TCA, Oxidation: 0.3 M $Br_3CN$ | $NH_2$-Supelco, 250 × 4.6 mm i.d. | Acetonitrile/phosphate buffer (90/10, v/v), Flow: 1.5 ml/min | Ex: 375 nm, Em: 450 nm | Tallaksen et al.[i] |

[a] B. Botticher and D. Botticher, *Int. J. Vit. Nutr. Res.* **56**, 155 (1986).

[b] A. L. Bailey and P. M. Finglas, *J. Micronutr. Anal.* **7**, 147 (1980).

[c] A. Laschi-Loquerie, S. Vallas, and J. Viollet, *Internat. J. Vitam. Nutr. Res.* **62**, 248 (1992).

[d] L. Bettendorff, C. Grandfils, and C. de Ricker, *J. Chromatogr.* **382**, 297 (1986).

[e] J. K. Fellman, W. E. Artz, and P. D. Tassinari, *J. Food Sci.* **47**, 20481 (1982).

[f] J. Bontemps, P. Philippe, and L. Bettendorff, *J. Chromatogr.* **307**, 283 (1984).

[g] W. Weber and H. Kervitz, *Eur. J. Clin. Pharmacol.* **28**, 213 (1985).

[h] C. Hasselmann, D. Franck, and P. Grimm, *J. Micronutr. Anal.* **5**, 269 (1989).

[i] C. M. E. Tallaksen, T. Bohmer, and H. Bell, *J. Chromatogr.* **564**, 127 (1991).

Abbreviations: TCA, Trichloroacetic acid; THF, tetrahydrofuran; Ex, excitation; Em, emission.

TABLE III

HPLC Systems for Total Thiamin Using Fluorimetric Detection and Postcolumn Derivatization

| Compound determined | Sample | Extraction | Column | Mobile phase | UV detection | Ref. |
|---|---|---|---|---|---|---|
| Thiamin Riboflavin | Foods | HCl Clarase | $C_{18}$ $\mu$Bondapak, 10 $\mu$m 300 × 4.6 mm i.d. | $CH_3OH/H_2O$ (40/60, v/v) + 5 $\mu M$ PIC $B_6$ Flow: 1.5 ml/min Oxidation: 0.1 m$M$ $K_3Fe(CN)_6$, 0.375 $M$ NaOH Flow: 1 ml/min | Ex: 360 nm Em: 425 nm | Wimalarasi and Wills[a] |
| Thiamin | Blood CSF Milk | $HClO_4$, 1.2 mol/liter Acid phosphatase, 0.4 U/mg pH 5.2 | $C_{18}$ $\mu$Bondapak, 10 $\mu$m 300 × 3.9 mm | $CH_3OH/H_2O$ (45/55, v/v) Citrate buffer, 0.05 mol/liter, pH 4 Sodium octanesulfonate, 10 mmol/liter Flow: 1.2 ml/min Oxidation: $K_3Fe(CN)_6$, 2.5 mmol/liter; NaOH, 3 mmol/liter Flow: 0.3 ml/min | Ex: 367 nm Em: 435 nm | Wielders and Hink[b] |
| Thiamin | Plasma | $HClO_4$ Acid phosphatase | $C_{18}$ Nucleosil 125 × 4 mm | Acetonitrile/$HClO_4$, octane sulfonic acid Flow: 2 ml/min Oxidation: $K_3Fe(CN)_6$/NaOH Flow: 1 ml/min | Ex: 365 nm Em: 435 nm | Mascher and Kikuta[c] |

| Analyte | Sample | Extraction | Column | Mobile phase / Oxidation | Detection | Reference |
|---|---|---|---|---|---|---|
| Thiamin | Urine | HCl, 0.1 $M$ | $NH_2$-Lichrosorb, 5 $\mu$m 250 × 3.2 mm i.d. | $CH_3OH$/diethyl ether (22/88, v/v) Oxidation: $K_3Fe(CN)_6$/NaOH (4/96, v/v) Flow: 0.85 ml/min | Ex: 365 nm | Roser et al.[d] |
| Thiamin Riboflavin | Cereal | $H_2SO_4$, 0.1 $N$ Mylase, 2 g/100 ml | $C_{18}$ $\mu$Bondapak, 10 $\mu$m 300 × 4.6 mm i.d. | $CH_3OH/H_2O/C_2H_5CO_2H$ (35.6/63/4/1, v/v) + 0.005 $M$ hexane sulfonic acid Flow: 0.8 ml/min Oxidation: $K_3Fe(CN)_6$, 30 g/100 ml; NaOH; 3.75 $N$ Flow: 0.25 ml/min | ND | Mauro and Wetzel[e] |
| Thiamin | Rice | Taka-Diastase | $C_{18}$ Nucleosil, 5 $\mu$m 150 × 4 mm i.d. | $NaClO_4$, 0.01 $M$ Phosphate buffer, 0.15 $M$, pH 2.2 Flow: 0.6 ml/min Oxidation: $K_3Fe(CN)_6$ (0.1%)/ NaOH (12%) Flow: 0.6 ml/min | Ex: 375 nm Em: 435 nm | Ohta et al.[f] |
| Thiamin | Food | HCl, 0.1 $N$ Mylase, 10% or Taka-Diastase, 10% (w/v) | | Oxidation: $K_3Fe(CN)_6$/NaOH (4/96, v/v) | Ex: 365 nm Em: 435 nm | AOAC[g] |

[a] P. Wimalasiri and R. B. H. Wills, *J. Chromatogr.* **318**, 412 (1985).
[b] J. P. Wielders and C. J. K. Hink, *J. Chromatogr.* **277**, 145 (1983).
[c] H. Mascher and C. Kikuta, *J. Pharm. Sci.* **82**, 56 (1993).
[d] R. L. Roser, A. H. Andrist, and W. H. Harrington, *J. Chromatogr.* **146**, 43 (1978).
[e] D. J. Mauro and D. L. Wetzel, *J. Chromatogr.* **299**, 281 (1984).
[f] H. Ohta, T. Baba, and Y. Suzuki, *J. Chromatogr.* **284**, 281 (1984).
[g] AOAC, Association of Official Analytical Chemists.

this salt has been found to be twice that of potassium hexacyanoferrate.[9] Wielders and Hink[2] reported an enhancement of the fluorescence intensity of thiochrome by using methanol in the mobile phase. The limits for detection in food were between 1 and 30 ng/injection and up to 1 pg/injection[10] in biological products. The analytical recoveries reported were between 87 and 105%.[2,10–14] The intraassay coefficient of variation was between 1.65 and 6.1%[2,9–12,15] and the interassay coefficient of variation was between 1.2 and 6.7%.[2,9,10,12,15] A good agreement was obtained with the AOAC methods[14,16] and the microbiological method.[8,9]

The advantage of precolumn derivatization versus postcolumn derivatization is the decreased peak broadening and hence better resolution with the former technique. Another advantage is that it requires no additional experimental device. A pump, mixing coil, and incubator are needed in postcolumn derivatization. The disadvantage of precolumn derivatization is the instability of thiochrome and its phosphate esters at pH values below 8.0; the elution solvent must have a pH of at least 8.0 and can affect the lifetime of the column. Other disadvantages are the yield of the reaction (about 67%) and the possible presence in the sample extract of an unknown amount of antioxidant, which can react with the oxidizing reagent.

## Conclusion

For multivitamin products and pharmaceutical preparations, methodologies based on UV detection can be performed. They are less sensitive but sample preparation is simpler. For biological and food products, however, techniques based on thiochrome formation (precolumn or postcolumn derivatization) are needed. We have developed in our laboratory[9] a simple technique applicable both to biological media and food products, using precolumn derivatization and the same extraction step as for the microbiological technique.

[9] A. Laschi-Loquerie, S. Vallas, and J. Viollet, *Int. J. Vitam. Nutr. Res.* **62**, 248 (1992).
[10] L. Bettendorff, C. Grandfils, and C. de Ricker, *J. Chromatogr.* **382**, 297 (1986).
[11] C. M. E. Tallaksen, T. Bohmer, and H. Bell, *J. Chromatogr.* **564**, 127 (1991).
[12] R. L. Roser, A. H. Andrist, and W. H. Harrington, *J. Chromatogr.* **146**, 43 (1978).
[13] P. Wimalasiri and R. B. H. Wills, *J. Chromatogr.* **318**, 412 (1985).
[14] H. Ohta, T. Baba, and Y. Suzuki, *J. Chromatogr.* **284**, 281 (1984).
[15] D. J. Mauro and D. L. Wetzel, *J. Chromatogr.* **299**, 281 (1984).
[16] J. K. Fellman, W. E. Artz, and P. D. Tassinari, *J. Food Sci.* **47**, 2048 (1982).

# [8] Determination of Thiamin and Its Phosphate Esters in Human Blood, Plasma, and Urine

By Chantal M. E. Tallaksen, Thomas Bøhmer, Jan Karlsen, and Helge Bell

Thiamin is a small, water-soluble molecule (molecular weight 337). It is present in human tissues and body fluids as unphosphorylated thiamin (T) and as the phosphorylated monophosphate (TP), diphosphate (TPP), and triphosphate (TPPP) (Fig. 1). Thiamin and TP are found extra- and intracellularly, while TPP and TPPP are found only intracellularly.[1,2] Approximately 30 mg of thiamin is estimated to be stored in various body tissues, 80% of the total amount as TPP, but the relative amount of each compound depends on the organ. Not more than 0.8–1% of the total amount is found in whole blood. Thiamin and TP are present in plasma and cerebrospinal fluid (CSF); T, TP, and TPP are found in erythrocytes (also in leukocytes and platelets); while T alone is found in urine.

Thiamin is phosphorylated directly to TPP by thiamin diphosphokinase, and TPP is dephosphorylated to TP via thiamin diphosphatase.[3] Thiamin monophosphate is further dephosphorylated to T by thiamin monophosphatase. Thiamin pyrophosphate-adenosine triphosphate-phosphoryltransferase catalyzes the phosphorylation of TPP into TPPP, and thiamin-triphosphatase (EC 3.6.1.28) catalyzes the conversion of TPPP to TPP. There are thiamin-specific enzymes, and the reactions need $Mg^{2+}$ as a cofactor.[3]

## Determination

### Previous Methods

The first methods used to determine thiamin concentration were microbiological assays. They used the growth of thiamin-dependent protozoa or bacteria as a measuring index. These assays are still used today,[4–6] but the

[1] R. E. Davies and G. C. Icke, *Adv. Clin. Chem.* **23,** 93 (1983).

[2] H. McIlwain and H. S. Bachelard, *in* "Biochemistry and the Central Nervous System," 5th Ed. Churchill Livingstone, Edinburgh, 1985.

[3] R. H. Haas, *Am. Rev. Nutr.* **8,** 483 (1988).

[4] S. L. Tobias, J. Van der Westhuyzen, R. E. Davies, G. C. Icke, and P. M. Attkinson, *S. Afr. Med. J.* **76,** 299 (1989).

[5] R. S. Wang and C. Kies, *Plant Food Hum. Nutr.* **41,** 337 (1991).

[6] A. A. Olkowski and S. R. Gooneratne, *Int. J. Vitam. Nutr. Res.* **62,** 34 (1992).

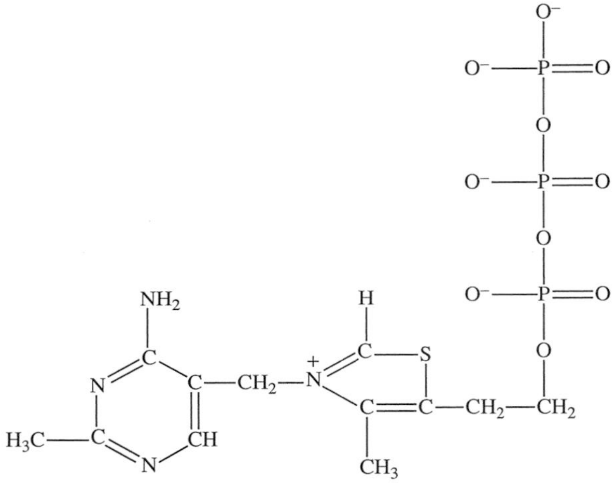

FIG. 1. Thiamin triphosphate (TPPP).

analyses are time consuming, the sensitivity of the assays is not always sufficient for analyses of human body fluids, and they do not allow a separate determination of each thiamin compound. There are also several functional tests for the estimation of TPP concentration. Among them are the glucose tolerance test,[7] and the erythrocyte transketolase assay.[8] The transketolase assay became a standard test used to estimate thiamin deficiency and is still widely used.[9-11] The thiochrome method[12] introduced the oxidation of thiamin to the fluorescent thiochrome, which was then determined by fluorometry.[13]

The use of high-performance liquid chromatography (HPLC) allowed separation of the thiamin compounds under analysis.[14] Several methods were established from 1979 to 1989, but it remained difficult to reconcile criteria of both specificity and sensitivity. Either the sensitivity of the method was too low for detection in human plasma,[14-17] or T had to be

[7] M. K. Horwitt and O. Kreisler, *J. Nutr.* **37,** 411 (1949).

[8] M. Brin and M. Tai, *J. Nutr.* **71,** 273 (1960).

[9] I. Darnton Hill and A. S. Truswell, *Med. J. Austr.* **152,** 5 (1990).

[10] L. J. Smidt, F. M. Cremin, L. E. Grivetti, and A. J. Clifford, *J. Gerontol.* **46,** M16 (1991).

[11] D. A. Gans and A. E. Harper, *Am. J. Clin. Nutr.* **53,** 1471 (1991).

[12] M. Fujiwara and K. Matsui, *Anal. Chem.* **25,** 811 (1948).

[13] H. B. Burch, O. A. Bessey, R. H. Love, and O. H. Lowry, *J. Biol. Chem.* **198,** 477 (1952).

[14] C. J. Gubler and B. C. Hemming, *Methods Enzymol.* **62,** 63 (1979).

[15] K. Ishii, K. Sarai, H. Sanemori, and T. Kawasaki, *Anal. Biochem.* **97,** 191 (1979).

[16] M. Amin and J. Reusch, *J. Chromatogr.* **390,** 448 (1987).

[17] H. Iwata, T. Matsuda, and H. Tonomura, *J. Chromatogr.* **450,** 317 (1988).

determined independently of TP, TPP, and TPPP.[16,18–25] We established a new method,[26] which combined the advantages of high sensitivity (detection limit, 1 nmol of T per liter of plasma) and determination of the four compounds in one chromatographic run.

*Proposed Conditions for Routine Analysis*

*Assay System*

Perkin-Elmer (Norwalk, CT) Series 400 solvent-delivery system
Perkin-Elmer LS1 fluorescence detector (excitation wavelength set at 375 nm and emission measured at 450 nm)
Perkin-Elmer LC1 100 laboratory computer integrator

*Columns*

Supelcosil $NH_2$ column (250 × 4.6 mm i.d.) from Supelco (Supelco Park, Bellefonte, PA)
Guard column (Supelco LC NH, 20 × 4.6 mm i.d.)
Guardpack inserts [$C_{18}$ Waters (Milford, CT) guardpack holder and insert]

*Solvents*

Phosphate buffer, 85 m$M$, pH 7.5
Acetonitrile (ACN)
Buffer/ACN, 90:10 (v/v), subsequently 60:40 (v/v)
*Procedure.* The flow rate is 1.2 ml/min, resulting in a pressure of 55–110 bar. The column is stabilized with the solvent for 2.5–4 min before injection of the sample. Thiamin is eluted after 4 min, and the solvent is then changed to 60:40 (v/v) (step gradient). Thiamin monophosphate, TPP, and TPPP

[18] H. Sanemori, H. Ueki, and T. Kawasaki, *Anal. Biochem.* **107,** 451 (1980).
[19] J. Schrijver, A. J. Speek, J. Klosse, H. Van Rijn, and W. H. P. Schruers, *Ann. Clin. Biochem.* **19,** 52 (1982).
[20] L. G. Warnock, *Anal. Biochem.* **126,** 394 (1982).
[21] M. Kimura, T. Fujita, and Y. Itokawa, *Clin. Chem.* **28,** 29 (1982).
[22] J. P. M. Wielders and C. H. R. J. K. Mink, *J. Chromatogr.* **277,** 145 (1983).
[23] M. Baines, *Clin. Chem. Acta* **153,** 43 (1985).
[24] L. Bettendorf, C. Grandfils, C. de Rycker, and E. Schoffeniels, *J. Chromatogr.* **382,** 297 (1986).
[25] B. Bötticher and D. Bötticher, *Int. J. Vitam. Nutr. Res.* **57,** 273 (1987).
[26] C. M. E. Tallaksen, T. Bøhmer, H. Bell, and J. Karlsen, *J. Chromatogr.* **564,** 127 (1991).

are eluted after 13, 17, and 20 min, respectively. In blood samples, an unidentified peak is eluted after 8–10 min (Fig. 2).

The molarity of the buffer is particularly important because of interaction with ACN. At a higher molarity (>85 m$M$), the ACN and the buffer do not mix homogeneously at 90:10 (v/v) and the column cannot be stabilized. At a lower molarity, peaks are broader, retention times higher, and the column unstable. The optimal molarity for the phosphate buffer is 82–85 m$M$.

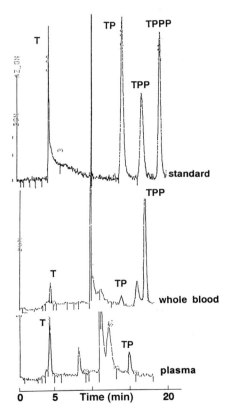

Fig. 2. Elution profile for a standard solution, a whole-blood sample, and a plasma sample. Peaks: T, thiamin; TP, thiamin monophosphate; TPP, thiamin diphosphate; TPPP, thiamin triphosphate. Injection at time 0. [Modified from *J. Chromatogr.* **564,** C. M. E. Tallaksen *et al.,* Concomitant determination of thiamin and its phosphate esters in human blood and serum, Copyright 1991 with kind permission of Elsevier Science–NL, Sara Burgerhartstraat 25, 1055 KV Amsterdam, The Netherlands.]

*Preparation of Solvents.* Stock solutions with buffer are kept refrigerated for no longer than 3 weeks. Prior to use, the buffer is filtered through a Millipore (Bedford, MA) filter (0.22-$\mu$m pore size), and the mobile phase is degassed for 20 min with helium before analysis.

## Derivatization

Cyanogen bromide, 97% (Aldrich, Steinheim, Germany): A 0.3 $M$ solution of cyanogen bromide in ACN is used

*Comments.* The conversion of thiamin compounds into fluorescent derivatives is obtained by alkaline oxidation, which converts thiamin into the fluorescent thiochrome. Cyanogen bromide is chosen as it has been described as the most efficient oxidizing agent.[12,27] As it is unstable in solution, weighed aliquots of the solid compound are kept refrigerated in glass tubes, and a fresh solution is prepared on each day of analysis. The solution is perfectly stable for at least 8 hr at room temperature. Prior to injection on the column, 200-$\mu$l samples are derivatized by addition of 20 $\mu$l of the cyanogen bromide solution followed by an equal volume of 1 $M$ sodium hydroxide. After derivatization, the samples are kept refrigerated and protected from the light, and stay stable for at least 6 hr. A 100-$\mu$l aliquot of the derivatized sample is injected on the column.

It is essential to add cyanogen bromide to the sample before alkalinization, as thiamin is destroyed in alkaline solutions. Samples treated with sodium hydroxide before cyanogen bromide are used as blanks, as done earlier by Iwata *et al.*[17]

## Standards

Thiamin hydrochloride, thiamin monophosphate chloride, cocarboxylase (Sigma, St. Louis, MO)
Thiamin triphosphate (gift from Takeda Chemicals, Osaka, Japan)
Thiazole–[2-[14]C]thiamin (specific activity, 918 mBq/mmol), (Amersham International, Amersham, U.K.)
Standard stock solutions ($10^{-3}$ $M$) are prepared in 0.01 $M$ hydrochloric acid and stored at $-20°$. Working solutions are prepared fresh by diluting the stock solutions to the required concentrations with deionized and filtered water. The currently used standard concentration is $2 \times 10^{-9}$ $M$ for each compound.

Because the standard solution is not stable over months, new aliquots of the standard mixture are prepared fresh every second month. Cleland's

[27] T. Nishimune, S. Ito, M. Abe, M. Kimoto, and R. Hayashi, *J. Nutr. Sci. Vitaminol.* (*Tokyo*) **34**, 543 (1988).

reagent (dithiothreitol) is added to the aliquots to improve storage conditions, but it does not improve the stability of the standards.

Recovery studies are done with [$^{14}$C]thiamin and have proved satisfactory. The linearity of the standards is tested within the physiological range and the correlation coefficients of the regression lines are all above 0.99.

*Preparation of Samples.* Venous blood is collected from the cubital vein in heparin tubes for whole-blood samples and in serum tubes (containing a clot activator) for serum samples. Whole-blood samples are hemolysed by freezing at $-20°$. Serum samples are stored at $-20°$ until further treatment. For sample preparation, 1 ml of hemolysed blood is diluted with 1.5 ml of acidified water (pH approximately 4.5).

Deproteinization is done by addition of 250 $\mu$l of 2.44 $M$ (40%) trichloroacetic acid (TCA) to the 2.0-ml serum samples and 315 $\mu$l to the diluted whole-blood samples, so that the final concentration of TCA in the samples is 5%. After thorough mixing, the samples are left for 1 hr in darkness. After centrifugation (20 min, 2000 $g$, 4°) 1 ml of supernatant of serum or hemolysate is transferred to fresh glass tubes, and TCA is extracted twice with 5 vol of water-saturated diethyl ether. A 400-$\mu$l aliquot is subsequently filtered through Millipore Ultrafri MC-10,000 NMWL filter units in a Sorvall centrifuge at 5000 $g$ for 20 min at 4°.

Urine samples are prepared in the following way. Thiamin is extracted from the urine by use of a CBA sorbent (carboxylic acid, 1 ml; Analytichem International, Harbor City, CA). The sorbent is pretreated with 2 ml of methanol and 2 ml of distilled water. One milliliter of acidified urine is then added. After washing with 2 ml of distilled water, 2 ml of methanol, and again with 2 ml of water, the thiamin is eluted with 2 ml of the following eluent heated to 65–75°: 2 $M$ potassium chloride (pH 1.5–2.0)–methanol (60/40, v/v). The method affords specific extraction of unphosphorylated thiamin with a recovery of 97.5 (4.9)%. The eluate is subsequently derivatized and analyzed.

*Comments.* Samples of cerebrospinal fluid could be prepared as plasma samples. Deproteinization is probably not necessary in most cases, but the procedure does not affect the results.

Heparinized whole blood that was not used for analysis was frozen within 2 hr of collection to avoid dephosphorylation by plasma enzymes.[28] Once deproteinized, samples were kept frozen until analysis. Storage of a pooled sample was checked at regular intervals for 1 year; analysis showed that plasma samples could be kept for at least 6 months, whereas whole-blood samples started to show a decrease in TPP after 3 months.

[28] J. Y. Thom, R. E. Davies, and G. C. Icke, *Int. J. Vitam. Nutr. Res.* **55,** 269 (1985).

TABLE I

THIAMIN, THIAMIN MONOPHOSPHATE, AND THIAMIN DIPHOSPHATE IN PLASMA AND WHOLE BLOOD[a]

| Parameter | Concentration (nmol/liter) | | | | | | |
|---|---|---|---|---|---|---|---|
| | Plasma T | Plasma TP | Blood T | Blood TP | Blood TPP | CSF T | CSF TP |
| Mean | 7.1 | 5.8 | 24 | 5.7 | 85 | 17 | 28 |
| SD | 1.6 | 2.6 | 11 | 2.2 | 18.1 | 8.3 | 8.6 |
| 95% CI | 0.7 | 1.2 | 4.9 | 1.0 | 8.5 | 2.5 | 2.6 |
| Geometric mean | 6.9 | 5.4 | 22 | 5.3 | 83 | 15 | 27 |
| Range | 4.5–11 | 2.8–11 | 8.6–56 | 2.4–12 | 55–125 | 5.7–40 | 15–50 |

[a] Concentrations of thiamin, TP, and TPP in 34 healthy subjects, and of T and TP in CSF in 44 healthy subjects. Mean, standard deviation (SD), 95% confidence interval (95% CI), geometric mean, and corresponding range calculated as the antilog of [log mean ± (2 log SD)] are given.

*Detection Limit.* On the basis of a signal-to-noise ratio of 3, the detection limit was 13–16 fmol of minimal absolute amount per aliquot for the different compounds.

*Results.* The subjects used in the reference group for plasma and whole-blood values were healthy, overnight-fasted volunteers (17 female, 17 male; mean age, 40.3 ± 8.5 years).[29] Forty-four patients (20 female, 24 male; mean age, 48.0 ± 14.7 years; range, 21–83 years) with lower back pain, but otherwise healthy, constituted the reference group for thiamin concentration in cerebrospinal fluid.[30] They were not overnight fasted.

We did not find any TPPP in plasma, whole blood, or CSF. Thiamin triphosphate is found in brain tissue and skeletal muscle, but we have not done any tissue studies. We therefore used only T, TP, and TPP as standards for the usual analyses during the last 3 years.

The concentrations of T, TP, and TPP showed a skew distribution. The distribution was, however, log-normal, and the use of a geometric mean appeared therefore better suited for clinical purposes, the reference range being defined by: antilog of [log mean ± (2 log SD)]. It establishes a clinically useful cutoff value for plasma T and TP concentrations, which was not available with the mere mean ± 2 SD (Table I).

## Concluding Remarks

After using this method for 6 years, we have found it reliable and easy to use, and have determined that it shows a satisfactory reproducibility for clinical use, with an intraassay coefficient of variation of, respectively, 5.4,

[29] C. M. E. Tallaksen, Ph.D. thesis. University of Oslo, Oslo, Norway, 1992.
[30] C. M. E. Tallaksen, T. Bøhmer, and H. Bell, *Am. J. Clin. Nutr.* **56,** 559 (1992).

4.8, 6.2, and 11.6 for T, TP, TPP, and TPPP, calculated on the basis of 10 analyses of the same sample on 1 day. Occasionally, however, we have had difficulties in obtaining identical new columns, and have been obliged to try a different batch. The variation of packing material for HPLC columns may be regarded as a parameter to take into account when using HPLC methods. Regular use of standards will partly eliminate this problem.

We have also observed that a stable ambient temperature improves the stability of the column.

There are several advantages to this method. The use of precolumn derivatization, and of a buffer at pH 7.5, make the working conditions easier than in previously described assays.[31,32] There is only one mobile phase, as opposed to the method of Bettendorf et al.,[24] in which the mobile phase is changed from buffer–tetrahydrofuran for the elution of T to buffer–methanol for the elution of the other compounds. The detection limit is 13–16 fmol, which is much lower than in the method described by Vanderslice and Huang,[33] and necessary for detection of thiamin in human samples.

### Acknowledgments

This work was supported by Vinmonopolet A.S. The authors thank Berit Falch for excellent technical assistance, and Maureen Dixon for kind secretarial assistance.

[31] J. Bontemps, P. Philippe, L. Bettendorf, J. Lombet, G. Dandrifosse, and E. Schoffeniels, *J. Chromatogr.* **307,** 283 (1984).
[32] M. Kimura and Y. Itokawa, *J. Chromatogr.* **332,** 181 (1985).
[33] J. T. Vanderslice and M. J. A. Huang, *J. Micronutr. Anal.* **2,** 189 (1986).

## [9] Determination of Thiamin and Thiamin Phosphates in Whole Blood by Reversed-Phase Liquid Chromatography with Precolumn Derivatization

*By* J. Gerrits, H. Eidhof, J. W. I. Brunnekreeft, and J. Hessels

### Introduction

Thiamin is present in four different forms: thiamin (T), thiamin monophosphate (TMP), thiamin diphosphate (TDP), and thiamin triphosphate (TTP). The role of TPP as a coenzyme in carbohydrate metabolism and the role of some phosphorylated form of thiamin in nerve conduction is well established.[1]

Several chromatographic methods for the determination of thiamin and its phosphate esters have been published.[2-10] These methods can be divided into pre- and postcolumn derivatization techniques with or without hydrolysis of thiamin phosphate esters. All these methods are cumbersome and in some cases poisonous chemicals are used. Here we present a simple, inexpensive, and sensitive method with automated precolumn derivatization of thiamin and its phosphate esters. The method is optimized for the simultaneous determination of thiamin and its phosphate esters and has previously been described in detail.[11] The thiochrome phosphate esters are stable and column life is extended.

## Materials and Methods

### Chemicals

Thiamin chloride hydrochloride, perchloric acid, $K_3Fe(CN)_6$, NaOH, and phosphoric acid are obtained from Merck (Darmstadt, Germany). Thiamin monophosphate (TMP) and thiamin diphosphate (TDP) are obtained from Sigma Chemical Co. (St. Louis, MO). Thiamin triphosphate (TTP) was donated by Dr. J. Schrijver (CIVO-TNO, Zeist, The Netherlands). Methanol and dimethylformamide (DMF) are obtained from Labscan (Dublin, Ireland) and *tert*-butylammonium hydroxide (TBAH) is from Fluka (Buchs, Switzerland).

### Stock Solutions and Working Standard Solutions

Stock solutions of thiamin chloride hydrochloride, TMP, TDP, and TTP are prepared fresh before use in 0.1 $M$ HCl, in concentrations of 3.0 m$M$.

[1] G. Rindi, *Acta Vitaminol. Enzymol.* **4,** 59 (1982).
[2] M. Kimura, T. Tujita, S. Nishida, and Y. Itokawa, *J. Chromatogr.* **188,** 417 (1980).
[3] J. Schrijver, A. J. Speek, J. A. Klosse, H. J. M. van Rijn, and W. H. P. Schreurs, *Ann. Clin. Biochem.* **19,** 52 (1982).
[4] M. Kimura and Y. Itokawa, *Clin. Chem.* **29,** 2073 (1983).
[5] J. P. M. Wielders and Chr. J. K. Mink, *J. Chromatogr.* **277,** 145 (1983).
[6] M. Baines, *Clin. Chim. Acta* **153,** 43 (1985).
[7] K. Ishii, K. Sarai, H. Sanemori, and T. Kawasaki, *Anal. Biochem.* **97,** 191 (1979).
[8] J. Bontemps, P. Philippe, L. Bettendorff, J. Lombet, G. Dandrifosse, E. Schoffeniels, and J. Crommen, *J. Chromatogr.* **307,** 283 (1984).
[9] J. Bontemps, L. Bettendorff, J. Lombet, C. Grandfils, G. Dandrifosse, E. Schoffeniels, F. Nevejans, and J. Commen, *J. Chromatogr.* **295,** 486 (1984).
[10] L. Bettendorff, C. Grandfils, C. de Rycker, and E. Schoffeniels, *J. Chromatogr.* **282,** 297 (1986).
[11] J. W. I. Brunnekreeft, H. Eidhof, and J. Gerrits, *J. Chromatogr.* **89** (1989).

Thiamin and its phosphates are dried with $P_2O_5$ overnight before preparing the stock solution. The purity of thiamin and its phosphates is evaluated by high-performance liquid chromatography (HPLC) and ultraviolet (UV) absorbance (molar extinction coefficient: $\varepsilon_{248 \text{ nm}; 0.1 N \text{ HCl}} = 13.400$).[12] The purity of TTP is 83%.

In our hands the fluorescence response factors of TTP, TDP, and TMP are almost the same. Therefore, it is possible to use the response factor of TDP to calculate the TTP concentration of the sample, because it is difficult to obtain TTP. The concentrations of these stock solutions are checked by the method of Penttinen.[12] Our results fully agree with those of Penttinen.

Working standard solutions are prepared by diluting the stock solution 1:2000 with distilled water. Working standard solutions are kept at $-20°$. These solutions are stable for 2 months.

The oxidation reagent for the method with manual derivatization is a freshly prepared aqueous solution of 12.1 m$M$ $K_3Fe(CN)_6$ and 3.35 $M$ NaOH. The oxidation reagent for the automated derivatization method is a freshly prepared solution of 15.2 m$M$ $K_3Fe(CN)_6$ in methanol–water (1:1) and 5.0 $M$ NaOH in water.

Two solutions of perchloric acid (PCA) are made: PCA-1, 7.2% and PCA-2, 7.2% in 0.25 $M$ NaOH.

## Mobile Phases

Solvent A is prepared as follows: 31.9 g of $K_2HPO_4 \cdot 3H_2O$ is dissolved in 800 ml of distilled water and 120 ml of methanol, and 0.6 ml of 0.5 $M$ TBAH and 15 ml of DMF are added.

Distilled water (1 liter) is added and then concentrated $H_3PO_4$ to a final pH of 7.0; the final volume is 1 liter. Solvent B is 70% (v/v) methanol. The solvents are filtered on 0.45-$\mu$m (pore size) membrane filters (RC 55; Schleicher & Schuell, Keene, NH) before use.

## Sample Preparation and Derivatization Procedure

Venous blood samples are collected in evacuated tubes containing lithium heparin as anticoagulant. The samples are immediately stored at $-20°$ in polystyrene tubes (55 × 11 mm). After thawing the working standard and the blood samples, 1 ml of PCA-1 (4°) is added to 1 ml of blood and 0.1 ml of distilled water; 1 ml of PCA-2 (4°) is added to 1 ml of distilled water and 0.1 ml of standard solution.

For recovery samples, a 100-$\mu$l standard solution is added instead of distilled water.

---

[12] H. K. Penttinen, *Acta Chem. Scand. B* **30**(7), 659 (1976).

After vigorous mixing, the tubes are placed on ice. After 10 min the tubes are again vigorously mixed and placed on ice for another 5 min. The mixtures are then centrifuged for 15 min at 2000 g at 4°. The sample preparation procedure is carried out in a dark room with subdued light. The supernatants are filtered through a Seraclear filter.

A. *Manual Derivatization.* The filtrates are pipetted in 1-ml portions in polystyrene tubes. When room temperature is reached, 100 $\mu$l of methanol is added to all tubes. During the mixing, 200 $\mu$l of oxidation reagent is added (pH > 12). After about 30 sec, 100 $\mu$l of 1.4 $M$ phosphoric acid is added, and the contents of the tubes are mixed. The final pH of the prepared samples and standards is 6.9 ± 0.2. The samples and standards are stable for at least 24 hr at room temperature when protected from light.

Before injection the samples are filtered through 0.2-$\mu$m (pore size) syringe filters (Dynagard 200-2-200 ME; Microgon, Inc., Laguna Hills, CA). A 50-$\mu$l sample is injected by a Rheodyne valve (model 7125; Rheodyne, Cotati, CA).

B. *Automated Derivatization.* The precolumn derivatization is performed by a Hitachi-AS400 autosampler (Merck). The filtrates are pipetted into insert vials (300 $\mu$l) in exact 100-$\mu$l portions. Just before injection, 20 $\mu$l of oxidation reagent, 20 $\mu$l of 5 $M$ NaOH solution, and 10 $\mu$l of 3.4 $M$ phosphoric acid are added to each vial. After every addition step, the sample is mixed. The injection volume is 75 $\mu$l. The sample tray of the autosampler is protected from light.

*Chromatographic Analysis*

A Microspher $C_{18}$ analytical column (3-$\mu$m particles; column, 100 × 4.6 mm) and a corresponding guard column are used (Chrompack, Middelburg, The Netherlands). The mobile phase is pumped at 1.0 ml/min by a low-pressure gradient pump (model 600; Waters, Milford, MA). With the help of a system controller (model 600 E; Waters) the gradient given in Table I is set up. Total analysis time is 15 min, including the stabilization of the

TABLE I
GRADIENT ELUTION PATTERN

| Time (min) | Flow rate (ml/min) | Solvent A (%) | Solvent B (%) | Curve shape |
|---|---|---|---|---|
| 0 | 1.0 | 100 | 0 | Linear |
| 2 | 1.0 | 80 | 20 | Linear |
| 6 | 1.0 | 80 | 20 | Linear |
| 7 | 1.0 | 100 | 0 | Linear |

column. Detection is performed by a spectrofluorimeter model RF 535 (Shimadzu, Tokyo, Japan) equipped with a 12-$\mu$l flow-through cell. The excitation is set at 367 nm; the emission is detected at 435 nm. The resulting signal is displayed and computed by an integrator DATA module 745 (Waters).

The calculation of the concentration of thiamin and its phosphate esters is based on an external standard and checked by recovery results.

### Quality Control Materials

For internal quality control, we use donor blood anticoagulated with citrate, heparin, or EDTA. For external quality control, we use lyophilized anticoagulated blood (SKZL, Nijmegen, The Netherlands). For stability of the various materials, see Table II. Lyophilized anticoagulated blood from SKZL may also be used for standardization purposes. Consensus values are available for this material.

### Results and Discussion

### Chromatographic Experiences

Bontemps et al.[8] have shown that thiamin, TMP, TDP, and TTP can be separated by gradient elution in the reversed-phase mode using 25 m$M$ phosphate buffer, pH 8.4, and methanol as mobile phase. This method is suitable for the measurement of thiamin compounds in excitable tissues.

Our aim was to develop a simple method for the simultaneous determination of the four different forms of thiamin in whole blood. First, we developed an isocratic method; however, this method was cumbersome because of the long stabilizing time of the column, owing to aspecific peaks

TABLE II

STABILITY OF THIAMIN PYROPHOSPHATE IN WHOLE AND LYOPHILIZED BLOOD AND IN STANDARDS AT DIFFERENT TEMPERATURES

| Material | Temperature[a] | | | |
|---|---|---|---|---|
| | $-20°$ | $4°$ | $20°$ | $37°$ |
| Heparin or EDTA whole blood | >6 months | <3 days | <4 hr | nd |
| Lyophilized anticoagulated blood | nd | >3 years | nd | >5 months |
| Standard in 0.01 $N$ HCl | <2 months | nd | nd | nd |

[a] nd, Not determined.

eluting at the end of the analysis. Second, we set up a new gradient method, which is a modification of the method of Bontemps.[8]

We carried out some experiments with different concentrations of $K_3Fe(CN)_6$ and NaOH; our results were in agreement with the results of Bontemps[8] (results not shown). Addition of methanol as a modifier influenced the result. Methanol increased the formation of the thiochromes (Thc), which resulted in higher fluorescence. Penttinen[12] suggested that the different fluorescence intensities of thiochromes were due to the different amounts of nonfluorescent oxidation products, especially the disulfide derivative. We varied the methanol concentration to obtain a stable, reproducible signal. Too much methanol produced asymmetric peaks.

Our final concentration of methanol was 12%. This resulted in well-shaped peaks and reproducible retention times: $K' = 1.7, 2.11, 3.26,$ and 6.88 for ThcTP, ThcDP, ThcMP, and Thc, respectively. The stability of the chromatographic system is shown by repeated injections of a standard sample (Fig. 1). The corresponding data are shown in Table III. Figure 2 shows a typical chromatogram of a blood sample injected three times.

Addition of TBAH as modifier resulted in longer retention times of the ThcTP, ThcDP, and ThcMP and a shorter retention time for Thc.

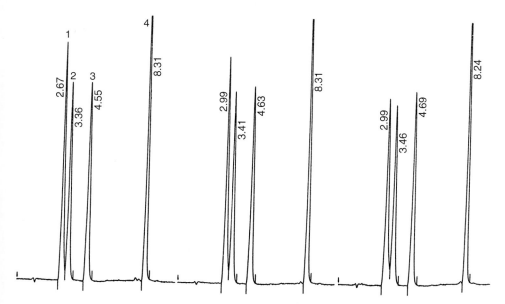

FIG. 1. Elution profile for a standard solution injected three times. Peaks: 1, TTP; 2, TDP; 3, TMP; and 4, thiamin.

TABLE III
CHROMATOGRAPHIC PARAMETERS[a]

| Compound | $N(\frac{1}{2})$ | $N(4.4\%)$ | $R_s$ |
|---|---|---|---|
| T | $13.1 \times 10^3$ | $11.3 \times 10^3$ | |
| TMP | $3.3 \times 10^3$ | $2.2 \times 10^3$ | |
| TDP | $2.0 \times 10^3$ | $1.3 \times 10^3$ | TDP–TMP : 3.4 |
| TTP | $1.7 \times 10^3$ | $1.3 \times 10^3$ | TTP–TDP : 1.2 |

[a] $N(\frac{1}{2}) = 5.53[t_r/W(\frac{1}{2})]^2$; $N(4.4\%) = 25[t_r/W(4.4\%)]^2$; $R_s = (t_{r1} - t_{r2})/[W_1(\frac{1}{2}) + W_2(\frac{1}{2})]$; $t_r$, retention time; $W(4.4\%)$, peak width at 4.4% of peak height; and $R_s$, resolution factor.

Dimethylformamide addition resulted in a better separation of the thiochrome esters.

## Column Lifetime and pH Influence on Stability of Samples

Although the fluorescence intensity is pH dependent and reaches a plateau at pH >8 according to Ishii *et al.*,[7] we used a buffer with pH 7.0. At this pH our signal was 20% lower in accordance with the results of Ishii *et al.*[7] Postcolumn addition of NaOH could increase the signal; however, we skipped this step because of dilution of the sample and peak broadening. Column life at pH 7.0 is extended and a better reproducibility is possible. Thiochrome and its phosphates are stable at this pH for at least 24 hr at

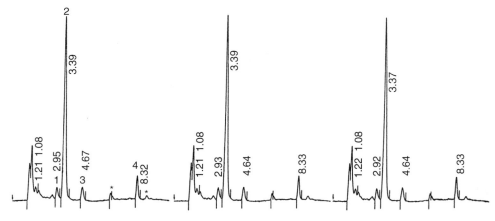

FIG. 2. Elution profile for a human blood sample injected three times. Peaks: 1, TTP; 2, TDP; 3, TMP; and 4, thiamin. Unknown impurities are indicated by an asterisk.

TABLE IV
COEFFICIENTS OF VARIATION IN BLOOD SAMPLES[a]

| | Intraassay precision | | | | Interassay precision | |
|---|---|---|---|---|---|---|
| | Sample | | Spiked sample | | Sample | |
| Compound | Concentration (nM) (mean ± SD) | CV (%) | Concentration (nM) (mean ± SD) | CV (%) | Concentration (nM) (mean ± SD) | CV (%) |
| TTP | 1.54 (±0.24) | 16 | 95.0 (±0.36) | 0.4 | 2.9 ±0.7 | 24 |
| TDP | 119.8 (±0.76) | 0.6 | 261.1 (±1.54) | 0.6 | 178 ±5.4 | 3.0 |
| TMP | 4.1 (±0.76) | 12 | 135.7 (±0.94) | 0.7 | 7.4 ±1.2 | 16 |
| Thiamin | 5.2 (±0.27) | 5 | 123.4 (±1.24) | 1.0 | 6.8 ±0.8 | 12 |

[a] Intraassay ($n = 10$) and interassay ($n = 10$) coefficients of variation (CV).

room temperature. At higher pH the oxidized compounds deteriorated continuously.

## Temperature Influence on Stability of Samples

After venipuncture samples should be stored at $-20°$ as soon as possible. At room temperature thiamin and its phosphates in whole blood are not stable after 4 hr (Table II). However, in lyophilized blood the compounds are very stable (Table II) and are therefore suitable for use as external quality control material. The whole procedure of sample derivatization should be done at constant temperature; the efficiency of the thiochrome formation varies with temperature. After the oxidation step, phosphoric acid is added after about 30 sec. Samples with a final pH of 6.9 ± 0.2 at room temperature and protected from light are stable for at least 24 hr.

## Precision, Accuracy, and Linearity

The precision results are given in Table IV. The intraassay coefficient of variation was calculated from the analyses of 10 whole-blood samples and a spiked sample. The interassay coefficient of variation was based on the analysis of 1 patient sample in 10 batches. The accuracy of the method was investigated by recovery experiments in different patient samples. The results are shown in Table V. The linearity of the method for TDP, TMP, and T goes up to 1000 nM. Considering a signal-to-noise ratio of 5 as the detection limit, the sample injected should contain at least 0.5 nM each of thiamin and phosphate esters. This corresponds to an absolute amount of about 7 fmol for each component injected. The calculation of the final results of the samples was based on external standards.

TABLE V
RECOVERY RESULTS

| Compound | $n^a$ | Recovery (mean $\pm$ SD) (%) | CV (%) |
|----------|------|------------------------------|--------|
| TTP      | 4    | 101 ($\pm$7.4)               | 7.3    |
| TDP      | 10   | 99 ($\pm$4.2)                | 4.2    |
| TMP      | 10   | 102 ($\pm$2.5)               | 2.4    |
| Thiamin  | 10   | 90 ($\pm$7.5)                | 8.3    |

[a] $n$, Number of samples analyzed.

*Reference Values*

Samples ($n$ = 65) were obtained from healthy volunteers. Reference values (mean $\pm$ SD) were as follows: TDP, 120 ($\pm$17.5); TMP, 4.1 ($\pm$1.6); T, 4.3 ($\pm$1.9) n$M$. For TTP all measurements were below 4.0 n$M$.

Conclusion

The results presented in this chapter show that the method is simple, inexpensive, and easy to perform. The parameters that affect oxidation have been optimized for manual and automated derivatization. The chromatographic conditions have been optimized with respect to stability of the signal, column life, time of analysis, reproducibility of the results, separation of thiochrome and its phosphates, and automated precolumn derivatization. The clinical importance of TTP is not yet clear.[1]

We encountered a few patients with low TDP and relatively high TTP. In most cases these patients were alcoholics. The precise role of TTP in the clinical setting still remains to be elucidated.

Acknowledgments

The authors are grateful to Dr. J. Schrijver for providing TTP and to A. Paus for typing the manuscript. The authors thank Elsevier Science for permission to adapt this article from *J. Chromatogr.* **491,** 89 (1989).

# [10] High-Performance Liquid Chromatography Determination of Total Thiamin in Human Plasma

By HERMANN J. MASCHER and CHRISTIAN KIKUTA

Thiamin lends itself to clinical and therapeutic use in the prevention and treatment of vitamin $B_1$ deficiencies. Thiamin (T) is present endogenously, mainly in the form of various phosphates, of which thiamin diphosphate (pyrophosphate) predominates in plasma. For relative bioavailability studies, determination of total T is performed after separation of the different T phosphates by phosphatase. The literature contains only a few studies on the detection of T in rat or human blood[1,2] or human plasma,[3] especially determination in connection with bioavailability studies.

## Assay Method

### Principle

After cleavage of T mono- and diphosphates by acid phosphatase, total T is determined by ion-pair reversed-phase high-performance liquid chromatography (HPLC) after protein precipitation with postcolumn oxidation of T resulting thiochrome. On-line detection is done by fluorescence.

### Materials

Thiamin diphosphate (TDP; pyrophosphate), thiamin monophosphate (TMP) chloride, and T chloride are obtained from Sigma (Munich, Germany). Acid phosphatase (2 units/mg, grade II) is from Boehringer GmbH (Mannheim, Germany). All other reagents and solvents used are analytical grade.

### Instrumentation

An HP-1090M liquid chromatograph (Hewlett Packard, Palo Alto, CA), with an F-1000 fluorescence detector (excitation, 365 nm; emission, 435 nm; Merck-Hitachi, Darmstadt, Germany), is used. An LC-420 pump (Kontron, Zurich, Switzerland) is used for postcolumn derivatization. The derivatizing

---

[1] M. Kimura and Y. Itokawa, *J. Chromatogr.* **332,** 181 (1985).
[2] J. W. I. Brunnekreeft, H. Eidhof, and J. Gerrits, *J. Chromatogr.* **491,** 89 (1989).
[3] B. Bötticher and D. Bötticher, *Int. J. Vit. Nutr. Res.* **57,** 273 (1987).

reagent is mixed with the eluate from the HPLC column in a mixing-T (Lee Company, Westbrook, CT), and the reaction takes place in a Teflon capillary (o.d., 1/16 in.; i.d., 0.3 mm; length, 10 m; volume, ~0.7 ml) having a three-dimensional configuration. The chromatograms are evaluated with the aid of a CI-10B integrator (LDC, Shannon, Ireland).

*Chromatographic Conditions*

The mobile phase for ion-pair reversed-phase chromatography is 25% acetonitrile : 75% aqueous phase (10 m$M$ perchloric acid : 10 m$M$ octanesulfonic acid) (v/v). The flow rate is 2.0 ml/min. The chromatographic separation takes place on a Nucleosil 120 5 $C_{18}$ column (125 × 4 mm; SRD Pannosch, Vienna, Austria).

*Postcolumn Derivatization*

The reaction mixture, 0.8 g of $K_3[Fe(CN)_6]$, in 1 liter of 3 $M$ sodium hydroxide solution, is pumped via the mixing-T and mixed with the eluate from the analytical column. The flow rate is 1.0 ml/min and the reaction is performed at 30°.

*Preparation of Calibration Samples*

The calibration samples are composed of 80% TDP and 20% T (e.g., 80 ng of TDP spiked with 20 ng of T) in accordance with the ratio given previously.[1] Pooled plasma samples are spiked with various quantities of this mixture to give concentrations of 33.7, 67.3, 134.6, 291.7, 583.4, and 1077.1 ng of T base per milliliter of pooled plasma. After analysis of a random selection of plasma samples from subjects after administration of T, only the range of 0 ng/ml (native value) to 134.6 ng/ml (T base added) is used for validation, because this concentration range is not exceeded. For the purpose of calculation, peak area is plotted against concentration. For calculation of the endogenous values for total T, spiked water samples (e.g., 80% TDP and 20% T) are used. These samples are also used for calculating the determination limit.

*Sample Preparation*

One milliliter of plasma is mixed with 0.15 ml of 3 $M$ perchloric acid by vortexing, and centrifuged at 20° for 3 min at >1500 $g$. Subsequently, 0.3 ml of buffer solution (1 $M$ sodium acetate/acetic acid with 2.4 g of NaOH/100 ml, pH 4.6) is added to 0.5 ml of the clear supernatant and mixed. Enzyme solution (0.1 ml of a solution consisting of 10 mg of acid phosphatase/ml of water) is then added, and the sample is incubated for

16 hr at 40°. After incubation, 0.15 ml of 3 $M$ perchloric acid is added, and the sample is mixed with a vortex before being centrifuged at 20° for 3 min at >1500 $g$. Finally, 20 $\mu$l of the clear supernatant is injected onto the column for HPLC analysis.

## Assay Results

### Phosphate Separation

The effectiveness of phosphate separation was examined by comparing TMP/TDP and T plasma admixtures. Systematic investigations on pure-substance solutions of TMP and TDP with enzymatic separation have already been reported by Defibaugh,[4] who found that with pure-substance solutions, numerous enzymes were suitable. We chose acid phosphatase, and our investigations demonstrated that 16 hr at 40° was sufficient to separate all of the TMP and TDP and leave only T after the admixture of plasma. After 3 hr, the phosphates had been only partially separated. In these preliminary investigations, the admixtures were made at various concentrations: TDP in the 30- to 1000-ng/ml plasma range (six different concentrations) and TMP in the 17- to 270-ng/ml range (three different concentrations).

The slopes of the validation lines after conversion to T base were within ±5% of the slopes of the validation lines after the admixture of T. The validation line relevant for the analysis of the samples from the subjects then consisted of 80% TDP and 20% T, which was then separated at 40° for 16 hr.

### Postcolumn Derivatization

As we have shown previously,[5] T may be transformed into highly fluorescent thiochrome, either by oxidation with $K_3[Fe(CN)_6]$ or by electrochemical oxidation. Because this coulometric oxidation must take place under highly basic conditions, occasional clogging of the graphite electrode (Coulochem, Bedford, MA) may occur after pumping in the NaOH solution. We therefore used oxidation with $K_3[Fe(CN)_6]$, which permitted a complete transformation under the reaction conditions.

During the investigations of the plasma samples from the subjects, an aqueous solution of T (corresponding to ~70 ng/ml of plasma) was injected at regular intervals as a means of monitoring the chromatographic condi-

[4] P. W. Defibaugh, *J. Assoc. Off. Anal. Chem.* **70**, 514 (1987).
[5] C. Kikuta and H. Mascher, Poster CLC 89, 13th International Symposium on Column Liquid Chromatography, June 1989, Stockholm M/TU-P-141, Sweden.

TABLE I
PRECISION IN PLASMA[a]

| Concentration of total thiamin in plasma (ng/ml) | CV (±%) |
|---|---|
| Endogenous value in pool–plasma | ±6.32 |
| 33.7 | ±2.74 |
| 67.3 | ±4.51 |
| 134.6 | ±1.60 |
| 291.7 | ±2.44 |
| 583.4 | ±1.82 |
| 1077.1 | ±3.04 |

[a] $n = 3$.

tions and the postcolumn derivatization: number $(n) = 82$, peak area = 91,245 ± 3108 [±3.4% coefficient of variation (CV)].

## Validation Results

Together with the phosphatase separation and the sample preparation work outlined previously, the validation yielded the results shown in Table I prior to the commencement of analysis of the samples taken from the subjects. The correlation coefficient with $n = 21$ was 0.9997 over a range of 34–1077 ng of total T base per milliliter of spiked, pooled plasma.

After initial analysis of samples from subjects, we decided to use only the validation concentration range from 0 to 134.6 ng/ml for all further work. The slopes of the validation lines on the 3 days were 1252.3, 1287.4, and 1268.5 (mean, 1269.4 ± 1.4%) with correlation coefficients of 0.9950, 0.9931, and 0.9914, respectively. The precision and reproducibility are shown in Table II. The endogenous value for total T base in pooled plasma

TABLE II
VALIDATION RESULTS DURING ANALYSIS[a]

| Concentration (ng/ml) | Precision (CV %)[b] | | | Reproducibility over 3 days (CV %) |
|---|---|---|---|---|
| | Day 1 | Day 2 | Day 3 | |
| Endogenous value | 4.3 | 4.0 | 2.0 | 1.8 |
| 33.7 | 2.7 | 2.5 | 0.2 | 2.7 |
| 67.3 | 3.2 | 1.4 | 1.3 | 4.3 |
| 134.6 | 1.5 | 1.4 | 0.9 | 2.6 |

[a] Concentration range for analysis of volunteer samples.
[b] Each day n = 3 at each point.

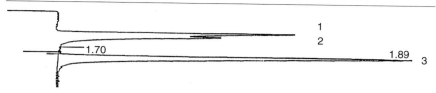

FIG. 1. Chromatographic separation of thiamin diphosphate (1), thiamin monophosphate (2), and thiamin (3; $t_R$: 1.89 min).

was 32 ng/ml. After the diet low in vitamin $B_1$, but before drug administration, the corresponding value for the plasma samples from the subjects ($n = 16$) was ~7 ng/ml (±25% CV). The calculated determination limit (signal-to-noise ratio, 3:1) was 2 ng/ml.

*Selectivity of Method*

Both TDP and TMP had distinctly shorter retention times than T (Fig. 1). Figure 2 shows chromatograms of a solution of pure substance, a plasma validation sample, and samples from the subjects after the administration of T.

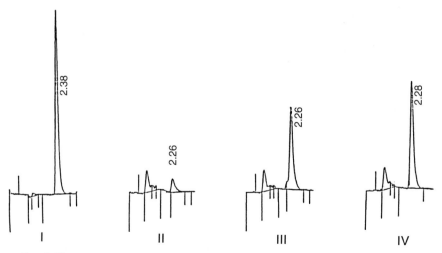

FIG. 2. Chromatograms of thiamin samples ($t_R$: ~2.3 min): (I) aqueous test solution, ~800 pg injected; (II) plasma sample from subject before vitamin $B_1$ administration (7.1 ng/ml); (III) plasma sample from subject after vitamin $B_1$ administration (41.2 ng/ml); and (IV) pooled plasma sample (validation plasma, 50.9 ng/ml).

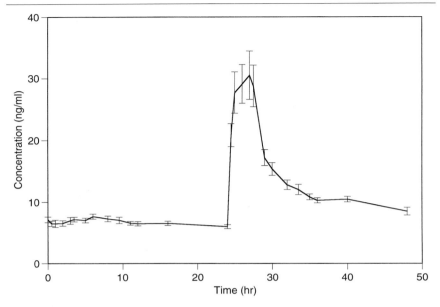

FIG. 3. Mean values of plasma concentrations ($\pm$ standard error of the mean, $n = 16$) of total T after oral administration of vitamin $B_1$ test tablets. Time 0–24 hr characterizes the placebo run-in phase. The verum application took place at $t = 24$ hr.

## Assay Method Used for Oral Bioavailability Study

### Subject Selection

The subjects are 15 females and 1 male between 21 and 36 years of age (mean, $27.3 \pm 4.4$ years), with body weights between 43.7 and 73.0 kg (mean, $60.5 \pm 7.3$ kg) and heights between 155 and 181 cm (mean, $167.9 \pm 6.8$ cm). Health status is confirmed by complete medical history, physical examination, and laboratory tests carried out at baseline. Drug screening and testing for hepatitis B surface antigens and human immuno-deficiency virus antibodies are also carried out.

### Diet

The subjects are hospitalized for 3 days, during which time they receive a diet low in vitamin $B_1$. On day 2, the circadian pattern for total T is determined with blood samples taken at the same time as on the day of drug administration (day 3).

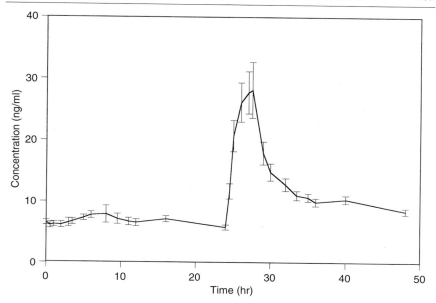

FIG. 4. Mean values of plasma concentrations ($\pm$ standard error of the mean, $n = 16$) of total T after oral administration of reference dragees. Time 0–24 hr characterizes the placebo run-in phase. The verum application took place at $t = 24$ hr.

## Drug Administration

On the 3rd day, the subjects receive 200 mg of thiamin hydrochloride (T $\times$ HCl) in the morning on an empty stomach. The reference product is vitamin $B_1$ pills (100 mg of T $\times$ HCl) and the test product is vitamin $B_1$ tablets (200 mg of T $\times$ HCl). The subjects receive either two pills of the reference product (corresponding to 200 mg of T $\times$ HCl) or one tablet of the test product (corresponding to 200 mg of T $\times$ HCl). Drug administration takes place in randomized two-way cross-over, with a 7-day washout period.

TABLE III
PHARMACOKINETIC PARAMETERS

| Vitamin $B_1$ formulation | $AUC_{0-24}$ (ng $\times$ hr/ml) | $AUC_{24-48}$ (ng $\times$ hr/ml) | $AUC_{24-48}-AUC_{0-24}$ (ng $\times$ hr/ml = net AUC) | Net $C_{max}$ (ng/ml) | Net $t_{max}$ (hr) |
|---|---|---|---|---|---|
| Test tablets | 159.3 $\pm$ 28.9 | 330.0 $\pm$ 84.7 | 170.7 $\pm$ 75.7 | 28.81 $\pm$ 16.22 | 2.34 $\pm$ 0.89 |
| Reference dragees | 157.8 $\pm$ 41.1 | 314.3 $\pm$ 101.5 | 156.5 $\pm$ 70.2 | 28.08 $\pm$ 16.75 | 2.56 $\pm$ 0.85 |

*Blood Samples*

Whole-blood samples (10 ml) are taken from a forearm vein into glass tubes containing heparin. The plasma is extracted and deep frozen at $-20°$ prior to analysis. The times at which samples are taken on days 2 and 3 (drug administration on day 3) are identical. Samples are taken at the following times: 0, 0.5, 1, 2, 3, 3.5, 5, 6, 8, 9.5, 11, 12, 16, and 24 hr. The 0-hr value is for the sample taken immediately prior to drug administration.

*Data Analysis*

The values of the area under the curve (AUC) of the plot of concentration versus time for days 2 and 3 are determined for each subject, and the differences between day 3 and day 2 are calculated (equals net AUC). The maximum concentration ($C_{max}$) values are obtained by subtracting the day 2 plasma level of endogenous T from the day 3 value. The values of time to reach $C_{max}$ ($t_{max}$) are determined on day 3 and correspond to the $C_{max}$ values of the measured total T levels. The calculations are performed with analysis of variance (ANOVA), Westlake (symmetrical confidence intervals), and Tukey (nonparametric signed-rank test), at the 95% significance level.

Bioavailability Results

To our knowledge, few data from human subjects and no studies on the circadian pattern for total T in plasma or whole blood have been published. It is interesting to note the relatively constant plasma values on the day before drug administration ($t = 0$–24 hr); these values indicate that a suitable diet low in vitamin $B_1$ had been chosen for this study. Furthermore, it is clear that the time of day had virtually no effect on the values of T in plasma obtained, the majority of which were ~6–8 ng/ml [$t = 0$–24 hr (Figs. 3 and 4)]. Both preparations show virtually identical bioavailability, with similar AUC, $C_{max}$, and $t_{max}$ values (Table III).

Conclusion

The method described allows complete separation of the various thiamin phosphates present in plasma and detection of the total T fluorimetrically by HPLC and postcolumn oxidation. The determination limit in plasma is ~2 ng of total T per milliliter of plasma. A comparative bioavailability test on two vitamin $B_1$ preparations was carried out by this method. To obtain native values for total T over the daily profile, the subjects had blood samples taken at identical intervals of time on the day preceding drug administration and received the same diet.

# [11] Cyanogen Bromide-Based Assay of Thiamin

*By* DAVID T. WYATT and RICHARD E. HILLMAN

## Introduction

Thiamin has a wide biological role. It is an essential coenzyme for oxidative metabolism and fatty acid synthesis[1] and may have noncoenzymatic functions as a neurotransmitter.[2] Analysis of thiamin levels by a thiochrome method involves three basic steps: (1) separation of thiamin from the original specimen, (2) dephosphorylation of the phosphate esters, and (3) oxidation of thiamin to thiochrome for final extraction and fluorometric measurement. This chapter describes the thiochrome method we used to develop a low-cost thiamin assay[3] sufficiently sensitive for pediatric application in both whole blood and cerebral spinal fluid, and the normative ranges[4] for those biological fluids.

## Reagents

Prepare trichloroacetic acid (TCA), potassium acetate, and sodium hydroxide as 0.32, 4.0, and 3.75 $M$ solutions, respectively. Prepare potato acid phosphatase to 1 mg/ml and cyanogen bromide to 0.47 $M$ just prior to use (use ventilation hood and gloves). Prepare a 3.8 p$M$ (3 $\mu$g/l) solution of quinine sulfate in 0.05 $M$ sulfuric acid as a calibration standard. Prepare stock solutions of thiamin hydrochloride, thiamin monophosphate, or thiamin diphosphate in 0.02 $M$ HCl at 3000 n$M$; dilute to 300, 150, and 75 n$M$ for standard curves. Distilled water may be used as a blank for all curves.

Standards show excellent linearity over the range of 0 to 3000 n$M$. The lower limit of detection is 10 n$M$. The precision of standards is shown in Table I. The between-run coefficients of variation (CV) for the slope and the intercept of the standard curve over 45 weeks are 5.2 and 10.5%, respectively.

## Sample Preparation

Whole-blood samples (0.5 ml) are frozen ($-20°$) overnight. Thaw at room temperature and pipette 100-$\mu$l aliquots into four tubes; add 500 $\mu$l of

[1] A. Lehninger, *in* "A Short Course in Biochemistry," pp. 215–235. Worth Publishers, New York; 1973.
[2] R. F. Butterworth, *Neurochem. Int.* **4**, 449 (1982).
[3] D. T. Wyatt, M. Lee, and R. E. Hillman, *Clin. Chem.* **35**, 2173 (1989).
[4] D. T. Wyatt, D. Nelson, and R. E. Hillman, *Am. J. Clin. Nutr.* **53**, 530 (1991).

TABLE I
PRECISION OF STANDARDS[a]

| Thiamin (nM) | Coefficient of variation (%) | |
| --- | --- | --- |
|  | Within run[b] | Between run[c] |
| 0 | 2.6 | 9.0 |
| 75 | 1.8 | 6.1 |
| 150 | 1.9 | 5.0 |
| 300 | 0.5 | 4.5 |

[a] Reprinted with permission from D. T. Wyatt, M. Lee, and R. E. Hillman, *Clin. Chem.* **35**(10), 2173 (1989).
[b] The CVs are based on thiamin diphosphate ($n = 6$ per concentration).
[c] The CVs are based on 45 weekly standard curves.

TCA to each tube; mix; centrifuge at 8750 $g$ for 12 min at room temperature; transfer 500 $\mu$l of supernatant to a glass tube. Add 400 $\mu$l of potassium acetate buffer and 100 $\mu$l of potato acid phosphatase and incubate overnight (18–20 hr) at 37°. Mix 200 $\mu$l of cyanogen bromide solution into each tube. After precisely 5 min, add 100 $\mu$l of NaOH; mix in 1.4 ml of 2-methyl-1-propanol (isobutanol); centrifuge at 700 $g$ for 5 min at room temperature; transfer 1.2 ml of the isobutanol (upper) layer to tubes for fluorometric reading. Omit cyanogen bromide from one and phosphatase from another of the four aliquots to determine the blank and the nonphosphorylated thiamin, respectively. Subtract the fluorescence value of the blank from that of both the total and nonphosphorylated thiamin before calculating the nanomolar value from the standard curve. Calculate the phosphorylated thiamin by subtracting the nonphosphorylated value from the total thiamin.

Trichloroacetic acid provides essentially complete deproteinization with only brief mixing. Potato acid phosphatase yields much better dephosphorylation than either wheat-germ acid phosphatase or $\alpha$-amylase. Duration of oxidation with cyanogen bromide is important. There is a steady increase in fluorescence up to 60 min, followed by a decline over the next 60 min. We stop oxidation after 5 min (via addition of NaOH) for several reasons: this exposure gives the least decline in standard fluorescence (quenching); discrimination between samples is as good as with much longer exposure; reproducibility is excellent; and the final thiamin concentrations are in the range reported with other methods.

Precision and Accuracy

For whole blood, the within-run CVs are 3.6% (total) and 4.6% (nonphosphorylated). The long-term between-run CVs (control blood over 40 weekly assays) are 6.0% (total) and 15.5% (nonphosphorylated). Recoveries for low (72 n$M$), intermediate (143 n$M$), and high (286 n$M$) additions are 93, 104, and 109%, respectively. Samples from patients receiving penicillin derivatives, phenytoin, carbamazepine, furosemide, hydrochlorothiazide plus triamterene, or acetazolamide show high net fluorescence and falsely elevated calculated thiamin levels. Amoxicillin falsely lowered the calculation (high blank). No interference is seen in samples from patients with renal failure, heart failure, or hyperbilirubinemia; or from those receiving theophylline, digitalis, heparin, or captopril.

The total and nonphosphorylated thiamin values are moderately higher in blood samples from heparinized vs EDTA tubes: total thiamin, 145 (±45 SD) n$M$ vs 129 (±50 SD) n$M$; nonphosphorylated thiamin, 61 (±24) n$M$ vs 45 (±27) n$M$. These differences were not due to differences in the blank or background fluorescence.

Overnight freezing decreases blank fluorescence (22 ± 7.9% SD; $p <$ 0.0001) and increases thiochrome fluorescence (13 ± 13% SD; $p < 0.001$) for a combined increase in calculated total thiamin of 58 n$M$ (mean increase, 41 ± 29% SD; $p < 0.0001$) above the value seen when the sample was analyzed prior to freezing. Repeated thawing and freezing produces no further changes. Samples are stable for 36 hr at room temperature, for 7 days at 4°, and for 24 months at −20°. After 36 months at −20°, total thiamin was relatively stable but nonphosphorylated thiamin increases substantially, probably owing to spontaneous dephosphorylation.

For basic thiamin screening, for which only a total thiamin determination is needed, 200 samples could be analyzed weekly at a cost of about $0.25/ sample. Sample size could be as small as 100 μl and could be done by finger or heel stick.

## Reference Ranges

We analyzed 146 blood samples from healthy adults, 323 blood samples from children, and 208 cerebrospinal fluid (CSF) samples. Blood samples came from children on growth hormone therapy (29), healthy siblings of diabetic patients (47), healthy black adolescents (22), and 225 residual blood specimens from the Children's Hospital of Wisconsin (Milwaukee, WI) pediatric laboratory. In all cases, samples met the following criteria: the patient was in good nutritional health with no history of chronic malabsorption and was receiving no multivitamins or medication known to interfere with the assay; if newborn, the patient was full term and of the appropriate weight for gestational age; and the sample had been refrigerated within 4 hr of collection and was less than 5 days old.

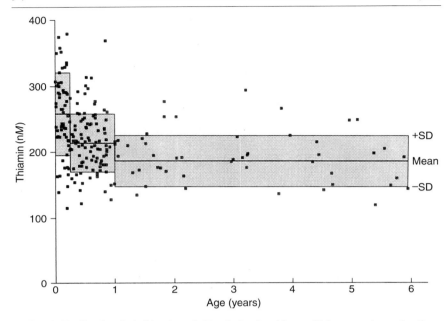

FIG. 1. Decline in whole-blood total thiamin levels with age. Values are shown for three age groups: ≤0–3, 3–12, and >12 months. For clarity, the age axis is extended only through 6 years. [Reprinted with permission from D. T. Wyatt, D. Nelson, and R. E. Hillman, *Am. J. Clin. Nutr.* **53**, 530 (1991). Copyright © 1991 *Am. J. Clin. Nutr.*]

The decline in whole-blood total thiamin is shown in Fig. 1. There is a strong negative correlation for both total and phosphorylated thiamin with age ($r = -0.387$ and $-0.417$, respectively, with $p < 0.0001$ for each). The greatest decline occurs in the first 3 months of age, with a smaller additional decrease from 3 to 12 months. Thereafter, thiamin levels are comparable to adults (Table II). The decline occurs relatively independently of the changes in hematocrit characteristically seen in infancy. There are no differences by sex at any age or by race if less than 10 years of age. If older than age 10, however, total and phosphorylated thiamin decline in blacks [for whites ($n = 51$), total $= 197 \pm 36$ and phosphorylated $= 131 \pm 28$; for blacks ($n = 31$), total $= 153 \pm 27$ and phosphorylated $= 91 \pm 23$; $p < 0.0001$ for both comparisons]. In another group of 101 adolescents and 34 adolescents paired with their parents,[5] we confirmed this racial difference. Thiamin levels in white adolescents are comparable to those in white adults;

[5] B. A. Cromer, D. T. Wyatt, L. A. Brandstaetter, S. Spadone, and H. R. Sloan, *J. Pediatr. Gastroenterol. Nutr.* **9**, 502 (1989).

TABLE II
WHOLE-BLOOD THIAMIN LEVELS IN INFANTS, CHILDREN, AND ADULTS[a]

| Sample | Thiamin level (nM) | | |
|---|---|---|---|
| | Total | Nonphosphorylated | Phosphorylated |
| All pediatric ($n = 323$) | $210 \pm 53$ | $69 \pm 23$ | $140 \pm 42$ |
| $\leq$0–3 months ($n = 64$)[b] | $258 \pm 63$ | $81 \pm 26$ | $177 \pm 48$ |
| 3–12 months ($n = 100$)[b] | $214 \pm 44$ | $66 \pm 20$ | $148 \pm 33$ |
| >12 months ($n = 159$)[b] | $187 \pm 39$ | $67 \pm 21$ | $120 \pm 31$ |
| Adult ($n = 146$) | $191 \pm 32$ | $62 \pm 17$ | $130 \pm 24$ |

[a] Reprinted with permission from D. T. Wyatt, D. Nelson, and R. E. Hillman, *Am. J. Clin. Nutr.* **53**, 530 (1991). Copyright © 1991 *Am. J. Clin. Nutr.*

[b] Significant difference among the three pediatric age groups for total, nonphosphorylated, and phosphorylated thiamin by one-way analysis of variance (ANOVA). ($p < 0.001$ for all comparisons; means $\pm$ SD).

levels in black adolescents are comparable to those in black adults; and whites of any age older than 10 years have higher total and phosphorylated thiamin levels than blacks. In the latter study, there was no correlation between thiamin level and dietary intake, socioeconomic status, and geographic locale.

Adjustment of calculated thiamin levels for differences in sample hematocrit is unnecessary, because only 8–10% of the variance in either total or phosphorylated thiamin is attributable to hematocrit. Adjustment based on red blood cell (RBC) volume also introduces large distortions in both the normal decline with age and between samples of widely varying hematocrit. Although white blood cell (WBC) count contributes 10–20% of the variance of both total and phosphorylated thiamin, correction for WBC mass is not feasible. Because most subjects have a WBC count between 6000 and 15,000,[6] RBC mass will usually contribute more than 75% of the thiamin. Caution may be needed when interpreting a low–normal thiamin level in a blood sample with leukocytosis.

The decline in CSF total thiamin concentration is shown in Fig. 2. As with whole blood, there is a strong negative correlation with age for both total and phosphorylated ($r = -0.335$ and $-0.316$, respectively, with $p < 0.0001$ for each). A two-phase decline can be constructed, which was somewhat slower than that seen in whole blood (Table III).

[6] L. D. Frenkel and J. A. Bellanti, Development and function of the lymphoid system. *In* "Children Are Different: Developmental Physiology" (T. R. Johnston, W. M. Moore, and J. E. Jeffries, eds.), pp. 177–183. Ross Laboratories, Columbus, Ohio, 1978.

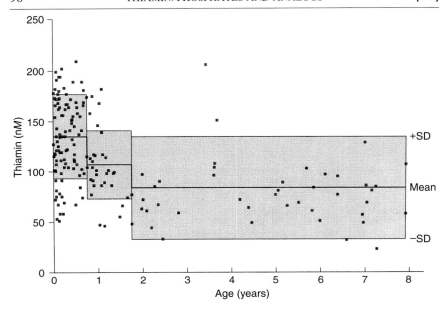

FIG. 2. Decline in CSF total thiamin levels with age. Values are shown for three age groups: ≤0–9, 9–18, and >18 months. For clarity, the age axis is extended only through 8 years. [Reprinted with permission from D. T. Wyatt, D. Nelson, and R. E. Hillman, *Am. J. Clin. Nutr.* **53,** 530 (1991). Copyright © 1991 *Am. J. Clin. Nutr.*]

TABLE III

CEREBROSPINAL FLUID LEVELS IN INFANTS AND CHILDREN[a]

| | Thiamin level (nM) | | |
|---|---|---|---|
| Sample | Total | Nonphosphorylated | Phosphorylated |
| All pediatric ($n = 208$) | 113 ± 50 | 40 ± 26 | 74 ± 38 |
| ≤0–9 months ($n = 107$)[b] | 135 ± 42 | 47 ± 27 | 87 ± 30 |
| 9–18 months ($n = 30$)[b] | 107 ± 34 | 30 ± 21 | 77 ± 21 |
| >18 months ($n = 71$)[b] | 84 ± 51 | 32 ± 24 | 52 ± 44 |

[a] Reprinted with permission from D. T. Wyatt, D. Nelson, and R. E. Hillman, *Am. J. Clin. Nutr.* **53,** 530 (1991). Copyright © 1991 *Am. J. Clin. Nutr.*

[b] Significant difference among the three pediatric age groups for total, nonphosphorylated, and phosphorylated thiamin by one-way ANOVA ($p < 0.001$ for all comparisons; means ± SD).

Because of these shifts, age-specific norms should be used for determining thiamin status in infancy. After the age of 12 and 18 months, a common range is applicable for whole blood and CSF, respectively, because levels are equal to those in adults.

# [12] Isotopically Labeled Precursors and Mass Spectrometry in Elucidating Biosynthesis of Pyrimidine Moiety of Thiamin in *Saccharomyces cerevisiae*

*By* KEIKO TAZUYA, KAZUKO YAMADA, and HIROSHI KUMAOKA

The biosynthetic pathway of the pyrimidine moiety of thiamin in *Saccharomyces cerevisiae*, a eukaryote, is different from that of prokaryotes. In prokaryotes, the pyrimidine moiety is synthesized from 5-aminoimidazole ribonucleotide, an intermediate in purine biosynthesis. This has been demonstrated by investigating the nutritional requirement of adenine- and thiamin-requiring mutants of *Salmonella typhimurium*[1] and by tracer experiments with radioactive compounds in *Escherichia coli*[2] and *S. typhimurium*.[3]

We investigated pyrimidine biosynthesis in *S. cerevisiae* by gas chromatography–mass spectrometry (GC–MS) analysis, using a stable isotope-labeled tracer.[4,5] These experiments showed that in *S. cerevisiae*, the pyrimidine moiety is synthesized from histidine and pyridoxine.

## Methods

### Extraction of Thiamin

*Saccharomyces cerevisiae* is grown in a synthetic medium containing a tracer for 7 hr at 30°. The medium consists of 9 g of glucose, 6 g of $(NH_4)_2HPO_4$, 1 g of $KH_2PO_4$, 1 g of sodium citrate, 0.25 g of $MgSO_4 \cdot 7H_2O$, 20 $\mu$g of biotin, 0.5 mg of calcium pantothenate, 10 mg of inositol, 1 mg of pyridoxine, 0.4 mg of $ZnSO_4 \cdot 7H_2O$, 25 $\mu$g of $CuSO_4 \cdot 5H_2O$, and 0.15 mg of $FeSO_4(NH_4)_2SO_4 \cdot 6H_2O$ per 1 liter of water. The pH is adjusted to 5.0 with $H_3PO_4$. The cells are harvested by centrifugation (8000 $g$ for 7

[1] P. C. Newell and R. G. Tucker, *Biochem. J.* **106**, 279 (1968).
[2] K. Yamada and H. Kumaoka, *J. Nutr. Sci. Vitaminol.* **29**, 389 (1983).
[3] B. Estramareix and S. David, *Biochem. Biophys. Res. Commun.* **134**, 1136 (1986).
[4] K. Tazuya, K. Yamada, and H. Kumaoka, *Biochim. Biophys. Acta* **990**, 73 (1989).
[5] K. Tazuya, C. Azumi, K. Yamada, and H. Kumaoka, *Biochem. Mol. Biol. Int.* **36**, 883 (1995).

min at 4°). After preparing the cell extract, phosphorylated thiamin in the extract is converted to free thiamin with Taka-Diastase (Sankyo Company, Ltd., Tokyo, Japan) and the extract is subjected to Amberlite CG-50 column chromatography ($H^+$, 15 × 100 mm) to separate thiamin.

### Isolation of 4-Amino-5-(ethylthio)methyl-2-methylpyrimidine (Pyrimidine)

The thiamin-containing eluate from the Amberlite CG-50 column is evaporated to dryness and the residue dissolved in 2 ml of distilled water. The solution is adjusted to pH 4.8 with 1 $M$ NaOH and 2 ml of ethanol and 2 ml of ethanethiol. The mixture is then heated at 100° for 15 hr in a sealed tube. After cooling, 10 ml of distilled water is added and the solution is acidified with concentrated HCl, then extracted with dichloromethane to remove ethanethiol. To the remaining aqueous solution is added 0.5 ml of 10 $M$ NaOH to extract the pyrimidine with dichloromethane (four times, 20 ml each time). The dichloromethane solutions are pooled and then dried over with sodium sulfate. The solvent is finally evaporated off *in vacuo*. The resulting residue is dissolved with 100 $\mu$l of ethyl acetate and subjected to GC–MS.

### Gas Chromatography–Mass Spectrometry

Pyrimidine is analyzed using a JEOL (Tokyo, Japan) mass spectrometer JMS-DX 303 and DB-5 capillary column (30 m × 0.25 mm i.d.; film thickness, 0.25 $\mu$m; J & W Scientific, Folsom, CA). The oven temperature is programmed to increase from 120 to 260° at 8°/min. Injection port and separator temperatures are 280°. Samples are assayed by mass spectrometry mode and selected ion mode, using an ionization energy of 70 eV and an ion source temperature of 250°. The percentage of isotopic enrichment is calculated using the data from samples extracted from cells growing with stable isotope and from cells growing with nonlabeled compound. The mass spectrum of pyrimidine is shown in Fig. 1. Analysis of the fragmentation of pyrimidine is based on that reported by White and Rudolph.[6]

### Participation of Histidine in Biosynthesis of Pyrimidine

The amide nitrogen atom of glutamine is incorporated into N-3 and amino nitrogen (amino-N) at C-4 of the pyrimidine moiety of thiamin in *S. cerevisiae*.[7] However, addition of Casamino Acids to the medium decreases incorporation of the amide nitrogen atom of glutamine into the

---

[6] R. H. White ad F. B. Rudolph, *Biochemistry* **18**, 2632 (1979).
[7] K. Tazuya, M. Tanaka, M. Morisaki, K. Yamada, and H. Kumaoka, *Biochem. Int.* **14**, 769 (1987).

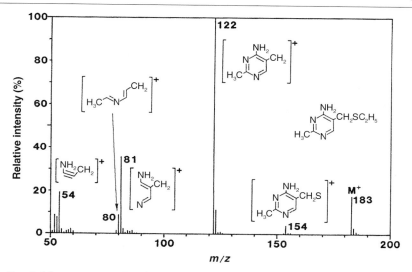

FIG. 1. Mass spectrum of pyrimidine. (Reprinted from *Biochim. Biophys. Acta*, **990**, K. Tazuya, K. Yamada, and H. Kumaoka, Incorporation of histidine into the pyrimidine moiety of thiamin in *Saccharomyces cerevisiae*, 73–79. Copyright 1989, with kind permission of Elsevier Science–NL, Sara Burgerhartstraat 25, 1055 KV Amsterdam, The Netherlands.)

pyrimidine.[8] This suggests that another amino acid in Casamino Acids is the direct precursor.

To determine the direct precursor, the competitive effects of $^{14}N$-labeled amino acids on the incorporation of $^{15}NH_4Cl$ into pyrimidine are investigated. Ammonium salts in the medium, $(NH_4)_2HPO_4$ and $FeSO_4(NH_4)_2 SO_4$, are substituted for $^{15}NH_4Cl$ and $FeSO_4 \cdot 7H_2O$. The concentration of $^{15}NH_4Cl$ added to the medium is 5 g/liter. Aspartate, asparagine, glutamate, and glutamine are added individually to the medium, while other amino acids are divided into five groups and added as shown in Table I. The amino acids are added to the medium at a concentration of 1 m$M$.

Two nitrogen atoms of $^{15}NH_4Cl$ are incorporated effectively into pyrimidine as shown in Fig. 2 and Table I, group A. The fragmentation patterns of pyrimidine, as shown in Fig. 2, suggest that two $^{15}N$ atoms are incorporated into the amino nitrogen atom and N-3 of pyrimidine. Fragment $m/z$ 54, which contains the amino-N atom of pyrimidine, is shifted to $m/z$ 55. One of the $^{15}N$ atoms is incorporated into the amino-N atom of pyrimidine. The fragment ion at $m/z$ 81 shifts only one mass unit. This fragment ion contains N-1 and amino-N atoms of pyrimidine. The incorporation of $^{15}N$ into the amino-N atom is thus confirmed. These results suggest that the

[8] K. Tazuya, M. Morisaki, K. Yamada, and H. Kumaoka, *Biochem. Int.* **16**, 955 (1988).

TABLE I

INCORPORATION OF $^{15}$N ATOM OF $^{15}$NH$_4$Cl ADDED TO MEDIUM INTO
PYRIMIDINE SYNTHESIZED BY *Saccharomyces cerevisiae*

| Group | Amino acids added to medium | Incorporation (%) | | | |
|---|---|---|---|---|---|
| | | $N_0$ | $N_1$ | $N_2$ | $N_3$ |
| A | None | 15.2 | 9.3 | 66.8 | 8.7 |
| B | Asn | 23.4 | 26.0 | 47.0 | 3.6 |
| C | Asp | 14.1 | 11.1 | 59.2 | 15.5 |
| D | Gln | 19.2 | 36.1 | 39.7 | 5.0 |
| E | Glu | 8.9 | 13.4 | 63.4 | 14.4 |
| F | Met, Cys, Arg, Lys | 21.2 | 16.7 | 58.8 | 3.4 |
| G | Val, Leu, Ile | 15.6 | 12.5 | 63.7 | 8.2 |
| H | Phe, Tyr, Trp | 11.0 | 10.5 | 64.0 | 14.6 |
| I | Ala, Gly, Ser, Hyp | 10.6 | 18.2 | 59.7 | 11.6 |
| J | Thr, His, Pro | 84.7 | 15.3 | 0.0 | 0.0 |
| K | Pro | 11.4 | 4.6 | 68.9 | 15.1 |
| L | Thr | 19.7 | 9.4 | 58.4 | 12.5 |
| M | His | 83.1 | 16.9 | 0.0 | 0.0 |

Reproduced from K. Tazuya, M. Morisaki, K. Yamada, and H. Kumaoka, *Biochem. Int.* **16,** 955 (1988), with permission.

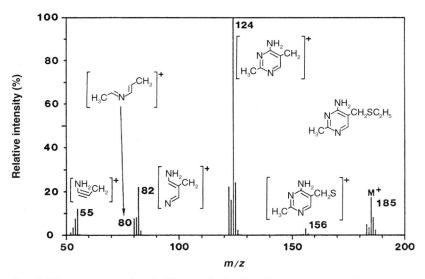

FIG. 2. Mass spectrum of pyrimidine synthesized by *S. cerevisiae* grown in the presence
of $^{15}$NH$_4$Cl. [Reproduced from K. Tazuya, M. Morisaki, K. Yamada, and H. Kumaoka,
*Biochem. Int.* **16,** 955 (1988), with permission.]

shift of $m/z$ 81 to $m/z$ 82 is due to the incorporation of $^{15}$N into the amino-N of pyrimidine but not into N-1 of pyrimidine. Consequently, the nitrogen atom of $^{15}$NH$_4$Cl is incorporated into the N-3 and amino-N atoms of pyrimidine.

Competitive experiments (Table I, groups B–J) show that only group J amino acids [histidine, threonine, proline (His, Thr, Pro)] inhibit the $^{15}$N incorporation into pyrimidine. Further investigation with histidine, threonine, and proline (Table I, groups K–M) reveals that nitrogen incorporation into pyrimidine is inhibited by histidine.

## Incorporation of L-[1,3-$^{15}$N$_2$]Histidine into Pyrimidine

To confirm the direct incorporation of histidine, *S. cerevisiae* is grown in the presence of L-[1,3-$^{15}$N$_2$]histidine. The two nitrogen atoms of the imidazole ring in histidine are incorporated directly into pyrimidine (Fig. 3). The base peak $m/z$ 122 is shifted to $m/z$ 124. The shift of the fragment ion at $m/z$ 54 to 55 shows that one of the nitrogen atoms of imidazole in histidine is incorporated into the amino-N of pyrimidine. The fragment ion at $m/z$ 81 contains the N-1 and amino-N atoms of pyrimidine. The one nitrogen of histidine is incorporated into this fragment. These results show

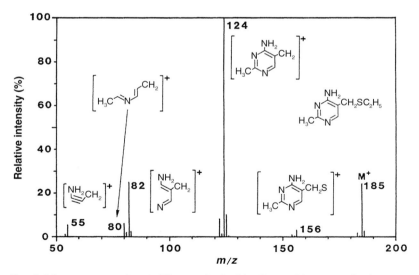

FIG. 3. Mass spectrum of pyrimidine synthesized by *S. cerevisiae* grown in the presence of DL-[1,3-$^{15}$N]histidine. (Reprinted from *Biochim. Biophys. Acta*, **990**, K. Tazuya, K. Yamada, and H. Kumaoka, Incorporation of histidine into the pyrimidine moiety of thiamin in *Saccharomyces cerevisiae*, 73–79. Copyright 1989, with kind permission of Elsevier Science–NL, Sara Burgerhartstraat 25, 1055 KV Amsterdam, The Netherlands.)

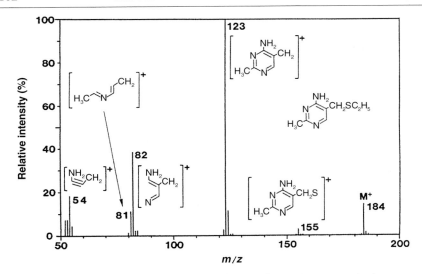

FIG. 4. Mass spectrum of pyrimidine synthesized by *S. cerevisiae* grown in the presence of $^{15}NH_4Cl$ with histidine and without pyridoxine. [Reproduced from K. Tazuya, K. Yamada, and H. Kumaoka, *Biochem. Mol. Biol. Int.* **30**, 893 (1993), with permission.]

that the nitrogen atom of histidine is incorporated into the amino-N atom but not the N-1 of pyrimidine. Therefore, the other nitrogen of histidine is incorporated into the N-3 of pyrimidine.

In another investigation with L-[$^{15}N$]aspartate, the label is incorporated into the N-3 of pyrimidine. This suggests that the amino-N of aspartate is incorporated into the N-3 of pyrimidine via the N-1 of the imidazole ring in histidine. Furthermore, the radioactive tracer experiment with L-[2-$^{14}C$]histidine showed that the C-2 of the imidazole ring in histidine was incorporated into pyrimidine.[4] However, the $\alpha$-amino-N atom of histidine was not incorporated into pyrimidine.[4] These findings show that the unit of N-1, C-2, and N-3 atoms of the imidazole ring in histidine is incorporated into the unit of N-3, C-4, and amino-N atoms of pyrimidine.

## Participation of Pyridoxine in Biosynthesis of Pyrimidine

The investigation with $^{15}NH_4Cl$ shows that only the two nitrogens are incorporated into pyrimidine (Fig. 2). This result suggests that the incorporation of $^{15}NH_4Cl$ into N-1 of pyrimidine is inhibited by an N-containing compound in the medium. The synthetic medium consists of a carbon source, inorganic salts, and vitamins (biotin, pantothenate, and pyridoxine) that have nitrogen atoms. The nitrogen of pantothenate originates from

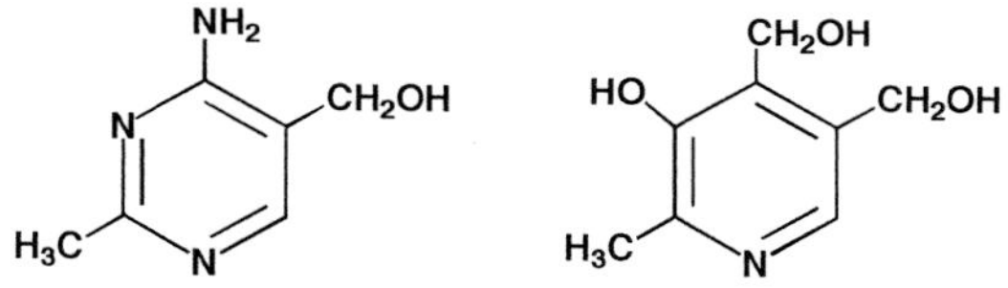

FIG. 5. Structures of 4-amino-5-hydroxymethyl-2-methylpyrimidine (1) and pyridoxine (2). [Reproduced from K. Tazuya, C. Azumi, K. Yamada, and H. Kumaoka, *Biochem. Mol. Biol. Int.* **33,** 769 (1994), with permission.]

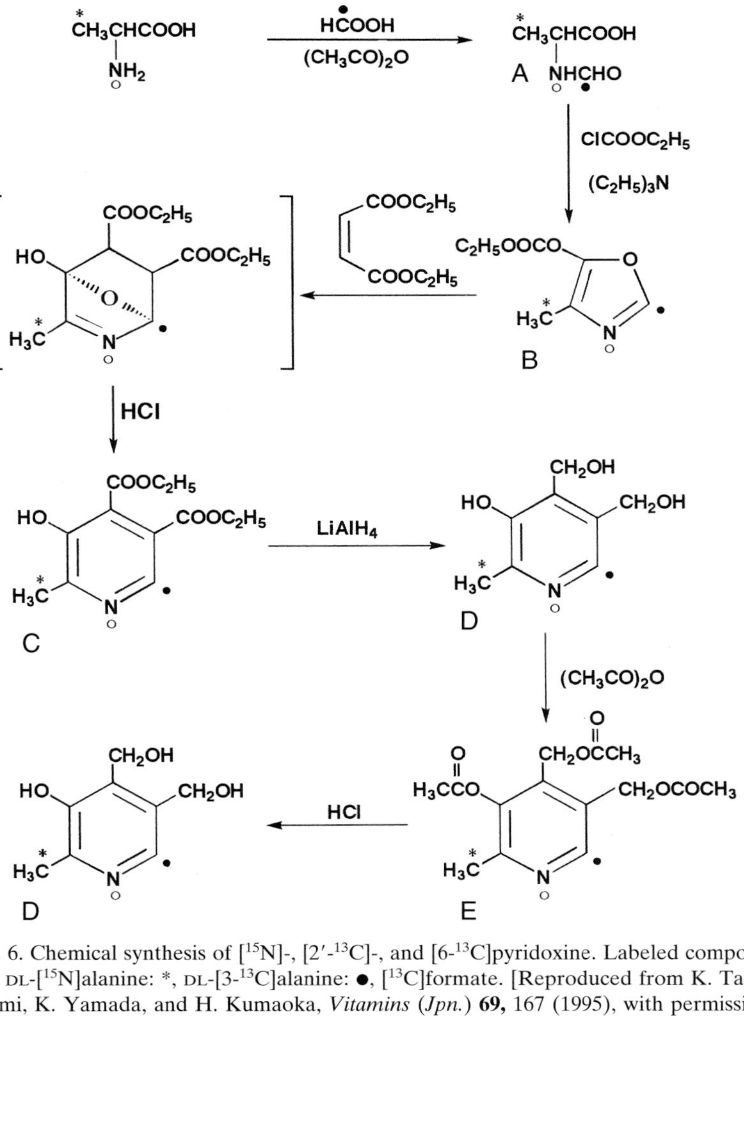

FIG. 6. Chemical synthesis of [$^{15}$N]-, [2'-$^{13}$C]-, and [6-$^{13}$C]pyridoxine. Labeled compounds used: °, DL-[$^{15}$N]alanine: *, DL-[3-$^{13}$C]alanine: ●, [$^{13}$C]formate. [Reproduced from K. Tazuya, C. Azumi, K. Yamada, and H. Kumaoka, *Vitamins (Jpn.)* **69,** 167 (1995), with permission.]

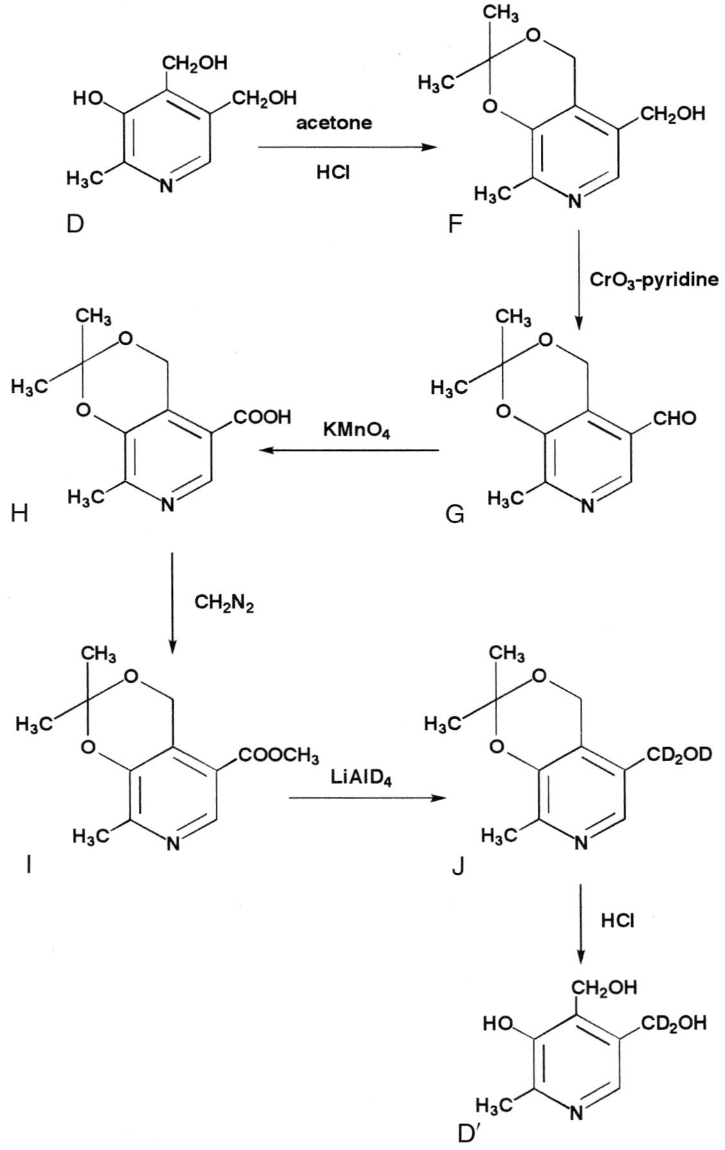

FIG. 7. Chemical synthesis of [5'-$^2$H$_2$]pyridoxine. [Reproduced from K. Tazuya, C. Azumi, K. Yamada, and H. Kumaoka, *Vitamins (Jpn.)* **69,** 167 (1995), with permission.]

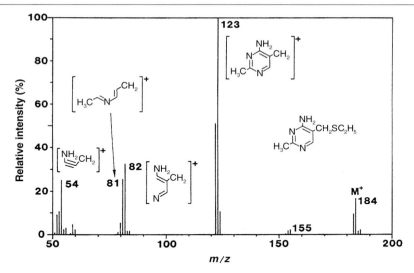

FIG. 8. Mass spectrum of pyrimidine synthesized by *S. cerevisiae* grown in the presence of [$^{15}$N]pyridoxine. [Reproduced from *Biochem. Mol. Biol. Int.* **33**, 769 (1994), with permission.]

aspartate and that of biotin from L-alanine and *S*-adenosylmethionine, which have no relation with pyrimidine biosynthesis. These facts suggest the participation of pyridoxine in pyrimidine biosynthesis. In an $^{15}NH_4Cl$-containing medium, to which L-$^{14}$N-histidine is added to inhibit $^{15}$N incorporation into the N-3 and amino-N of pyrimidine and pyridoxine is excluded, *S. cerevisiae* does grow. As shown in Fig. 4, the nitrogen of $^{15}NH_4Cl$ is incorporated efficiently into pyrimidine. The fragment $m/z$ 54 is not shifted, whereas fragments $m/z$ 80 and 81 are shifted to $m/z$ 81 and 82, respectively. The common nitrogen of both fragment ions is the N-1 of pyrimidine. The nitrogen atom of $^{15}NH_4Cl$ is incorporated into the N-1 of pyrimidine by the exclusion of pyridoxine from the medium. This strongly suggests that the origin of the N-1 of pyrimidine is pyridoxine.

*Incorporation of Labeled Pyridoxine into Pyrimidine*

As shown in Fig. 5, the structures of pyrimidine and pyridoxine are similar. It was hypothesized that the unit of the six atoms, C-2′, C-2, N-1, C-6, C-5, and C-5′, of pyrimidine moiety of thiamin may be derived from the same positions in pyridoxine. To confirm this hypothesis, the labeled pyridoxines are chemically synthesized as shown in Figs. 6 and 7.[9]

[9] K. Tazuya, C. Azumi, K. Yamada, and H. Kumaoka, *Vitamins (Jpn.)* **69**, 167 (1995).

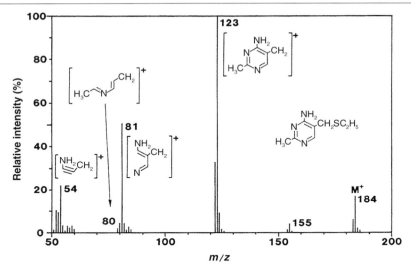

FIG. 9. Mass spectrum of pyrimidine synthesized by *S. cerevisiae* grown in the presence of [2′-¹³C]pyridoxine. [Reproduced from *Biochem. Mol. Biol. Int.* **33,** 769 (1994); with permission.]

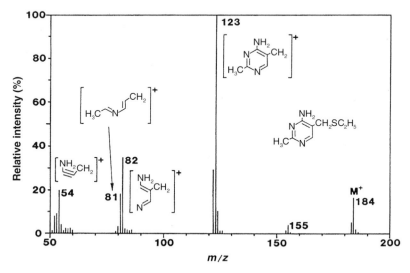

FIG. 10. Mass spectrum of pyrimidine synthesized by *S. cerevisiae* grown in the presence of [6-¹³C]pyridoxine. [Reproduced from K. Tazuya, C. Azumi, K. Yamada, and H. Kumaoka, *Biochem. Mol. Biol. Int.* **36,** 883 (1995), with permission.]

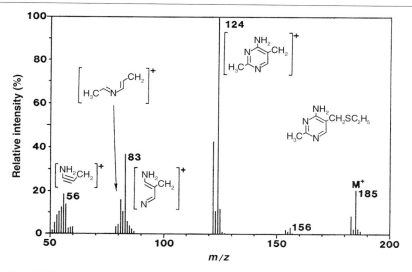

FIG. 11. Mass spectrum of pyrimidine synthesized by *S. cerevisiae* grown in the presence of [5'-²H₂]pyridoxine. [Reproduced from K. Tazuya, C. Azumi, K. Yamada, and H. Kumaoka, *Biochem. Mol. Biol. Int.* **36,** 853 (1995), with permission.]

The result of incorporation of [¹⁵N]pyridoxine into pyrimidine is shown in Fig. 8. The ¹⁵N incorporation moves the peaks M⁺ and $m/z$ 122 to $m/z$ 184 and $m/z$ 123, respectively. The nitrogen atom of pyridoxine is incorporated into pyrimidine efficiently. The fragment ions at $m/z$ 80, which contains N-1 of pyrimidine, and $m/z$ 81, which has N-1 and the amino-N atom at C-4 of pyrimidine, are shifted to $m/z$ 81 and 82, respectively. These shifts are due to the incorporation of [¹⁵N]pyridoxine into the N-1 of pyrimidine.

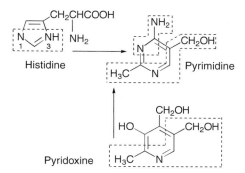

FIG. 12. Precursors of the pyrimidine moiety of thiamin in *S. cerevisiae.*

The nitrogen atom of pyridoxine is the direct precursor of the N-1 atom of pyrimidine.

The mass spectrum of pyrimidine biosynthesized by *S. cerevisiae* with [2′-$^{13}$C]pyridoxine is shown in Fig. 9. The shift of the $M^+$ and $m/z$ 122 to $m/z$ 184 and $m/z$ 123 shows the incorporation of C-2′ of pyridoxine into pyrimidine. The fragment ion at $m/z$ 80 has disappeared and the intensity of $m/z$ 81 is higher than that of the standard. The fragment peak at $m/z$ 81 is a sum of the fragment ions at $m/z$ 80 + 1 (isotope peak of $m/z$ 80) and $m/z$ 81. The $m/z$ 80 peak is shifted to $m/z$ 81, while the $m/z$ 81 peak is not shifted; the two peaks, therefore, overlap. The fragment ion at $m/z$ 81 contains N-1, C-4, C-5, C-5′, and C-6 atoms and an amino group at C-4. The fragment ion at $m/z$ 80 has N-1, C-2′, C-2, C-5, C-5′, and C-6 atoms of pyrimidine. The difference between the two ions is C-2′, C-2, C-4, and an amino group at C-4. The fragment ion at $m/z$ 54, which contains C-4 and the amino group at C-4, is not shifted. These results suggest that the C-2′ carbon of pyridoxine is incorporated into the C-2 or C-2′ of pyrimidine. These findings and the structural resemblance suggest that the C-2′ of pyridoxine is incorporated into the C-2′ of the pyrimidine.

The result of incorporation of [6-$^{13}$C]pyridoxine into the pyrimidine is shown in Fig. 10. The C-6 of pyridoxine is incorporated into $M^+$ and the fragments of $m/z$ 122, 81, and 80, respectively. The common carbon atoms of fragments $m/z$ 80 and 81 are C-5, C-5′, and C-6. The label is not incorporated into the fragment ion at $m/z$ 54, which contains C-4, C-5, and C-5′ atoms in pyrimidine. From the structural differences between these ions, $m/z$ 81, 80, and 54, it is decided that the C-6 of pyridoxine must be incorporated into the C-6 atom of pyrimidine.

The result of the investigation with [5′-$^2$H$_2$]pyridoxine is shown in Fig. 11. The label is incorporated into all of the fragment. The two hydrogen atoms shared in common between the fragments are the hydrogen atoms at C-5′ of pyrimidine. The labeled hydrogen atoms of pyridoxine are incorporated into the hydrogen atoms at C-5′ of pyrimidine.

In light of these findings, it is concluded that the unit of C-2′, C-2, N, C-6, C-5, and C-5′ atoms of pyridoxine is incorporated into the unit of C-2′, C-2, N-1, C-6, C-5, and C-5′ of pyrimidine. As shown in Fig. 12, the pyrimidine moiety of thiamin is therefore synthesized from histidine and pyridoxine in *S. cerevisiae*.

# [13] Thiamin Transporters in Yeast

By Akio Iwashima, Kazuto Nosaka, Hiroshi Nishimura,
and Fumio Enjo

The yeast *Saccharomyces cerevisiae* has been known to take up thiamin from the extracellular environment, although this organism can synthesize the vitamin *de novo*. Since Suzuoki[1] demonstrated for the first time that thiamin enters yeast cells by an active transport system in cell suspensions, several studies have been made on thiamin transport in various types of living cells.[2] Among eukaryotic cells, yeast has the most active transport system for thiamin. Although thiamin transport is widely found in nature, the structure of thiamin transporters has not been well characterized.

## Thiamin Transport in Yeast Cells

### Measurement Technique

Yeast cells are washed once with cold water and are suspended in 50 m$M$ potassium phosphate buffer, pH 5.0, containing 0.1 $M$ glucose. [$^{14}$C]Thiamin is obtained from Amersham International (Amersham, U.K.); the specific activity varies from 699 to 899 MBq/mmol. For measurement of uptake, cells suspended in the above buffer at a final $A_{560\,nm}$ between 0.05 and 0.1 are preincubated at 37° for 15 min with constant shaking. The uptake is initiated by adding [$^{14}$C]thiamin at a final concentration of 1 $\mu M$; the incubation is continued at 37° with constant shaking. At appropriate intervals, 1-ml samples are removed and filtered rapidly on a membrane filter (TM-1, 0.65-$\mu$m pore size; Toyo Roshi Co., Tokyo, Japan) on a filtration apparatus connected to a water aspirator. The filters are washed with 10 ml of 50 m$M$ potassium phosphate buffer (pH 5.0), air dried, and counted in 10 ml of Bray's scintillation cocktail in a Packard (Downers Grove, IL) Tri-Carb liquid scintillation spectrometer to a 1% standard counting error. The rate of [$^{14}$C]thiamin uptake at 37° is expressed as nanomoles of [$^{14}$C]thiamin taken up per milligram dry weight after subtracting the uptake at 0° from that at 37°. A standard curve of $A_{560\,nm}$ versus dry weight of *S. cerevisiae* is used to express the results per milligram of cells.

---

[1] Z. Suzuoki, *J. Biochem.* **42**, 27 (1955).
[2] C. R. Rose, *Biochim. Biophys. Acta* **947**, 335 (1988).

TABLE I
EFFECT OF INHIBITORS ON THIAMIN TRANSPORT IN YEAST CELLS

| Inhibitor | Concentration (mM) | Relative rate of transport |
|---|---|---|
| None | — | 1.00 |
| Sodium arsenate | 10 | 0.63 |
| KCN | 2 | 0.51 |
| 2,4-Dinitrophenol | 0.4 | 0.35 |
| N-Ethylmaleimide | 0.1 | 0.07 |
| Dicyclohexylcarbodiimide | 0.02 | 0.17 |
| Diethylstilbestrol | 0.02 | 0.08 |

## Characterization of Reaction

The initial rate of [$^{14}$C]thiamin uptake by *S. cerevisiae* grown at exponential phase in thiamin-free Wickerham's synthetic medium[3] is approximately 10-fold higher than that by commercially available bakers' yeast. The uptake by the former cells is linear for about 2 min, whereas that by the latter cells is for about 20 min. Thiamin accumulation is mediated by a single active transport system operating with a pH optimum of 4.5–5.0 and an apparent $K_m$ value for thiamin of 0.18 $\mu M$. Cells are able to accumulate thiamin to 10,000 times the extracellular concentration. The uptake is inhibited by arsenate, cyanide, 2,4-dinitrophenol, and N-ethylmaleimide (Table I). The inhibitors of yeast H$^+$-ATPase, such as dicyclohexylcarbodiimide and diethylstilbestrol, are also inhibitory. Glucose is the most effective energy source; relative rates of uptake with 0.1 $M$ glucose and 0.1 $M$ ethanol are 1:0.3, respectively.

Table II summarizes the effects of various thiamin analogs on the rate of thiamin uptake. The transport system is more specific for the chemical structure of the pyrimidine moiety of the thiamin molecule than that of the thiazole moiety; in particular, the amino group attached to C-4 appears to play an important role in the binding of thiamin. Actually, 2-methyl-4-amino-5-hydroxymethylpyrimidine alone was taken up by the yeast thiamin transport system.[4] O-Benzoylthiamin disulfide is the strongest competitive inhibitor among the thiamin analogs tested, with an apparent $K_i$ value of 1.8 nM. The inhibitory action of this thiamin disulfide probably depends on its high solubility in the lipid component of the yeast cell membrane, where the thiamin transporter may be integrated.

[3] L. J. Wickerham, *U.S. Dept. Agric. Tech. Bull.* **1029,** 1 (1951).
[4] A. Iwashima, Y. Kawasaki, and Y. Kimura, *Biochim. Biophys. Acta* **1022,** 211 (1990).

TABLE II
EFFECT OF THIAMIN ANALOGS ON THIAMIN TRANSPORT IN YEAST CELLS

| Thiamin analog | Concentration ($\mu M$) | Inhibition (%) |
|---|---|---|
| Modification of thiazole moiety | | |
| Pyrithiamin | 10 | 94.0 |
| Chlorothiamin | 10 | 92.1 |
| Dimethialium | 10 | 90.8 |
| Hydroxyethylthiamin | 10 | 90.0 |
| Thiamin acetic acid | 10 | 93.7 |
| Modification of pyrimidine moiety | | |
| 2-Northiamin | 10 | 46.9 |
| Oxythiamin | 10 | 0 |
| Effect of pyrimidine moiety | | |
| 2-Methyl-4-amino-5-hydroxymethylpyrimidine | 10 | 87.6 |
| 2-Methyl-4-oxy-5-hydroxymethylpyrimidine | 10 | 0 |
| Effect of thiazole moiety | | |
| 4-Methyl-5-($\beta$-hydroxyethyl)thiazole | 10 | 1.4 |
| 3,4-Dimethyl-5-($\beta$-hydroxyethyl)thiazole | 10 | 4.9 |
| Modification of quaternary nitrogen | | |
| Thiamin disulfide | 1.0 | 73.6 |
| $O$-Benzoylthiamin disulfide | 0.1 | 92.9 |

## Thiamin-Binding Proteins

### Measurement Technique

Two thiamin-binding proteins have been found in *S. cerevisiae*[5]: one is a soluble thiamin-binding protein (sTBP) in the periplasmic space, and the other is bound to plasma membrane (mTBP). The thiamin-binding activities of these proteins can be assessed by a standard equilibrium dialysis method as follows. The dialysis bag containing thiamin-binding protein in a total volume of 1.0 ml is equilibrated with 500 ml of 50 m$M$ potassium phosphate buffer, pH 5.0, containing 0.1 $\mu M$ [$^{14}$C]thiamin (699 to 899 MBq/mmol) at 4° for 20 hr with constant stirring. The amount of thiamin bound to protein is calculated by subtracting the radioactivity in the outer solution of the dialysis bag from that in the bag. Because free thiamin and thiamin phosphates in crude samples compete with the radioactive thiamin, these samples should be dialyzed against the buffer for 5 hr prior to the activity assay. The thiamin-binding activity is expressed as picomoles of the amount of thiamin bound per milligram of protein. Assay methods other than the

[5] A. Iwashima, H. Nishimura, and Y. Nose, *Biochim. Biophys. Acta* **557**, 460 (1979).

usual equilibrium dialysis method have been developed for thiamin-binding proteins from microorganisms.[6,7]

## Purification of Soluble Thiamin-Binding Protein

*Saccharomyces cerevisiae* is grown in 10 liters of thiamin-free Wickerham's synthetic medium at 30°. Cells are harvested at late exponential phase and washed twice with distilled water, then subjected to cold osmotic shock treatment as follows. Washed yeast cells (18 g of wet weight) are suspended in 300 ml of 0.1 $M$ Tris-HCl, pH 8.0, containing 0.9 $M$ NaCl, 1 m$M$ 2-mercaptoethanol, and 0.5 m$M$ EDTA. The suspension is gently shaken at 30° for 20 min, then centrifuged for 10 min at 6500 $g$ at 4°. The pellet is resuspended in 300 ml of ice-cold 0.5 m$M$ MgCl$_2$, gently stirred for 1 hr in the cold, and centrifuged for 10 min at 6500 $g$ at 4°. The supernatant fluid containing the released protein is adjusted to pH 7.0 with one-tenth volume of 0.5 $M$ potassium phosphate buffer, pH 7.0. The osmotic shock fluid (335 ml) is applied to a DEAE-cellulose column (1.5 × 15 cm) previously equilibrated with 50 m$M$ potassium phosphate buffer, pH 7.0, and washed with 300 ml of the same buffer. Elution is carried out with a linear gradient consisting of 100 ml of the buffer in the mixing flask and an equal volume of buffer containing 0.2 $M$ KCl in the reservoir. The flow rate is 30 ml per hour and the fraction volume is 4 ml. The thiamin-binding protein is reproducibly eluted at 0.1–0.11 $M$ KCl. The DEAE-cellulose fractions are combined and dialyzed twice against 2.5 liters of distilled water. The dialysate is concentrated on an Amicon (Danvers, MA) Ultrafilter (XM 100A), washed with water several times, and finally dissolved in 1.8 ml of 50 m$M$ potassium phosphate buffer, pH 7.0.

It has been reported that the so-called constitutive acid phosphatase localized in the yeast periplasmic space is repressed by exogenous thiamin.[8] Because both thiamin transport and the activity of sTBP are repressed by pregrowth of yeast cells in thiamin-containing medium, the activity of the thiamin-repressible acid phosphatase is also measured throughout the purification of sTBP. As shown in Table III, the specific activities of the purified protein represent an overall purification of 11.6-fold for thiamin-binding protein and 10.5-fold for acid phosphatase, indicating that the sTBP and the thiamin-repressible acid phosphatase are copurified.

[6] F. R. Leach and C. A. C. Carraway, *Methods Enzymol.* **62,** 87 (1979).
[7] T. Nishimune and R. Hayashi, *Methods Enzymol.* **62,** 91 (1979).
[8] M. E. Schweingruber, R. Fluri, K. Maundrell, A.-M. Schweingruber, and E. Dumermuth, *J. Biol. Chem.* **261,** 15877 (1986).

TABLE III
PURIFICATION OF SOLUBLE THIAMIN-BINDING PROTEIN FROM YEAST

| Fraction | Protein (mg) | Thiamin-binding activity | | | Acid phosphatase activity | | |
|---|---|---|---|---|---|---|---|
| | | Total (pmol) | Specific (pmol/mg) | Yield (%) | Total (units) | Specific (units/mg) | Yield (%) |
| Osmotic shock | 12.2 | 10,100 | 829 | 100 | 63.1 | 5.2 | 100 |
| DEAE-cellulose | 1.02 | 6,890 | 6,750 | 68.2 | 45.5 | 44.7 | 72.2 |
| Ultrafiltration | 0.66 | 6,290 | 9,580 | 62.3 | 35.7 | 54.4 | 56.6 |

## Purification of Membrane-Bound Thiamin-Binding Protein

Plasma membranes are prepared, according to the procedure of Ongjoco et al.[9] with some modifications, from S. cerevisiae (10 g of wet weight) grown at 30° for 16 hr in thiamin-free Wickerham's synthetic medium. After harvesting, the cells are washed once with cold distilled water. The modified procedure is as follows: yeast cells washed with 100 ml of 50 mM EDTA are suspended in 90 ml of 50 mM EDTA and 0.25 M 2-mercaptoethanol, and allowed to stand for 30 min at room temperature. The cells are resuspended in 80 ml of a sorbitol–EDTA solution (1 M sorbitol and 0.1 M EDTA) and then Zymolyase-100T (0.1 mg/ml) is added to the cell suspension, followed by a 1-hr incubation at 37°. The resultant yeast protoplasts are centrifuged for 10 min at 750 g at 4°. The yeast protoplasts washed once with the chilled sorbitol–EDTA solution are suspended in 15 ml of a buffer, pH 7.5, containing 0.25 M sucrose, 10 mM imidazole, 1 mM EDTA, 2 mM 2-mercaptoethanol, 0.02% (w/v) sodium azide, and 1 mM phenylmethylsulfonyl fluoride, and homogenized with 30 ml of glass beads (0.5 mm in diameter) by seven 1-min bursts (alternating with 2-min rests) in a Bead-Beater (Biospec Products, Bartlesville, OK). After removal of the glass beads by filtration, the homogenate is centrifuged twice for 3 min at 2300 g. The supernatant (crude particulate fraction) is cleared of mitochondria by the procedure of Fuhrmann et al.[10] by acidification to pH 5.0, followed by two centrifugation steps for 5 min each at 9000 g and 4°, leaving precipitates. The precipitates are resuspended in 50 mM potassium phosphate buffer, pH 5.0, to give the acid precipitates described in Table IV. The acidic supernatant is readjusted to pH 7.5 and centrifuged for 1 hr at 123,000 g and 4°. The pellet of the purified membranes is suspended in a cold appropriate buffer or water by using a Dounce homogenizer. As

[9] R. Ongjoco, K. Szkutnicka, and V. P. Cirillo, J. Bacteriol. 169, 2926 (1987).
[10] F. G. Fuhrmann, C. Boehm, and A. P. R. Theuvenet, Biochim. Biophys. Acta 433, 583 (1976).

TABLE IV
PURIFICATION OF PLASMA MEMBRANE-BOUND THIAMIN-BINDING PROTEIN
FROM YEAST

| | | Thiamin-binding activity | | |
| Fraction | Protein (mg) | Total (pmol) | Specific (pmol/mg) | Yield (%) |
| --- | --- | --- | --- | --- |
| Crude particulates | 1034 | 3728 | 3.6 | 100 |
| Acid precipitates | 712 | 1498 | 2.1 | 40.2 |
| Plasma membranes | 13.2 | 2013 | 152.5 | 54.0 |

shown in Table IV, the specific thiamin-binding activity of the plasma membrane represents a purification of 42-fold from the crude particulates, with a recovery of 54%.

## Properties

The purified preparation of sTBP is a glycoprotein with an apparent molecular weight of 140,000 as estimated by sodium dodecyl sulfate (SDS)–polyacrylamide gel electrophoresis. The apparent $K_d$ of the binding of thiamin is 29 n$M$, which is about sixfold lower than the apparent $K_m$ of thiamin transport (Table V). The optimal pH for the binding was 5.5. Although pyrithiamin inhibited the binding of thiamin as well as transport, $O$-benzoylthiamin disulfide did not show similarities in the effect on the binding and transport *in vivo*.[5] Furthermore, a thiamin transport mutant

TABLE V
COMPARISON OF PROPERTIES OF YEAST THIAMIN-BINDING PROTEINS

| | Thiamin-binding activity | |
| Property | sTBP | mTBP |
| --- | --- | --- |
| $K_d$ for thiamin | 29 n$M$ | 0.11 $\mu M$ |
| pH optimum | 5.5 | 5.0 |
| Effect of thiamin analog | | |
|   Pyrithiamin | Inhibited | Inhibited |
|   $O$-Benzoylthiamin disulfide | No change in activity | Inhibited |
| Control | Repressible by thiamin | Repressible by thiamin |
| Thiamin transport mutant | Same as that from parent strain | 1.1% of that from parent strain |

of *S. cerevisiae* (PT-R$_2$) isolated as a strain resistant to pyrithiamin[11] still contains the sTBP in amounts comparable to the wild-type parent strain. These findings strongly suggest that sTBP is not an essential component of the thiamin transport system. It has been further demonstrated that sTBP is identical to the thiamin-repressible acid phosphatase in *S. cerevisiae* encoded by *PHO3*.[12] This conclusion is supported by the following findings: (1) exogenous thiamin leads to the parallel disappearance of both activities; (2) both activities are copurified throughout the purification procedure (Table III); (3) both activities are detected at the same position on nondenaturing polyacrylamide gel electrophoresis; (4) the thiamin-binding activity is found to be deficient in thiamin-repressible acid phosphatase-defective mutants; and (5) sTBP is demonstrated to be encoded by the *PHO3* gene by genetic analysis.[13] The most striking characteristic of the thiamin-repressible acid phosphatase (sTBP) is its high affinity for thiamin phosphates. It shows $K_m$ values of 1.6 and 1.7 $\mu M$ at pH 5.0 for thiamin monophosphate and thiamin pyrophosphate, respectively. These $K_m$ values are two to three orders of magnitude lower than those (0.61 and 1.7 m$M$) for *p*-nitrophenyl phosphate. It is therefore highly probable that sTBP plays a physiological role in the hydrolysis of thiamin phosphates in the periplasmic space prior to the uptake of their thiamin moieties by *S. cerevisiae*. This is further supported by genetic evidence showing that *PHO3* mutant cells of *S. cerevisiae* have markedly reduced activity of the uptake of [$^{14}$C]thiamin phosphates in contrast to the parent yeast cells.[13]

However, mTBP shows a thiamin-binding activity with an apparent $K_d$ of 0.11 $\mu M$ at an optimal pH of 5.0, which are almost the same values as those of the thiamin transport system *in vivo*. The membrane thiamin-binding activity is inhibited by both pyrithiamin and *O*-benzoylthiamin disulfide as well as thiamin transport. In addition, the binding activity is also repressible by exogenous thiamin and is largely absent from the PT-R$_2$ mutant strain (Table V). These results indicate that mTBP may be directly involved in the transport of thiamin in *S. cerevisiae*.

## Photoaffinity Labeling of Yeast Thiamin Transport System

### Preparation of 4-Azido-2-nitrobenzoylthiamin

4-Amino-2-nitrobenzoic acid (5.2 g, 28.5 mmol) is dissolved in 12 *M* HCl (45 ml); NaNO$_2$ (3.19 g, 46.2 mmol) dissolved in water (15 ml) is added

[11] A. Iwashima, H. Nishino, and Y. Nose, *Biochim. Biophys. Acta* **330**, 222 (1973).

[12] K. Nosaka, H. Nishimura, and A. Iwashima, *Biochim. Biophys. Acta* **967**, 49 (1988).

[13] K. Nosaka, Y. Kaneko, H. Nishimura, and A. Iwashima, *FEMS Microbiol. Lett.* **60**, 55 (1989).

portion-wise with stirring over 1 hr at $-5$ to $0°$; acetic acid (42.7 ml) is added at $-5$ to $0°$; NaN$_3$ (3.20 g, 49.2 mmol) dissolved in water (12 ml) is slowly added drop-wise over 1 hr at $0-5°$. The reaction mixture is diluted with cold water (56 ml), stirred for another 15 min at $0-5°$, and further diluted with cold water (200 ml); the resulting precipitate is collected by suction filtration and dried over P$_2$O$_5$ in a desiccator to give pale yellow crystals of 4-azido-2-nitrobenzoic acid weighing in total 5 g (yield, 84.2%), mp 177–179° (decomposed). In this reaction sequence, the addition of NaN$_3$, and the succeeding procedures are carried out in subdued light.

4-Azido-2-nitrobenzoic acid (0.9 g, 4.32 mmol) is heated with SOCl$_2$ (3 ml) for 20 min under reflux, concentrated in vacuum at $50°$, and traces of SOCl$_2$ are removed by addition of benzene and alternate evaporation in vacuum, to leave crystalline 4-azido-2-nitrobenzoyl chloride, which is then used without purification. 4-Methyl-5-($\beta$-hydroxyethyl)thiazole (0.65 g, 4.54 mmol) is added to the chloride, the mixture is refluxed in benzene (2 ml) in the presence of pyridine (0.5 ml) for 1 hr, concentrated in vacuum at $50°$, and the crystalline residue is recrystallized from ethanol to give 4-methyl-5-[2-(4-azido-2-nitrobenzoyl)ethyl]thiazole (1.25 g; yield, 86.8%), mp 94–97°. Further recrystallization from ethanol gives colorless needles, mp 98–99°.

A mixture of 2-methyl-4-amino-5-bromomethylpyrimidine hydrobromide (1.7 g, 6.01 mmol) and 4-methyl-5-[2-(4-azido-2-nitrobenzoyl)ethyl]-thiazole (2.0 g, 6.00 mmol) in $n$-butanol (5 ml) is heated for 75 min under reflux and filtered while hot; the residue is washed with ethanol and then well washed with acetone to leave a straw-colored powder of 4-azido-2-nitrobenzoylthiamin bromide hydrobromide (0.94 g; crude yield, 25.4%), mp 204–205° (decomposed). Recrystallization from ethanol gives colorless crystals, mp 207–208° (decomposed). 4-Azido-2-nitrobenzoylthiamin (ANBT), thus prepared as a photoreactive thiamin derivative, irreversibly inactivates both membrane-bound thiamin-binding activity and thiamin transport in yeast cells.[14] [$^3$H]ANBP is then synthesized from [$^3$H]2-methyl-4-amino-5-bromomethylpyrimidine hydrobromide, which has been tritiated by catalytic exchange at DuPont–New England Nuclear (Boston, MA), and 4-methyl-5-[2-(4-azido-2-nitrobenzoyl)ethyl]thiazole according to the procedure described in this section.

*Photoaffinity Labeling of Thiamin-Binding Component in Yeast Plasma Membrane with [$^3$H]ANBT*

The cell suspension ($8.74 \times 10^6$/ml) in 1 liter of 50 m$M$ potassium phosphate buffer, pH 5.0, with or without 1 m$M$ thiamin, is irradiated with

---

[14] K. Sempuku, H. Nishimura, and A. Iwashima, *Biochim. Biophys. Acta* **645,** 226 (1981).

1 $\mu M$ [³H]ANBT (specific activity, 6500 dpm/nmol) for 15 min at 4° under a Toshiba (Tokyo, Japan) black light lamp (40 W) at a distance of 25 cm from the reaction vessel. Following the photoaffinity labeling of the cells, the plasma membranes are prepared by the same procedure as previously described. The specific labeling of [³H]ANBT to the membrane is calculated as 6.17 nmol/mg of protein. The obtained plasma membranes are subjected to SDS–polyacrylamide slab gel (linear gradient 7.5–22.5%) electrophoresis. The gels are sliced (2 mm) and digested by overnight incubation in 1 ml of 15% (v/v) $H_2O_2$ at 55–60° and the radioactivities in the gels are counted in 10 ml of scintillation fluid. The photoaffinity labeling of yeast plasma membranes with [³H]ANBT results in the specific covalent modification of a membrane component with an apparent molecular mass of 6–8 kDa.

However, no specific incorporation of the radioactivity is found in the plasma membrane from the cells of the PT-$R_2$ mutant strain.[15] This suggests that the small membrane component photolabeled with [³H]ANBT takes part in thiamin binding by a thiamin transporter in the yeast plasma membrane. It is not reasonable, though, to rule out the possibility that this component could be a proteolytic degradation product of a larger membrane polypeptide.

## Conclusion

We have isolated a thiamin transporter gene, *THI10*, from a yeast genomic library by the complementation of a mutant of *S. cerevisiae* defective in thiamin transport. The nucleotide sequence of this gene contains an open reading frame of 1794 bp encoding a 598-amino acid polypeptide with a calculated molecular weight of 66,903, and it has been submitted to the DDBJ/EMBL/GenBank Data Bank with Accession No. D55634. The deduced amino acid sequence of THI10 protein has been found to be identical with that of a protein of unknown function encoded by *L8083.2* gene on chromosome XII (Accession No. U19027) and the sequence is similar to those of allantoin and uracil permeases encoded by *DAL4* (30.2% identity) and *FUR4* (27.9% identity), respectively.

---

[15] H. Nishimura, K. Sempuku, Y. Kawasaki, K. Nosaka, and A. Iwashima, *FEBS Lett.* **255**, 154 (1989).

## [14] *In Vitro* Systems for Studying Thiamin Transport in Mammals

### By Gianguido Rindi and Umberto Laforenza

### Introduction

In mammals the transport of low (physiological) concentrations of thiamin ($<2\ \mu M$) is a carrier-mediated, mainly energy-dependent process that, *in vitro,* can be studied in preparations with different levels of anatomical complexity (intact tissues, isolated cells, and membrane vesicles). Each of these allows different aspects of the transport mechanism to be investigated. Thus intact tissue preparations (everted jejunal sacs or rings, human small intestinal biopsies, brain slices) can be used to study biotransformations of thiamin related to transport. Isolated cell preparations both from normal (hepatocytes, enterocytes, erythrocytes) and pathological tissues (Ehrlich ascites tumor and neuroblastoma cells) can also be used, according to the particular problem to be assessed. Choroid plexus and placenta preparations, as well as blood–brain barrier, can give insight into the modalities of thiamin transfer from the plasma through complex biological structures.

By contrast, membrane preparations (small intestinal brush border or basolateral membrane vesicles, brain microsacs, erythrocyte ghosts) are well suited for characterizing pure membrane thiamin transport mechanism without the interference of intracellular thiamin metabolism.

The results of studies *in vitro* using low thiamin concentrations are summarized in Table I. High concentrations of thiamin are mainly transferred prevailingly by simple diffusion (see Rindi[1]). Table I shows that the specific transport mechanism displays affinities for thiamin of the same order of magnitude ($K_m$ range from $10^{-6}$ to $10^{-7}\ M$) and similar molecular specificities. The transport mechanism is inhibited by the same thiamin analogs (especially pyrithiamin). By contrast, transport capacity appears to vary markedly from one tissue to another, suggesting that the number or efficiency of the carrier sites may differ according to the type and function of tissue. It is only in intact tissue or isolated cell preparations that metabolic energy is needed and that uptake is associated with thiamin pyrophosphorylation.

[1] G. Rindi, *Acta Vitaminol. Enzymol.* **6,** 47 (1984).

TABLE I

CARRIER-MEDIATED THIAMIN TRANSPORT BY DIFFERENT MAMMAL TISSUE PREPARATIONS[a]

| Preparation | Species | Kinetic constants | | Activators | Inhibitors | Peculiarities | References[b] |
|---|---|---|---|---|---|---|---|
| | | $K_m$ ($\mu M$) | $J_{max}$ | | | | |
| Everted jejunal sacs | Rat | 0.634 | 0.173 nmol/100 mg · 2 min | Na$^+$ | Metabolic; Th analogs; anoxia; ouabain | Electroneutral process | 1 |
| | Mouse | 0.60 | 1.98 nmol/ml · 30 min | Na$^+$ | Metabolic; Li$^+$; anoxia | | 2 |
| Everted small intestinal rings | Rat | 0.16–0.63 | | | Pyrithiamin | | 3 |
| | Rat | 0.59 | | | Metabolic; anoxia; Th analogs | No Th phosphorylation | 4 |
| | Rat | 0.42 | 1.6 nmol/g w.t. · 15 min | Na$^+$ | Metabolic; Th analogs; choline | Th phosphorylation | 5 |
| Bioptic duodenal mucosal tissue | Human | 0.43 | 0.23 nmol/100 mg · 3 min | Na$^+$ | Metabolic; anoxia; Th analogs | | 6 |
| | Human | 4.4 | 0.23 nmol/100 mg · 6 min | | Metabolic; Th analogs | | 7 |
| Brain slices | Rat | 0.36 | | | Metabolic; Th analogs; anoxia; ouabain | Probably with phosphorylation | 8 |
| Kidney slices | Rat | 77[c] | 231 nmol/g w.t. · 30 min[c] | | Ethanol | Th phosphorylation | 9 |
| Isolated hepatocytes | Rat | 310 | 0.7 $\mu$mol/ml · 5 min | Na$^+$ | Metabolic; Th analogs; anoxia; ouabain | Th phosphorylation | 10 |
| | Rat | 34.1[d] | 20.8 pmol/10$^5$ cells · 30 sec[d] | Na$^{+d}$ | Th analogs[d]; choline[d]; acetylcholine[d] | Th phosphorylation | 11 |
| | Rat | 1.26[d] | 1.21 pmol/10$^5$ cells · 30 sec[d] | | | | |

(continued)

TABLE I (continued)

| Preparation | Species | $K_m$ ($\mu M$) | $J_{max}$ | Activators | Inhibitors | Peculiarities | References[b] |
|---|---|---|---|---|---|---|---|
| | | Kinetic constants | | | | | |
| Erythrocytes | Human | 0.1–0.12 | 0.01 pmol/mg protein · min | | Th analogs | Th phosphorylation; electroneutral | 12 |
| Isolated enterocytes | Rat | 0.18 | 1.38 pmol/mg protein · min | | Rotenone | | 13 |
| Kidney tubular cells | Rabbit | 28.8[d] | 11.4 nmol/g protein · 30 min[e] | | Ethanol | Thiamin phosphorylation | 14 |
| Ehrlich ascites tumor cells | Mouse | 0.043 | 0.71 pmol/mg protein · min | Na⁺ | Metabolic; Th analogs | | 15 |
| Neuroblastoma cells | Mouse | 0.035[f] 800[g] | | ATP | Metabolic; veratridine; batracotoxin | Na⁺ independent; Th phosphorylation; membrane potential dependent | 16 |
| Choroid plexus | Rabbit | 0.41 | | | | No Th phosphorylation | 17 |
| Blood–brain barrier (in vivo) | Rat | 0.4[b:18] | 17.6 pmol/g · min[b:18] | | Th analogs[b:20] | Carrier-mediated[b:18] energy independent[b:19] | 18, 19, 20 |
| Placenta | Human | 1.29[h] | 1.49 nmol/min[h] | | | | 21 |
| Erythrocyte ghosts | Human | 0.16–0.51 | 0.36 pmol/mg protein · min | | Th analogs | | 12 |
| Brush border membrane vesicles (small intestine) | Rat | 0.8 | 0.35 pmol/mg protein · 4 sec | | Unlabeled Th; Th analogs | Na⁺ independent; no Th phosphorylation | 22 |
| Basolateral membrane vesicles (small intestine) | Rat | 1.38 | 1.93 pmol/mg protein · 4 sec | ATP | Unlabeled Th; Th analogs; ouabain; frusemide; vanadate | Na⁺,K⁺-ATPase dependent; no Th phosphorylation; electroneutral | 23 |

| | | | | | | | |
|---|---|---|---|---|---|---|---|
| Basolateral membrane vesicles (liver) | Rat | $1.41^i$ $28.6^j$ | 1.75 nmol/mg protein · 5 sec$^d$ 36.6 nmol/mg protein · 5 sec$^j$ | $H^+$ | $NH_4^+$; unlabeled Th; imipramine; choline | Th/$H^+$ antiport; electroneutral | 24 |
| Brush border membrane vesicles (small intestine) | Rat | 6.23 | 14.9 pmol/mg protein · 6 sec | $H^+$ | $NH_4^+$; Th; Th analogs; guanidine; imipramine; clonidine; harmaline | Th/$H^+$ antiport; electroneutral | 25 |
| Brain microsacs | Rat | 2.36 | 4.53 pmol/mg protein · 4 sec | | Th analogs; guanidine; $Mg^{2+}$, $Ca^{2+}$ | $Na^+$ independent | 26 |

[a] Th, Thiamin; w.t., wet tissue.

[b] Key to references: (1) A. M. Hoyumpa Jr., H. M. Middleton III, F. A. Wilson, and S. Schenker, Gastroenterology 68, 1218 (1975); (2) R. C. de Angelis, Arq. Gastroent. (S. Paulo) 14, 135 (1977); (3) U. Ventura, G. Ferrari, R. Tagliabue, and G. Rindi, Life Sci. 8, 699 (1969); (4) T. Komai, K. Kawai, and H. Shindo. J. Nutr. Sci. Vitaminol. 20, 163 (1974); (5) T. Akiyama, H. Wada, and K. Miyaji, Mie Med. J. 31, 349 (1981); (6) A. M. Hoyumpa Jr., R. Strickland, J. J. Sheehan, G. Yarborough, and S. Nichols. J. Lab. Clin. Med., 99, 701 (1982); (7) U. Laforenza, C. Patrini, C. Alvisi, A. Faelli, A. Licandro, and G. Rindi, Am. J. Clin. Nutr., in press (1997); (8) Y. Nose, A. Iwashima, and H. Nishino, in "Thiamine" (C. J. Gubler, M. Fujiwara, and P. M. Dreyfus, eds.), p. 157. John Wiley & Sons, New York, 1976; (9) M. A. Mahajan and M. Acara, J. Pharmacol. Exp. Ter. 268, 1311 (1994); (10) C.-P. Chen, J. Nutr. Sci. Vitaminol. 24, 351 (1978); (11) K. Yoshioka, Biochim. Biophys. Acta 778, 201 (1984); (12) D. Casirola, C. Patrini, G. Ferrari, and G. Rindi, J. Membr. Biol. 118, 11 (1990); (13) V. Ricci and G. Rindi, Arch. Int. Physiol. Biochim. 100, 275 (1992); (14) M. A. Mahajan, M. Acara, and M. Taub. J. Pharmacol. Exp. Ther. 268, 1316 (1994); (15) S. Yamamoto, S. Koyama, and T. Kawasaki, J. Biochem. 89, 809 (1981); (16) L. Bettendorff and P. Wins. J. Biol. Chem. 269, 14379 (1994); (17) R. Spector, Am. J. Physiol. 230, 1101 (1976); (18) J. Greenwood, E. R. Love, and O. E. Pratt, J. Physiol. (London) 327, 95 (1982); (19) J. Greenwood and O. E. Pratt, J. Physiol. (London 317, 65P (1981); (20) J. Greenwood, P. J. Luthert, O. E. Pratt, and P. L. Lantos, Brain Res. 399, 148 (1986); (21) J. Dancis, D. Wilson, I. A. Hoskins, and M. Levitz, Am. J. Obstet. Gynecol. 159, 1435 (1988); (22) D. Casirola, G. Ferrari, G. Gastaldi, C. Patrini, and G. Rindi, J. Physiol. (London) 398, 329 (1988); (23) U. Laforenza, G. Gastaldi, and G. Rindi, J. Physiol. (London) 468, 401 (1993); (24) R. H. Moseley, P. G. Vashi, S. M. Jarose, C. J. Dickinson, and P. A. Permoad, Gastroenterology 103, 1056 (1992); (25) U. Laforenza and G. Rindi, Pflügers Arch. 423, R13 (1993); (26) U. Laforenza and G. Rindi, Unpublished results (1996).

[c] Calculated from data on Fig. 3 of the cited paper.

[d] Low- and high-affinity entry process.

[e] Calculated from data of Fig. 3 of the cited paper.

[f] High-affinity carrier.

[g] Low-affinity carrier.

[h] Calculated from data of Fig. 1 of the cited paper.

[i] High-affinity, low-capacity process.

[j] Low-affinity, high-capacity process.

Preparation and Characterization Procedures

*Isolated Rat Enterocytes*

*Preparation.* Enterocytes are isolated from the small intestine of male adult Wistar albino rats (about 350 g body weight), after 15–16 hr of fasting with water *ad libitum,* according to the procedure of Weiser[2] with minor modifications (preparation time, 1 hr). The entire small intestine is removed from two animals, which have been decapitated to avoid mucus secretion caused by anesthesia.[3] The intestine is then rinsed with a cold and oxygenated (95% $O_2$ and 5% $CO_2$) solution containing 154 m$M$ NaCl and 1 m$M$ dithiothreitol (DTT).

The washed small intestine is filled with oxygenated solution A containing 96 m$M$ NaCl, 1.5 m$M$ KCl, 27 m$M$ sodium citrate, 0.2 m$M$ phenylmethylsulfonyl fluoride (PMSF), bovine serum albumin (BSA, 10 mg/ml), 5.6 m$M$ $K_2HPO_4$–$KH_2PO_4$ (pH 7.3), and incubated in 100 ml of the same solution at 37° for 10 min. After removal of the content, the intestine is filled with solution B [140 m$M$ NaCl, 1.5 m$M$ EDTA, 0.5 m$M$ DTT, 0.2 m$M$ PMSF, BSA (10 mg/ml), 16 m$M$ $Na_2HPO_4$–$NaH_2PO_4$ (pH 7.3)] and incubated in 100 ml of the same oxygenated solution for 3 min at 37°. The intestine is then gently fingered for 2 min. The content with enterocytes is collected in 30 ml of cold solution C [137 m$M$ NaCl, 5.2 m$M$ KCl, 0.6 m$M$ $CaCl_2$, 0.8 m$M$ $MgSO_4$, 0.2 m$M$ PMSF, 10 m$M$ D-glucose, 5 m$M$ glutamine, BSA (10 mg/ml), 3 m$M$ $Na_2HPO_4$–$NaH_2PO_4$ (pH 7.3)] and filtered through a 250-$\mu$m (pore size) mesh nylon filter. The intestine is filled again with solution B and processed as described. The filtrate of the lumenal content is then added to the previous filtrate.

The cellular suspension is filtered again (in an ice-cold bath) through a 100-$\mu$m (pore size) mesh nylon filter and then centrifuged at 50 $g_{av}$ for 2 min at 4° in a Beckman TJ-6 centrifuge (TH-4 rotor; Beckman, Fullerton, CA). The pellet is suspended in 4–5 ml of solution C and then diluted up to 30 ml with solution C to reach a final protein content of 3 mg/ml.

The protein content is determined according to Lowry *et al.,*[4] with BSA as a standard.

*Characterization.* Cell viability, routinely assessed by trypan blue dye exclusion,[5] is 95 ± 1.7% (mean ± SEM of 31 different preparations).[6]

[2] M. M. Weiser, *J. Biol. Chem.* **248,** 2536 (1973).
[3] M. Watford, P. Lund, and H. A. Krebs, *Biochem. J.* **178,** 589 (1979).
[4] O. H. Lowry, N. J. Rosebrough, A. L. Farr, and R. J. Randall, *J. Biol. Chem.* **193,** 265 (1951).
[5] H. Baur, S. Kasperek, and E. Pfaff, *Hoppe-Seyler's Z. Physiol. Chem.* **356,** 827 (1975).
[6] V. Ricci and G. Rindi, *Arch. Int. Physiol. Biochim.* **100,** 275 (1992).

Further characterization of isolated enterocytes is obtained by measuring the homogeneity of the cell dispersion, $O_2$ consumption, ATP, and phosphocreatine contents and ATP/ADP, phosphocreatine/creatine, and lactate/pyruvate ratio values.[6]

Transport efficiency can be assessed by measuring valine uptake according to the method of Kimmich.[7] L-[U-$^{14}$C]valine (Amersham International plc, Amersham, UK; specific activity, 2.48 MBq/mmol) taken up by normoenergized (normal) enterocytes is three times higher than that of deenergized with rotenone.[6]

## Brush Border Membrane Vesicles of Rat Small Intestine

Brush border membrane vesicles (BBMVs) make up purified plasma membrane preparations derived from the lumenal side of the epithelial cells characterized by digestive and absorptive properties. These are taken from rat small intestine and prepared at 0–4° following Biber *et al.*[8] with minor modifications.

*Preparation.* The mucosal scraping from the duodenum and the jejunum (first tract of the small intestine, 60 cm) of six adult Wistar albino rats is homogenized for 45 sec, at the maximal speed, in a Waring blender-type homogenizer (Braun Multimix MX 32; Braun AG, Kronberg/Taunus, Germany) in 360 ml of a solution containing 60 m$M$ D-mannitol, 1 m$M$ ethylene glycol-$O,O'$-bis(2-aminoethyl)-$N,N,N',N'$-tetraacetic acid (EGTA), 2.5 m$M$ Tris-HCl (pH 7.1). After 1 min, 4.5 ml of 1 $M$ MgCl$_2$ (12.5 m$M$, final concentration) is added, and the suspension is homogenized again as described. After 15 min, the homogenate is centrifuged at 4500 $g_{av}$ for 15 min in a Sorvall RC 5 supercentrifuge (SS-34 rotor; Du Pont Instruments, Newton, CT). The supernatant is separated and centrifuged at 16,000 $g_{av}$ for 30 min. The pellet obtained is suspended in 30 ml of a solution containing 60 m$M$ D-mannitol, 5 m$M$ EGTA, 12 m$M$ Tris-HCl (pH 7.1) and homogenized (10 strokes) with a Teflon–glass Potter-Elvehjem homogenizer (Kontes, Vineland, NJ) and, after dilution with another 30 ml of the same solution, 0.72 ml of 1 $M$ MgCl$_2$ (final concentration, 12 m$M$) is added. After 15 min at 2°, the homogenate is centrifuged at 4500 $g_{av}$ for 15 min, and the resulting supernatant centrifuged at 16,000 $g_{av}$ for 30 min at 4°. The pellet, suspended in 18 ml of preloading solution (usually 280 m$M$ D-mannitol, 2 m$M$ MgSO$_4$, and a 20 m$M$ concentration of an appropriate HEPES–MES–Tris buffer) is centrifuged again (16,000 $g_{av}$ for 30 min at 4°). The new pellet (BBMVs) is, finally, suspended with 500–600 $\mu$l of preloading solution

[7] G. A. Kimmich, *Biochemistry* **9**, 3659 (1970).

[8] J. Biber, B. Stieger, W. Haase, and H. Murer, *Biochim. Biophys. Acta* **647**, 169 (1981).

(final protein concentration,[4] 12–14 mg/ml), stored in the cold for 2 hr, and preincubated for 30 min at 25° (to load the vesicles) before uptake studies.

*Purity and Characteristics.* The purity of BBMVs is estimated by evaluating the degree of enrichment in the marker enzymes saccharase[9] and alkaline phosphatase[10] as compared to that of the initial mucosal homogenate. Enrichments are $11.3 \pm 0.8$ and $13.4 \pm 1.0$, respectively (mean $\pm$ SEM of at least eight different preparations).[11]

For the study of thiamin uptake, it is better to use $Mg^{2+}$- than $Ca^{2+}$-precipitated BBMVs, because the former are less leaky and dissipate ionic gradients more slowly,[12,13] even if the latter can have higher purity.

Conventional electron microscopy of the final vesicle preparation shows microvillous membranes, almost exclusively sealed and right-side-out oriented, structurally intact, and virtually uncontaminated by cytosol or other subcellular components.

The transport efficiency of BBMVs, evaluated by determining the time course profile of D-[U-$^{14}$C]glucose (Amersham International plc; specific activity, 0.31 MBq/mmol) uptake,[14] shows a typical overshoot at about 30 sec, indicating the suitability of the preparation for transport studies.

### Basolateral Membrane Vesicles of Rat Small Intestine

The basolateral membrane represents the contralumenal region of the plasma membrane of small intestinal epithelial cells, and it contains about 95% of the total $Na^+,K^+$-ATPase.[15] Basolateral membrane vesicles (BLMVs) are prepared according to the method of Schron *et al.*,[16] on the basis of ultracentrifugation of discontinuous sucrose gradients.

*Preparation.* The entire procedure is carried out at 0–4°, making sure that the solutions are prepared daily to minimize bacterial contamination.

For each BLMV preparation, the mucosal scraping from six to eight small intestines of adult Wistar albino rats is suspended in 360 ml of the isolation medium [250 m$M$ sucrose, 1 m$M$ EDTA, 0.1 m$M$ PMSF, 12 m$M$ Tris, 16 m$M$ HEPES (pH 7.5)] and homogenized for 3 min at the maximal

[9] A. Dahlqvist, *Anal. Biochem.* **7**, 18 (1964).

[10] H. Murer, E. Ammann, J. Biber, U. Hopfer, *Biochim. Biophys. Acta* **433**, 509 (1976).

[11] U. Laforenza and G. Rindi, Unpublished observations (1996).

[12] M. Kessler, O. Acuto, C. Storelli, H. Murer, M. Müller, and G. Semenza, *Biochim. Biophys. Acta* **506**, 136 (1978).

[13] I. Sabolic and G. Buckhardt, *Biochim. Biophys. Acta* **772**, 140 (1984).

[14] D. Casirola, G. Ferrari, G. Gastaldi, C. Patrini, and G. Rindi, *J. Physiol.* (*London*) **398**, 329 (1988).

[15] A. K. Mircheff and E. M. Wright, *J. Membr. Biol.* **28**, 309 (1976).

[16] C. M. Schron, R. G. Knickelbein, P. S. Aronson, and J. W. Dobbins, *Am. J. Physiol.* **253**, G404 (1987).

speed in a Waring blender-type homogenizer (Braun Multimix MX 32; Braun AG). The homogenate is centrifuged at 3000 $g_{av}$ for 10 min in a Sorvall RC 5 supercentrifuge (SS-34 rotor; Du Pont Instruments).

After removal of the pellet, the supernatant is centrifuged at 41,000 $g_{av}$ for 30 min and the resulting pellet is kept and the second supernatant discarded. The white, soft pellet, which lays on top of a brownish pellet well fixed to the walls of the tube, is removed by mild stirring by hand, resuspended with about 40 ml of isolation medium, and homogenized (20 strokes) with a Teflon–glass Potter-Elvehjem homogenizer (Kontes). The homogenate is diluted with a further 40 ml of isolation medium and centrifuged again at 41,000 $g_{av}$ for 30 min. Both white, soft pellets (crude membranes) are resuspended in 1 ml of 60% (w/v) sucrose solution and diluted with an equal volume of 60% (w/v) sucrose (final homogenate volume, 18 ml). Density gradients were prepared by stratifying in the following order from bottom to top: 2 ml of crude membrane homogenate, 5 ml of 36% (w/v) sucrose solution, and 4 ml of 31% (w/v) sucrose solution. All the gradient solutions are prepared in 0.5 m$M$ EDTA and 10 m$M$ Tris–HEPES, pH 7.5. The gradients are then centrifuged at 200,000 $g_{av}$ for 2 hr in a Sorvall Ultra Pro 80 ultracentrifuge (TH 641 rotor; Du Pont Instruments).

After ultracentrifugation, a BLMV-rich layer is obtained at the interface between 31 and 36% (w/v) sucrose. The membranes are collected, diluted with 60 ml of a vesicle-suspending medium [100 m$M$ D-mannitol, 2 m$M$ MgSO$_4$, 10 mM Tris–HEPES (pH 7.5)], and centrifuged at 48,000 $g_{av}$ for 30 min in a Sorvall RC 5 supercentrifuge (SS-34 rotor). The resulting pellet is resuspended in about 500–600 $\mu$l of vesicle-suspending medium (final protein concentration,[4] 12–14 mg/ml) and immediately used.

*Purity and Characteristics.* Purity is evaluated by determining the enrichment of the marker enzymes, in different subcellular fractions, with respect to that of the initial mucosal homogenate.[17] The enrichment values (means ± SEM of at least five different preparations) are (in parentheses are subcellular fractions): K$^+$-stimulated phosphatase[10] (BLMVs), 22.2 ± 0.7; saccharase[9] (BBMVs), 0.28 ± 0.06; succinate dehydrogenase[18] (mitochondria), 0.24 ± 0.07; NADPH-cytochrome $c$ reductase[19] (microsomes), 0.79 ± 0.05.

Percentage and sidedness of BLMVs, calculated according to Boumendil-Podevin and Podevin[20] from the Na$^+$,K$^+$-ATPase activity,[21] are

[17] U. Laforenza, G. Gastaldi, and G. Rindi, *J. Physiol. (London)* **468,** 401 (1993).

[18] R. J. Pennington, *Biochem. J.* **80,** 649 (1961).

[19] G. L. Sottocasa, B. Kuylenstierna, L. Ernster, and A. Bergstrand, *J. Cell Biol.* **32,** 415 (1967).

[20] E. F. Boumendil-Podevin and R. A. Podevin, *Biochim. Biophys. Acta* **735,** 86 (1983).

[21] R. Marin, T. Proverbio, and F. Proverbio, *Biochim. Biophys. Acta* **858,** 195 (1986).

(means $\pm$ SEM of eight different preparations): 63.8 $\pm$ 8.4% inside out; 5.0 $\pm$ 2.0% right side out, and 31.2 $\pm$ 4% open (sheets).[17]

The functional efficiency (and the purity) of BLMVs are evaluated from the profile of the time course of D-[U-$^{14}$C]glucose (Amersham International plc; specific activity, 0.31 MBq/mmol) transport.[14] The profile lacks the overshoot typical of BBMVs.

## Human Erythrocyte Resealed Ghosts

The term "resealed ghosts" describes a cell membrane preparation, essentially devoid of intracellular structures, that is prepared from erythrocytes by hypotonic hemolysis followed by restoration of the original membrane structure and function. The possibility of changing (and varying) the original cell content with appropriate solutions makes resealed ghosts a particularly suitable preparation for uptake studies.

*Preparation.* In accordance with Schwoch and Passow,[22] with minor modifications, right-side-out, pink ghosts are prepared from human erythrocytes (RBCs) (less than 10 days old; obtained from a blood bank) stored at 4° in a citrate buffer [142 m$M$ D-glucose, 0.2 m$M$ adenine, 70 m$M$ NaH$_2$PO$_4$, and 100 m$M$ citrate buffer (pH 7.4)]. After centrifugation at 12,000 $g_{av}$ for 10 min at 0°, the supernatant and the buffy coat are removed by suction, and the RBCs are washed three times with cold saline (166 m$M$ NaCl) and centrifuged as described. The RBC suspension [50% (v/v) in saline] at 0° is mixed at a ratio of 1:9 with hemolysis medium (1.2 m$M$ acetic acid and 4 m$M$ MgSO$_4$, unbuffered) at 0°.

After 5 min of slow stirring, the pH of the suspension (5.9 < pH < 6.2) and tonicity are restored by the addition of 1 vol of reversal medium (240 m$M$ EDTA sodium salt, pH 8.1) at 0° (final pH: 6.9 < pH < 7.2) under slow stirring. After 10 min, the RBC suspension is centrifuged at 20,000 $g_{av}$ for 10 min at 0°. The pellet is routinely washed three times by resuspending with cold resealing medium [140 m$M$ NaCl, 5 m$M$ KCl, 2 m$M$ MgSO$_4$, 15 m$M$ Tris-HCl (pH 7.35)] and centrifuging at 20,000 $g_{av}$ for 5 min at 0°. At this stage, any other medium can be used to load the ghosts, depending on the nature of experimental uptake studies.

The resealing process is performed by incubating the ghosts suspended in resealing medium at a ratio of 1:5 at 37°, for 45–60 min. Resealed ghosts are prepared just before use and stored at 0–4°.

*Characterization.* As an index of the extent of resealing, the hemoglobin content of ghosts is determined by a cyanmethemoglobin method (hemoglo-

[22] G. Schwoch and H. Passow, *Mol. Cell Biochem.* **2,** 197 (1973).

bin kit; Boehringer GmbH, Mannheim, Germany). It is found to be about 3% of that of the initial RBCs.[23]

*Brain Microsacs*

The term "brain microsacs"[24,25] describes a cell-free preparation, enriched in sealed membrane vesicles derived from pre- and postsynaptic elements, that is suitable for uptake studies.[25–29] The procedure described here can utilize the whole brain of rat or specific brain regions, and is essentially that of Harris and Allan.[25]

*Preparation.* The brain from one rat killed by decapitation is rapidly removed and manually homogenized with a glass–glass homogenizer (10–12 strokes in a Dounce tissue grider; Kontes) in 4 ml of ice-cold homogenizing medium [145 m$M$ NaCl, 5 m$M$ KCl, 1 m$M$ MgCl$_2$, 1 m$M$ CaCl$_2$, 10 m$M$ D-glucose, and 10 m$M$ HEPES–Tris (pH 7.5)]. The homogenate is filtered through a double layer of gauze, diluted with 4 ml of homogenizing medium, and centrifuged at 900 $g_{av}$ for 15 min at 2°. After removal of the supernatant, the pellet is resuspended in 8 ml of homogenizing medium, and centrifuged again as described. The pellet obtained is resuspended in 7–8 ml of homogenizing medium to a final protein concentration[4] of 12–14 mg/ml.

*Purity and Characteristics.* The microsac preparation is characterized by use of conventional electron microscope techniques. Morphological examination reveals sealed vesicles from pre- and postsynaptic elements with few free mitochondria, nuclei, and myelin-like membrane structures, and no intact cells.

Uptake Measurement Procedures

*Thiamin Uptake by Enterocytes*

Thiamin uptake by rat small intestinal mucosa is a "tandem" process,[30] whereby a membrane-located specific transport mechanism[14,17,31] is associ-

[23] D. Casirola, C. Patrini, G. Ferrari, and G. Rindi, *J. Membr. Biol.* **118,** 11 (1990).
[24] J. W. Daly, E. McNeal, C. Partington, M. Neuwirth, and C. R. Creveling, *J. Neurochem.* **35,** 326 (1980).
[25] R. A. Harris and A. M. Allan, *Science* **228,** 1108 (1985).
[26] A. M. Allan and R. A. Harris, *Mol. Pharmacol.* **29,** 497 (1986).
[27] L. Bettendorff, P. Wins, and E. Schoffeniels, *Biochem. Biophys. Res. Commun.* **171,** 1137 (1990).
[28] L. C. Daniell, *Neuropharmacology,* **30,** 539 (1991).
[29] L. C. Daniell, *Anal. Biochem.* **202,** 239 (1992).
[30] R. M. Wohlhueters and P. G. W. Plagemann, *Int. Rev. Cytol.* **64,** 171 (1980).
[31] U. Laforenza and G. Rindi, *Pflügers Arch.* **423,** R13 (1993).

ated with intracellular metabolic trapping (phosphorylation).[32] The thiamin transport mechanism in the membrane can be studied in isolated entero-cytes by using a short incubation time (30 sec) and deenergized cells, whereas the trapping mechanism can be studied by using a long incubation time (30 min) and normoenergized cells.[6] Three types of enterocytes are used: (1) normoenergized (normal) cells, (2) deenergized cells, preincu-bated (see as follows) with 25 $\mu M$ rotenone in 0.5% (v/v) ethanol (final concentrations), and (3) control normoenergized cells, preincubated in a medium containing 0.5% (v/v) ethanol.[6]

Four milliliters of enterocytes (about 1.5 mg/ml protein concentration) suspended in solution C (see description of enterocyte preparation) is preincubated in a small plastic flask for 10 min at 37° in a thermostatic shaker (70 oscillations/min). Incubation is started by adding [$^3$H]thiamin (Amersham International plc; specific activity, 27.75 GBq/mmol) to each flask at a final concentration of 0.5 $\mu M$ for time course experiments, and at increasing concentrations from 0.12 to 7.5 $\mu M$ for kinetic studies. At appropriate time intervals, thiamin uptake by enterocytes is evaluated by the rapid filtration of aliquots of 50 $\mu l$ of cell suspension directly pipetted on nitrocellulose filter (MFS, Dublin, CA; 0.65-$\mu$m pore diameter). The filter has been previously saturated with unlabeled thiamin by extensive washing with a fresh solution, containing 300 m$M$ D-mannitol, 0.5 m$M$ unlabeled thiamin, and 10 m$M$ Tris–HEPES, pH 7.5.[14] The filters are then rapidly rinsed with 10 ml of solution C. After the addition of 7 ml of Hydroluma (Lumac, Schaesberg, The Netherlands) the amount of [$^3$H]thia-min taken up by the enterocytes and retained on the filters is determined radiometrically by means of a liquid scintillation counter (Packard Instru-ments Co., Inc., Downers Grove, IL) and expressed as a picomoles of [$^3$H]thiamin per milligram of protein. All the uptake values are corrected for the volume of adherent water, which is determined in each experiment by [$^{14}$C]dextran[33] (Du Pont–NEN Products, Cologno Monzese, Italy; spe-cific activity, 30 MBq/g) and separate small flasks.

*Thiamin Uptake by Small Intestinal Brush Border, Basolateral Membrane Vesicles, and Cerebral Microsacs*

Incubation (at 25 or 37°) is started by adding 10 $\mu l$ of vesicle suspension to 90 $\mu l$ of a solution containing 2 m$M$ MgSO$_4$, a 20 m$M$ concentration of the appropriate HEPES–MES–Tris buffer and D-mannitol, eventually reaching 300 mos$M$ osmolarity, and [$^3$H]thiamin (Amersham International

[32] G. Ferrari, C. Patrini, and G. Rindi, *Pflügers Arch.* **393**, 37 (1982).
[33] F. A. Wilson and L. L. Treanor, *Biochim. Biophys. Acta* **406**, 280 (1975).

plc; specific activity, 27.75 GBq/mmol). Tritiated thiamin (0.25 or 1 $\mu M$) can be used in binding, inhibition, or time course studies, and in increasing concentrations from 0.25 to 10 $\mu M$ in kinetic studies. Incubation is stopped at selected time intervals by pipetting 100 $\mu l$ of vesicle suspension into 3 ml of ice-cold stop solution (150 m$M$ NaCl and 1 m$M$ Tris–HEPES, pH 7.5). The amount of [$^3$H] thiamin taken up by the vesicles is evaluated with a rapid filtration procedure,[12,14] using nitrocellulose filters (MFS; 0.65-$\mu$m pore diameter) previously saturated with unlabeled thiamin as described previously for enterocytes. The filters with the vesicles are rapidly rinsed with 5 ml of ice-cold stop solution, and the amount of radioactivity taken up by the vesicles retained on the filter is measured radiometrically as previously described. In each experiment appropriate blanks are prepared to evaluate the radioactivity of [$^3$H]thiamin not specifically adsorbed by the filter, the values are subtracted from those for total radioactivity retained on the filter. For incubation times ranging from 4 to 15 sec, a short-time incubation apparatus (Struma; Innovativ-Labor AG, Adliswill, Switzerland) is used.

### Thiamin Uptake by Erythrocytes and Ghosts

Incubation of RBCs and ghosts is carried out at 37° according to Casirola *et al.*[23] Washed RBCs are used after dilution with a suspending solution containing 140 m$M$ NaCl, 5 m$M$ KCl, 2 m$M$ MgSO$_4$, 15 m$M$ Tris-HCl (pH 7.35) at a hematocrit value of 0.5.

Ghosts, stored in the previously mentioned solution at a ratio of 1 : 5, are sedimented by centrifuging at 20,000 $g_{av}$ for 5 min. They are then resuspended in the same solution at a hematocrit value of 0.5.

Incubation is started by adding 100 $\mu l$ of RBCs or ghost suspension to 100 $\mu l$ of the suspending medium containing appropriate concentrations of [$^3$H]thiamin (Amersham International plc; specific activity, 27.75 GBq/ mmol). Normal concentrations are as follows: 0.2 $\mu M$ for studies of time course, membrane potential, and metabolic inhibitor or structural analog influence; from 0.15 to 15 $\mu M$ for kinetic studies. Incubations are stopped by transferring the samples to an ice-cold bath and adding 1 ml of ice-cold thiamin-free medium to each sample. Erythrocytes or ghosts are immediately centrifuged at 10,000 $g_{av}$ (Beckman microfuge 12; Beckman) for 30 and 60 sec, respectively, and washed three times with cold thiamin-free suspending medium to remove [$^3$H]thiamin completely from the water adherent to the cells. Ten minutes after the addition of 0.5 ml of 10% (w/v) trichloroacetic acid to the final RBC or ghost pellets, the samples are centrifuged as described. After the addition of 7 ml of Hydroluma (Lumac) to 0.5 ml of the supernatant the total amount of [$^3$H]thiamin taken up by RBCs or ghosts is determined radiometrically by means of a liquid scintillation counter (Packard Tri-Carb 2000 CA analyzer, Packard Instruments Co., Inc.).

The results obtained are expressed as picomoles of [³H]thiamin per microliter of intracellular water, determined according to the method of Speizer et al.[34]

## Determination of Thiamin Compounds in Tissues

The content of thiamin and its phosphoesters (endogenous and exogenous) in enterocytes and erythrocytes is determined by a high-performance liquid chromatography (HPLC) method[35] after homogenization, deproteinization, and trichloroacetic acid extraction of the cell pellets.

Enterocytes (40–50 mg of protein) are homogenized with a Teflon–glass Potter-Elvehjem homogenizer (Kontes) in the cold, with 1–2 ml of 5% (w/v) trichloroacetic acid, and, after 10 min, centrifuged at 20,000 $g_{av}$ for 10 min at 2°. The supernatant is extracted three times with 5 ml of diethyl ether to eliminate trichloroacetic acid.

Erythrocytes (usually 1–3 ml of washed and centrifuged cells) are deproteinized with 70% (w/v) trichloroacetic acid [final concentration, 8% (w/v)], centrifuged for 10 min at 20,000 $g_{av}$, and the supernatant extracted as described with diethyl ether to eliminate trichloroacetic acid.

Eighty microliters of the trichloroacetic acid-free supernatant is oxidized to fluorescent thiochromes with 50 $\mu$l of ferricyanide alkaline solution [4.3 m$M$ potassium ferricyanide in 15% (w/v) NaOH]. The oxidized sample (20 $\mu$l) is injected into a PRP-1 reversed-phase analytical column (150 × 4.1 mm i.d., 5-$\mu$m particle size; Hamilton, Reno, NV). The mobile phase [containing 25 m$M$ tetrabutylammonium hydrogen sulfate, 4% (v/v) tetrahydrofuran, and 50 m$M$ Na$_2$HPO$_4$] is prepared and buffered to pH 9.0 just before use. The flow rate is 0.5 ml/min and the fluorescence is detected on a Perkin-Elmer LS-30 spectrofluorometer (Perkin-Elmer, Beaconsfield, UK; wavelengths were $\lambda_{ex}$ = 365 nm; $\lambda_{em}$ = 433 nm). Before each analysis a standard solution prepared with a mixture of thiamin and its phosphoesters, each at a concentration of 0.25 $\mu M$, is processed. The retention times of three thiamin derivatives were 9.5 min for thiochrome, 11.5 min for thiochrome monophosphate, and 21 min for thiochrome pyrophosphate. Compared with the results obtained by Bettendorff et al.,[35] present results show an inversion in the elution order of thiochrome and thiochrome

[34] L. Speizer, R. Haugland, and H. Kutchai, *Biochim. Biophys. Acta* **815**, 75 (1985).
[35] L. Bettendorff, M. Peeters, C. Jouan, P. Wins, and E. Schoffeniels, *Anal. Biochem.* **198**, 52 (1991).

monophosphate, while the retention time of thiochrome pyrophosphate is approximately two times higher.

Contents are expressed as picomoles per milligram of protein for entero-cytes, and as femtomoles per $10^6$ cells for erythyrocytes.

## [15] Cofactor Designing in Functional Analysis of Thiamin Diphosphate Enzymes

*By* ALFRED SCHELLENBERGER, GERHARD HÜBNER, and HOLGER NEEF

Introduction

Thiamin diphosphate (ThDP) has evolved as a cofactor of enzymes catalyzing the splitting and resynthesis of C–C bonds. It reacts exclusively with substrates of the general structure R–CO–X, where X as a cationic leaving group (i.e., $CO_2$ or R–CHO) is replaced by another positively charged residue (Y). Among the best studied ThDP enzymes are pyruvate decarboxylase (PDC, EC 4.1.1.1) with $Y = H^+$, transketolases (TK, EC 2.2.1.1) and acetolactate synthase (ALS, EC 4.1.3.18), with carbonyl compounds as Y, or the $\alpha$-oxoacid dehydrogenase complexes, e.g., pyruvate dehydrogenase complex (PDH, EC 1.2.2.2), with a cyclic disulfide (lipoic acid) as Y (Fig. 1, Table I).

Figure 1 (structures **1–4**) summarizes the general course of ThDP-catalyzed reactions. Apart from the simple X–Y transfer mechanism (Table I) another type of reaction also starts at the level of the carbanion/enamine intermediate **2**. In this case oxidation of **2** by a second cofactor (FAD, lipoic acid) yields (by hydrolysis of **4**) the corresponding carbonic acids as phosphates [pyruvate oxidase (POX), EC 1.2.3.3] or as CoA derivatives (PDH).

In all ThDP enzymes the first substrate contains, besides the activated carbonyl group and the cationic leaving group X, an alkyl residue R, which defines the substrate specificity of the ThDP enzyme. The second substrate Y characterizes the enzyme species (operational specificity) as shown in Table I.

Thiamin diphosphate with its two cooperating heterocycles early on provoked interest in its synthesis. It was, therefore, the first coenzyme in which nearly all atoms and groups expected to be of functional significance were modified (Fig. 2, Table II). As a result of these synthesis studies,

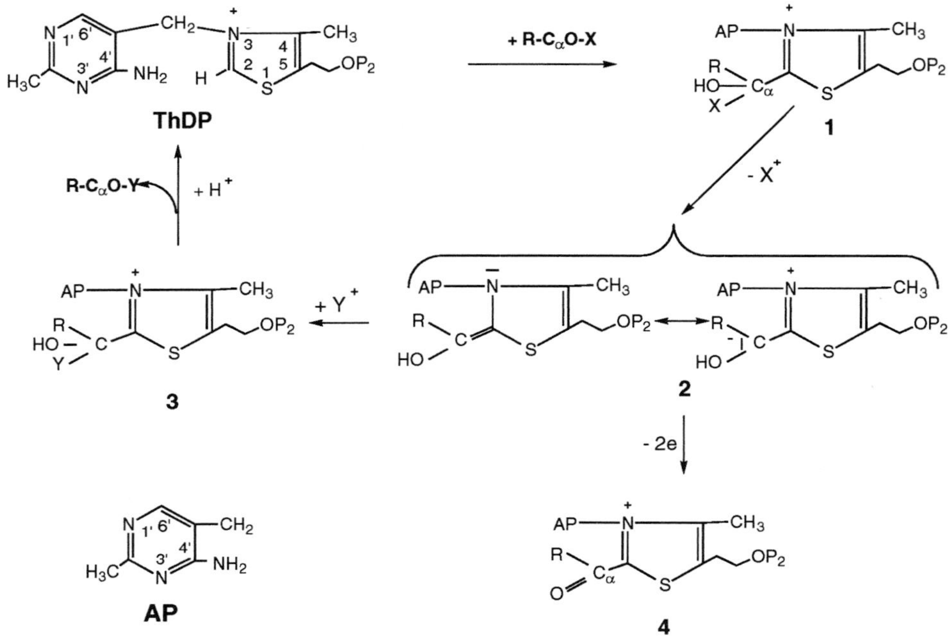

FIG. 1. Thiazolium catalysis of the carbonyl-activated X–Y exchange by ThDP enzymes. X and Y are cationic groups in the (irreversible) reaction R–CO–X + Y → R–CO–Y + X. AP, Aminopyrimidine component of ThDP.

made long before genetic engineering came into fashion,[1] not only groups and atoms with a direct significance for the catalytic mechanism, but also steric conditions for coenzyme action (the so-called V-conformation), were published as essentials of ThDP catalysis.[2]

These experiments have shown that the method of cofactor designing is an approach by which distinct chemical modifications in defined regions of enzymes can be introduced. It is not just limited to those analogs that are bound as active cocatalysts to the proteins. Methods have been developed that describe the binding properties of inactive analogs, helping to elucidate which groups are involved in binding and the specific influence of molecular parameters on the binding mechanism of the cofactors.

[1] A. Schellenberger, G. Oske, and W. Rödel, *Hoppe-Seyler's Z. Physiol. Chem.* **329,** 149 (1962).
[2] A. Schellenberger, *Angew. Chem. Int. Ed.* **6,** 1024 (1967).

TABLE I
THIAMIN DIPHOSPHATE AS COENZYME OF CATION-TRANSFER REACTIONS OF CARBONYL METABOLITES

| Enzyme[a] | R—CO—$\mid$X | Y | R—CO—$\mid$Y | X |
|---|---|---|---|---|
| PDC | $CH_3$—CO—$\mid CO_2^-$ | $H^+$ | $CH_3$—CO—$\mid$H | $CO_2$ |
| TK | $CH_2OH$—CO—$\mid$CHOH—$R^1$ | $R^2$—$\overset{\delta+}{C}HO$ | $CH_2OH$—CO—$\mid$CHOH—$R^2$ | $R^1$—CHO |
| ALS | $CH_3$—CO—$\mid CO_2^-$ | $CH_3$—$\overset{\delta+}{C}O$—$CO_2^-$ | $CH_3$—CO—$\mid\overset{\overset{\textstyle OH}{\mid}}{\underset{\underset{\textstyle CH_3}{\mid}}{C}}$—$CO_2^-$ | $CO_2$ |
| PDH | $CH_3$—CO—$\mid CO_2^-$ | $\overset{\delta+}{S}$—S | $CH_3$—CO—$\mid$S    S$^-$ | $CO_2$ |

[a] PDC, Pyruvate decarboxylase; TK, transketolase; ALS, acetolactate synthase; PDH, pyruvate dehydrogenase complex.

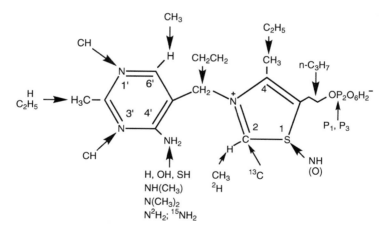

Exchange of $Mg^{2+}$ :  $Mn^{2+}$, $Ca^{2+}$, $Co^{2+}$, $Ni^{2+}$, $Zn^{2+}$, $Cd^{2+}$

FIG. 2. Thiamin diphosphate and modified areas.

# TABLE II
## THIAMIN DIPHOSPHATE ANALOGS AS COFACTORS OF THIAMIN DIPHOSPHATE ENZYMES

| Compound | Synthesis (Ref.)[c] | PDC[a] $K_D$ ($K_I$) (M) | PDC Activity Percentage | PDC Ref. | TK[b] $K_D$ ($K_I$) (M) | TK Activity (%) Hydroxypyruvate + ferricyanide | TK Activity (%) X-5-P + R-5-P[f] Percentage | TK Ref. | PDH[c,d] $K_D$ ($K_I$) (M) | PDH Activity Percentage | PDH Ref. | POX[d] $K_D$ ($K_I$) (M) | POX Activity Percentage | POX Ref. |
|---|---|---|---|---|---|---|---|---|---|---|---|---|---|---|
| ThDP | 1 | Irreversible | 100 | | $3 \times 10^{-8}$ ($K_m$); $5 \times 10^{-7}$ ($K_m$); $9.7 \times 10^{-7}$ ($K_D$) | 100 | 100 | 16 | $1 \times 10^{-7}$; $1 \times 10^{-6}$; $1.7 \times 10^{-5}$; $1.4 \times 10^{-6}$ | 100[c]; 100[d] | 20; 21 | $6 \times 10^{-6}$ | 100 | 27 |
| **ThDP analogs** | | | | | | | | | | | | | | |
| 4'-Hydroxy- | 2 | Irreversible | 0 | 2 | | | | | $2 \times 10^{-8}$; $1.5 \times 10^{-6}$; $1.0 \times 10^{-6}$[h] | 0[c]; 0[c] | 20, 22; 21 | | | |
| 4'-Mercapto- | 2 | | 0 | 2 | | | | | | | | | | |
| 4'-Methylamino- | 2 | | 0 | 2 | $1.4 \times 10^{-5}$ | 215 | 0 | 17 | $4.1 \times 10^{-5}$ | 0[c] | 23, 24 | | | |
| 4'-Dimethylamino- | 2 | | 0 | 2 | $3 \times 10^{-6}$ ($K_m$); $1 \times 10^{-5}$ ($K_m$) | 0 | 0 | 16 | $8.5 \times 10^{-5}$ | 0[c] | 23, 24 | | | |
| 4'-Deamino- | 3 | Reversible | 0 | 3 | | 0 | 0 | 10, 17 | $2.3 \times 10^{-6}$; $2.2 \times 10^{-6}$[h] | 0[c] | 21 | | | |
| 4'-$^{15}$N-amino- | 4 | Irreversible | 100 | 4 | | | | | | | | | | |
| 2'-Ethyl- | 5 | Irreversible | 48 | 5 | $5.6 \times 10^{-6}$ | 220 | 65 | 17 | $6 \times 10^{-7}$ | 125[c] | 23 | | | |
| 2'-Demethyl- | 3 | Reversible | 0 | 3 | | 20 | 0 | 17 | $2 \times 10^{-5}$ | 34[c] | 23 | | | |
| 6'-Methyl- | 6 | Reversible | 0 | 6 | $1.4 \times 10^{-5}$ | 131 | 50 | 17 | | 0[c] | 22, 23 | | | |
| 6'-Methyl-4'-hydroxy- | 5 | Reversible | 0 | 5 | | | | | | | | | | |
| 6'-Methyl-4-demethyl- | 5 | Irreversible | 22 | 5 | $1.6 \times 10^{-5}$ | 24 | 0 | 17 | $6 \times 10^{-6}$ | 43[c] | 22, 23 | | | |
| 5'-Ethylene- | 7 | Reversible | 0 | 7 | | | | | | | | | | |
| N'-Pyridyl- | 8 | Reversible | 65 | 10 | $8 \times 10^{-8}$ ($K_m$); $4 \times 10^{-7}$ ($K_m$) | 100 | 100 | 10, 16 | $9.1 \times 10^{-5}$; $7.3 \times 10^{-6}$[h] | 70[d] | 21 | | | |

| $N^3$-Pyridyl- | 9 | | 0 | 10 | $1.3 \times 10^{-9}$ ($K_m$) | 0 | 0 | 0 | |
|---|---|---|---|---|---|---|---|---|---|
| | | | | | $1.3 \times 10^{-9}$ ($K_m$) | | | | |
| 1-Aza-(Imidazolium-ThDP) | 11 | Irreversible | 0 | 11 | | $2.8 \times 10^{-6}$ $>10^{-4}$ $>10^{-4h}$ | 10, 16 | $0^c$ $0^f$ | 25 21 |
| 1-Methylaza-(1-methyl-imidazolium-ThDP) | 3 | Irreversible | 0 | 3 | | | | | |
| 2-Methyl- | 5 | Reversible | 0 | 5 | | | | | |
| 4-Ethyl- | 5 | Irreversible | 33 | 5 | | | | | |
| 5-Demethyl- | 6 | Reversible | 24 | 6 | $2.6 \times 10^{-5}$ | | | 185 | $0^c$ 23 |
| 5-Hydroxypropyl- | 5 | Reversible | 0 | 5 | | | 7 18 | | |
| 2-(1-Hydroxyethyl)- | 12 | Reversible | 0 | 14 | | | | | |
| 2-Oxothiazolidine- | 13 | $g$ | | | $2.6 \times 10^{-8}$ | $\sim 10^{-6}$ | 0 19 | $\sim 90^c$ | 26 |
| 2-Thiothiazolidine- | 13 | $g$ | | | | | | $<2 \times 10^{-7}$ $2 \times 10^{-8}$ | 0 28 0 29 |

$^a$ From *Saccharomyces cerevisiae*. This PDC binds ThDP irreversibly. Binding of ThDP analogs was proofed by Sephadex filtration.

$^b$ From bakers' yeast. This TK has two different active centers of equal catalytic activity.

$^c$ From pigeon breast muscle. The $E_1$ component of this PDH complex has two catalytic centers.

$^d$ From *E. coli*.

$^e$ Key to references: (1) 4'-Hydroxythiamin: H. N. Rydon, *Biochem. J.* **48**, 383 (1954) (phosphorylation and separation according to 4'-deaminothiamin); (2) A. Schellenberger and K. Winter, *Hoppe-Seyler's Z. Physiol. Chem.* **344**, 16 (1966); (3) H. Neef, K.-D. Kohnert, G. Hübner, and A. Schellenberger, *Acta Biol. Med. Germ.* **31**, 525 (1973) (4) G. Hübner, H. Neef, G. Fischer, and A. Schellenberger, *Z. Chem.* **15**, 221 (1975); (5) A. Schellenberger, K. Winter, G. Hübner, R. Schwaiberger, D. Helbig, S. Schumacher, R. Thieme, G. Bouillon, and K.-P. Rädler, *Hoppe-Seyler's Z. Physiol. Chem.* **346**, 123 (1966); (6) A. Schellenberger, I. Heinroth, and G. Hübner, *Hoppe-Seyler's Z. Physiol. Chem.* **348**, 506 (1967); (7) A. Schellenberger, H. Hanke, and G. Hübner, *Hoppe-Seyler's Z. Physiol. Chem.* **349**, 517 (1968); (8) H. Neef, R. Golbik, B. Fahlbusch, and A. Schellenberger, *Liebigs Ann. Chem.* **1990**, 913; (9) A. Schellenberger, K. Wendler, P. Creutzburg, and G. Hübner, *Hoppe-Seyler's Z. Physiol. Chem.* **348**, 501 (1967); (10) R. Golbik, H. Neef, G. Hübner, St. König, B. Seliger, L. Meshalkina, G. A. Kochetov, and A. Schellenberger, *Bioorg. Chem.* **19**, 10 (191); (11) A. Schellenberger, H. Thieme, and G. Hübner, *Z. Chem.* **9**, 62 (1969); (12) H. Holzer, U. Goedde, and J. Ullrich, *Biochem. Biophys. Res. Commun.* **5**, 447 (1961); (13) J. A. Gutowski and G. E. Lienhard, *J. Biol. Chem.* **251**, 2863 (1976); (14) A. Schellenberger, V. Müller, K. Winter, and G. Hübner, *Hoppe-Seyler's Z. Physiol. Chem.* **344**, 244 (1966); (15) R. Kluger, G. Gish, and G. Kauffman, *J. Biol. Chem.* **259**, 8960 (1984); (16) L. E. Meshalkina, R. Golbik, R. A. Usmanov, H. Neef, A. Schellenberger, and G. A. Kochetov, *Dokl. AN USSR* **307**, 486 (1989) (in Russian); (17) R. A. Usmanov, H. Neef, M. G. Pustynnikov, H. Neef, R. A. Usmanov, A. Schellenberger, and G. A. Kochetov, *Biokhimiya* **51**, 1003 (1986) (in Russian); (18) M. G. Pustynnikov, H. Neef, R. A. Usmanov, A. Schellenberger, and G. A. Kochetov, *Biochem. Int.* **10**, 479 (1985); (19) D. S. Shreve, M. P. Holloway, J. C. Haggerty, III, and H. Z. Sable, *J. Biol. Chem.* **258**, 12405 (1983); (20) L.S Khailova, L. G. Korochkina, and S. E. Severin, *Ann. N.Y. Acad. Sci.* **573**, 36 (1989); (21) J. Hennig, G. Kern, H. Neef, H. Bisswanger, and A. Schellenberger, IVth Joint Symposium of Biochem. Soc. of the USSR and the GDR, Kiev, 1977, Abstracts, p. 24; (24) S. E. Severin, L. S. Khailova, H. Neef, G. Hübner, and A. Schellenberger, *Ukr. Biokhim. Zh.* **48**, 503 (1976) (in Russian); (25) L. S. Khailova, unpublished (1996); (26) G. Hübner, H. Neef, A. Schellenberger, R. Bernhardt, and L. S. Khailova, *FEBS Lett.* **86**, 6 (1978); (27) T. A. O'Brien, R. Blake II, and R. B. Gennis, *Biochemistry* **16**, 3105 (1977); (28) T. A. O'Brien and R. B. Gennis, *J. Biol. Chem.* **255**, 3302 (1980); (29) L. A. Shaw-Goldstein, R. B. Gennis, and C. Walsh, *Biochemistry* **17**, 5605 (1978).

$^f$ X-5-P, Xylulose 5-phosphate; R-5-P, ribose 5-phosphate.

$^g$ From Ref. 15. In case of wheat germ PDC this compound is not a transition state analog.

$^h$ In the presence of pyruvate.

## Cofactor Analogs with Thiamin Diphosphate Enzymes

### Modifications of Thiamin Diphosphate

Synthesis of ThDP analogs has sometimes proven relatively easy (e.g., 4'-hydroxy-ThDP[3]) but in other cases is extremely difficult (e.g., 4'-deamino-ThDP[4]). In principle the selective exchange of protein side chains via genetic engineering as the corresponding countertest will be the method of choice if data regarding the protein structure are available. However, chemical modifications of cofactors allow the intervention into distinct and limited areas, which are not restricted by genetic mechanisms. Like complementation and control of genetic information, cofactor designing has proved therefore of inestimable value.

Presently it is only PDC listed in Table II where all analogs have been tested. In the case of TK and PDH, most data refer to the AP (aminopyrimidine component of ThDP) moiety. Within the last 20 years, the techniques for analyzing and interpreting the experimental data have evolved considerably. Therefore earlier data are discussed in this chapter only to the extent that allows reliable interpretation.

Information obtained by modification of the ThDP structure can contribute to the understanding of the catalytic mechanism only on the condition that, contrary to the results of genetic engineering techniques, a correct fixation of the coenzyme within the intact active center is established. Proving this situation (in the absence of activity!) is difficult and needs careful analysis by suitable and sensitive optical (e.g., fluorescence) signals or by comparison of structural data from X-ray analysis (see Methods in Cofactor Analysis). Another technique that reveals the functional role of structural elements utilizes isotope exchange. In this case the effect on the binding mechanism is excluded and the measured rate constants provide reliable evidence of the participation of the modified area in the catalytic mechanism.

In the aminopyrimidine part of ThDP, all nitrogens ($N^1$, $N^3$, and the amino group) have been eliminated or replaced by other groups. Moreover, the substituents in the 2'-, 5'-, and 6'-positions, respectively, have been modified as shown in Fig. 2.

The thiazolium moiety as the real center of catalytic activity has also been studied extensively. Sulfur was substituted for nitrogen and all the other positions (2, 4, and 5) were modified as much as the experimental

---

[3] S. Mizuhara, R. Tamura, and H. Arata, *Proc. Jpn. Acad.* **27,** 302 (1951); S. Mizuhara and P. Handler, *J. Am. Chem. Soc.* **76,** 571 (1954).

[4] A. Schellenberger, W. Rödel, and H. Rödel, *Hoppe-Seyler's Z. Physiol. Chem.* **339,** 122 (1964).

conditions allowed. In some cases synthesis of the thiamin analog was successful, but the conversion into the corresponding pyrophosphate failed on account of the increased lability of the heterocycle (e.g., oxazolium analog).

## Modification of Metal Ion

It has proved much easier in the case of PDC, TK, and PDH to substitute the metal ion ($Mg^{2+}$, $Ca^{2+}$) as the second cofactor. As a result of these experiments, it could be shown that generally divalent ions (with the exception of SH reagents such as $Cu^{2+}$ or $Hg^{2+}$) can substitute for the native metal ($Mg^{2+}$), but the resulting active enzymes have in some cases a considerably decreased binding stability for the coenzyme. This finding has been confirmed by the crystal structures of TK, PDC, and POX, demonstrating that the metal functions as an anchor between the pyrophosphate and the protein.

## Methods in Cofactor Analysis

### Test of Activity

Active analogs are formed if the essential binding groups and mechanistic structures are preserved. Modification or elimination of the binding groups is reflected in a decrease of the affinity ($K_d$, $K_i$). If this modification influences exclusively the binding mechanism a complete or partial compensation of this effect should be possible by increasing the cofactor concentration. As an example $N^{1'}$-pyridyl-ThDP (missing N-3'; Table II) delivers full activity with the apoenzymes of TK and PDH at slightly increased concentrations. This result demonstrates that the N-3' atom of the pyrimidine part plays no role in the catalytic mechanism of the coenzyme.[5]

If the cofactor modification results exclusively in a restriction of the catalytic mechanism, these analogs have the same binding constants ($K_d$, $K_i$) as the native coenzyme, but show reduced catalytic properties ($k_{cat}$). 4'-Hydroxy-ThDP (missing amino group, Table II), a typical example, binds to the apoproteins of PDC, TK, and PDH with the same affinity as the native cofactor, but the complexes are inactive as a result of the missing (essential) amino group. Often both effects are observed simultaneously.

---

[5] R. Golbick, H. Neef, G. Hübner, S. König, B. Seliger, L. Meshalkina, G. A. Kochetov, and A. Schellenberger, *Bioorg. Chem.* **19**, 10 (1991).

If the coenzyme binds reversibly to the active site (as in the case of PDH[6] or TK[7]), the affinity of inactive analogs can be measured in competition with the native coenzyme. In Table II most of the $K_i$ values were estimated in this way. If the native coenzyme binds irreversibly to the protein, as in the case of PDC[2] or POX,[8] estimations of the binding properties of ThDP analogs are more difficult. In these cases, a qualitative estimation of the binding capacity was possible by measuring the reduced incorporation rate of ThDP in the presence of the analogs. Table III shows the influence of pyrophosphate and thiazolium pyrophosphates on the ThDP incorporation rate.[9] In agreement with X-ray studies the essential part of the coenzyme in the cofactor-binding mechanism is the pyrophosphate–metal anchor. Therefore, no influence on the enzyme recombination is observed, if this part is missing (thiamin, Table III).

In the case of PDH (*Escherichia coli*), another interesting effect was found when the coenzyme affinity was measured in the absence and presence of the substrate: The affinity of ThDP (and also of other active ThDP analogs) increases considerably (Table II) in the presence of pyruvate.[10] This result suggests that intermediates of the catalytic mechanism show a higher affinity for the apoenzyme than the coenzyme itself. In some cases both the coenzyme and the (inactive) analog bind irreversibly to the active site. This must be assumed when the amount of active enzyme depends on the sequence of the addition of active and inactive cofactors to the protein.

*Binding Constants by Spectroscopic Methods*

Binding of ThDP and analogs results generally in modifications of the circular dichroism (CD; chirality) or fluorescent properties (tryptophan quenching) of the protein. Binding constants are usually measured in the case of PDC[11] and POX[12] by using a CD peak at 263 nm (maximum). In the case of TK, a CD peak at 280 nm has proved to be the method of choice.[13] With PDH the observed modifications of the CD spectrum are too small. In this case a quenching of the inner fluorescence as the result

[6] F. Horn and H. Bisswanger, *J. Biol. Chem.* **258,** 6912 (1983).

[7] P. C. Heinrich, H. Steffen, P. Janser, and O. Wiss, *Eur. J. Biochem.* **30,** 553 (1972).

[8] B. Risse, G. Stempfer, G. Schuhmacher, R. Rudolph, and R. Jaenicke, *Protein Sci.* **1,** 1699 (1992).

[9] S. Eppendorfer, S. König, R. Golbik, H. Neef, K. Lehle, R. Jaenicke, A. Schellenberger, and G. Hübner, *Hoppe-Seyler's Z. Physiol. Chem.* **374,** 1129 (1993).

[10] J. Hennig, G. Kern, H. Neef, H. Bisswanger, and G. Hübner, in preparation (1997).

[11] J. Ullrich and A. Wollmer, *Hoppe-Seyler's Z. Physiol. Chem.* **352,** 1635 (1971).

[12] D. Proske, Diploma work. University of Halle, Halle, Germany, 1995.

[13] G. A. Kochetov and R. A. Usmanov, *FEBS Lett.* **9,** 265 (1970).

TABLE III
INFLUENCE OF PYROPHOSPHATE ANCHOR ON COENZYME-BINDING RATE OF PYRUVATE
DECARBOXYLASE[a]

| ThDP analog | Activity (%) | Affinity | Rate constant of ThDP incorporation $(min^{-1})$ |
|---|---|---|---|
| (structure: pyrimidine–$CH_2$–thiazolium with $OP_2O_6H_2^-$, $NH_2$, $CH_3$, $H_3C$) | 100 | Full | 7.4 |

Influence on ThDP incorporation rate after addition of:

| ThDP analog | Activity (%) | Affinity | Rate constant of ThDP incorporation $(min^{-1})$ |
|---|---|---|---|
| (structure: pyrimidine–$CH_2$–thiazolium with $OH$, $NH_2$, $CH_3$, $H_3C$) | 0 | No | 7.4 (no) |
| $HOP_2O_6H_2^-$ | 0 | Comp. | 5.4 |
| (structure: $H-N^+$ thiazolium with $CH_3$ and $OP_2O_6H_2^-$, $H$, $S$) | 0 | Comp. | 4.4 |
| (structure: $H_3C-N^+$ thiazolium with $CH_3$ and $OP_2O_6H_2^-$, $H$, $S$) | 0 | Comp. | 3.2 |
| (structure: phenyl–$CH_2-N^+$ thiazolium with $CH_3$ and $OP_2O_6H_2^-$, $H$, $S$) | 0 | Comp. | 2.0 |
| (structure: pyrimidinone–$CH_2$–thiazolium with $P_2O_6H_2^-$, $CH_3$, $H_3C$, $O$, $N$–$H$) | 0 | Comp. (high-affinity range) | (0.04) |

[a] *Saccharomyces cerevisiae.* Recombination measured at pH 6.0.

of coenzyme binding has proved useful (after correction of the inner-filter effect of the cofactor).[14]

*Proving Cofactor Binding by X-Ray Analysis*

Thiamin diphosphate shows a high degree of conformational flexibility as a result of the single bonds around the methylene bridge. Different

[14] F. Horn and H. Bisswanger, *J. Biol. Chem.* **258**, 6912 (1983).

conformers of the coenzyme have been found[15] and calculated,[16] therefore, depending on the specific surroundings (so-called V-, S-, and F-conformations[15]), that differ by only small enthalpy differences. For this reason the active-site conformation of ThDP has played a central role in many functional discussions.

This conformation problem could be solved when the first X-ray structures of ThDP-dependent enzymes were published. As proposed much earlier from experiments with the 6'-methyl-ThDP analog,[2] the V-conformation (characterized by a small distance of 3.5 Å between the AP amino group and the $C^2$–H bond of the thiazolium structure) was found to be identical in all crystal structures and must be considered as a strongly conserved structural and functional element of all ThDP enzymes.

The most convincing proof of the correct binding of a coenzyme analog would also result from the X-ray analysis of cofactor-modified enzymes. Up to now it has been possible only in the case of TK to crystallize the enzyme after incorporation of analogs modified in the structure of the AP component. The X-ray data of these enzyme analogs confirm that ThDP, as well as analogs such as $N^1$- and $N^3$-pyridyl-ThDP or 6'-methyl-ThDP (Table II), bind precisely as V-conformers within the active site of TK.[17]

## Results

### Thiazolium

*Chemical Modifications.* The thiazolium system allows modifications without a complete loss of activity only in position 4. Elimination of the 4-methyl group (4-H ThDP) reduces considerably the cofactor affinity and activity (PDC and TK), while the 4-ethyl analog produces 32% activity in the case of PDC (Table II).

Modifications of the 5-ethylene "spacer" (*n*-propylene) as well as an enlargement of the methylene bridge (ethylene) have generally produced inactive analogs.

An interesting result was obtained with the (inactive) 2-methyl analog (2-methyl-ThDP): a complete lack of affinity for the active site of PDC was detected.

*Isotope Modifications.* Experiments with $^{31}$P nuclear magnetic resonance (NMR) have confirmed an extremely restricted mobility of the pyrophos-

[15] J. Pletcher and M. Sax, *J. Am. Chem. Soc.* **94,** 3998 (1972).
[16] R. Friedemann and W. Uslar, *J. Mol. Struct.* (*Theor. Chem.*) **181,** 401 (1988).
[17] S. König, A. Schellenberger, H. Neef, and G. Schneider, *J. Biol. Chem.* **269,** 10879 (1994).

phate group.[18] This finding agrees with the anchor function found by X-ray analysis and with the pyrophosphate exchange experiments, shown in Table III.

Labeling the $C^2$-position with $^{13}C$ ($[^{13}C^2]$ThDP) allowed the finding that deprotonation of the $C^2$–H bond (yielding the ylid form of ThDP as a permanently existing structure) by interaction with the apoprotein of PDC can be excluded.[19] As demonstrated in Fig. 3, no change in the chemical shift is observed when free and enzyme-bound ThDP are compared. Moreover it could be shown that the addition of pyruvamide as an artificial activator of PDC causes no alteration of the $^{13}C$ peak.

The question of how the slow $C^2$–H deprotonation of free ThDP (Table IV) fits into the general rate profile of ThDP enzymes (the deprotonation rate of the free coenzyme must be increased by four orders of magnitude in the enzyme reaction) could be considered by measuring the influence of the protein surrounding of the active site and of the AP component on the H/D exchange rate of the $C^2$–H bond.[19] This rate enhancement as a key reaction of ThDP enzymes could be followed in $^2H_2O$ by analyzing the $C^2$–H deprotonation rate within the active site of PDC (*Saccharomyces cerevisiae*) by $^1H$ NMR. $^2H$-incorporation in the $C^2$-position (very slow with free coenzyme) was stopped after different time intervals by denaturation with DCl/trichloroacetate and centrifugation of the denatured protein. As shown in Table IV, the protein architecture together with the AP conjunction (Fig. 5) is responsible for the high increase in H/D exchange observed in the enzyme reaction.

## Aminopyrimidine

Some positions of the aminopyrimidine (AP) component can be modified without loss of activity, but the results differ with respect to the tested enzymes more than in the case of the thiazolium ring.

Elimination of the 2′-methyl group (2′-H-ThDP) yields low, but modest enlargement (2′-ethyl-ThDP) high activities with PDC and TK.

The functional significance of the two ring nitrogens was analyzed via the two corresponding pyridyl analogs of ThDP. Substitution of the 3′-N atom by a CH group ($N^1$-pyridyl-ThDP) did not significantly affect the catalytic efficiency as coenzyme of PDC, TK, or PDH. A decrease of affinity was measured in the case of PDC (loss of irreversible binding) and PDH.[5] All of these results are in contrast to the analog with a missing N-1′ atom

---

[18] G. Hübner, S. König, and K. D. Schnackerz, *FEBS Lett.* **314,** 101 (1992).

[19] D. Kern, G. Kern, H. Neef, M. Killenberg-Jabs, K. Tittmann, C. Wikner, G. Schneider, and G. Hübner, *Science* **275,** 67 (1997).

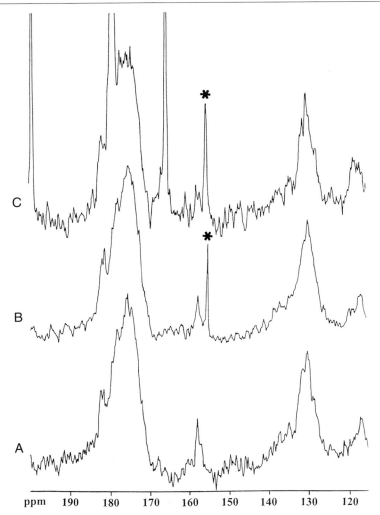

FIG. 3. Proof of the protonation state of the C-2 atom of ThDP. Expansion of $^1$H-decoupled $^{13}$C NMR spectra of unlabeled (A) and [$^{13}$C$^2$]ThDP-labeled PDC (*S. cerevisiae*) in the absence (B) and presence (C) of 100 m$M$ pyruvamide as artificial PDC activator. The signal of [$^{13}$C$^2$]ThDP is marked with an asterisk and shows the same chemical shift as the protonated C-2 atom of ThDP. (The strong signals in spectrum C are from pyruvamide in its dehydrated and hydrated states.) [Reprinted from *Science* **275,** 68 (1997).]

TABLE IV

PSEUDO-FIRST-ORDER RATE CONSTANTS OF $C^2$-
DEPROTONATION OF FREE AND ENZYME-BOUND
THIAMIN DIPHOSPHATE[a]

| Assay | $k$ (sec$^{-1}$) |
|---|---|
| Free ThDP | $9.5 \times 10^{-4}$ |
| Pyruvamide-activated PDC | $>6 \times 10^{+2}$ |
| Free DAThDP | $1.3 \times 10^{-3}$ |
| Pyruvamide-activated PDC with DAThDP as coenzyme analog | $4 \times 10^{-5}$ |

[a] At pH 6.0 and 4°. Experiments with 4'-deamino-ThDP (DAThDP) show the essential function of the amino group in ThDP enzymes.

($N^{3'}$-pyridyl-ThDP), which was found to be completely inactive in all of the tested ThDP enzymes.[5]

If measured with PDC, methylation of the 6'-position (6'-methyl-ThDP) also generates an inactive enzyme. To understand this finding we considered (by molecular modeling) different conformational states of ThDP. In one special case (later described as the V-conformation[15]) a steric twisting of the two ring planes was detected as a result of a steric contact of the two methyl groups in the 6'- and 4-positions. A proof of this assumption by elimination of the 4-methyl group (6'-methyl-4-H-ThDP) has produced an analog with 22% activity.[2] The steric constraint, derived from this experiment, is PDC specific. In the case of TK, 6'-methyl-ThDP was found active and 6'-methyl-4-H-ThDP inactive (possibly a result of the missing 4-methyl group).[20]

The 4'-amino group has been found to be very sensitive with respect to any modifications. Substitutions (mono- and dimethylamino) as well as modifications (hydroxy, mercapto) or elimination (4'-H-ThDP) produced inactive analogs in the case of PDC (sometimes with high affinity to the apoenzyme, e.g., 4'-hydroxy-ThDP). The only exception has been found with TK, where the 4'-methylamino analog shows 20% activity with hydroxypyruvate and glycolaldehydes as substrates.[20] These differences observed in the functional properties of special coenzyme analogs (such as 6'-methyl-ThDP) with different apoenzymes show that the (apparent) agreement of the X-ray structures (V-conformations) of different ThDP enzymes allows, nevertheless, functional differences due to the specific architecture and mobility of the different active sites.

[20] R. A. Usmanov, H. Neef, M. G. Pustynnikov, A. Schellenberger, and G. A. Kochetov, *Biochem. Int.* **10,** 479 (1985).

The essential function of the amino group could also be demonstrated by measuring a (direct) $^{15}$N isotope effect with the $4'$-$^{15}$NH$_2$ analog.[21] This finding has demonstrated, once more, that the amino group (together with the N-$1'$ atom) takes part in the rate-limiting step of the PDC reaction.


Some Mechanistic Conclusions

Considering the fact that Langenbeck's first studies on amino compounds as model catalysts of cocarboxylase (the ThDP coenzyme of PDC) started in 1930,[22] it seems remarkable that the ThDP enzymes still present a considerable number of unsolved problems.

As we have shown in Fig. 3, the ylid structure as a "permanently dissociated state" within the active site can be excluded. Moreover, the aminopyrimidine part obviously plays a crucial role in the building of the coenzyme–substrate complex by increasing the dissociation rate of the $C^2$–H bond by four orders of magnitude.[19] This demonstrates that the ylid formation is part of a process, which itself must be strongly influenced by the interaction of the N-1 atom with a conserved glutamate side chain of the protein.

Another open problem refers to the mechanism by which the essential $N^{1'}$-$4'$-amino "conjunction" influences the catalytic reaction of the thiazolium moiety. Only this heterocycle (and not the imidazolium structure; see above) has a dissociable $C^2$–H bond (p$K_a$ 18 as a result of the low electron level), which—as in the case of cyanide—is necessary to catalyze the X-elimination step. Thiamin and cyanide can simulate (at pH values $>8.5$) the catalytic function of ThDP enzymes. This suggests the chemically and sterically (V-conformation) expensive ThDP structure to be designed by nature mainly for reducing the functional pH value to the physiologic level between pH 6 and 7.

All X-ray structures that have been solved in the case of PDC,[23] TK,[24] and POX[25] have confirmed the results obtained with the two pyridine analogs of ThDP.[5] Without the hydrogen bond formed between N-$1'$ and a conserved glutamate side chain, no catalysis is observed even if a stable and correct fixation and conformation of the analogs have been proved by X-ray analysis.[17] This special aspect is missing up to now in most of the

[21] G. Hübner, H. Neef, G. Fischer, and A. Schellenberger, *Z. Chem.* **374,** 221 (1975).

[22] W. Langenbeck and R. Hutschenreuter, *Z. Anorg. Allg. Chem.* **188,** 1 (1930).

[23] F. Dyda, W. Furey, S. Swaminathan, M. Sax, B. Farrenkopf, and F. Jordan, *Biochemistry* **32,** 6165 (1993).

[24] Y. Linqvist, G. Schneider, U. Ermler, and M. Sundström, *EMBO J.* **11,** 2373 (1993).

[25] Y. A. Muller and G. A. Schultz, *Science* **259,** 965 (1993).

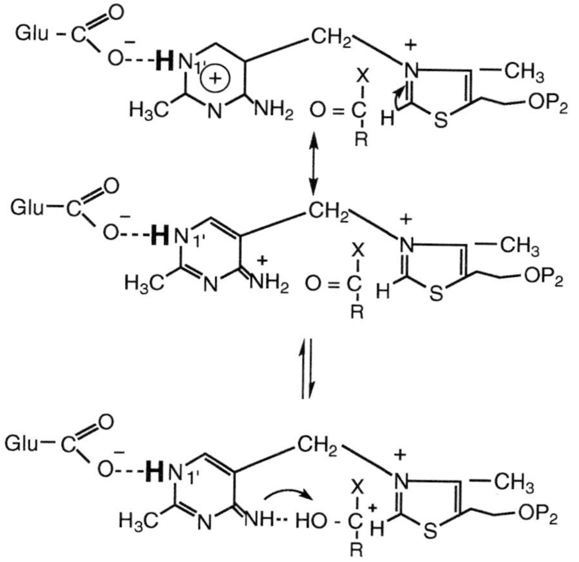

FIG. 4. The $N^{1'}$-4'-amino conjunction increases as protonator of the substrate carbonyl group the electrophilic substitution in the $C^2$-position.

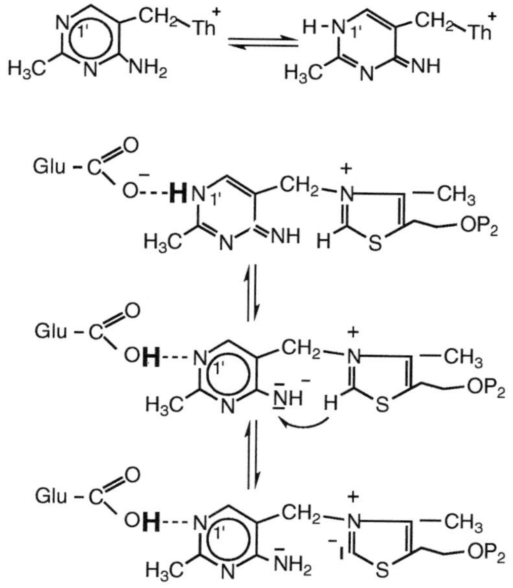

FIG. 5. The $N^{1'}$-4'-amino conjunction increases as deprotonator of the $C^2$-H bond with the substrate addition at the deprotonated C-2 atom.

usual attempts to explain the catalytic role of ThDP. Two models could explain the function of the $N^1$-amino conjunction.

The first model starts from the aminopyrimidine structure of ThDP. $N^{1'}$-protonation would produce a considerable decrease in the electronic level of the pyrimidine ring and consequently a strong acidification of the amino group (Fig. 4). In the catalytic mechanism, this protonation power could be used to increase the reactivity of the substrate carbonyl group. The resulting carbonium ion would function consequently as an electrophile for the C-2 atom as shown in Fig. 4.

The second attempt to understand the influence of the $N^{1'}$–protein interaction starts with the assumption that the aminopyrimidine structure exists within the active site as the tautomeric imino structure (Fig. 5). In this case a deprotonation of the $HN^{1'}$ structure by the (conserved) glutamate residue would increase the basic character of the 4'-N atom considerably. This basic center in the direct vicinity of the $C^2$–H bond (3.5 Å) could explain not only the cocatalytic function of the 4'-amino group, but also the direct influence on dissociation rate as found by the H/D exchange experiments (Table IV).

# [16] Enzymatic Preparation of Derivatives of Thiamin: O-β-Galactosylthiamin and O-α-Glucosylthiamin

*By* Yukio Suzuki and Kei Uchida

Glycosylated thiamin had previously not been synthesized chemically or known to be biologically synthesized. Two novel derivatives of thiamin, identified as 5'-O-(β-D-galactopyranosyl)thiamin (thiamin β-galactoside) and 5'-O-(α-D-glucopyranosyl)thiamin (thiamin α-glucoside), have now been found to be formed from o-nitrophenyl-β-D-galactopyranoside and thiamin in 33% (v/v) CH$_3$CN by *Aspergillus oryzae* β-galactosidase,[1] and also from dextrin and thiamin by successive actions of *Bacillus stearothermophilus* cyclomaltodextrin glucanotransferase (CGTase) and *Rhizopus* glucoamylase,[2] respectively. These two derivatives have been isolated, identified, and characterized, as described in this chapter.

---

[1] Y. Suzuki and K. Uchida, *Biosci. Biotech. Biochem.* **58**, 1273 (1994).

[2] Y. Suzuki and K. Uchida, *BioFactors* **5**, 44 (1995). Abstracts presented in the 2nd International Congress on Vitamins and Biofactors in Life Science; Y. Suzuki and K. Uchida, *Biosci. Biotech. Biochem.* **61**, in press (1997).

Thiamin $\beta$-Galactoside

*Preparation*

*Assay.* Thiamin compounds are analyzed by the method of Kimura *et al.*[3] with slight modifications, using high-performance liquid chromatography (HPLC). The mobile phase [0.04 $M$ sodium phosphate buffer (pH 6.0)] is pumped in a tube (inside diameter, 0.23 mm) at a flow rate of 1 ml/min by a Waters Associates (Milford, CT) chromatography pump model 6000 A. A sample of a solution containing thiamin compounds is injected by a Waters injector model U6K onto a column [Asahipack GS-320 (7.6 × 500 mm); Asahi Chemical Industries Ltd., Tokyo, Japan]. Thiamin compounds in the effluent are first measured at a wavelength of 280 nm with a Waters UV absorbance detector model 441, connected to a Shimadzu (Tokyo, Japan) Chromatopac model C-R1B. A 0.01% (w/v) $K_3[Fe(CN)_6]$–15% (w/v) NaOH solution is then introduced at 1 ml/min by a Waters 501 G pump and mixed in a stainless steel mixing coil (0.51 × 1500 mm) with the column effluent to convert thiamin compounds into fluorescent compounds. The fluorescent compounds are measured with a Shimadzu fluorescence HPLC monitor model RF-530 (excitation, 375 nm; emission, 435 nm), connected to a Shimadzu Chromatopac model C-R1B and recorded graphically. The peak–area method is used for measurement.

*Paper Chromatography.* A suitable amount of the fluid is applied as a band on Toyo (Tokyo, Japan) filter paper No. 50 (40 × 40 cm), and developed by ascent in a solvent system of *n*-butanol–pyridine–water (6 : 4 : 4, v/v/v) (solvent A). After drying, the chromatogram is sprayed with a mixture of ethanol and an alkaline $K_3[Fe(CN)_6]$ solution (1 : 1, v/v). The alkaline solution contains 200 ml of 1% (w/v) $K_3[Fe(CN)_6]$ aqueous solution, 26.4 ml of 30% (w/v) NaOH, and 13.6 ml of distilled water. Thiamin and its compounds on a chromatogram appear as blue fluorescent bands under an ultraviolet lamp (3650-Å filter).

*Reagents*

$\beta$-Galactosidase from *A. oryzae* (Kohjin Co., Ltd., Tokyo, Japan)
$\beta$-Galactosyl donor: *o*-Nitrophenyl-$\beta$-D-galactopyranoside
Acceptor: Thiamin hydrochloride
Buffer: Sodium phosphate buffer, pH 5.0

*Enzymatic Reactions.* A reaction mixture (1.5 ml, adjusted to pH 5.0) containing 60 mg of *o*-nitrophenyl-$\beta$-D-galactopyranoside as donor, 169 mg of thiamin hydrochloride as acceptor, 0.5 ml of 0.2 $M$ sodium phosphate

---

[3] M. Kimura, T. Fujita, S. Nishida, and Y. Itokawa, *J. Chromatogr.* **188,** 417 (1980).

buffer (pH 5.0), 0.5 ml of $CH_3CN$, $\beta$-galactosidase from *A. oryzae* (390 units), and distilled water to make a total volume of 1.5 ml is incubated at 20° for 48 hr in the dark. After incubation, 1 vol of diethyl ether is added, and the mixture is shaken to redissolve $CH_3CN$ in the diethyl ether layer, and then centrifuged (10,000 $g$, 10 min, 4°). In the water layer, two new thiochrome reaction-positive spots with lower $R_f$ values than that of thiamin are detected in addition to thiamin on a paper chromatogram. These spots are not formed in the reaction mixture without *o*-nitrophenyl-$\beta$-D-galacto-pyranoside and without enzyme.

*Isolation.* The reaction is run on a large scale with 42.2 g of thiamin hydrochloride, 15.0 g of *o*-nitrophenyl-$\beta$-D-galactopyranoside, 125 ml of 0.2 $M$ sodium phosphate buffer (pH 5.0), 125 ml of $CH_3CN$, 97,500 units of *A. oryzae* $\beta$-galactosidase, and distilled water to make a total volume of 375 ml (pH adjusted to 5.0) at 20° for 48 hr in the dark. After being heated in a boiling water bath for 10 min to stop the enzymatic action, the reaction mixture is centrifuged (10,000 $g$, 20 min, 4°). To the supernatant solution is added 1 vol of diethyl ether and the mixture is stirred at 0° for 30 min to redissolve $CH_3CN$ in the diethyl ether layer. The water layer is concentrated at room temperature *in vacuo* to remove diethyl ether, and the concentrate (200 ml) is applied to the first preparative paper chromatog-raphy (PPC) with 400 sheets of Toyo filter paper No. 50 (40 × 40 cm) in solvent A, using 8 paper-developing boxes. Examination of the paper chromatogram, after a spray of alkaline $K_3[Fe(CN)_6]$ solution, reveals two purplish fluorescent bands besides that of thiamin [major and very minor compounds (**I** and **II**)] having lower $R_f$ values than that of thiamin on the paper chromatogram. Each band position corresponding to compounds **I** and **II** on all paper chromatograms, in which all strips at both edges of the chromatogram are made visible with an alkaline $K_3[Fe(CN)_6]$ solution, is cut out, and each band is extracted with acetic acid aqueous solution (pH 4.2). The amounts of compounds **I** and **II** in the extract are analyzed by HPLC (**I**, 1.95 g; **II**, 0.15 g as thiamin hydrochloride; 13.3% yield based on *o*-nitrophenyl-$\beta$-D-galactopyranoside). The extract of the major compound (**I**), after it is concentrated below 30° *in vacuo,* is put on an activated Vitachange (Wako Pure Chemical Industries, Ltd., Osaka, Japan) column (5.6 × 16 cm) at 4°. After the column is washed with water to remove sugars, compound **I** is eluted with a 25% (w/v) KCl aqueous solution at 4°. To the effluent, which contains compound **I** and contaminants such as thiamin and compound **II**, is added ethanol to a final concentration of 75% (v/v) at 0°. The resultant precipitate is removed by filtration. The filtrate is concentrated and lyophilized. The powdered preparation is extracted three times with 85% (v/v) ethanol. After concentration, the alcoholic extract is put on the second PPC with a solvent system of *n*-butanol–acetic

acid–water (2:1:1, v/v/v) (solvent B), to separate compound **I** from contaminants (thiamin and compound **II**). After appropriate sectioning and
elution, the compound **I** solution is concentrated, and passed through a
Toyopearl HW-40S gel column (2.6 × 70 cm), eluted with water, and
lyophilized. The purified powdered preparation of compound **I** is crystallized several times from methanol. The crystalline preparation is dried *in
vacuo* on $P_2O_5$ (yield, 0.77 g). The melting point and fast atom bombardment–mass spectrometry (FAB–MS) data are as follows: mp (decomposition): 143–147°; $[M + H]^+$ ion at 427 (thiamin: $[M + H]^+$ ion at 265).

## Identification and Characterization

$R_f$ values (0.32, 0.33, 0.39) of the isolated compound **I** on PPC in three
solvent systems [solvent A, B, and *n*-propanol–water–1 *M* sodium acetate
buffer (pH 5.0) (7:2:1, v/v/v) (solvent C)] are different from those of
thiamin (0.48, 0.55, 0.58), thiamin monophosphate (0.19 in solvent C), and
thiamin diphosphate (0.06 in solvent C). On enzymatic hydrolysis with
β-galactosidase from *Escherichia coli* or *A. oryzae,* compound **I** is completely hydrolyzed to a sugar and a thiochrome reaction-positive substance,
which has $R_f$ values identical to those of thiamin on paper chromatograms.
The sugar component in the hydrolysate is confirmed as galactose on a
Kieselgel 60 plate (E. Merck, Darmstadt, Germany) developed with a
solvent system of *n*-propanol–2% (v/v) $NH_4OH$ (2:1, v/v), by HPLC using
a Waters carbohydrate analysis column developed with acetonitrile–water
(90:10, v/v), and also by the Waters assay kit of D-galactose using
D-galactose dehydrogenase and NAD. The molar ratio of thiamin and
galactose liberated is around 1:1 from compound **I**. No reducing activity
for compound **I** is found by the method of Nelson.[4] The ultraviolet (UV)
absorption spectra of compound **I** in 0.1 *N* HCl, 0.1 *M* sodium phosphate
buffer (pH 6.8), and 0.1 *N* NaOH solution show the positions of maxima
and minima to be similar to those of thiamin. The infrared (IR) spectrum
of compound **I** indicates absorption peaks at 3700 to 2500 $cm^{-1}$ (alcoholic
hydroxyl group in sugar moiety, O–H), 1655 $cm^{-1}$ ($N–H_3$), 1605 $cm^{-1}$ (C=N
and C=C), 1220 $cm^{-1}$ (CN), and 1100 to 1000 $cm^{-1}$ (sugar hydroxyl group,
C–O). Figure 1 shows the $^1H$ nuclear magnetic resonance (NMR) spectrum
of compound **I** (a Varian spectrometer VXR-500 at 500 MHz). From the
spectrum in $D_2O$, the following signals (ppm) are assigned to the protons
of thiamin based on the data reported[5]: 2.55 (thiazole $–CH_3$, H-9, s, 3H),
2.61 (pyrimidine $–CH_3$, H-12, s, 3H), 3.31 (inner $–CH_2–$ of side chain, H-10,

[4] N. Nelson, *J. Biol. Chem.* **153**, 375 (1944).
[5] H. Z. Sable and J. E. Biaglow, *Proc. Natl. Acad. Sci. U.S.A.* **54**, 808 (1965).

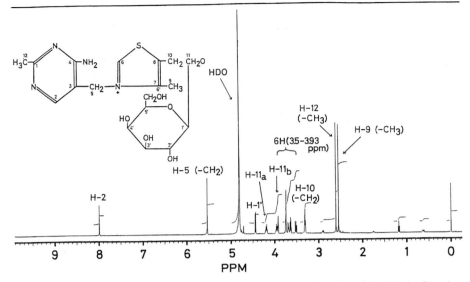

FIG. 1. $^1$H NMR spectrum of compound **I** in D$_2$O. [From Y. Suzuki and K. Uchida, *Biosci. Biotech. Biochem.* **58**(7), 1273 (1994), with permission of the Japan Society for Bioscience, Biotechnology, and Agrochemistry.]

t, 2H), 4.19 (–O–CH$_a$H$_b$)– side chain, H-11$_a$, m, 1H), 3.95 (–O–CH$_a$H$_b$– side chain, H-11$_b$, m, 1H), 5.54 (bridge –CH$_2$–, H-5, s, 2H), and 8.00 (aromatic proton in pyrimidine ring, H-2, s, 1H). Additional signals are as follows: signals of ring protons of the sugar moiety (3.50–3.93 ppm, 6H) and one anomeric proton signal of a β-galactosidic linkage (4.44, 1H, d, $J = 7.9$ Hz). These spectral data suggest that compound **I** is a β-galactoside of thiamin. To confirm the fine structure of compound **I**, carbon-13 chemical shifts of compound **I** in D$_2$O are compared with those of thiamin (Table I). Signal assignments are based on the data reported for methyl-β-D-galacto-pyranoside[6] and thiamin[7] in D$_2$O. Comparing the thiamin carbons of compound **I** with those of the parent thiamin, it is apparent that the galactosylation site in compound **I** is the CH$_2$OH group in the 4-methyl-5-hydroxyethylthiazolium moiety, because the appended C-11 signal is shifted downfield by a sizable 7.80 ppm with a concomitant small upfield shift of C-10 in the hydroxyethyl group and with only smaller effects on other thiamin carbons. Thus, compound **I** is identified as 5′-*O*-(β-D-galactopyranosyl) thiamin.

[6] P. E. Pfeffer, K. M. Valentine, and F. W. Parrish, *J. Am. Chem. Soc.* **101**, 1265 (1979).
[7] R. E. Echols and G. C. Levy, *J. Org. Chem.* **39**, 1321 (1974).

TABLE I
$^{13}$C NMR CHEMICAL SHIFTS OF THIAMIN AND COMPOUNDS **I** AND **I′** IN D$_2$O[a]

| Carbon no. | Thiamin (ppm) | Compound **I**[b] (ppm) | Compound **I′**[b] (ppm) |
|---|---|---|---|
| Thiamin carbon | | | |
| C-1 | 164.38 | 164.86 | 164.19 |
| C-2 | 145.72 | 146.78 | 145.57 |
| C-3 | 107.21 | 107.05 | 106.95 |
| C-4 | 164.13 | 163.91 | 163.89 |
| C-5 | 51.00 | 50.99 | 50.74 |
| C-6 | 155.72 | 155.45 | 155.69 |
| C-7 | 143.95 | 143.94 | 143.72 |
| C-8 | 137.67 | 137.10 | 137.16 |
| C-9 | 12.10 | 12.11 | 11.90 |
| C-10 | 30.30 | 28.04 | 27.59 |
| C-11 | 61.33 | 69.13 | 67.03 |
| C-12 | 22.04 | 22.31 | 21.85 |
| Transferred hexose carbon | | | |
| C-1 | | 103.74 | 99.18 |
| C-2 | | 71.71 | 71.89 |
| C-3 | | 73.75 | 73.88 |
| C-4 | | 69.62 | 70.40 |
| C-5 | | 76.20 | 73.21 |
| C-6 | | 61.97 | 61.46 |

[a] **I**, Thiamin β-galactoside. **I′**, thiamin α-glucoside.
[b] From Y. Suzuki and K. Uchida, *Biosci. Biotech. Biochem.* **58**(7), 1273 (1994), with permission of the Japan Society for Bioscience, Biotechnology, and Agrochemistry.

Commercial thiamin hydrochloride has a thiamin-specific odor and a strong tongue-pricking taste, but thiamin β-galactoside is odorless, scarcely stimulative, and mildly sweet.

## Thiamin α-Glucoside

### Preparations

*Assay.* Thiamin compounds are analyzed by the method of Kimura *et al.*[3] with slight modifications, using high-performance liquid chromatography, as described in the section on thiamin β-galactoside.

*Paper Chromatography.* Paper chromatography is done in a Toyo filter paper No. 50 (40 × 40 cm) with a solvent system of *n*-butanol–acetic acid–water (2:1:1, v/v/v) (solvent A). After drying, thiamin and its compounds on chromatograms are detected as blue fluorescent bands under an ultraviolet lamp (3650-Å filter), by spraying with a mixture of ethanol and alkaline $K_3[Fe(CN)_6]$ solution (1 : 1, v/v). The alkaline solution contains 200 ml of 1% (w/v) $K_3[Fe(CN)_6]$ aqueous solution, 26.4 ml of 30% (w/v) NaOH, and 13.6 ml of distilled water.

*Reagents*

Cyclomaltodextrin glucanotransferase (CGTase) from *B. stearother-mophilus* (Hayashibara Biochemical Laboratories, Inc., Okayama, Japan) and its homogeneous preparation purified by the method of Kitahata and Okada[8]

Glucoamylase from *Rhizopus* sp. (Toyobo Co., Ltd., Osaka, Japan)

$\alpha$-Glucosyl donor: Dextrin (Pine-Dex No. 1, DE 8; Matsutani Chemical Kogyo Co. Ltd., Itami, Japan

Acceptor: Thiamin hydrochloride

Calcium dichloride

Buffer: 0.1 $M$ sodium acetate buffer, pH 5.5

*Enzymatic Reaction.* A reaction mixture (10 ml, adjusted to pH 5.5) containing 0.5, 1.0, or 1.5 g of dextrin, 0.5 g of thiamin hydrochloride, 5 ml of 0.1 $M$ sodium acetate buffer (pH 5.5), 5.5 mg of $CaCl_2$, 680 units of a homogeneously purified preparation of CGTase from *B. stearothermophi-lus,* and distilled water to make a total volume of 10 ml is incubated at 37° for 72 hr in the dark. After incubation, many new thiochrome reaction-positive spots with lower $R_f$ values than that of thiamin are detected besides thiamin on a paper chromatogram. These spots are not formed in the reaction without dextrin and without enzyme. The formation of these new spots from thiamin with 15% (w/v) dextrin is larger than those with 5 or 10% (w/v) dextrin. Furthermore, two thiochrome reaction-positive spots [major and very minor compounds (**I′** and **II′**)] having lower $R_f$ values than that of thiamin on a paper chromatogram are formed when the reaction mixture incubated for 72 hr is digested with 30 units of glucoamylase from *Rhizopus* sp. at 37° for 20 hr. These results show that these spots are transglucosylated derivatives of thiamin.

*Isolation.* A reaction mixture (500 ml, adjusted to pH 5.5) containing 75 g of dextrin, 25 g of thiamin hydrochloride, 250 ml of 0.1 $M$ sodium acetate buffer (pH 5.5), 275 mg of $CaCl_2$, $34 \times 10^3$ units of *B. stearother-mophilus* CGTase, and distilled water to make a total volume of 500 ml is incubated at 37° for 72 hr in the dark. After the pH is adjusted to 4.5, the reaction mixture is heated in a boiling water bath for 10 min in the dark and centrifuged. The supernatant solution is incubated with 1500 units of glucoamylase from *Rhizopus* sp. (Toyobo Co., Ltd.) at 37° for 20 hr in the

[8] S. Kitahata and S. Okada, *J. Jpn. Soc. Starch Sci.* **29,** 7 (1982).

dark. After being heated in a boiling water bath for 10 min, the mixture is applied to the first preparative paper chromatography with 500 sheets of Toyo filter paper No. 50 (40 × 40 cm) in solvent A, using 8 paper-developing boxes. Two purplish fluorescent bands corresponding to major and very minor compounds (I' and II') in addition to that of thiamin are observed on the paper chromatograms after spraying with alkaline $K_3[Fe(CN)_6]$. The band corresponding to compound I' is cut out and eluted with acetic acid aqueous solution, pH 4.2. The amount of compound I' in the effluent is analyzed by HPLC (7.50 g as thiamin hydrochloride). The effluent is concentrated below 30° *in vacuo,* and then lyophilized. The preparation of compound I' dissolved in water is applied to an activated Vitachange column (5.6 × 16 cm) at 4°. After the column is washed with water to remove sugars, compound I' is eluted with a 25% (w/v) KCl aqueous solution at 4°. To the effluent containing compound I' and contaminants (i.e., thiamin, compound II', and a large amount of KCl) is added ethanol to a final concentration of 75% (v/v) at 0°. The resultant precipitate is removed by filtration. The filtrate is concentrated and lyophilized. The powdered preparation is extracted three times with 85% (v/v) ethanol. The alcoholic extract, after concentration, is subjected to the second PPC with solvent A. After appropriate sectioning and elution, the compound I' solution is concentrated and lyophilized. The preparation, after dissolved in 85% (v/v) ethanol, is decolorized with activated carbon, concentrated, and then lyophilized. To the powdered preparation is added a small amount of methanol, and the insoluble materials are removed. The filtrate is concentrated and lyophilized. The purified powdered preparation of compound I' is crystallized several times from *n*-butanol–methanol solution. The crystalline preparation is dried *in vacuo* on $P_2O_5$ (yield, 2.38 g). The melting point and FAB–MS data are as follows: mp (decomposition): 184–186°; $[M + H]^+$ ion at 427 (thiamin: $[M + H]^+$ ion at 265).

## Identification and Characterization

On treatment with pig liver α-glucosidase,[9] compound I' can be completely hydrolyzed to a sugar and a thiochrome reaction-positive substance that has $R_f$ values identical to those of thiamin on paper chromatograms. This does not occur with β-glucosidase from almond. The sugar component in the hydrolysate is confirmed as glucose on a Kieselgel 60 plate developed with a solvent system of *n*-propanol–2% (w/v) $NH_4OH$ (2:1, v/v) and by HPLC using a Waters carbohydrate analysis column developed with $CH_3CN$–water (90:10, v/v). The molar ratio of thiamin and glucose liber-

[9] K. Uchida and Y. Suzuki, *Agric. Biol. Chem.* **38,** 195 (1974).

ated is around $1:1$ in compound $\mathbf{I}'$. No reducing activity for compound $\mathbf{I}'$ is found by the method of Nelson.[4] The UV absorption spectra of compound $\mathbf{I}'$ in 0.1 $N$ HCl, 0.1 $M$ sodium phosphate buffer (pH 6.8), and 0.1 $N$ NaOH solution show the positions of maxima and minima to be similar to those of thiamin. Examination of the $^1$H NMR spectrum of compound $\mathbf{I}'$ (in $D_2O$, ppm) shows the following signals (Fig. 2): 2.57 (thiazole $-CH_3$, H-9, s, 3H), 2.64 (pyrimidine $-CH_3$, H-12, s, 3H), 3.34 (inner $-CH_2-$ of side chain, H-10, t, 2H), 3.82 ($-O-CH_aH_b-$ side chain, H-11$_a$, m, 1H), 4.00 ($-O-CH_aH_b-$ side chain, H-11$_b$, m, 1H), 5.58 (bridge $-CH_2-$, H-5, s, 2H), and 8.04 (aromatic proton in pyrimidine ring, H-2, s, 1H). Additional signals are as follows: signals of ring protons of the sugar moiety (3.36–3.81 pm, 6H) and one anomeric proton signal of an $\alpha$-glucosidic linkage (4.97, 1H, d, $J = 4.0$ Hz). These data suggest that compound $\mathbf{I}'$ is an $\alpha$-glucoside of thiamin. To confirm the fine structure of compound $\mathbf{I}'$, carbon-13 chemical shifts of compound $\mathbf{I}'$ in $D_2O$ are compared with those of thiamin (Table I). Signal assignments are based on the data reported for methyl-$\alpha$-D-glucopyranoside[6] and thiamin[7] in $D_2O$. It is apparent that the glucosylation site in compound $\mathbf{I}'$ is the $CH_2OH$ group in the 4-methyl-5-hydroxyethylthiazolium moiety, because the appended C-11 signal is displaced downfield by a sizable 5.70 ppm with a concomitant small upfield shift of C-10 in the

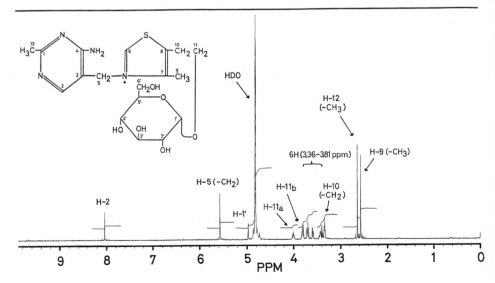

FIG. 2. $^1$H NMR spectrum of compound $\mathbf{I}'$ in $D_2O$. [From Y. Suzuki and K. Uchida, *Biosci. Biotech. Biochem.* **58**(7), 1273 (1994), with permission of the Japan Society for Bioscience, Biotechnology, and Agrochemistry.]

hydroxyethyl group and only smaller effects on other thiamin carbons. Thus, compound **I′** is identified as 5′-*O*-(α-D-glucopyranosyl)thiamin.

Thiamin α-glucoside is odorless and mildly sweet with no stimulative tongue-pricking taste. When a 0.05 *M* aqueous solution (adjusted to pH 7.0) of thiamin hydrochloride, compound **I** (thiamin β-galactoside), or compound **I′** (thiamin α-glucoside) is allowed to stand overnight at room temperature, much of the odor is produced from the neutralized thiamin hydrochloride solution after 2 hr, but both neutralized thiamin glycoside solutions are odorless even after 24 hr. Compounds **I** and **I′** show about 50 and 80% activities of the equivalent moles of thiamin hydrochloride, respectively, when the use of each glycosylthiamin by thiamin-deficient male rats fed on a semisynthetic diet is examined in terms of its effects on the growth curve, food intake, liver weight, and hepatic thiamin content.

## Acknowledgments

The authors thank the Japan Society for Bioscience, Biotechnology, and Agrochemistry for permission to adapt this article from Y. Suzuki and K. Uchida, *Biosci. Biotech. Biochem.* **58**(7), 1273 (1994).

# Section III

# Lipoic (Thioctic) Acid and Derivatives

# [17] High-Performance Liquid Chromatography Methods for Determination of Lipoic and Dihydrolipoic Acid in Human Plasma[1]

By JENS TEICHERT and RAINER PREISS

## Introduction

α-Lipoic acid (1,2-dithiolane-3-pentanoic acid) and its reduced form dihydrolipoic acid (6,8-dimercaptooctanoic acid) are physiologically occurring substances. The oxidized state of the lipoic acid is characterized by a disulfide-containing ring. It is mainly known as a cofactor of α-keto acid dehydrogenases. Lipoic acid is also used as a therapeutic agent in the therapy of diabetic polyneuropathy and chronic and toxic liver disorders. To date there is little information concerning the pharmacokinetics of lipoic acid in humans. The metabolic fate in humans was not investigated until recently, although lipoic acid metabolism in the rat was described in 1974.[2]

The redox potential of the dihydrolipoic acid/lipoic acid couple is −0.32 V. Dihydrolipoic acid is a potent sulfhydryl reductant and therefore its isolation from plasma samples is difficult. Traces of oxygen and transition metals cause rapid oxidation in solution. We have tried to determine both lipoic and dihydrolipoic acid in human plasma.[3] All preparations were carried out under inert gas, and all reagents were thoroughly bubbled with helium gas. Dihydrolipoic acid was detected in plasma samples of six healthy volunteers. Calibration of dihydrolipoic acid in plasma samples was impossible because of its rapid oxidation. Therefore, concentrations of dihydrolipoic acid were calculated by using the standard curves of lipoic and dihydrolipoic acid in solution.

G. J. Handelman et al.[4] measured lipoic and dihydrolipoic acid in tissue culture medium simultaneously by using dual mercury amalgam electrodes in the series mode. The mercury surface-specific reaction is much more specific than the reaction on the glassy carbon surface. Therefore, the former is commonly used for the measurement of disulfides and thiols.

[1] Portions of the method are reprinted from *J. Chromatogr. B*, **672**, J. Teichert and R. Preiss, Determination of lipoic acid in human plasma by high-performance liquid chromatography with electrochemical detection, pp. 277–281, Copyright 1995 with kind permission from Elsevier Science–NL, Sara Burgerhartstraat 25, 1055 KV Amsterdam, The Netherlands.
[2] E. H. Harrison and D. B. McCormick, *Arch. Biochem. Biophys.* **160**, 514 (1974).
[3] J. Teichert and R. Preiss, *Int. J. Clin. Pharmacol. Ther. Toxicol.* **30**, 511 (1992).
[4] G. J. Handelman, D. Han, H. Tritschler, and L. Packer, *Biochem. Pharmacol.* **47**, 1725 (1994).

However, standard curves display linearity only at low nanomolar concentrations because large amounts of dihydrolipoic acid overwhelm the redox capacity. The detection limit for this method was found to be 0.05 nmol (10.3 ng) of lipoic acid and 0.01 nmol (2.08 ng) of dihydrolipoic acid. The Hg/Au electrochemical detection had not been adapted to measure plasma levels of lipoic and dihydrolipoic acid until now.

We have developed a method to measure levels of lipoic acid in human plasma.[5] Dihydrolipoic acid is not detected separately but detected as its corresponding disulfide. Oxidation occurs during sample preparation. By using a new lot of the crude enzyme Alcalase, some interfering compounds were detected. To prevent this, we have substituted subtilisin for the Alcalase. Both standard curves display equal nonlinear relationships of peak area to sample quantity.

### Principle of Detection

Both reduced and oxidized lipoic acid can be detected by using a glassy carbon electrode at +1.1 V. The high potential is required for the multielectron process of oxidation of lipoic acid. The identity of the highly oxidized species is unknown. The oxidation of dihydrolipoic acid involves a two-electron transfer depending on the flow and the condition of the electrode. The oxidized species is lipoic acid. Therefore, the sensitivity for lipoic acid is approximately fourfold greater than for dihydrolipoic acid under these conditions.

#### Chemicals

R,S-α-Lipoic acid (Asta Medica, Frankfurt/Main, Germany)
Alcalase, subtilisin, trichloroacetic acid, trisodium citrate dihydrate, sodium chloride, sodium phosphate dibasic dihydrate (Merck, Darmstadt, Germany)
Acetonitrile, methanol, water, potassium dihydrogen phosphate, sodium phosphate monobasic monohydrate (J. T. Baker, Deventer, The Netherlands)

### Standard Solutions

Lipoic acid is dissolved in an equimolar solution of sodium hydroxide in ultrapure water to obtain a stock solution of 0.2 mg/ml. Calibration standards in the concentration range of 0.008–20 μg/ml are prepared by appropriate dilution of the stock solution in water.

[5] J. Teichert and R. Preiss, *J. Chromatogr. B* **672,** 277 (1995).

The buffer used for enzymatic hydrolysis is prepared as follows: 147.4 mmol of sodium chloride, 5.6 mmol of potassium chloride, 5.48 mmol of disodium phosphate dibasic dihydrate, and 0.32 mmol of sodium phosphate monobasic monohydrate are dissolved in 1 liter of ultrapure water (pH 7.4). One milligram of subtilisin is dissolved in 12.5 ml of buffer and diluted with water to a total volume of 25 ml.

## Preparation of Samples

For calibration, plasma samples are spiked with the diluted stock solution and frozen in 1-ml aliquots. Eight concentrations are measured over the whole calibration range. Trisodium citrate dihydrate is used as anticoagulant. After thawing, 1 ml of plasma is diluted with 0.5 ml of enzyme solution and hydrolyzed at 37° for 10 min. The reaction mixture is shaken slightly and 1 ml of water and 1 ml of 0.2 $M$ trichloroacetic acid are added.

## Extraction Procedure

A BAKERBOND spe Phenyl cartridge (1 ml) is preconditioned with 3 ml of methanol and 3 ml of water. The mixture is applied to the cartridge, washed with 5 ml of water, and dried under full vacuum for 5 min. The samples are eluted with 1 ml of methanol and evaporated to dryness by a stream of nitrogen at 50°. The residue is dissolved in 200 $\mu$l of the mobile phase and injected into a high-performance liquid chromatography (HPLC) system.

## Column and Mobile Phase

The HPLC system consists of a pump (model 64; Knauer, Berlin, Germany), a Knauer manual loop sample injector with a 20-$\mu$l sample loop, an electrochemical detector (model 656; Metrohm, Herisau, Switzerland) equipped with a glassy carbon electrode, and a Knauer HPLC software package version 2.11 for integration of peaks. Chromatographic separations are performed with a Nucleosil 120-$C_{18}$ 5-$\mu$m (Macherey-Nagel, Düren, Germany) column (250 × 4 mm i.d.). A precolumn (30 × 4 mm i.d.) with an integrated guard column (5 × 4 mm i.d.) packed with 5-$\mu$m Nucleosil $C_{18}$ is placed in front of the analytical column. The guard column is replaced frequently to maintain good column efficiency. The mobile phase consists of 227.5 g of acetonitrile, 31.5 g of 2-propanol, and 674.5 g of 0.05 $M$ potassium dihydrogen phosphate, which is adjusted to pH 2.5 with phosphoric acid (20°). The flow rate is 1 ml/min.

## Standards and Instrument Response

The electrode is set at a potential of +1.1 V. The system experiences a gradual decrease in sensitivity with time. External lipoic acid standards are injected daily as a system suitability test. Refilling of the reference electrode with electrolyte or polishing the glassy carbon electrode with aluminum oxide powder remedies this problem of decreasing sensitivity. Quality control samples (QCs) are prepared prior to the beginning of the study. Human plasma is spiked at low, middle, and high lipoic acid concentrations (LQCs, MQCs, and HQCs). The QCs are used to validate the assay and to monitor the performance of the assay during the study. Quality controls and samples are processed in batch.

## Standard Curve and Protein Binding

Lipoic acid displays a nonlinear relationship between the quantity injected and the peak area. Addition of large quantities of lipoic acid overwhelms the binding capacity of plasma proteins and causes linearity (Fig. 1). The enzyme subtilisin hydrolyzes the proteins and releases the lipoic acid bounded by hydrophobic interactions. The commonly used denaturation of proteins by acid (e.g., trichloroacetic acid) is not favored because preliminary experiments showed an unsatisfactory recovery. Nevertheless, a minority of the lipoic acid added is not recoverable by hydrolysis with subtilisin.

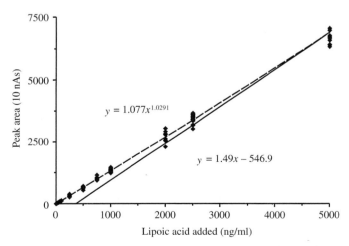

FIG. 1. Standard curve for lipoic acid in the nonlinear range (– – –). It displays linearity at high nanomolar or micromolar concentrations (————). (*Note:* The part of the scale greater than 5000 ng/ml is not shown.)

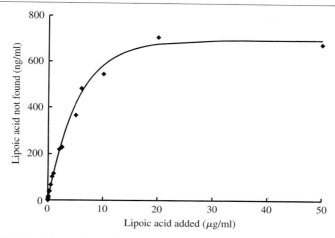

FIG. 2. Lipoic acid not released by action of subtilisin. Saturation of protein binding causes linearity of the standard curve in the high concentration range.

The relationship between the part of lipoic acid not recovered and the quantities added is shown in Fig. 2. For calculation, the $x$ values of the nonlinear standard curve are subtracted from the $x$ values of the standard curve in the linear range. Therefore, the linear curve is translated through the origin. Both standard curves obtained after hydrolysis by Alcalase and subtilisin are used for calculation.

In biological systems the carboxylic acid group of lipoic acid forms a peptide linkage to the $\varepsilon$-amino group of protein-bound lysine. Both enzymes Alcalase and subtilisin are not able to release lipoic acid from the $\varepsilon$-amino group of the lysine.[6] One way to release lipoic acid from the $\varepsilon$-amino group of lysine residues is by nonenzymatic hydrolysis under acidic or alkaline conditions. We hydrolyzed plasma samples of seven healthy volunteers with concentrated hydrochloric acid at 120° for 6 hr. The concentration of lipoic acid extracted with hexane–chloroform is in the range of 12.3–43.1 ng/ml (mean, 29.1 ± 8.8 ng/ml). The recovery of this method is about 30%. The lipoic acid content of nine plasma samples estimated after enzymatic hydrolysis (subtilisin) followed by solid-phase extraction is in the range of 1.4–11.6 ng/ml (mean, 5.6 ± 3.5 ng/ml). It seems the higher content after acid hydrolysis is due to the cleavage of the peptide linkage. However, lipoic acid occurs in tissues largely bound to the lysine residue, but in plasma the amount bound in this manner seems to be small.

[6] A. Mattulat and W. Baltes, *Z. Lebensm. Unters. Forsch.* **194,** 326 (1992).

## Extraction Recovery and Assay Reproducibility

The recovery of lipoic acid from plasma is determined at three different concentrations. Known amounts of lipoic acid are added to drug-free plasma, and the area responses of lipoic acid are compared to those obtained by direct injection of standard solutions containing equivalent amounts of lipoic acid. The recoveries (mean ± SD) at 25, 250, and 2000 ng/ml are 81.7 ± 2.4, 82.1 ± 5.2, and 81.5 ± 2.5%, respectively.

Assay reproducibility is assessed at the same concentrations. The intra-day precision, as measured by relative standard deviation (RSD), is 2.9, 6.4, and 3.1%, respectively. Within-day reproducibility is calculated from six replicate analyses. The corresponding values for between-day analysis are 9.3, 11.6, and 8.3%, respectively. Day-to-day reproducibility is determined from analyses on six different days.

## Storage Stability of Lipoic Acid in Plasma

Plasma samples spiked with lipoic acid are stored at −70° and analyzed at different times over a 6-month period. Areas of lipoic acid in these

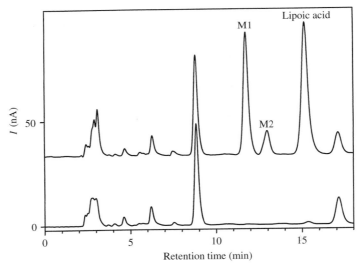

Fig. 3. Chromatograms of plasma (1 ml) extract before (*bottom*) and 1 hr after (*top*) oral administration of lipoic acid to a volunteer. The top chromatogram shows the concomitant generation of metabolites (M1 and M2). They are unidentified at the time of writing. The detectability under these conditions indicates that the dithiolane ring is unaltered and that a two-chain shortened analog is probably produced. Peak M2 consists of two substances. The chromatographic system is not capable of separating them. The peak at the retention time of 17 min is not dihydrolipoic acid. The current range of the detector is 10 nA.

samples are compared to those from freshly prepared controls. The average recovery from 6-month spiked plasma is 103.6% at the 25-ng/ml level and 97.7% at the 1000-ng/ml level. Two plasma samples are measured after a storage period of 2 years. The recovery at the 5- and 10-μg/ml level is 109.4 and 104.9%, respectively.

## Discussion

The single-dose kinetics of lipoic acid following oral and intravenous administration was studied in 12 healthy volunteers and 12 patients with diabetic polyneuropathy. The sensitivity of the assay is adequate to evaluate the pharmacokinetic parameters of lipoic acid in humans (Fig. 3). The minimum detectability of lipoic acid is 0.45 pmol (about 0.1 ng) corresponding to 4.5 pmol/ml of human plasma by choosing a signal-to-noise ratio of 3:1. In a cell suspension, we have measured lower quantities of lipoic acid owing to the lack of interfering substances. Dihydrolipoic acid was also detected in this suspension (Fig. 4). After storage in the eluent for 24 hr, the decrease in dihydrolipoic acid and the increase in lipoic acid concentrations were monitored. The present assay for measuring lipoic acid, specifically in plasma, is simple, robust, and easy to automate. Calibration was

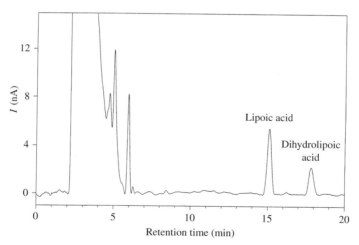

FIG. 4. HPLC chromatogram of lipoic and dihydrolipoic acid in a cell suspension. The analysis shows 0.2 ng of lipoic acid and approximately 0.5 ng of dihydrolipoic acid. Although the quantity of dihydrolipoic acid injected on the HPLC is greater, the electrode shows a much greater sensitivity for lipoic acid than for dihydrolipoic acid. The current range of the detector is 0.5 nA.

carried out using an external standard. Therefore, it is necessary to analyze quality control samples every day.

## Summary

This assay method was applied to determine plasma levels of lipoic acid in humans. The method consists of enzymatic hydrolysis to release the protein-bound lipoic acid, solid-phase extraction, and electrochemical detection at a potential of $+1.1$ V. Previous methods did not provide adequate sensitivity for these studies or required procedural modifications for detection of low levels of plasma lipoic acid. The chromatographic system is capable of separating lipoic acid from dihydrolipoic acid. Both reduced and oxidized lipoic acid can be detected. Therefore, oxidation of dihydrolipoic acid must be prevented. In the described procedure, we do not prevent oxidation and the whole content is measured as lipoic acid. The method does not detect lipoic acid covalently bound to lysine. The detection limit for this method is 1 ng of lipoic acid per milliliter of plasma.

# [18] Analysis of Lipoic Acid by Gas Chromatography with Flame Photometric Detection

*By* HIROYUKI KATAOKA, NORITOSHI HIRABAYASHI, and MASAMI MAKITA

## Introduction

Lipoic acid is a naturally occurring cofactor reported in a diverse group of microorganisms[1] and a variety of plant and animal tissues.[2] It serves as an acyl carrier in the oxidative decarboxylation of $\alpha$-keto acids such as pyruvate and $\alpha$-ketoglutarate[2-4] and as an aminomethyl carrier in glycine-cleavage enzyme systems.[5,6] Lipoic acid has been shown to provide potent

[1] A. A. Herbert and J. R. Guest, *Arch. Microbiol.* **106**, 259 (1975).
[2] L. J. Reed, *in* "Comprehensive Biochemistry" (M. Florkin and E. H. Stotz, eds.), Vol. 14, p. 99. Elsevier, New York, 1966.
[3] U. Schmidt, P. Grafen, K. Altland, and H. W. Goedde, *in* "Advances in Enzymology" (F. F. Nord, ed.), Vol. 32, p. 423. Wiley Interscience, New York, 1969.
[4] L. J. Reed and M. L. Hackert, *J. Biol. Chem.* **256**, 8971 (1990).
[5] K. Fujiwara, K. Okamura, and Y. Motokawa, *Arch. Biochem. Biophys.* **197**, 454 (1979).
[6] G. Kikuchi and K. Hiraga, *Mol. Cell Biochem.* **45**, 137 (1982).

antioxidant abilities against attack by free radicals[7–9] and an inhibitory effect against human immunodeficiency virus (HIV) replication,[10] and an interplay between lipoic acid and glutathione in the protection against lipid peroxidation and metal toxicity has been demonstrated.[7,11–13] A number of *in vitro* and *in vivo* studies also suggest that lipoic acid is able to recycle other natural antioxidants such as ascorbic acid and $\alpha$-tocopherol, and a dietary supplement of lipoic acid is effective for resistance of tissues to lipid peroxidation[14] and maintaining the life span of the mouse.[15] Moreover, lipoic acid is used extensively in the treatment of various diseases, such as alcoholic liver disease,[16] mushroom poisoning,[17] metal poisoning,[12,13,18] diabetes,[19] glaucoma,[20] and neurodegenerative disorder.[21]

The determination of lipoic acid has been carried out by microbiological assay,[22] colorimetric assay,[23] enzyme immunoassay,[24] high-performance liquid chromatography (HPLC),[25–27] gas chromatography (GC),[28,29] and GC–mass spectrometry.[30–32] However, many of these methods are not satisfac-

[7] A. Bast and G. R. M. M. Haenen, *Biochim. Biophys. Acta* **963,** 558 (1988).

[8] Y. J. Suzuki, M. Tsuchiya, and L. Packer, *Free Radical Res. Commun.* **15,** 255 (1991).

[9] B. C. Scott, O. I. Aruoma, P. J. Evans, C. O'Neil, A. Van der Vliet, C. E. Cross, H. Tritschler, and B. Halliwell, *Free Radical Res.* **20,** 119 (1994).

[10] A. Baur, T. Harrer, G. Jahn, J. R. Kalden, and B. Fleckenstein, *Klin. Wochenschr.* **69,** 722 (1991).

[11] L. Muller and H. Menzel, *Biochim. Biophys. Acta* **1052,** 386 (1990).

[12] Z. Gregus, A. F. Stein, F. Varga, and C. D. Klaassen, *Toxicol. Appl. Pharmacol.* **114,** 88 (1992).

[13] D. Han, H. J. Tritschler, and L. Packer, *Biochem. Biophys. Res. Commun.* **207,** 258 (1995).

[14] V. Kagan, S. Khan, C. Swanson, A. Shvedova, E. Serbinova, and L. Packer, *Free Radical Biol. Med.* **9**(s), 15 (1990).

[15] M. Podda, H. J. Tritschler, H. Ulrich, and L. Packer, *Biochem. Biophys. Res. Commun.* **204,** 98 (1994).

[16] A. W. Marshall, R. S. Graul, M. Y. Morgan, and S. Sherlock, *Gut* **23,** 1088 (1982).

[17] R. Plotzker, D. Jensen, and J. A. Payne, *Am. J. Med. Sci.* **283,** 79 (1982).

[18] P. Ou, H. J. Tritschler, and S. P. Wolff, *Biochem. Pharmacol.* **50,** 123 (1995).

[19] V. Burkart, T. Koike, H. H. Brenne, Y. Imai, and H. Kolb, *Agents Actions* **83,** 60 (1993).

[20] A. A. Filina, N. G. Davydova, and E. M. Kolmirtseva, *Vestn. Oftal'mol.* **109,** 5 (1993).

[21] H. Altenkirch, H. Stoltenburg-Didinger, M. Wagner, J. Herrmann, and G. Walter, *Neurotoxicol. Teratol.* **12,** 619 (1990).

[22] A. A. Herbert and J. R. Guest, *Methods Enzymol.* **XVIII,** 269 (1970).

[23] M. Koike and K. Suzuki, *Methods Enzymol.* **XVIII,** 292 (1970).

[24] A. I. MacLean and L. G. Bachas, *Anal. Biochem.* **195,** 303 (1991).

[25] S. C. Howard and D. B. McCormick, *J. Chromatogr.* **208,** 129 (1981).

[26] J. Teichert and R. Prieß, *J. Chromatogr. B* **672,** 277 (1995).

[27] D. Han, G. J. Handelman, and L. Packer, *Methods Enzymol.* **251,** 315 (1995).

[28] R. H. White, *Anal. Biochem.* **110,** 89 (1981).

[29] J. C. H. Shih and S. C. Steinsberger, *Anal. Biochem.* **116,** 65 (1981).

[30] R. H. White, *Biochemistry* **19,** 15 (1980).

tory owing to lack of sensitivity and specificity, time-consuming procedures, expensive sample preparation, and the need for expensive equipment. We have described a selective and sensitive method for analyzing lipoic acid by GC with flame photometric detection (FPD).[33] In this chapter, a method for the derivatization and FPD–GC analysis of lipoic acid in biological and food samples is described, and optimum conditions and typical problems encountered in the development and application of the method are discussed.

Principle

The derivatization process is shown in Fig. 1. After reduction of lipoic acid with sodium borohydride (NaBH$_4$), the reduced form, dihydrolipoic acid, is converted into its (S,S)-diethoxycarbonyl (DEOC) methyl ester derivative by the ethoxycarbonylation with ethyl chloroformate (ECF) and the subsequent methylation with hydrogen chloride in methanol (HCl–methanol). The derivative is measured selectively and sensitively by FPD–GC, using a DB-210 capillary column (J & W, Folsom, CA).

*Reagents*

Sodium borohydride (NaBH$_4$): 100 mg/ml in 0.1 $M$ NaOH, stored at 4°, stable for at least 2 weeks

Ethyl chloroformate (ECF): stored at 4°

Hydrogen chloride in methanol (HCl–methanol): 0.25 $M$, stored at 4°

Bovine serum albumin (fraction V, BSA): 50 mg/ml in 2.5 $M$ KOH, stored at 4°

2-Mercaptoethanol: 1 mg/ml in distilled water, stored at 4°

DL-$\alpha$-Lipoic acid: 1 mg/ml in methanol, stored at 4°, used after dilution with distilled water

(S,S)-Dimethoxycarbonyllipoic acid [(S,S)-DMOC lipoic acid]: 1 mg/ml in methanol, stored at 4°, used as an internal standard (IS) after dilution with distilled water. (S,S)-DMOC lipoic acid is prepared as follows: 10 mg of lipoic acid is dissolved in 1 ml of NaBH$_4$ (50 mg/ml in 0.05 $M$ NaOH) and then incubated at 60° for 10 min. To the reaction mixture is added 0.3 ml of methyl chloroformate, and the mixture is shaken for 20 min at room temperature. The reaction

[31] K. J. Pratt, C. Carles, T. J. Carne, M. J. Danson, and K. J. Stevenson, *Biochem. J.* **258**, 749 (1989).

[32] A. Mattulat and W. Baltes, *Z. Lebensm. Unters. Forsch.* **194**, 326 (1992).

[33] H. Kataoka, N. Hirabayashi, and M. Makita, *J. Chromatogr.* **615**, 197 (1993).

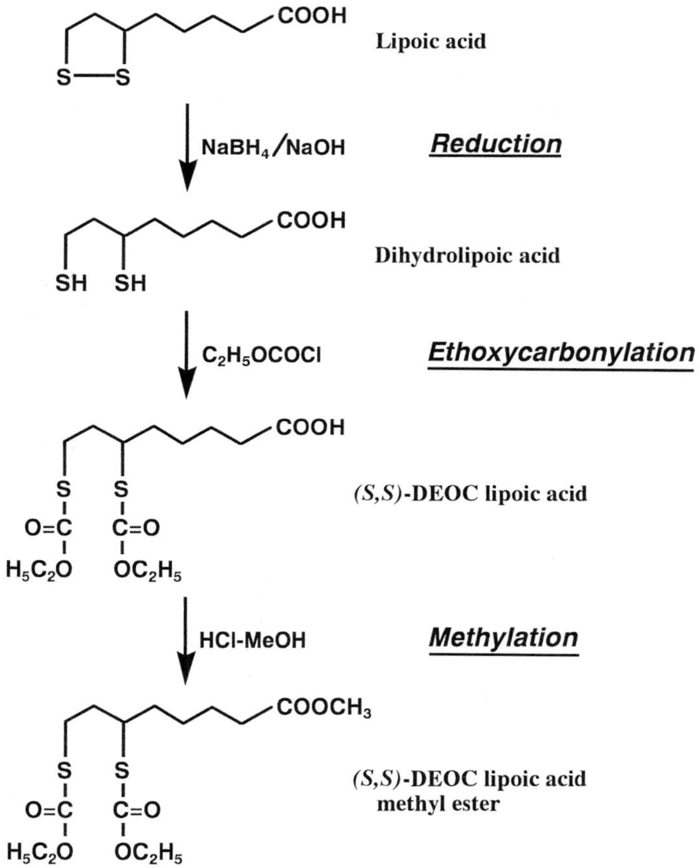

FIG. 1. Reduction and derivatization of lipoic acid.

mixture is acidified to pH < 1 with 2 *M* HCl and then extracted twice with 3 ml of *n*-pentane. The combined *n*-pentane extracts are washed with 1 ml of 0.2 *M* HCl and evaporated to dryness at 60°, and the residue is then reconstituted with methanol to a concentration of 1 mg/ml.

Sample Preparation

*Tissue Samples*

Immediately after dissection, each organ is removed and frozen, and stored at −20° until use. Each pooled tissue sample is chopped, and an aliquot (ca. 1 g) is homogenized in 4 vol of distilled water with an ultra-disperser, and then made up to 5 ml with distilled water. After an aliquot (0.2 ml) of tissue homogenate is placed in a polypropylene tube (75 × 9 mm i.d.), 0.8 ml of BSA (50 mg/ml in 2.5 $M$ KOH) is added, and the mixture is then hydrolyzed.

*Bacterial Samples*

After an aliquot (5–50 mg) of freeze-dried bacterial cells is placed in a polypropylene tube, 0.8 ml of BSA (50 mg/ml in 2.5 $M$ KOH) and 0.2 ml of distilled water are added and the mixture is then hydrolyzed.

*Food Samples*

For liquid food samples, an aliquot (0.05–0.1 g) is placed in a polypropylene tube and treated in the same manner as bacterial samples. For solid food samples, an aliquot (ca. 1 g) is treated in the same manner as tissue samples.

*Base Hydrolysis and Extract of Samples*

The samples obtained by the previously described methods are hydrolyzed for 3 hr at 110° under vacuum in a Pico-Tag workstation (Waters Assoc., Milford, CT). To the resulting hydrolysate are added 0.05 ml of 2-mercaptoethanol (1 mg/ml) and 0.4 ml of 6 $M$ HCl, in an ice bath, and the solution is transferred to 10-ml Pyrex glass tube with a polytetrafluoro-ethylene (PTFE)-lined screw cap. The mixture is extracted twice with 2 ml of dichloromethane, and the pooled dichloromethane extracts are evaporated to dryness. The residue is then dissolved in 1 ml of 0.01 $M$ NaOH and used for the analysis as a hydrolysate sample.

*Comments*

Because the majority of lipoic acid in biological and food samples is present as a protein-bound form with an amide linkage to the ε-amino group of a lysyl residue,[6,34,35] the release of lipoic acid by hydrolysis of the sample is necessary. Although lipoic acid tends to become oxidized to its thiosulfinate or thiosulfonate during hydrolysis, this oxidation is far more resistant to base hydrolysis than to acid hydrolysis. The optimum concentration of KOH for base hydrolysis of a sample is 2–3 $M$. Furthermore, the degradation of lipoic acid during hydrolysis can be reduced by adding BSA

[34] M. Koike, L. J. Reed, and W. R. Carroll, *J. Biol. Chem.* **235,** 1924 (1960).
[35] D. T. Chuang, C. W. C. Hu, L. S. Hu, W. L. Niu, D. E. Myers, and R. P. Cox, *J. Biol. Chem.* **259,** 9277 (1984).

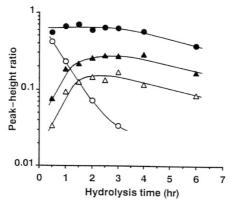

FIG. 2. Effect of BSA on the base hydrolysis of standard lipoic acid and mouse liver. (○) Lipoic acid (−BSA); (●) lipoic acid (+BSA); (△) mouse liver (−BSA); (▲) mouse liver (+BSA). (Reprinted from *J. Chromatogr.* **615**, H. Kataoka, N. Hirabayashi, and M. Makita, 197, Copyright 1993 with kind permission of Elsevier Science–NL, Sara Burgerhartstraat 25, 1055 KV Amsterdam, The Netherlands.)

(Fig. 2) and the optimum concentration of BSA is 2–6% (w/v). Maximum recovery of lipoic acid from sample is obtained by hydrolysis at 110° for 2–4 hr (Fig. 2), because a part of the released lipoic acid is lost by oxidation during hydrolysis.

When the base hydrolysate is acidified, 6 $M$ HCl should be added, taking care not to generate heat. The lipoic acid released by the base hydrolysis is quantitatively extracted into dichloromethane from aqueous acid medium and the solvent is easily evaporated. Although lipoic acid tends to be oxidized during this extraction, this oxidation can be obviated by adding 2-mercaptoethanol to the sample; the recovery of lipoic acid by one extraction is higher than 94%. In the recovery of the lower organic layer, care should be taken not to suck up the upper aqueous layer.

Derivatization Procedure

*Reduction with NaBH₄*

To the hydrolysate sample is added 0.05 ml of NaBH₄ (100 mg/ml solution) and the mixture is incubated at 60° for 5 min. After cooling, the reaction mixture is used as the sample for derivatization.

*(S,S)-Diethoxycarbonylation and Methylation*

To the NaBH₄ reduction sample are added 0.05 ml of 2 $M$ NaOH and 0.05 ml of ECF and the mixture is shaken with a shaker set at 300 rpm (up and down) for 3 min at room temperature. After the reaction mixture is acidified to pH <1 by addition of 0.2 ml of 2 $M$ HCl, taking care not to foam, 0.1 ml of IS solution (1 $\mu$g/ml) is added and the mixture is extracted with 3 ml of $n$-pentane. After the $n$-pentane extract is evaporated to dryness at 60° under a stream of nitrogen, 0.3 ml of 0.25 $M$ HCl–methanol is added to the residue and the mixture is incubated at 80° for 10 min. After the solvent is evaporated to dryness at 80° under a stream of nitrogen, the residue is dissolved in 0.1 ml of ethyl acetate and 1 $\mu$l of this solution is injected into the GC system.

*Comments*

Although standard lipoic acid can be easily gas chromatographed as its methyl ester derivative, it is difficult to detect the lipoic acid in biological and food samples as this derivative because a part of lipoic acid forms mixed disulfides[36] and disulfide polymers[37] in these samples. Therefore, these sulfides must be reduced to thiols and then derivatized with appropriate reagents for GC analysis. By reduction of lipoic acid and these disulfide compounds with NaBH$_4$, dihydrolipoic acid is efficiently derivatized. This reduction is accomplished within 2 min at 60° by using 5 mg of NaBH$_4$ in aqueous alkaline medium. Formation of gas and foaming during the NaBH$_4$ reduction of biological and food samples can be reduced by adding one drop of $n$-hexanol, a surface-active agent.

The ethoxycarbonylation of dihydrolipoic acid is accomplished within 2 min in aqueous alkaline medium by shaking at room temperature. The resulting ($S,S$)-DEOC lipoic acid is quantitatively extracted into $n$-pentane in acidic medium and the recovery by one extraction is higher than 95%. The ($S,S$)-DEOC lipoic acid in aqueous medium is stable at room temperature for at least 2 hr, but subsequently decomposes slowly. Therefore, ($S,S$)-DEOC lipoic acid should be extracted into $n$-pentane as soon as possible. Subsequent methylation of carboxyl group is accomplished within 5 min at 80° by using 0.25 $M$ HCl–methanol. The resulting ($S,S$)-DEOC methyl ester derivative of lipoic acid is stable in ethyl acetate at 4° for at least 2 months. The previously described derivative preparation, including the reduction step, can be performed within 30 min.

($S,S$)-DMOC lipoic acid used as an IS behaves similarly to lipoic acid during derivatization. However, when the IS is added before reduction or ethoxycarbonylation, loss of the IS and appearance of another peak due to partial substitution from the ($S,S$)-DMOC form to the ($S$)-MOC and

[36] L. J. Reed, B. G. DeBush, C. S. Hornberger, and I. C. Gunsalus, *J. Am. Chem. Soc.* **75**, 1271 (1953).
[37] R. C. Thomas and L. J. Reed, *J. Am. Chem. Soc.* **78**, 6148 (1956).

(S)-EOC forms are observed. Therefore, the IS should be added after decomposition of excess $NaBH_4$ by acidification. The $C_7$ or $C_9$ homologs of lipoic acid used in a previous report[28] are desirable for an IS if they are easily available.

## Gas Chromatography

The derivatives of lipoic acid and IS are separated and measured by using a GC system equipped with a fused-silica capillary column (15 m × 0.53 mm i.d., 1.0-$\mu$m film thickness) of cross-linked DB-210 (J & W, Folsom, CA) and a flame photometric detector (S-filter). The operating conditions are as follows: column temperature, programmed at 5°/min from 200 to 250°; injection and detector temperature, 260°; nitrogen flow rate, 10 ml/min.

As shown in Fig. 3A, the derivatives of lipoic acid and IS are eluted as a single and symmetrical peak within 6 min. The derivatives are stable and no decomposition is observed during GC analysis.

## Linearity, Reproducibility, Sensitivity, and Specificity

The calibration graph obtained both from logarithmic plots of the peak–height ratio against the IS and the amount of lipoic acid gives a straight

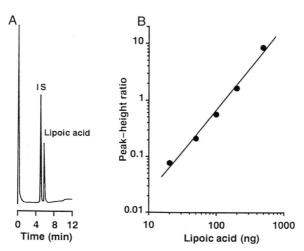

Fig. 3. (A) Standard chromatogram by FPD–GC and (B) calibration graph for lipoic acid. GC conditions are as follows: column, DB-210 (15 m × 0.53 mm i.d., 1.0-$\mu$m film thickness); column temperature, programmed at 5°/min from 200 to 250°; injection and detector temperature, 260°; nitrogen flow rate, 10 ml/min. Calibration graph is constructed from both logarithmic plots.

line in the range of 20–500 ng (Fig. 3B). The regression line is log $y$ = 1.464 log $x$ − 3.104 ($r$ = 0.9950, $n$ = 15). The relative standard deviations in each point are 0.5–4.5% ($n$ = 3). The derivative provides an excellent FPD response. The minimum detectable amount of lipoic acid injected to give a signal three times as high as the noise is ca. 50 pg. Lipoic acid can be analyzed without any influence from coexisting substances such as nonsulfur amino acids, amines, phenols, and fatty acids, because the S-mode FPD is highly selective for sulfur-containing compounds and lipoic acid is selectively extracted and derivatized as previously described.

## Applications

The method described in the previous section can be successfully applied to milligram levels of animal tissue, bacterial cells, or food samples. Typical gas chromatograms obtained from 50 mg of mouse liver and 5 mg of *Escherichia coli* by FPD and flame ionization detection (FID) are shown in Fig. 4. Although it is difficult to determine by FID–GC because of the interfering peaks and low sensitivity (Fig. 4C and D), the lipoic acid in

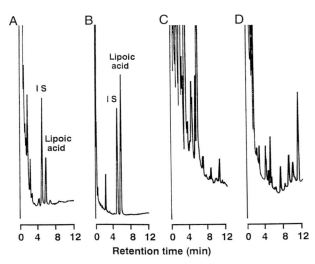

FIG. 4. Gas chromatograms demonstrating selectivity and sensitivity of the analysis of lipoic acid in biological samples. (A and C) Mouse liver (50 mg); (B and D) *Escherichia coli* (5 mg). The derivatized samples are analyzed by FPD–GC (A and B) and FID–GC (C and D). The GC conditions are described in the caption to Fig. 3. (Reprinted from *J. Chromatogr.* **615**, H. Kataoka, N. Hirabayashi, and M. Makita, 197, Copyright 1993 with kind permission of Elsevier Science–NL, Sara Burgerhartstraat 25, 1055 KV Amsterdam, The Netherlands.)

TABLE I
LIPOIC ACID CONTENT IN MOUSE TISSUES,
BACTERIAL CELLS, AND FOODS

| Sample | Content[a] ($\mu$g/g) |
| --- | --- |
| Mouse tissue[b] | |
| Brain | 0.83 ± 0.03 |
| Heart | 2.03 ± 0.07 |
| Lung | 0.80 ± 0.03 |
| Liver | 1.23 ± 0.03 |
| Spleen | 0.52 ± 0.03 |
| Pancreas | 0.84 ± 0.05 |
| Kidney | 1.54 ± 0.06 |
| Testis | 1.06 ± 0.05 |
| Muscle | 0.78 ± 0.02 |
| Bacterial cell[b] | |
| *Aerobacter aerogenes* | 10.62 ± 0.62 |
| *Azotobacter vinelandii* | 34.96 ± 2.39 |
| *Bacillus subtilis* | 17.65 ± 1.38 |
| *Clostridium perfringens* | ND[c] |
| *Escherichia coli* | 23.45 ± 0.44 |
| *Micrococcus lysodeikticus* | 4.34 ± 0.18 |
| *Pseudomonas fluorescens* | 16.25 ± 0.25 |
| *Saccharomyces cerevisiae* | 1.54 ± 0.01 |
| Food | |
| Milk | ND |
| Egg white | ND |
| Egg yolk | 1.24 ± 0.12 |
| Chicken | 0.91 ± 0.04 |
| Pork | 1.07 ± 0.09 |
| Beef | 2.36 ± 0.07 |
| Yellowtail | 0.75 ± 0.01 |
| Cuttlefish | 0.55 ± 0.03 |

[a] Mean ± SD ($n = 4$).
[b] Reprinted from *J. Chromatogr.* **615,** H. Kataoka, N. Hirabayashi, and M. Makita, 197, Copyright 1993 with kind permission of Elsevier Science–NL, Sara Burgerhartstraat 25, 1055 KV Amsterdam, The Netherlands.
[c] ND, Not detectable.

these samples can be analyzed by FPD–GC without any such interference from matrix substances (Fig. 4A and B). The detection limit of lipoic acid in these samples is ca. 10 ng/g.

The overall recoveries of lipoic acid added to mouse tissue and bacterial cell samples are 50–60%, and the relative standard deviations are 0.5–5.3% ($n = 4$). However, the recoveries of lipoic acid added to hydrolysates are higher than 83%. Therefore, these findings suggest that 20–30% of free lipoic acid is lost during hydrolysis of the sample.

Table I shows the results obtained from the analysis of lipoic acid in mouse tissues, several bacterial cell samples, and commercial foods by the method previously developed and described. These analytical data represent the total contents containing free and bound forms of lipoic acid and dihydrolipoic acid in these samples, because the sample is hydrolyzed with KOH and reduced with $NaBH_4$. Lipoic acid is widely distributed in mouse tissues, bacterial cells, and foods. Bacterial cells contain high concentrations of lipoic acid. In food samples, lipoic acid contents are high in those derived from animals, but the levels in those from plants are low or not detectable (data not shown).

Conclusion

Lipoic acid can be selectively and sensitively determined by FPD–GC, in which lipoic acid is converted into its $(S,S)$-DEOC methyl ester derivative after reduction to dihydrolipoic acid with $NaBH_4$. The detection limit of lipoic acid is ca. 50 pg injected, and the calibration graph is linear in the range of 20–500 ng. The best hydrolysis conditions for the samples are 110° for 3 hr using 2 $M$ KOH containing 4% (w/v) BSA. Using this method, total lipoic acid in the hydrolysates of biological and food samples can be determined without any interference from coexisting substances.

# [19] Biosynthesis of Lipoic Acid and Posttranslational Modification with Lipoic Acid in *Escherichia coli*

*By* Sean W. Jordan and John E. Cronan, Jr.

Introduction

α-Lipoic acid (systematic name 1,2-dithiolane pentanoic acid, also called 6,8-thioctic acid) is a cofactor covalently attached to the E2 subunits of pyruvate dehydrogenase (PDH), α-ketoglutarate dehydrogenase (KGDH) (oxoglutarate dehydrogenase), and branched-chain keto acid dehydrogenase (3-methyl-2-oxobutanoate dehydrogenase) complexes and to the glycine cleavage enzyme of *Escherichia coli* and other organisms. Attachment to such proteins is essential for the physiological function of lipoate. Lipoate is essentially an eight-carbon fatty acid modified by sulfur insertion at C-6

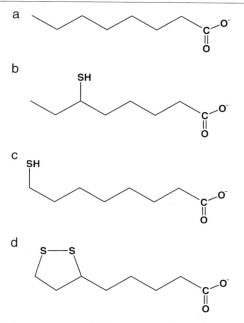

FIG. 1. Structures of (a) octanoate, (b) 6-thiooctanoate, (c) 8-thiooctanoate, and (d) lipoate.

and C-8 to give a thiolane ring (Fig. 1). Although the mechanism of action of lipoic acid as a cofactor is well studied, current knowledge of the biosynthesis of lipoic acid is extremely limited. We first present a brief review of lipoic acid synthesis, followed by a section on the enzymes that attach lipoic acid to the target proteins.

*Synthesis of Lipoic Acid*

Virtually all of the work on lipoic acid synthesis has used *E. coli*. Parry[1] and White[2] independently showed that administration of labeled octanoic acid to cultures of *E. coli* results in specific incorporation of the label into lipoic acid.[1,2] The insertions of sulfur at the C-6 and C-8 positions of lipoic acid proceed without removal of hydrogen from the C-7 or C-5 positions,[1,2] and the only hydrogen atoms lost from octanoic acid are those replaced by the sulfur atoms. These data are consistent with a lack of desaturation of the carbon chain during sulfur insertion, but could also result from an enzyme-bound intermediate that binds the hydrogen atoms abstracted from

[1] R. J. Parry, *J. Am. Chem. Soc.* **99**, 6464 (1977).
[2] R. H. White, *Biochemistry*, **19**, 15 (1980).

octanoate and replaces the atoms following the desaturation event. Insertion of sulfur at C-6 occurs with inversion of configuration at C-6.[2,3] White also tested various labeled substrates for incorporation into lipoic acid and concluded that the existence of hydroxylated intermediates in sulfur insertion is unlikely.[4]

Although the first mutants of *E. coli*, called *lipA*, defective in lipoic acid biosynthesis were isolated in the 1960s,[5,6] the cloning of the affected gene was not accomplished until 1991.[7,8] The original mutants have been used to bioassay lipoic acid,[9] but residual growth without lipoate and reversion to lipoic acid prototrophy were complications. Molecular cloning has allowed the construction of nonreverting null mutants in which the *lipA* gene is disrupted and that fail to grow without lipoate.[7] Strains with mutations in *lipA* were shown to utilize 8-thiooctanoate or 6-thiooctanoate to replace the requirement for lipoate for growth in minimal media,[10,11] implicating involvement of the *lipA* gene product in the insertion of the first sulfur into lipoate. A difficulty with these studies is that growth of *lipA* mutants on the monothiol acids required much higher concentrations of these compounds than did growth on lipoic acid. These findings raised the possibility that contaminating lipoic acid in these preparations (at levels below those detectable by chemical analysis) was the compound supporting growth. However, we have isolated mutants of an *E. coli lipA* strain that grow normally on lipoic acid, but that are unable to grow on 8-thiooctanoate.[12] Therefore, growth on 8-thiooctanoate was not due to lipoic acid contamination.

Proteins having sequences similar to that of LipA are found in both eukaryotic and prokaryotic organisms and thus elucidation of its enzymatic role should provide information applicable across the biological world. The sequence of the LipA protein is similar to that of biotin synthase,[8] the enzyme that introduces the sulfur atom of biotin. Biotin synthase is known to be an iron–sulfur protein and iron–sulfur protein signature sequences are also found in LipA. Therefore, the origin of the sulfur atoms and

[3] R. J. Parry, *J. Am. Chem. Soc.* **100**, 5243 (1978).
[4] R. H. White, *J. Am. Chem. Soc.* **102**, 6605 (1990).
[5] A. B. Vise and J. Lascelles, *J. Gen. Microbiol.* **48**, 87 (1967).
[6] A. A. Herbert and J. R. Guest, *J. Gen. Microbiol.* **53**, 363 (1968).
[7] T. J. Vanden Boom, K. E. Reed, and J. E. Cronan, Jr., *J. Bacteriol.* **173**, 6411 (1991).
[8] M. A. Hayden, I. Huang, D. E. Bussiere, and G. W. Ashley, *J. Biol. Chem.* **267**, 9512 (1991).
[9] A. A. Herbert and J. R. Guest, *Arch. Microbiol.* **106**, 259 (1975).
[10] M. A. Hayden, I. Y. Huang, G. Iliopoulos, M. Orozco, and G. W. Ashley, *Biochemistry* **32**, 3778 (1993).
[11] K. E. Reed and J. E. Cronan, Jr., *J. Bacteriol.* **175**, 1325 (1993).
[12] T. W. Morris, K. E. Reed, and J. E. Cronan, Jr., *J. Biol. Chem.* **269**, 16091 (1994).

mechanism of sulfur insertion seem likely to be common between biotin synthesis and lipoic acid synthesis. Unfortunately, the biotin synthase reaction is complex in that it requires numerous cofactors and has an extremely low turnover number *in vitro*. Thus, sulfur insertion in lipoate biosynthesis may involve a mechanism of similar complexity.

### Attachment of Lipoic Acid: Lipoate Ligases of Escherichia coli

Enzymes that attach lipoate to lipoylated proteins are called lipoate ligases. The structures of the lipoylated enzymes in general and of their lipoyl domains in particular are highly conserved throughout biology (e.g., mammalian and plant lipoyl domains are specifically modified when expressed in bacteria). The first enzyme shown to attach lipoic acid to the lipoyl domains of the 2-oxo acid dehydrogenases uses ATP to activate lipoic acid to lipoyl-AMP followed by transfer of the lipoyl moiety to the ε-amino group of a specific lysine residue centrally located within the domain to give the modified (and thus active) dehydrogenase. *Escherichia coli* contains such an enzyme (LplA protein) that functions primarily in the utilization of exogenously supplied lipoic acid. The *E. coli lplA* gene has been cloned, the protein overexpressed, and purified to homogeneity.[12] The LplA enzyme is the only ligase of known sequence. The enzyme has been purified to homogeneity and has become a valuable reagent in the study of lipoylated enzymes of bacteria and eukaryotes.[13] *Escherichia coli* strains with null mutations in the *lplA* gene retain the ability to attach endogenously synthesized lipoate to the 2-oxo acid dehydrogenases, indicating the presence of a second lipoate–protein ligase.[14] This second lipoate–protein ligase is dependent on the *lipB* gene.[14] *Escherichia coli* strains with null mutations in both *lipB* and *lplA* are completely deficient in protein lipoylation, even when supplied with exogenous lipoate. We have determined that the lipoate donor of the *lipB*-dependent ligase is the lipoyl-acyl carrier protein (lipoyl-ACP).[15] Therefore, octanoyl-ACP, a known intermediate of fatty acid biosynthesis, seems a likely precursor in the biosynthesis of lipoate.

The following is a sensitive assay for the detection of *in vitro* lipoylation via either the *lipB*-dependent ligase or LplA. This assay relies on the use of the nonlipoylated apo-PDH complex purified from a *lipB lplA* strain, a strain completely deficient in protein lipoylation,[14] and thus lacks the en-

[13] S. Ravindran, G. A. Radke, J. R. Guest, and T. E. Roche, *J. Biol. Chem.* **271**, 653 (1996).
[14] T. M. Morris, K. E. Reed, and J. E. Cronan, Jr., *J. Bacteriol.* **177**, 1 (1995).
[15] L. J. Reed, F. R. Leach, and M. Koike, *J. Biol. Chem.* **232**, 123 (1957).

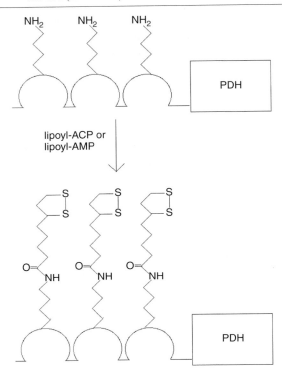

FIG. 2. PDH-coupled assay. Apo-PDH is converted to active lipoylated PDH by lipoate–protein ligase, using either lipoyl-ACP or lipoyl-AMP as the lipoate donor.

zyme activities that require lipoate as a cofactor (Fig. 2). Lipoylation of the purified PDH complex is detected by assay of the ligation reaction product for PDH activity. This method has proven useful in the detection and purification of the *lipB*-dependent ligase[12] and should prove useful in the elucidation of the steps involved in the conversion of octanoyl-ACP to lipoyl-ACP. This assay is essentially that used many years ago with the PDH of *Streptococcus* (now *Enterococcus*) *faecalis*.[16] This organism is unable to synthesize lipoate and when grown in the absence of lipoate accumulates the apo form of PDH. At first glance *E. faecalis* would seem equivalent to a *lipA* null mutant of *E. coli*. However, this is not the case. Whereas *E. faecalis* accumulates the apo form of PDH, *E. coli lipA* strains modify the PDH complex with octanoate (in place of lipoate), resulting in inactive

[16] Manuscript submitted (1997).

proteins that cannot be activated.[17] The *lipB*-dependent ligase seems primarily responsible for this enigmatic modification, although LplA can also perform this modification. Therefore, production of apo-PDH in *E. coli* requires mutational inactivation of both lipoate ligases.

## Methods

### Purification of Apo-Pyruvate Dehydrogenase Complexes from Escherichia coli

Nonlipoylated PDH complex is purified from strain TM136 (*lplA*::Tn*10 lipB*::Tn*1000dkan*). This strain lacks lipoate–protein ligase activity and the PDH complex purified from this strain is not lipoylated and therefore lacks PDH activity.[12] Strain TM136 is grown in a 200-liter fermenter in 2XYT [tryptone (16 g/liter), yeast extract (10 g/liter), sodium chloride (5 g/liter)] supplemented with 5 mM sodium acetate, 5 mM sodium succinate, tetracycline (15 $\mu$g/ml), and kanamycin (50 $\mu$g/ml). The fermenter is inoculated with a 3-liter inoculum and allowed to grow to stationary phase (approximately 28 hr, reaching a final cell density of about $5 \times 10^8$ cells/ml) to give 400 g of wet cell paste.

All purification steps are performed at 4°. Cell paste (150 g) is suspended in 110 ml of 0.1 M potassium phosphate buffer (pH 7.0) and passed through a French pressure cell twice at 25,000 psi. The resulting lysate is centrifuged at 39,000 g for 30 min in a Beckman (Palo Alto, CA) JA-20 rotor to remove unbroken cells and cell debris. The supernatant is then centrifuged at 236,000 g in a Beckman 45Ti rotor for 8 hr to pellet the PDH complex. The resulting pellet is then resuspended in 60 ml of 50 mM potassium phosphate buffer, using a Dounce homogenizer. This suspension is cleared by centrifugation at 39,000 g in a Beckman JA-20 rotor for 30 min. This supernatant is then centrifuged again at 236,000 g for 8 hr in a Beckman 45Ti rotor. The resulting pellet is then resuspended in 32 ml of 50 mM potassium phosphate buffer (pH 7.0), using a Dounce homogenizer. This suspension is cleared once more by centrifugation in the Beckman JA-20 rotor. The resulting suspension has no detectable PDH activity in the absence of lipoylation. When fully lipoylated the preparations has a specific activity of 1.54 U/mg protein, where 1 unit is defined as the activity required to produce 1 $\mu$mol of acetyl-CoA in 1 min at 25°).[15] It should be noted that this method does not separate the PDH complex from the KGDH

---

[17] S. T. Ali, A. J. G. Moir, P. R. Ashton, P. C. Engel, and J. R. Guest, *Mol. Microbiol.* **4,** 943 (1990).

complex. The KGDH complex in these preparations also is not lipoylated and no KGDH activity is detectable in the absence of *in vitro* lipoylation.

*Assay of Apo-Pyruvate Dehydrogenase Complex.* To assay apo-PDH preparations the protein is first lipoylated via the ATP-dependent reaction using either purified LplA enzyme purified as described by Morris and co-workers[12] or a cell-free extract from strain SWJ49 (Δ*aceEF lipB*::Tn*1000d-kan* carrying plasmid pTM64, which encodes *lplA* under the control of the *lac* promoter). The reaction mixture (0.5-ml total volume) contains 1–500 $\mu$g of PDH/KGDH complex, 50 m$M$ sodium phosphate buffer (pH 7.0), 12.5 $\mu$g of lipoic acid, 5 m$M$ ATP, 1.5 m$M$ dithiothreitol, 5 m$M$ MgCl$_2$ plus either 400 $\mu$g of strain SWJ49 cell extract protein or 1 $\mu$g of purified LplA enzyme. Following incubation for 10 min at 37°, the ligated PDH is assayed for PDH activity.

## Pyruvate Dehydrogenase-Coupled Lipoate–Protein Ligase Assays

With both lipoate–protein ligases, activity is assayed by first lipoylating the E2 subunit of the PDH complex purified from a lipoate–protein ligase-deficient strain and then assaying the reaction product for PDH activity. We define 1 unit (U) of lipoate–protein ligase as the amount of ligase required to convert 1 unit of PDH from the nonlipoylated, nonactive form to the lipoylated active form in 1 min.

*Assay of ATP-Dependent LplA Ligase.* The ligation substrate is 0.2 U of apo-PDH in 10 m$M$ sodium phosphate buffer (pH 7.0) containing 0.3 m$M$ dithiothreitol. Either 5–10 ng of the purified enzyme or 50–200 $\mu$g of a cell-free extract protein is used. The LplA reaction mixture (50 $\mu$l, final volume) also contains lipoic acid (25 $\mu$g/ml), 5 m$M$ ATP, and 5 m$M$ MgCl$_2$. The reactions are brought to 50 $\mu$l with water and incubated for 10 min at 37°. The reactions are stopped by bringing the volume to 500 $\mu$l with 120 m$M$ Tris-HCl (pH 8.5) followed by assay for PDH activity in a total volume of 1 ml.

*Assay of lipB-Dependent Ligase.* The ligation substrate is 0.2 U of apo-PDH in 10 m$M$ sodium phosphate buffer (pH 7.0) containing 0.3 m$M$ dithiothreitol. The reaction (50 $\mu$l, final volume) also contains 8 $\mu$g of lipoyl-ACP and 10–100 $\mu$g of a cell-free extract of *E. coli* strain SWJ45 carrying pKR112,[11] which encodes *lipB* under the control of the *tac* promoter. The reactions are incubated for 10 min at 37°, then stopped by bringing the volume to 500 $\mu$l with 120 m$M$ Tris-HCl (pH 8.5). The reactions are then assayed for PDH activity in a total volume of 1 ml. The lipoyl-ACP is synthesized according to the method of Shen *et al.*[18] with 0.7 m$M$

[18] Z. Shen, D. Fice, and D. M. Byers, *Anal. Biochem.* **204,** 34 (1992).

lipoic acid substituted for the fatty acid substrate. The reaction product is precipitated by addition of 2 vol of acetone and resuspended in 20 m$M$ Tris-HCl (pH 7.0) at a concentration of 4 mg/ml. Completion of the lipoyl-ACP synthesis is tested either by gel-shift assay[18] or by the ability to lipoylate PDH via the *lipB*-dependent lipoate–protein ligase.

*Pyruvate Dehydrogenase Assay.* The PDH activity is assayed by modification of the method of Reed and Willms[19] using 3-acetylpyridine adenine dinucleotide (APAD) as the electron acceptor. To the 50-$\mu$l ligation mixture is added 730 $\mu$l of water, 120 $\mu$l of 1 $M$ Tris-HCl (pH 8.5), 30 $\mu$l of cysteine (12 mg/ml), 20 $\mu$l of 10 m$M$ thiamine PP$_i$ (TPP), 8.1 $\mu$l of 10 m$M$ coenzyme A (sodium salt), and 33 $\mu$l of a 4-mg/ml APAD solution. This mixture is divided in half (496 $\mu$l) and to one half was added 4 $\mu$l of 1 $M$ sodium pyruvate. The other half received 4 $\mu$l of water and served as the background control. The absorbance at 366 nm was followed for 1–2 min at 25°. The rate of increase of absorbance at 366 nm was multiplied by a factor of 79.37 to give the rate of reduced APAD formation in $\mu$mol/min. One unit of PDH activity is one $\mu$mol of reduced APAD formed per minute.

## Conclusion

Using this array we have demonstrated the presence of a lipoyl-ACP-dependent protein ligase activity. In wild-type strains lipoate–protein ligase activity can be detected with as little as 5 ng of cell-free extract protein, using either lipoyl-ACP or lipoate plus ATP-Mg$^{2+}$ as substrate for the lipoyl-ACP-dependent or LplA ligase, respectively. In extracts from *lplA* null mutants, ligase activity is detected only when lipoyl-ACP is the lipoyl source whereas in extracts from *lipB* null mutants ligase activity can be detected only by use of lipoic acid plus ATP-Mg$^{2+}$. However, when LplA is overexpressed from an inducible plasmid in *lipB* null strains, activity can be detected using lipoyl-ACP, indicating that LplA can also utilize lipoyl-ACP albeit at a reduced rate.[12]

The finding that the *lipB*-dependent ligase uses lipoyl-ACP as the lipoate donor indicates the likelihood that synthesis of lipoate involves the attachment of sulfur to the C-6 and C-8 carbons of octanoyl-ACP. If this is the case, successful *in vitro* synthesis of lipoate will require that octanoyl-ACP be converted to lipoyl-ACP. The detection of newly synthesized lipoyl-ACP will require a sensitive assay that readily distinguishes octanoyl-ACP from lipoyl-ACP.

---

[19] L. J. Reed and C. R. Willms, *Methods Enzymol.* **IX,** 247 (1966).

# [20] Lipoate Addition to Acyltransferases of $\alpha$-Keto Acid Dehydrogenase Complexes and H-Protein of Glycine Cleavage System

By Kazuko Fujiwara, Kazuko Okamura-Ikeda,
and Yutaro Motokawa

## Introduction

Lipoic acid is a sulfur-containing prosthetic group of the dihydrolipoamide acyltransferases of the pyruvate, $\alpha$-ketoglutarate, and branched-chain $\alpha$-keto acid dehydrogenase complexes (E2p, E2k, and E2b, respectively) and H-protein of the glycine cleavage system.[1–4] Lipoic acid is covalently bound to these proteins via an amide linkage between the carboxyl group of lipoic acid and the $\varepsilon$-amino group of a specific lysine residue. The lipoyl-lysine residue functions as a carrier of intermediates and reducing equivalents between the active sites of the components of the complexes. The attachment of lipoate to the proteins proceeds by two consecutive reactions [Eqs. (1) and (2)]:

$$\text{Lipoate} + \text{ATP} \rightarrow \text{lipoyl-AMP} + \text{PP}_i \qquad (1)$$
$$\text{Lipoyl-AMP} + \text{apoprotein} \rightarrow \text{lipoylated holoprotein} + \text{AMP} \qquad (2)$$

In mammals, the two reactions are catalyzed by separate enzymes in mitochondria. The first reaction is catalyzed by lipoate-activating enzyme in the presence of $\text{Mg}^{2+}$ ion[5] and the second reaction is catalyzed by lipoyl-AMP: $N^{\varepsilon}$-lysine lipoyltransferase (lipoyltransferase).[6] However, only one enzyme, lipoate protein ligase, catalyzes the activation and ligation of a lipoyl group in *Escherichia coli*.[7,8]

[1] L. J. Reed and M. L. Hackert, *J. Biol. Chem.* **265,** 8971 (1990).
[2] R. N. Perham, *Biochemistry* **30,** 8501 (1991).
[3] K. Fujiwara, K. Okamura, and Y. Motokawa, *Arch. Biochem. Biophys.* **197,** 454 (1979).
[4] K. Fujiwara, K. Okamura-Ikeda, and Y. Motokawa, *J. Biol. Chem.* **267,** 2001 (1992).
[5] J. N. Tsunoda and K. T. Yasunobu, *Arch. Biochem. Biophys.* **118,** 395 (1967).
[6] K. Fujiwara, K. Okamura-Ikeda, and Y. Motokawa, *J. Biol. Chem.* **269,** 16605 (1994).
[7] D. E. Brookfield, J. Green, S. T. Ali, R. S. Machado, and J. R. Guest, *FEBS Lett.* **295,** 13 (1991).
[8] T. W. Morris, K. E. Reed, and J. E. Cronan, Jr., *J. Biol. Chem.* **269,** 16091 (1994).

0076-6879/97 $25

## Lipoylation of Lipoyl Domains of E2 Components and H-Protein

We have purified lipoyltransferase from bovine liver mitochondria using lipoyl-AMP and apoH-protein as donor and acceptor of the lipoyl group, respectively.[6,9] Two isoforms were separated by chromatography on a hydroxylapatite column. The lipoyltransferase that eluted at about 220 m$M$ phosphate has been purified to homogeneity and termed lipoyltransferase II. Lipoyltransferase I eluted at about 200 m$M$ phosphate and has been partly purified; the final preparation still contains minor contaminants. The molecular masses and optimal pH values of lipoyltransferase I and II are both 40 kDa and pH 7.9, respectively. The $K_m$ values for lipoyl-AMP obtained with lipoyltransferase I and II are 13 and 16 $\mu M$ and for apoH-protein are 0.29 and 0.17 $\mu M$, respectively. The $V_{max}$ values of lipoyltransferase I and II are 135 and 144 nmol/min/mg of protein, respectively. Thus, the biochemical properties of both isoforms are similar.

We describe here the assay methods for and properties of the lipoylation reaction of lipoyl domains of mammalian acyltransferases (E2s) and H-protein, using purified lipoyltransferase I and II.

### Materials

Lipoyl-AMP is prepared as described by Reed et al.[10] and resolved before use by neutralizing with an NaOH solution. The concentration of lipoyl-AMP is determined spectrophotometrically with a molar extinction coefficient of 15.4 $\times$ 10$^3$ at 259 nm. Oligonucleotides are synthesized on an Applied Biosystems (Foster City, CA) 392 DNA/RNA synthesizer. Purification and assay of lipoyltransferase are carried out as described previously.[6,9] The final preparations of lipoyltransferase I and II show specific activities of 135 and 144 units/mg of protein, respectively. One unit of lipoyltransferase activity is defined as 1 nmol of H-protein lipoylated per minute. cDNAs for mature bovine H-protein[4] and the lipoyl domains of rat E2p (LE2p, amino acids 127–253 of mature E2p[11]), E2k (LE2k, amino acids 1–117 of mature E2k[12]), and bovine E2b (LE2b, amino acids 1–113 of mature E2b[13]) subcloned into pTZ18U are constructed as described by Fujiwara et al.[14] Bovine apoH-protein is prepared as described.[4]

[9] K. Fujiwara, K. Okamura-Ikeda, and Y. Motokawa, *Methods Enzymol.* **251,** 340 (1995).

[10] L. J. Reed, F. R. Leach, and M. Koike, *J. Biol. Chem.* **232,** 123 (1958).

[11] S. Matuda, K. Nakano, S. Ohta, M. Shimura, T. Yamanaka, S. Nakagawa, K. Titani, and T. Miyata, *Biochim. Biophys. Acta* **1131,** 114 (1992).

[12] K. Nakano, S. Matuda, T. Yamanaka, H. Tsubouchi, S. Nakagawa, K. Titani, S. Ohta, and T. Miyata, *J. Biol. Chem.* **266,** 19013 (1991).

[13] T. A. Griffin, K. S. Lau, and D. T. Chuang, *J. Biol. Chem.* **263,** 14008 (1988).

[14] K. Fujiwara, K. Okamura-Ikeda, and Y. Motokawa, *J. Biol. Chem.* **271,** 12932 (1996).

*Expression of Apoform of LE2p, LE2k, and LE2b in Escherichia coli*

Standard molecular biology techniques are carried out essentially according to Sambrook *et al.*[15] To construct expression vectors, cDNA for the lipoyl domain of E2p is synthesized by the polymerase chain reaction (PCR) employing a 5′ primer consisting of 5′ GCTCGGT<u>CATATG</u>**TCC**-TATCCCGTTCAGATG 3′ [*Nde*I site, Met codon, then nucleotides 337–354 (Ref. 11)], a 3′ primer consisting of 5′ GA<u>GGATCC</u>**TTAT-TA**GGGGGT 3′ [*Bam*HI, antisense stop codons, then nucleotides 717–712 (Ref. 11)], and template plasmid for LE2p synthesized as described previously. (The underlined nucleotides represent restriction sites that were introduced for cloning purposes. The nucleotides shown in boldface letters are initiator methionine codons introduced at the start site of the target sequences in 5′ primers and antisense stop codons in 3′ primers). For the cDNA for the lipoyl domain of E2k, a 5′ primer consisting of 5′ CGAGCTC-<u>CATATG</u>AATGATGTGATTACAGTC 3′ [*Nde*I site, Met codon, then nucleotides 199–216 (Ref. 12)], a 3′ primer consisting of 5′ GA<u>GGATCC</u>**T-TA**CATCATCTGA 3′ [*Bam*HI site, antisense stop codon, Met, then nucleotides 549–543 (Ref. 12)], and template plasmid for LE2k are employed. For the cDNA for the lipoyl domains of E2b, a 5′ primer consisting of 5′ GGCGGT<u>CATATG</u>**GGA**CAGATTGTTCAGTTC 3′ [*Nde*I site, Met codon, then nucleotides 193–210 (Ref. 13)], a 3′ primer consisting of 5′ AT<u>GGATCC</u>**TTA**CGCCAGTGTTTTCTGGCC 3′ [*Bam*HI site, antisense stop codon, then nucleotides 528–511 (Ref. 13)], and template plasmid for LE2b are employed. The nucleotide sequence of the PCR products is confirmed by DNA sequencing employing a 373A DNA sequencing system (Applied Biosystems). Each PCR product and expression vector, pET-3a,[16] are digested with *Nde*I and *Bam*HI and ligated. The expression plasmid is introduced into *E. coli* BL21(DE3)pLysS.[16] The cells are grown in 20 ml of M9ZB (1 g of $NH_4Cl$, 3 g of $KH_2PO_4$, 6 g of $Na_2HPO_4$, 4 g of glucose, 1 ml of 1 $M$ $MgSO_4$, 10 g of Bacto-tryptone, and 5 g of NaCl in 1 liter of water) containing ampicillin (25 $\mu$g/ml) and chloramphenicol (25 $\mu$g/ml) at 37°. When the $A_{600}$ of the culture reaches 0.8, isopropyl-$\beta$-thiogalactopyranoside is added to a final concentration of 0.4 m$M$, and cells are grown for an additional 4 hr and harvested.

*Purification of Recombinant Apoform of LE2p, LE2k, and LE2b*

All purification steps are performed at 4°. Each cell pellet obtained from 20 ml of the culture is frozen and thawed in 2 ml of 50 m$M$ Tris-HCl

[15] J. Sambrook, E. F. Fritsch, and T. Maniatis, "Molecular Cloning: A Laboratory Manual," 2nd Ed. Cold Spring Harbor Laboratory, Cold Spring Harbor, New York, 1989.
[16] F. W. Studier, A. H. Rosenberg, J. J. Dunn, and J. W. Dubendorff, *Methods Enzymol.* **185,** 60 (1990).

(pH 7.5), 1 mM EDTA, 40 $\mu M$ p-amidinophenylmethylsulfonyl fluoride (amidino-PMSF), leupeptin (10 $\mu$g/ml), 0.5 mM dithiothreitol. After sonication twice for 30 sec, the cell extract is obtained by centrifugation at 105,000 $g$ for 1 hr, diluted with an equal volume of buffer A [20 mM potassium phosphate buffer (pH 7.2), 0.2 mM dithiothreitol, 10 $\mu M$ amidino-PMSF], and applied to a DEAE-Sepharose CL-6B column (1 × 3 cm; Pharmacia Biotech, Piscataway, NJ) equilibrated with buffer A. The column is washed with three column volumes of buffer A and then developed with a 0–0.3 $M$ NaCl gradient in 80 ml of buffer A. Fractions containing the apolipoyl domain are located by nondenaturing polyacrylamide gel electrophoresis (PAGE) and sodium dodecyl sulfate (SDS)-PAGE. In this condition, LE2p, LE2k, and LE2b are eluted from the DEAE-Sepharose column at NaCl concentrations of about 11, 80, and 139 mM, respectively. Protein concentrations are estimated by the method of Lowry et al.[17] All of the final preparations of the lipoyl domains are 80–90% pure as assessed by SDS–PAGE.

## Lipoylation Reaction

The lipoylation reaction is carried out in a mixture of 8 $\mu$l containing 300 ng of the purified apolipoyl domain or apoH-protein, 40 mM potassium phosphate buffer (pH 7.8), bovine serum albumin (0.2 mg/ml), 10 $\mu M$ MnCl$_2$, 0.1 mM lipoyl-AMP, and lipoyltransferase. In some experiments 0.1 mM lipoic acid, 2 mM ATP, and 2 mM MgCl$_2$ are employed instead of lipoyl-AMP. After incubation at 37° for 60 min, 4 $\mu$l of 3× sample buffer [1× sample buffer is 62.5 mM Tris-HCl (pH 6.8), 10% (v/v) glycerol, 0.002% (w/v) bromphenol blue] is added to the mixture, and 10 $\mu$l of the mixture is subjected to 20% PAGE.[4] The proteins in the gel are stained (Silver Staining Kit, Protein; Pharmacia Biotech). Lipoylation is determined by the increase in mobility of the lipoylated domain on nondenaturing PAGE because the binding of the lipoyl group on the lysine residue neutralizes a positive charge on the domain.[4,6]

As shown in Fig. 1, lipoyl domains of E2s are well lipoylated with lipoyltransferase I and II, depending on the presence of lipoyl-AMP, suggesting that the lipoyltransferases are responsible for the lipoylation of E2 components in vivo. Lipoic acid plus MgATP cannot replace lipoyl-AMP. Fully lipoylated proteins are obtained with apolipoyl domains of E2p and E2k and apoH-protein employing $1.25 \times 10^{-3}$ unit of either lipoyltransferase I or II (Fig. 1A, B, and D), whereas more than four times this activity of lipoyltransferases is required to lipoylate the apolipoyl domain of E2b (Fig. 1C).

---

[17] O. H. Lowry, N. J. Rosebrough, A. L. Farr, and R. J. Randall, J. Biol. Chem. **193,** 265 (1951).

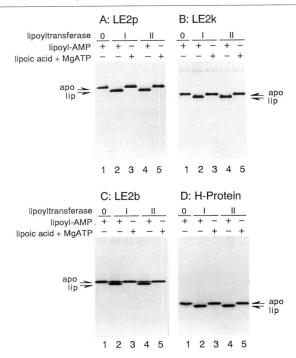

FIG. 1. Lipoylation of the lipoyl domains of the acyltransferase components and H-protein. The lipoyl domains of the acyltransferase component of the pyruvate (LE2p, A), $\alpha$-ketogluta-rate (LE2k, B), and branched-chain $\alpha$-keto acid (LE2b, C) dehydrogenase complexes and H-protein (D) are incubated with lipoyltransferase I (lanes 2 and 3), lipoyltransferase II (lanes 4 and 5), or without enzyme (lane 1) in the presence of lipoyl-AMP (lanes 1, 2, and 4) or lipoic acid plus MgATP (lanes 3 and 5). An enzyme activity of $1.25 \times 10^{-3}$ unit is used for LE2p, LE2k, and H-protein and of $5 \times 10^{-3}$ unit for LE2b. *apo* and *lip*, Apoform and lipoylated form, respectively.

## Role of Conserved Glutamic Acid and Glycine Residues in Lipoylation

Figure 2 shows the amino acid sequences around the lipoic acid attach-ment site of H-protein and various E2 components. Gly-43, Glu-56, and Gly-70 of bovine H-protein[18] are highly conserved among these proteins. The importance of Glu-56 and Gly-70 for the intramitochondrial lipoylation has been demonstrated by site-directed mutagenesis of these residues.[19] The roles of these conserved residues on E2 components are examined

[18] K. Fujiwara, K. Okamura-Ikeda, and Y. Motokawa, *J. Biol. Chem.* **265,** 17463 (1990).
[19] K. Fujiwara, K. Okamura-Ikeda, and Y. Motokawa, *FEBS Lett.* **293,** 115 (1991).

```
                                                    *
    rat       E2p       28  KEGEKISEGDLIAEVETDKATVGFESLEECYM  59
                        154 KVGEKLSEGDLLAEIETDKATIGFEVQEEGYL 185
    rat       E2k       25  AVGDAVAEDEVVCEIETDKTSVQVPSPANGII  56
    bovine    E2b       26  KEGDTVSQFDSICEVQSDKASVTITSRYDGVI  57
    bovine    H-protein 41  EVGTKLNKQEEFGALESVKAASELYSPLSGEV  72
```

FIG. 2. Comparison of the amino acid sequences surrounding the lipoic acid attachment site. The sequences of rat acetyltransferase (E2p),[11] rat succinyltransferase (E2k),[12] bovine acyltransferase of the branched-chain α-keto acid dehydrogenase complex (E2b),[13] and bovine H-protein[18] are compared. The lysine residue involved in lipoic acid attachment is marked with an asterisk. The shadowed amino acid residues show the conserved residues that were chosen for mutagenesis. The numbers on the right- and left-hand sides refer to the positions of the amino acids of the mature proteins.

employing purified lipoyltransferases and lipoyl domains of E2 components and their mutants.

*Materials*

[1-$^{14}$C]Octanoic acid (0.055 Ci/mmol) is purchased from Du Pont–New England Nuclear (Boston, MA). [$^{14}$C]Octanoyl-AMP is synthesized essentially as described[10] using 250 $\mu$Ci of [$^{14}$C]octanoic acid and isolated by high-performance liquid chromatography (HPLC) on an ODS-120T column (4.5 × 250 mm; Tosoh, Tokyo, Japan) with a linear gradient of 0.05 $M$ sodium phosphate, pH 5.5 (solvent A) and acetonitrile (solvent B). The gradient is developed from 10 to 45% B, and the elution of the product is monitored at 259 nm. Octanoyl-AMP is eluted at an acetonitrile concentration of 29%. Under the same condition, lipoyl-AMP is eluted at an acetonitrile concentration of 26%.

*Mutagenesis*

Site-directed mutagenesis is carried out to obtain the mutants LE2k-E40Q, LE2k-G27S, LE2k-G27N, LE2k-G54S, LE2k-G54N, and LE2b-Q41E by the method of Kunkel *et al.*[20] with synthetic oligonucleotides described previously.[14]

*Expression of Apoform of Mutant Lipoyl Domains in Escherichia coli*

To construct the expression vectors for LE2k-E40Q, LE2k-G27S, LE2k-G27N, LE2k-G54S, LE2k-G54N, and LE2b-Q41E, PCRs are carried out as described above for the wild type, employing the respective template synthesized by site-directed mutagenesis and 5' and 3' primers used for

---

[20] T. A. Kunkel, J. D. Roberts, and R. A. Zakour, *Methods Enzymol.* **154,** 367 (1987).

the PCR of the respective wild-type lipoyl domain of E2. The PCR products are ligated with pET-3a, introduced into *E. coli* BL21(DE3)pLysS, and expressed. Recombinant mutant lipoyl domains are purified with a DEAE-Sepharose column as mentioned previously for the wild type. LE2k-E40Q, LE2k-G27S, LE2k-G27N, LE2k-G54S, LE2k-G54N, and LE2b-Q41E are eluted from the column at NaCl concentrations of about 75, 80, 80, 75–80, 75, and 146 m$M$, respectively. All of the final preparations of the lipoyl domains are 80–90% pure as assessed by SDS–PAGE.

## Octanoylation of LE2s and Their Mutants

Although estimation of the radiolabeled lipoyl moiety incorporated into proteins is convenient for the quantitative determination of the protein lipoylation, it is not easy to obtain radiolabeled lipoic acid. We, therefore, employ [14C]octanoyl-AMP as an alternative substrate, because [14C]octanoic acid is available commercially. Octanoic acid is incorporated into apoH-protein from octanoyl-AMP as efficiently as lipoic acid.[6] Octanoylation reactions are carried out in the condition described for lipoylation except that 800 ng of each recombinant LE2 or its mutant is employed and 0.05 m$M$ [14C]octanoyl-AMP is substituted for lipoyl-AMP. After nondenaturing PAGE of the reaction products, the proteins in the gel are electrotransferred onto an Immobilon-P membrane (Millipore, Bedford, MA) in a semidry Sartoblot apparatus (Sartorius) and autoradiographed. Quantification is carried out by counting radioactivities of [14C]octanoic acid on the lipoyl domains with a BAS-1500 Mac Bio-Imaging analyzer (Fuji Photo Film, Tokyo, Japan). Radioactivities nonenzymatically bound to lipoyl domains without lipoyltransferases are subtracted.

## Role of Conserved Glutamic Acid Residue Situated Three Residues on N-Terminal Side from Lipoyllysine Residue

Comparison of the amino acid sequences surrounding the lipoic acid attachment site of E2s and H-protein and the fact that LE2b is lipoylated less efficiently than the other lipoyl domains led to an assumption that the presence of a glutamine residue instead of a glutamic acid residue at the position three residues to the N-terminal side of the lipoylation site is responsible for the low rate of lipoylation. To test the assumption, the glutamine residue has been substituted by glutamic acid by site-directed mutagenesis. As shown in Fig. 3 and Table I, the octanoylation rate of LE2b-Q41E is improved about eightfold as compared with the wild-type LE2b and becomes comparable with the octanoylation rate of LE2p and LE2k. The importance of the glutamic acid residue at this position for the octanoylation reaction is further confirmed by the replacement of Glu-40

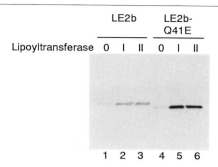

FIG. 3. Octanoylation of wild-type LE2b and mutant LE2b-Q41E. Wild-type LE2b and mutant LE2b-Q41E are incubated with lipoyltransferase I (lanes 2 and 5), lipoyltransferase II (lanes 3 and 6), or without enzyme (lanes 1 and 4). For each enzyme, $1 \times 10^{-4}$ unit of activity is used. After nondenaturing PAGE of the reaction products, proteins in the gel are transferred onto Immobilon-P and autoradiographed.

of LE2k by glutamine. The replacement reduced the octanoylation rate of LE2k-E40Q dramatically to 0.3% of the wild type (Fig. 4 and Table I). These results are consistent with the previous observations obtained with lipoyl-AMP and lipoyl domains synthesized with a reticulocyte lysate.[14] The lipoylation of LE2b-Q41E generated with the *in vitro* transcription–translation system was stimulated about 80-fold as compared with the wild-type LE2b[14], which is an ~10 times greater increase than that observed in octanoylation of isolated LE2b-Q41E presented in this study. The difference in substrate may be responsible for the discrepancy. The three-dimensional structure of the lipoyl domain of E2p from *E. coli* indicates that the glutamic acid residue located three residues to the N-terminal side of the

TABLE I
EFFECT OF AMINO ACID SUBSTITUTION ON OCTANOYLATION

| Lipoyl domain | Lipoylation (%) | |
| --- | --- | --- |
| | Lipoyltransferase I | Lipoyltransferase II |
| LE2b | 100 | 100 |
| LE2b-Q41E | 782 | 721 |
| LE2k | 100 | 100 |
| LE2k-E40Q | 0.3 | 0.3 |
| LE2k-G27S | 87 | 87 |
| LE2k-G27N | 101 | 102 |
| LE2k-G54S | 34 | 38 |
| LE2k-G54N | 1.1 | 1.5 |

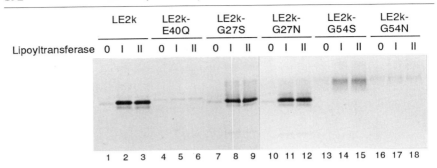

FIG. 4. Octanoylation of wild-type LE2k and its mutants. Wild-type LE2k and mutants LE2k-E40Q, LE2k-G27S, LE2k-G27N, LE2k-G54S, and LE2k-G54N are octanoylated with lipoyltransferase I (lanes 2, 5, 8, 11, 14, and 17), lipoyltransferase II (lanes 3, 6, 9, 12, 15, and 18), or without enzyme (lanes 1, 4, 7, 10, 13, and 16). An enzyme activity of $1 \times 10^{-4}$ unit is used. The autoradiogram is obtained as described in the caption to Fig. 3.

lipoylation site is situated at the end of the $\beta$-strand preceding the tight $\beta$ turn in which the lysine residue to be lipoylated is located.[21] The results obtained suggest that the glutamic acid residue plays an important role in the lipoylation reaction, possibly as a recognition signal for lipoyltransferase.

### Role of Conserved Glycine Residues

Glycine residues 16 residues to the N-terminal side and 11 residues to the C-terminal side of the lipoylation site are also highly conserved among E2 components and H-protein (Fig. 2). To investigate the role of the glycine residues in the lipoylation reaction, Gly-27 and Gly-54 of LE2k have been replaced by a serine or asparagine residue. The replacement of Gly-27 by serine or asparagine does not cause any effect on the rate of lipoylation (Fig. 4 and Table I), suggesting that the glycine residue 16 residues on the N-terminal side of the lipoyllysine has no role in the lipoylation reaction. However, the replacement of Gly-54 by a serine or asparagine residue causes the retardation of the mobility of the mutants LE2k-G54S and LE2k-G54N on nondenaturing PAGE (Fig. 4). The rates of octanoylation of LE2k-G54S and LE2k-G54N are reduced to less than 40 and 2%, respectively, of that of the wild type (Fig. 4 and Table I). The results obtained here are similar to that obtained with lipoyl-AMP.[14] Thus, the change in mobility on nondenaturing PAGE indicates that the amino acid substitution causes some conformational change in the lipoyl domain and suggests that

[21] J. D. F. Green, E. D. Laue, R. N. Perham, S. T. Ali, and J. R. Guest, *J. Mol. Biol.* **248,** 328 (1995).

the glycine residue 11 residues to the C-terminal side of the lipoyllysine residue plays an important role in the folding of the lipoyl domain. The effects of the mutation of Gly-54 on the octanoylation rate depend on the amino acid residue replaced, suggesting that the glycine residue does not play a direct role in the lipoylation and that a substituted bulky residue prevents lipoylation or causes an inappropriate conformational change of the domain. Interestingly, a glycine residue corresponding to Gly-54 of LE2k is also present in biotin enzymes.[22,23] The replacements of the glycine residue of *E. coli* biotin carboxyl carrier protein of acetyl-CoA carboxylase and the $\alpha$ subunit of human propionyl-CoA carboxylase with serine causes a less efficient biotinylation and make the proteins temperature sensitive, suggesting that the incompletely folded protein has been biotinylated.[22,23]

## Acknowledgments

We thank Dr. Sadayuki Matuda (Department of Biology, Kanoya National Institute of Fitness and Sports, Kanoya, Japan) for the generous gift of cDNAs for rat E2p and E2k. This investigation is supported in part by grants from the Ministry of Education, Science, and Culture of Japan.

[22] S.-J. Li and J. E. Cronan, Jr., *J. Biol. Chem.* **267**, 855 (1992).
[23] A. Leon-Del-Rio and R. A. Gravel, *J. Biol Chem.* **269**, 22964 (1994).

# [21] Lipoylation of Acyltransferase Components of 2-Oxo Acid Dehydrogenase Complexes

*By* JANET QUINN

## Introduction

To date five lipoate-containing proteins have been identified. These are the acyltransferase components of the three 2-oxo acid dehydrogenase complexes, protein X of the pyruvate dehydrogenase complex, and the H-protein of the glycine cleavage system. This chapter focuses on the methods available to study lipoylation of the acyltransferase proteins. Posttranslational modification of the H-protein is considered elsewhere.[1]

[1] K. Fujiwara, K. Okamura-Ikeda, and Y. Motokawa, *Methods Enzymol.* **279** [20], 1997 (this volume).

The 2-oxo acid dehydrogenase family comprises the pyruvate dehydrogenase (PDH), the 2-oxoglutarate dehydrogenase (OGDH), and the branched-chain 2-oxo acid dehydrogenase (BCOADH) multienzyme complexes.[2,3] Each occupies a key position in intermediary metabolism, catalyzing the lipoic acid-mediated oxidative decarboxylation of their respective 2-oxo acid substrates. The acyltransferase (E2) components of the complexes contain one or more lipoyl domains to which the lipoic acid moiety is covalently attached via amide linkage to the ε-amino group of a specific lysine residue. This lipoate–lysine moiety is central to the activity of the 2-oxo acid dehydrogenases, providing a swinging arm with sufficient rotational mobility to allow movement of reaction intermediates between different active sites within the complexes.[4]

There has been a long-standing interest into the lipoate protein ligases responsible for lipoate attachment and the mechanism of protein lipoylation. As early as 1958, Reed and colleagues demonstrated the presence of lipoate ligases in *Escherichia coli* and *Streptococcus* (*Enterococcus*) *faecalis*.[5,6] This was achieved by generating inactivated apo-protein PDH complexes with lipoamidase and subsequently assaying for reactivation of complex activity following incubation with lipoic acid, ATP, and protein extracts containing ligase activity. Reed and colleagues proposed that proteins are lipoylated in a two-step ATP-dependent reaction in which lipoic acid is first activated to form lipoyl-AMP, followed by transfer of the activated lipoyl group to the pa-protein.[6] This is analogous to the reaction sequence of the well-characterized biotin protein ligase of *E. coli*.[7]

Other than the demonstration of similar lipoate ligase activities in eukaryotes,[8,9] further studies were hindered by difficulty in obtaining adequate amounts of pa-protein substrate with which to follow lipoylation. More recently, developments in recombinant DNA technology have, however, allowed for the easy construction of subgenes encoding single lipoyl domains of *E. coli*,[10,11] *Bacillus stearothermophilus*,[12] and human PDH com-

[2] S. J. Yeaman, *Biochem. J.* **257**, 625 (1989).
[3] R. N. Perham, *Biochemistry* **30**, 8501 (1991).
[4] M. C. Ambrose and R. N. Perham, *Biochem. J.* **155**, 429 (1976).
[5] L. J. Reed, F. R. Leach, and M. Koike, *J. Biol. Chem.* **232**, 123 (1958).
[6] L. J. Reed, M. Koike, M. E. Levitch, and F. R. Leach, *J. Biol. Chem.* **232**, 143 (1958).
[7] J. E. Cronan, Jr., *Cell* **58**, 427 (1989).
[8] S. K. Mitra and D. P. Burma, *J. Biol. Chem.* **240**, 4072 (1965).
[9] J. N. Tsunoda and K. T. Yasunobu, *Arch. Biochem. Biophys.* **118**, 395 (1967).
[10] J. S. Miles and J. R. Guest, *Biochem. J.* **245**, 869 (1987).
[11] S. T. Ali and J. R. Guest, *Biochem. J.* **271**, 139 (1990).
[12] F. Dardel, L. C. Packman, and R. N. Perham, *FEBS Lett.* **264**, 206 (1990).

plexes.[13] Overexpression of these subgenes in *E. coli* generates a mixture of lipoylated and unlipoylated domains, owing to the high levels of expression of apo-domains exceeding the capacity of the host cells for lipoate biosynthesis. This ready availability of apo-domain substrate has stimulated renewed interest in protein lipoylation and as a consequence several methods have been developed with which to follow lipoate attachment. These include an electrophoretic assay, based on the observation that modified lipoyl domains have a higher mobility than apo-domains on nondenaturing polyacrylamide gel electrophoresis (PAGE),[11,14] and a radioassay monitoring the incorporation of $^3$H- or $^{35}$S-labeled lipoate into apo-proteins. Mass spectroscopy analysis has proved a useful tool in assessing whether recombinant lipoyl domains are correctly posttranslationally modified. This is important because modifications other than lipoylation, such as octanoylation, have previously been observed.[12–14] As a further test of correct posttranslational modification, an assay is available to assess whether recombinant lipoyl domains, once lipoylated, are functional substrates for reductive acetylation by the E1 enzyme component of the appropriate 2-oxo acid dehydrogenase complex.

The availability of apo-domain substrates has allowed for rapid progress to be made in the area of protein lipoylation. This chapter first describes the method used to generate substrate amounts of apo-lipoyl domain from the human PDH complex and the techniques utilized to study and characterize lipoylation of this and other apo-domains. The chapter concludes with a summary of what has been learned thus far of the mechanism of protein lipoylation of the acyltransferases.

## Purification of Human E2p Inner Lipoyl Domain

### Materials

The plasmid pGLIP-2T encodes a recombinant *S*-transferase (GST)–lipoyl domain fusion protein, the expression of which is under the control of the isopropyl-β-D-thiogalactopyranoside (IPTG)-inducible $P_{tac}$ promoter.[13] *Escherichia coli* strain XL1-Blue is used as host for pGLIP-2T.

### Method

The method used is essentially as described previously.[13] Overnight cultures of pGLIP-2T transformants (100 ml) are diluted into Luria broth/

---

[13] J. Quinn, A. G. Diamond, A. K. Masters, D. E. Brookfield, N. G. Wallis, and S. J. Yeaman, *Biochem. J.* **289,** 81 (1993).

[14] S. T. Ali, A. J. G. Moir, P. R. Ashton, P. C. Engel, and J. R. Guest, *Mol. Microbiol.* **4,** 943 (1990).

ampicillin (50 $\mu$g/ml) medium and grown until an $OD_{600}$ of 0.6 is achieved. Recombinant protein synthesis is induced on the addition of IPTG (0.1 m$M$). Following a 4-hr induction, cells are harvested (10,000 $g$, 10 min at 4°) and resuspended in 5 ml of buffer A [50 m$M$ Tris (pH 8.0), 25% (w/v) sucrose, 10 m$M$ EDTA, 0.1 m$M$ phenylmethylsulfonyl fluoride (PMSF)]. Lysozyme (80 mg) is added and cells incubated on ice for 30 min. Bacteria are collected by centrifugation (10,000 $g$, 10 min at 4°) and resuspended in 5 ml of buffer B (14 m$M$ $Na_2HPO_4$–8 m$M$ $KH_2PO_4$). Bacterial lysis is completed by rapid freezing in liquid nitrogen and thawing twice and the cell debris is removed by centrifugation (12,000 $g$, 10 min at 4°). It is sometimes necessary to treat the lysate with DNase I (10 mg) prior to centrifugation. The resulting supernatant is mixed with glutathione (GSH)–agarose beads[15] and incubated for 2 hr at 4° to allow coupling of the recombinant protein. Following adsorption, the beads are washed several times with buffer B and once with thrombin cleavage buffer [50 m$M$ Tris, pH 8.0), 150 m$M$ NaCl, 2.5 m$M$ $CaCl_2$, 0.1% (v/v) 2-mercaptoethanol]. To release the lipoyl domain, the GSH–agarose is resuspended in 2 ml of thrombin cleavage buffer and incubated with 25 units of thrombin (2331 units/mg; Calbiochem, La Jolla, CA) at room temperature for 30 min. To resolve the lipoylated and apo-domains, the protein preparation is dialyzed against 10 m$M$ ammonium acetate, pH 5.0, and fractionated by anion-exchange chromatography on a Mono Q FPLC, (fast protein liquid chromatography; Pharmacia, Piscataway, NJ) column. A linear ammonium acetate gradient (10–300 m$M$) resolves the domain preparation into two peaks that correspond to the lipoylated and unlipoylated forms of the domain. The lipoylated form has one less positively charged lysine side chain that the unlipoylated form and therefore elutes at a slightly higher ionic concentration. The final yield from a 1-liter culture is approximately 1 and 1.5 mg of lipoylated and unlipoylated domains, respectively. Supplementing the growth medium with lipoic acid (0.5 m$M$, added at the same time as IPTG) generates only the lipoylated form of the domain. This method can be easily scaled up to culture sizes of 10–20 liters. Methods for purification of *E. coli* and *B. stearothermophilus* recombinant lipoyl domains have also been described.[11,12]

## Assays for Lipoylation of Recombinant Lipoyl Domain

### *Electrophoretic Mobility on Nondenaturing Polyacrylamide Gel Electrophoresis*

*Principle.* Modified lipoyl domains migrate more slowly than their unmodified counterparts on nondenaturing polyacrylamide gels.[11,14] This tech-

---

[15] P. C. Simons and D. L. Vander Jagt, *Anal. Biochem.* **82**, 334 (1977).

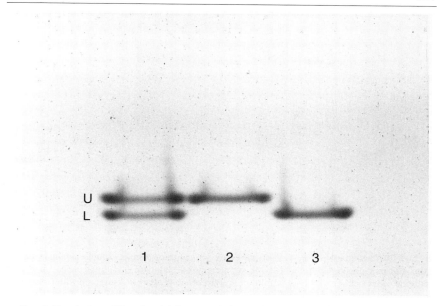

FIG. 1. Resolution of lipoylated (L) and unlipoylated (U) human lipoyl domains on nondenaturing PAGE. Lane 1 shows the domain preparation (3 $\mu$g) before chromatography; lanes 2 and 3, resolution of the two domain forms (1.5 $\mu$g) after anion-exchange chromatography on Mono Q. Following electrophoresis, proteins are visualized by staining with Coomassie Brilliant Blue. (Reprinted from Ref. 13, with permission.)

nique can be used to measure *in vivo* lipoylation following expression of wild-type or mutated lipoyl domain subgenes in *E. coli,* and to follow *in vitro* lipoylation following incubation of apo-domains with lipoate ligase preparations (see the next section).

*Method.* Nondenaturing PAGE is performed as previously described.[16] For the human domain, which has a relative molecular weight of approximately 12,000, a 5% (w/v) polyacrylamide stacking gel and 15% (w/v) polyacrylamide resolving gel are routinely used for analysis. Purified domain mixtures are added to Laemmli sample buffer[17] prepared without sodium dodecyl sulfate (SDS) and kept at 4° prior to loading. Proteins are visualized by staining with Coomassie Brilliant Blue and the relative amounts of lipoylated and unlipoylated domain determined by quantitative densistometry. Figure 1 illustrates the resolution of the two domain forms on nondenaturing PAGE (lane 1, Fig. 1) and their separation following anion-exchange chromatography (lanes 2 and 3, Fig. 1).

[16] B. J. Davis, *Ann. N.Y. Acad. Sci.* **121,** 404 (1964).
[17] U. K. Laemmli and M. Favre, *J. Mol. Biol.* **80,** 579 (1973).

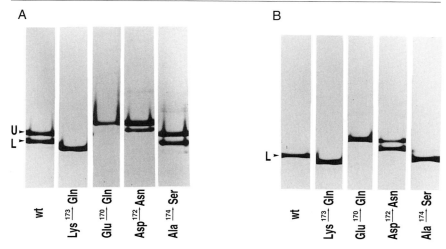

FIG. 2. Nondenaturing PAGE of wild-type and mutant human lipoyl domains purified from cultures unsupplemented (A) or supplemented (B) with [³H]lipoic acid. Each lane contains 3 μg of either wild-type or mutant lipoyl domain preparations stained with Coomassie Brilliant Blue. Mutant lipoyl domains are each designated by the amino acid change incorporated into the polypeptide. The positions of unlipoylated (U) and lipoylated (L) wild-type domains are indicated.

This assay was used to analyze the effect on protein lipoylation of mutating several conserved residues of the human lipoyl domain. Figure 2 illustrates the approximate amounts of unlipoylated and lipoylated domains generated from wild-type or mutated subgenes, expressed in cultures ± [³H]lipoic acid. Mutating the lipoyllysine (Lys[173] → Gln) and Glu[170] (→ Gln) residues of the human domain abolished lipoylation even in cultures supplemented with [³H]lipoic acid, because no radioactivity was detected in these domain species. In contrast, residues Asp[172] and Ala[174], which flank the lipoyllysine and are highly conserved in the primary structures of all known lipoyl domains, do not appear to be required for lipoylation of the human domain by the *E. coli* lipoylating system.

*Measuring Lipoylation in Vitro*

The assay described here has previously been used to analyze whether the unlipoylated human recombinant domain, generated on overexpression in *E. coli,* can be subsequently lipoylated on incubation with partially purified *E. coli* lipoate ligase.[13] It has also been used to follow purification of the *E. coli* lipoate protein ligase, using *E. coli* pa-lipoyl domain as substrate.[18,19]

*Principle.* To measure lipoate ligase activity by following either [³H] lipoic acid incorporation into apo-domain substrates or the shift in electrophoretic mobility of the apo-domain on nondenaturing PAGE.

*Preparation of [2-³H]Lipoic Acid.* This can be done commercially (Amersham, Amersham, U.K.) or can be synthesized using a catalyzed-exchange procedure involving metalation of lipoic acid with lithium at C-2, followed by treatment with tritiated water.[20,21]

*Assay Methodology.* Human pa-domain (0.2–0.5 μg) is incubated with ATP (80 μM), DL-lipoic acid or DL-[2-³H]lipoic acid (60 μM), MgCl₂ (3.2 mM), sodium phosphate buffer (25 mM, pH 7.0), and partially purified *E. coli* lipoate ligase (specific activity, 0.12 nmol of *E. coli* domain modified per minute per milligram of protein), in a final volume of 30 μl for 4–6 hr at 30°. The incubation time can be varied depending on the substrate and activity of the lipoate ligase preparation being tested (see Refs. 19 and 20). The reaction is terminated by heating at 70° for 1 min. Modified domains are detected by migration on nondenaturing PAGE and quantified by densitometry or by incorporation of DL-[2-³H]lipoic acid. For the latter method electrophoresed proteins are blotted onto Fluorotrans membrane (Pall Corp.) and exposed to Hyperfilm ³H (Amersham) for 2 weeks at room temperature. Figure 3a illustrates that after incubation with partially purified lipoate ligase, the majority of the human domain becomes lipoylated, and that this reaction requires both ATP and lipoic acid. When lipoylation is carried out using [2-³H]lipoic acid, radioactivity is detected only in the faster migrating form (Fig. 3b). Radioactive incorporation can also be assayed by acid precipitation of reaction mixtures onto filter disks, followed by scintillation counting.[19] However, it has been reported that the lipoate protein ligase activities obtained using this method are only one-tenth of those obtained by the electrophoretic assay.[19]

## Characterization of Recombinant Lipoyl Domain

### Mass Spectroscopy Analysis

Analysis of recombinant lipoyl domains by mass spectroscopy is a useful tool to ensure that the differences in mobility observed on nondenaturing

[18] D. E. Brookfield, J. Green, S. T. Ali, R. S. Machado, and J. R. Guest, *FEBS Lett.* **295,** 13 (1991).

[19] D. E. Green, T. W. Morris, J. Green, J. E. Cronan, Jr., and J. R. Guest, *Biochem. J.* **309,** 853 (1995).

[20] P. E. Pfeffer and L. S. Silbert, *J. Org. Chem.* **35,** 262 (1970).

[21] P. E. Pfeffer, L. S. Silbert, and J. M. Chrinko, *J. Org. Chem.* **37,** 1256 (1972).

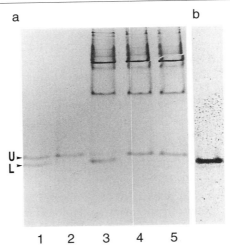

F<small>IG</small>. 3. Lipoylation *in vitro* of human apo-domain. (a) Purified apo-domain (lane 2) was incubated with lipoate ligase in the presence of both ATP and [³H]lipoic acid (lane 3), in the presence of ATP (lane 4), and in the presence of lipoic acid (lane 5). The positions of unlipoylated (U) and lipoylated (L) domains are indicated (lane 1). (b) Corresponding autoradiograph of lane 3. (Reprinted from Ref. 13, with permission.)

PAGE are a consequence of lipoylation and are not due to proteolysis. In addition, there have been several reports of modification of recombinant domains with octanoic acid (a biosynthetic precursor of lipoic acid), which can be detected using mass spectroscopy.[12–14]

*Method.* The human lipoyl domain is prepared for electrospray mass spectroscopy analysis by freeze drying from $NH_4HCO_3$ and water twice.[13] Samples are redissolved in methanol–water–acetic acid (50:49:1 by volume), to give a final concentration of 10–20 n$M$. Mass analysis is performed on a VG BioQ quadrupole mass spectrometer using myoglobin as the calibration standard. The faster migrating band on nondenaturing PAGE gave a clean spectrum showing a single component with a relative molecular weight of 11,650 ± 2.0 (SD), which corresponds well to the calculated mass of a nonlipoylated domain (11,652). Mass analysis of the slower migrating domain form gave a major component with a relative molecular weight of 11,842 ± 2, corresponding to the mass of a lipoylated domain. A minor component with a mass representative of an octanoylated domain was also detected in this domain preparation. Mass spectroscopy analysis of *E. coli*[14] and *B. stearothermophilus*[12] lipoyl domains has also been reported.

## Acetylation of Lipoyl Domains

*Principle.* Acetylation of lipoyl domains can also be used to verify that the slower migrating protein on nondenaturing PAGE is correctly modified with lipoic acid. In the presence of [2-[14]C]pyruvate only correctly lipoylated domains should be reductively acetylated by the E1 component of the PDH complex. This can be measured by monitoring the incorporation of [[14]C]acetyl groups.[22]

*Method.* Recombinant human apo-domain preparation (~2 $\mu$g) is incubated at 25° in 50 m$M$ NaPO$_4$ (pH 6.5)–1 m$M$ EDTA, containing 0.1 $\mu$g of human E1p, 0.4 m$M$ thiamine pyrophosphate (TPP) 0.5 m$M$ $N$-ethylmaleimide (NEM), and 0.2 m$M$ [2-[14]C]pyruvate (30,000 cpm/nmol), in a final volume of 20 $\mu$l. Following incubation for 30 min at room temperature, the reaction mixture is electrophoresed on 15% (w/v) nondenaturing gels and transferred onto Fluorotrans membrane (Pall Corp.). The membrane is stained for protein with Coomassie Brilliant Blue and exposed to Amersham Hyperfilm [14]C for 2 weeks at room temperature to allow detection of radiolabeled acetyl groups. Radioactive groups should be incorporated only into the slower migrating form of the domain on nondenaturing PAGE. In addition to verifying the domain to be correctly lipoylated, this result also suggests the recombinant lipoyl domain to be correctly folded as the E1 polypeptide is able to recognize the domain as a substrate. Adaptations of this procedure have been used to study reductive acetylation of *E. coli*[11] and *B. stearothermophilus*[12] domains.

## Discussion

The generation of substrate amounts of apo-lipoyl domains and the development of assays with which to follow protein lipoylation have led to a dramatic increase in our understanding of this important posttranslational modification event. A lipoyl protein ligase has been purified from *E. coli* by virtue of its ability to modify apo-domains of the *E. coli* PDH complex[20,23] and cloning of its structural gene (*lplA*) has been reported.[20] The ligase utilizes lipoic acid and ATP as substrates to lipoylate the *E. coli* PDH and OGDH complexes (*E. coli* do not possess a BCOADH complex), and also the H-protein of the *E. coli* glycine cleavage system, illustrating that the prokaryotic enzyme possesses both lipoic acid-activating and lipoylation activities. Surprisingly, genetic data strongly support the presence of two distinct lipoate protein ligases in *E. coli* and a two-pathway model of protein

[22] D. M. Bleile, M. L. Hackert, F. H. Pettit, and L. J. Reed, *J. Biol. Chem.* **256**, 514 (1981).
[23] T. W. Morris, K. E. Reed, and J. E. Cronan, Jr., *J. Biol. Chem.* **269**, 16091 (1994).

lipoylation has been proposed.[24] Molecular recognition by the *E. coli* lipoate ligase of the lipoyl domain substrates does not appear to require the highly conserved residues flanking the lipoyllysine (this work and Ref. 24). Instead, selection of the correct target lysine appears to be dependent on recognition of a novel structure in which the lipoyllysine residue is prominently displayed in the sharp β turn of the domain.[25] Finally, an enzyme from bovine mitochondria has been purified that can only utilize lipoyl-AMP as a substrate to modify bovine H-protein, and thus lacks the lipoic acid-activating present in the *E. coli* lipoate ligase.[20] Whether this reflects a true difference between prokaryotic and eukaryotic lipoylation enzymes has yet to be proven.

## Acknowledgments

I thank Professor S. J. Yeaman and Dr. A. G. Diamond for assistance and valuable discussions during the course of this work, Dr. A. K. Masters for labeled lipoate, Mr. J. M. Palmer for human E1p, and the Science and Engineering Research Council for financial support. The mass spectroscopy analysis detailed in this chapter was kindly performed by Dr. N. G. Wallis at the Cambridge Centre for Molecular Recognition (Department of Biochemistry, University of Cambridge).

[24] T. W. Morris, K. E. Reed, and J. E. Cronan, Jr., *J. Bacteriol.* **177,** 1 (1995).
[25] N. G. Wallis and R. N. Perham, *J. Mol. Biol.* **236,** 209 (1994).

## [22] Purification and Properties of Brain Lipoamidase

*By* JUN OIZUMI and KOU HAYAKAWA

### Introduction

Lipoamidase (lipoyl-X hydrolase; EC number not yet assigned), originally discovered in *Streptococcus faecalis* 10C1, is an enzyme that releases lipoic acid from pyruvate dehydrogenase complex in *Escherichia coli*.[1] The enzyme also hydrolyzes several lipoylamide compounds such as lipoamide, lipoylalanine, and lipoyllysine.[2] However, differences in the enzyme, lipoyl-X hydrolase, among bacterial species have already been indicated by Nakamura.[3] For example, the enzyme of *S. faecalis* R recognizes only

[1] L. J. Reed, M. Koike, M. E. Levitch, and F. R. Leach, *J. Biol. Chem.* **232,** 143 (1958).
[2] K. Suzuki and L. J. Reed, *J. Biol. Chem.* **238,** 4021 (1963).
[3] M. Nakamura, *Vitamins (Jpn.)* **26,** 217 (1962).

lipoyllysine and does not recognize lipoylglutamate, whereas the enzyme of *Corynebacterium bovis* 187 recognizes both lipoylglutamate and lipoyllysine.

In animals, Saito[4] has reported that the tissues of rabbit and humans are able to hydrolyze lipoamide, whereas Suzuki and Reed[2] have reported that bovine tissues (heart, liver, and kidney) cannot. Saito[5] has also reported the presence of lipoyl-X hydrolase (lipoamidase) in rat liver, which liberates lipoic acid from both lipoamide and $\alpha$-ketoglutarate dehydrogenase complex.

In 1985 Mesavage and Wolf[6] reported that lipoic acid and *p*-aminobenzoic acid were produced from lipoyl-*p*-aminobenzoic acid (LPAB) in human serum (i.e., LPAB hydrolase was present in human

Lipoyl-*p*-aminobenzoic acid

serum). Soon after, we developed the high-performance liquid chromatography (HPLC)–fluorimetric LPAB hydrolase assay method,[7] and reported that LPAB hydrolase activity in human serum was strongly correlated with biotinidase (BPAB hydrolase) activity ($r = 0.893$; $p < 0.01$), and that the correlation was stronger in female ($r = 0.936$) than in male specimens ($r = 0.881$).[8] Human serum LPAB hydrolase (lipoamidase) has also been shown to be identical to human serum biotinidase (BPAB hydrolase; $M_r$ 76,000).[9-11]

Unlike the case for human serum, lipoamidase (LPAB hydrolase) and biotinidase (BPAB hydrolase) were independently isolated from guinea pig liver: $M_r$ 60,000 for lipoamidase and $M_r$ 70,000 for biotinidase.[12] Isoelectric points of the liver microsomal lipoamidase and biotinidase of guinea pig liver were 5.7 and 6.3, respectively.[12] Although the enzyme from human liver has not yet been studied, lipoamidase and biotinidase activities of human cerebrum show the same p$I$ of 5.0,[9] and the p$I$ of human serum

[4] J. Saito, *Vitamins (Jpn.)* **21,** 359 (1960).
[5] J. Saito, *Vitamins (Jpn.)* **41,** 216 (1970).
[6] W. C. Mesavage and B. Wolf, *Am. J. Hum. Genet.* **37,** A13 (1985). [Abstract]
[7] K. Hayakawa and J. Oizumi, *J. Chromatogr.* **423,** 304 (1987).
[8] K. Hayakawa and J. Oizumi, *Enzyme* **40,** 30 (1988).
[9] J. Oizumi and K. Hayakawa, *Biochem. Biophys. Res. Commun.* **162,** 658 (1989).
[10] C. L. Garganta and B. Wolf, *Clin. Chim. Acta* **189,** 313 (1990).
[11] L. Nilsson and B. Kagedahl, *Biochem. J.* **291,** 545 (1993).
[12] J. Oizumi and K. Hayakawa, *Biochim. Biophys. Acta* **991,** 410 (1989).

lipoamidase/biotinidase has been shown to be 4.3.[13] These findings may be due to species differences.

Although species differences may be present, pig brain lipoamidase (LPAB hydrolase)[14] has been studied and purified to clarify lipoate-responsive neuronal diseases in humans. Lipoamidase (LPAB hydrolase) from pig brain is a glycoprotein enzyme of $M_r$ 140,000 and p$I$ 5.4.[14] The enzyme shows the unique characteristics of multiple hydrolase; that is, in spite of its inability to recognize the biotinylamide substrates, it recognizes and hydrolyzes LPAB, lipoamide, lipoyllysine, some physiological peptides, and acetylcholine.[14,15] Although the physiological role of pig brain lipoamidase has not yet clearly been determined, we have proposed its role to be that of a "deanchoring" enzyme in a signal-transducing system in pig brain.[16] In line with its importance in signal transduction, we have synthesized a novel lipoamidase substrate, lipoyl-6-aminoquinoline (LAQ).[17] This new substrate is also expected to be helpful in studying human diseases, such as neuronal brain diseases and cancer.[17]

## Assay Methods

### Lipoamidase Assay

*Principle.* The assay is based on the measurement of the rate of liberation of $p$-aminobenzoic acid (PAB) from the synthetic substrate lipoyl-$p$-aminobenzoic acid (LPAB). Although the photometric PAB assay of Bratton and Marshall[18] has been widely used, various interferences frequently occur. Therefore, the PAB assay by HPLC with fluorimetric detection has been devised[19] and applied to the assay of lipoamidase (LPAB hydrolase).[7]

*Instruments.* A Waters model 600E HPLC (Waters, Milford, MA) with a gradient elution unit is used. The column is a $50 \times 4.6$ mm i.d. stainless steel tube packed manually with spherical, $10$-$\mu$m silica gel particles chemically bonded with octadecylsilane (ODS) [Develosil ODS-10 (Nomura Chemical Co., Seto, Aichi, Japan); a prepacked column can now be obtained from the manufacturer]. The sample injection unit used is a diaphragm-type injector (model U6K; Waters) with a 2-ml sample-loading loop. A line filter

[13] J. Oizumi and K. Hayakawa, *BioFactors* **1**, 179 (1988).
[14] J. Oizumi and K. Hayakawa, *Biochem. J.* **266**, 427 (1990).
[15] J. Oizumi and K. Hayakawa, *Biochem. J.* **271**, 45 (1990).
[16] J. Oizumi and K. Hayakawa, *Mol. Biochem.* **115**, 11 (1992).
[17] K. Yoshikawa, K. Hayakawa, N. Katsumata, T. Tanaka, T. Kimura, and K. Yamauchi, *J. Chromatogr. B* **679**, 41 (1996).
[18] A. C. Bratton and E. K. Marshall, Jr., *J. Biol. Chem.* **128**, 537 (1939).
[19] K. Hayakawa and J. Oizumi, *J. Chromatogr.* **383**, 148 (1986).

and a miniguard column (10 × 4 mm i.d., packed with Develosil ODS-10 gel) (GL Sciences, Tokyo, Japan) are inserted between the injector and the column. Detection of PAB is carried out with a fluorimeter (F-3000; Hitachi Co., Tokyo, Japan), using a flow-through cell (cell volume, 18 μl): excitation wavelength, 276 nm; emission wavelength, 340 nm.

*p-Aminobenzoic Acid Assay.* A stepwide elution is carried out for PAB at 2.5 min. Solvent A (0.1 $M$ sodium phosphate buffer adjusted to pH 2.1 with orthophosphoric acid) is applied at a flow rate of 1.5 ml/min for 2.5 min, then solvent B (methanol, the washing solvent) is applied for 3 min at a flow rate of 3.0 ml/min. Solvent A is then applied for 4 min at the same rate and for 2.5 min at a flow rate of 1.5 ml/min, to return the column to the initial condition. This cyclic HPLC assay requires 12 min/sample.

*Reagents*

Sodium phosphate buffer: 0.1 $M$, pH 7.0, containing 1 m$M$ (452 mg/ liter) Na$_4$EDTA and 10 m$M$ (781 mg/liter) 2-mercaptoethanol
Lipoyl-*p*-aminobenzoic acid (LPAB)[8]: Synthesized LPAB is stored at 333 μmol/liter of chloroform at −80°
Acetonitrile
Methanol
Standard PAB solution: PAB (10 mg) is dissolved in 1 liter of distilled water (0.005 ml contains 365 pmol of PAB)

*Procedure.* To a 10 × 75 mm Pyrex test tube is added 0.03 ml of a chloroform solution of synthesized LPAB (10 nmol), and the solution is dried under a stream of nitrogen gas. Next, 0.09 ml of 0.1 $M$ sodium phosphate buffer containing 10 m$M$ 2-mercaptoethanol, 1 m$M$ EDTA, and 0.01 ml of purified enzyme or brain homogenates is added to make a final volume of 0.10 ml. The solution is thoroughly mixed, using a Vortex mixer (Scientific Industries, Inc., Bohemia, NY) and is incubated at 37°. The reaction is allowed to proceed for an appropriate time interval (10 hr) and is stopped by adding 0.2 ml of acetonitrile–methanol solution (1:1, v/v). After mixing and centrifuging (1500 $g$ for 15 min at 4°) a portion (0.005 ml) of the clear supernatant (0.3 ml) is injected into the HPLC system.

*Specific Activity.* The amount of PAB in 5 μl of injected sample (AP, amount of PAB) is first calculated using an equal volume of an external standard solution: AP (picomoles) = [peak height (mm) of PAB in 5 μl of injected sample]/[peak height (mm) of PAB in 5 μl of injected standard solution] × 365 pmol. Specific activity (picomoles per minute per milligram) = AP (picomoles) × [0.3 (ml)/0.005 (ml)]/[reaction time (minutes) × 0.01 (ml) × protein concentration of the enzyme solution (mg/ ml)]. Because the International Union of Biochemistry has defined 1 unit of enzyme as the amount that hydrolyzes 1 μmol of substrate (LPAB) per minute, we express the specific activity of LPAB hydrolase (picomoles per minute per milligram of protein) as microunits per milligram of protein. Protein is determined by the method of Lowry *et al.*[20] or by a bicinchoninic acid (BCA) protein assay kit from Pierce (Rockford, IL). Bovine serum albumin is used as a standard protein.

*Purification Procedure*

See Ref. 14 for details of this procedure.

*Reagents*

Nonidet P-40 (NP-40): Nonylphenyl polyethyleneglycol (9) ether
Glycerol
Acetone

*Step 1. Pig brain homogenates:* Pig brain is purchased from Pel-Freez Biologicals (Rogers, AR). Brain (82 g) is homogenized in 0.32 *M* sucrose containing 1 m*M* sodium phosphate, pH 7.4 (800 ml).

*Step 2. Ultracentrifugation:* The homogenate is centrifuged at 105,000 *g* for 90 min at 4°, and the resultant precipitate is dispersed in 0.05 *M* phosphate buffer (pH 6.0) containing 0.2 m*M* EDTA, 10% (v/v) glycerol, and 1% (v/v) NP-40 (600 ml).

*Step 3. CM-cellulose:* The dispersed membrane fractions are mixed with 500 ml of CM-cellulose and filtered through a gauze cloth. The filtrate (1.2 liters) is diluted to one-fifth of its original concentration and adjusted to 1 m*M* EDTA, 10% (v/v) glycerol, and 0.5% (v/v) NP-40 (pH 8).

*Step 4. DEAE-cellulose chromatography:* A portion (2 liters) of the diluted filtrate is loaded onto the DEAE-cellulose column (40 × 1.9 cm i.d.) and developed with a linear gradient of NaCl (from 0 to 0.5 *M*; total volume, 500 ml). The enzyme-containing fractions are collected and concentrated by an ultrafiltration membrane (PM-10; Amicon, Danvers, MA). This step is repeated twice.

*Step 5. Sephadex G-200 chromatography:* The concentrated DEAE-cellulose-enriched enzyme is then separated by Sephadex G-200 column (100 × 1.9 cm i.d.). The mobile phase is 0.05 *M* sodium phosphate buffer (pH 6.8) containing 1 m*M* EDTA, 0.1% (v/v) NP-40, and 10% (v/v) glycerol. The enzyme-containing fractions are collected and concentrated as described. This step is repeated eight times.

*Step 6. Sepharose CL-4B chromatography:* The concentrated enzyme is then separated on a Sepharose CL-4B column (100 × 1.9 cm i.d.) with the mobile phase as described, except for a nonionic detergent (NP-40)

---

[20] O. H. Lowry, N. J. Rosebrough, A. L. Farr, and R. J. Randall, *J. Biol. Chem.* **193,** 265 (1951).

TABLE I
PURIFICATION OF MEMBRANE LIPOAMIDASE (LPAB HYDROLASE) FROM PIG BRAIN[a]

| Step | Volume (ml) | Protein (mg) | Specific activity (microunits/mg) | Recovery (%) |
|------|------------|--------------|-----------------------------------|--------------|
| 1. Homogenate | 800 | 7792 | 14.2 | 100 |
| 2. Membrane fraction | 400 | 3656 | 27.8 | 92 |
| 3. CM-cellulose | 1000 | 2830 | 27.8 | 71 |
| 4. DEAE-cellulose | 800 | 1104 | 47.0 | 47 |
| 5. Sephadex G-200 | 320 | 826 | 55.7 | 41 |
| 6. Sepharose CL-4B | 450 | 329 | 122 | 36 |
| 7. WGA–agarose | 256 | 59 | 238 | 13 |
| 8. Serotonin HPLC | 14 | 1.54 | 8530 | 12 |

[a] Modified from J. Oizumi and K. Hayakawa, *Biochem. J.* **266,** 431 (1990), with kind permission from The Biochemical Society.

concentration of 0.5% (v/v). This gel-permeation chromatography is repeated five times, and the enzyme fractions are collected for concentration by PM-10.

*Step 7. Wheat-germ agglutinin chromatography:* The concentrated enzyme solution (25 ml) is dialyzed against 2 liters of a 0.5% (v/v) NP-40 solution containing 5 m$M$ sodium phosphate (pH 7.0), 10% (v/v) glycerol, and 100 m$M$ NaCl. The dialyzed enzyme solution is loaded onto a wheat germ agglutinin (WGA)–agarose column (40 × 10 mm i.d.). The bound enzyme is then specifically eluted by a linear gradient of 0–0.3 $M$ $N$-acetylglucosamine. The enzyme is eluted at an $N$-acetylglucosamine concentration of 0.06 $M$. The enzyme fractions are concentrated and dialyzed against a 0.5% (v/v) NP-40 solution containing 5% (v/v) acetone, 5% (v/v) glycerol, 1 m$M$ EDTA, 1 m$M$ 2-mercaptoethanol, and 5 m$M$ sodium phosphate (pH 7.0).

*Step 8. HPLC separation by serotonin column:* The previously described solution is loaded onto a serotonin HPLC column (150 × 4.0 mm i.d.; Seikagaku Co., Tokyo, Japan) and eluted with a linear gradient of 5–50 m$M$ phosphate for 30 min at a flow rate of 1.0 ml/min. The enzyme-containing solutions are concentrated and stored at −80°. The enzyme shows a specific activity of 8530 microunits/mg of protein. An outline of purification is given in Table I and also in Ref. 14.

*Alternative Purification Procedure*

The following alternative procedure is described in detail in Oizumi and Hayakawa.[21]

[21] J. Oizumi and K. Hayakawa, *Biomed. Chromatogr.* **3,** 274 (1989).

*Preparation of Arg-Phe-NH₂-Bound Gel:* Arg-Phe-NH₂ (5 mg; Peninsula Laboratories, Inc., Belmont, CA) is attached to 5 ml of Formyl-Cellulofine gel (Seikagaku Co.) according to the protocol of the supplier

*Alternative step 8. Arg-Phe-NH₂ column:* The partially purified enzyme solution from step 7 (WGA-chromatography step) is loaded onto an Arg-Phe-NH₂ column (40 × 10 mm i.d.). The column is preequilibrated with 0.1 $M$ sodium phosphate (pH 7.0) containing 0.3 $M$ NaCl, 1 m$M$ EDTA, 1 m$M$ 2-mercaptoethanol, and 5% (v/v) glycerol. The enzyme activity is eluted with the same solvent containing 1% (v/v) NP-40. This method yields the enzyme with a specific activity of 9520 microunits/mg of protein.[21]

## Properties

*Distribution of Lipoyl-p-aminobenzoic Acid Hydrolase in Pig Brain.* The cerebellum showed a higher specific activity (microunits per milligram) of 26.8 as compared to the cerebrum with 14.4 and the medulla with 7.6. More than 90% of lipoamidase activities were present in membrane subfractions. The nerve terminal button fraction seemed to have higher enzyme activity than did myelin membranes. In a human cerebrum autopsy (1 year 2 months, male), LPAB hydrolase activity was present in gray matter (19.6 microunits/mg) and was absent in white matter.[17]

*Chemical Characteristics of Brain Lipoamidase.* The relative molecular weight ($M_r$) of purified brain lipoamidase of pig was 140,000 as assessed by sodium dodecyl sulfate–polyacrylamide gel electrophoresis (SDS–PAGE), present as a single polypeptide.[14] Lipoamidase was specifically stained with periodic acid–Schiff (PAS) stain, which strongly indicated that lipoamidase was a glycoprotein enzyme. Treatment of the enzyme with either $N$-glycanase or glycopeptidase F shifted the enzyme protein in the direction of slightly smaller size (from $M_r$ 140,000 to 132,000). The amino acid composition of pig brain lipoamidase was similar to that of guinea pig liver lipoamidase ($M_r$ 60,000), and both lipoamidases are integral membrane protein enzymes.[12] Unexpectedly, the amino acid composition of acetylcholine receptor from *Electrophorus electricus* is similar to that of brain lipoamidase. The p$I$ value of purified pig brain lipoamidase was 5.4. The p$I$ values of lipoamidases as estimated from isoelectric focusing of the microsomes of guinea pig livers and of human cerebrum were 5.7 and 5.0, respectively. The pH optimum of brain lipoamidase using LPAB as substrate was a relatively broad range of pH 6.0–7.5. A similar optimum pH was also obtained when measuring the release of membrane-bound enzyme (5′-nucleotidase) from pig brain microsomes by this enzyme.[16] A relatively sharp pH optimum at pH 8.0 using LPAB as substrate has been obtained for lipoamidase from guinea pig livers.[12]

*Kinetic Properties and Substrate Specificity.* The $K_m$ values for LPAB and reduced-form LPAB were 31 and 12 $\mu M$, respectively. Biotinyl-*p*-aminobenzoate (BPAB) was not hydrolyzed. Smaller $K_m$ values were always obtained for the reduced forms of lipoyllysine and lipoamide. Competitive inhibition by lipoic acid and reduced lipoic acid (dihydrolipoic acid) against LPAB hydrolysis was also observed. The $K_i$ values for lipoate and dihydrolipoate were 1.8 and 0.40 m$M$, respectively. Competitive inhibition by biotin was not observed up to a concentration of 20 m$M$. The releasing reaction of 5′-nucleotidase from pig brain microsomes by brain lipoamidase was competitively inhibited ($K_i = 0.036$ m$M$) by lipoyllysine.[16] Brain lipoamidase showed typical characteristics of multiple hydrolase[15] (i.e., amide, ester, and peptide bonds were hydrolyzed). The enzyme specifically recognized the whole molecular structure of the substrate, whereas it loosely recognized the bond structure of the substrate (e.g., the dipeptide Asp-Phe was not hydrolyzed, whereas the methyl ester of Asp-Phe (aspartame) was). The exopeptidase activity was demonstrated by lipoamidase; however, peptides longer than the hexamer seemed not to be substrates. Lipoyl esters, which were electrically neutral, exhibited higher specificity with longer acyl groups. Molecular mass and molecular hydrophobicity (hydropathy) seemed to determine the substrate specificity. Lipoyllysine, acetylcholine, and oligopeptides were hydrolyzed at similar $K_m$ values (ca. 0.5 m$M$); however, acetylcholine was hydrolyzed at a rate 100 times higher. Although many similar specificities were found between electric eel acetylcholinesterase and lipoamidase, a distinctly different specificity was demonstrated with lipoyl compounds.

*Inhibitor.* The amidase reaction (LPAB hydrolysis) of lipoamidase showed characteristics of both serine- and thiol-type reactions (i.e., diisopropylfluorophosphate (DFP) and *p*-chloromercuribenzoate (PCMB) at 50 $\mu M$ inhibited 56 and 50%, respectively). Furthermore, 1,10-phenanthroline (metalloproteinase inhibitor) at 1 m$M$ inhibited LPAB hydrolysis (31%). LPAB hydrolysis of lipoamidase was inhibited 89 and 46% by Fe-EDTA (0.1 m$M$) and hemin (0.01 m$M$), respectively.[22] Acetylcholine hydrolysis by lipoamidase was inhibited almost completely by 0.01 m$M$ acetylcholinesterase inhibitors, except arecoline. The reaction of releasing 5′-nucleotidase from pig brain microsomes by brain lipoamidase was strongly inhibited by 0.05 m$M$ DFP and ebelactone B.[16] These results indicate that the active center of brain lipoamidase is of a complex type.

---

[22] J. Oizumi and K. Hayakawa, *BioFactors* **2,** 127 (1989).

*Effect of Phospholipids.* (See Ref. 23.) Approximately twofold activation of LPAB hydrolase activity occurred on the addition of phosphatidylethanolamine (at 5 mg/ml). However, phosphatidylserine, cardiolipin, and phosphatidic acid reduced the enzyme activity by approximately 80%. These characteristics of brain lipoamidase as an integral membrane enzyme are similar to adenylate cyclase, and in contrast to ATPase.

## Acknowledgment

This work was supported by the Ministry of Health and Welfare, Japan.

23 J. Oizumi and K. Hayakawa, *Experientia* **46,** 459 (1990).

# Section IV

# Pantothenic Acid, Coenzyme A, and Derivatives

# [23] Measurement of Pantothenic Acid and Hopantenic Acid by Gas Chromatography–Mass Spectroscopy

*By* Kiyoshi Banno

## Introduction

Hopantenic acid (HOPA), a natural homolog of pantothenic acid (PaA) containing $\gamma$-aminobutyric acid (GABA) in place of $\beta$-alanine, was first discovered in biological fluids by paper chromatography.[1,2] Its calcium salt and pharmacologic properties have been used to improve blood circulation and metabolism in the brain. There are many reports on the pharmacokinetics and assay of HOPA. Biserte and co-workers succeeded in identifying hopantenic acid as a GABA derivative by paper chromatography in human urine,[3] and also in renal and hepatic tissue of normal rats.[4–6] Because hopantenic acid is used pharmacologically it is important to be able to determine HOPA and PaA separately for pharmacokinetic studies.

Gas chromatography–mass fragmentography (GC–MF) has been used to assay HOPA after administration of calcium hopantenate.[7] Other methods for the assay of HOPA have also been reported, such as the GC–MF determination of pantoyllactone, which is a product of HOPA hydrolysis,[8] the colorimetric determination of GABA with sodium 1,2-naphthoquinone 4-sulfate,[9] and a high-performance liquid chromatographic (HPLC) method with 9-anthryldiazomethane as a fluorescent derivative.[10] A bioassay method using *Lactobacillus arabinosus*,[11] and the determination of trimethylsilylated pantoyllactone by GC–MF, were developed for the assay of

---

[1] P. Boulanger, G. Biserte, and F. Courtot, *Bull. Soc. Chim. Biol.* **34,** 366 (1952).

[2] A. G. Moiseenok, V. M. Kopelevich, M. A. Izraelit, and L. M. Shmuilovich, *Farmakal. Toksikol.* **36,** 489 (1973).

[3] G. Biserte, R. Plaquest, and P. Boulanger, *Bull. Soc. Chim. Biol.* **37,** 7 (1955).

[4] G. Biserte, P. Boulanger, A. Finot, M. Davril, E. Sacquet, and H. Charlier, *C.R. Acad. Sci.* **260,** 3219 (1965).

[5] P. Boulanger, G. Biserte, M. Davril, and M. Rache, *C.R. Acad. Sci.* **265,** 157 (1967).

[6] M. Davril, G. Biserte, and P. Boulanger, *Biochimie* **53,** 419 (1971).

[7] Y. Umeno, K. Nakai, E. Matsushima, and T. Marunaka, *J. Chromatogr.* **226,** 333 (1981).

[8] M. Anetai, T. Takahashi, H. Ogawa, and H. Kaneshima, *Hokkaidoritsu Eisei Kenkyushoho* **33,** 138 (1983).

[9] H. Terada, T. Hayashi, S. Kawai, and T. Ohno, *J. Chromatogr.* **130,** 281 (1977).

[10] T. Fukuyama, T. Maki, and M. Matsuoka, Japanese Patent Publication No. 86655 (1986). [Unexamined]

[11] H. R. Skeggs and L. D. Wright, *J. Biol. Chem.* **156,** 21 (1944).

PaA.[12] The methods described here, however, were not developed for the simultaneous determination of PaA and HOPA, and are not sensitive enough for pharmacokinetic studies.

The simultaneous rapid microanalysis for PaA and HOPA in biological samples and natural products by GC–MF with a wide-bore column and multiple ion detection (MID) is described here. We use 5-[(2,4-dihydroxy-3,3-dimethyl-1-oxobutyl)amino]pentanoic acid calcium salt as an internal standard (IS) to measure PaA and HOPA simultaneously in plasma samples from humans, monkeys, dogs, pigs, rabbits, mice, rats, chickens, and soft-shelled turtles; in brain samples of chicken and soft-shelled turtle; and also in natural products (rice, green tea, and dried yeast).

## Sample Preparation

### Materials

Calcium hopantenate and 5-[(2,4-dihydroxy-3,3-dimethyl-1-oxobutyl)amino]pentanoic acid calcium[13] salt used as an internal standard are synthesized at Tanabe Seiyaku Co., Ltd. (Osaka, Japan). The calcium pantothenate used is a reagent of the Pharmacopoeia of Japan. Bis(trimethylsilyl)trifluoroacetamide (BSTFA) for the silylation reagent is obtained from GL Sciences, Inc. (Tokyo, Japan). The ion-exchange resin (H type, MCI GEL CK08P) is obtained from Mitsubishi Chemical Corporation (Tokyo, Japan).

### Plasma Samples

Blood samples are collected in heparinized containers and centrifuged to obtain the plasma.

The plasma samples (1.0 ml) are diluted to 2.0 ml with distilled water, applied to 2.5 ml of ion-exchange resin (H type, MCI GEL CK08P, 17 cm × 10 mm i.d.), and eluted several times with 1 ml of distilled water to make 6 ml. Chloroform (3.0 ml) is added to the eluate and the mixture is shaken vigorously for 5 min. The aqueous layer (5.0 ml) is taken, and after addition of 6 $M$ hydrochloric acid (0.5 ml), ammonium sulfate (5.0 g), and ethyl acetate (20.0 ml), the mixture is shaken vigorously for 15 min and centrifuged at 1800 $g$ for 5 min at 4°. The supernatant is taken and filtered through a filter paper. The filtrate is extracted with ethyl acetate (20.0 ml) again and centrifuged at 1800 $g$ for 5 min at 4°. The supernatant

---

[12] P. Tarli, *Anal. Biochem.* **42,** 8 (1971).
[13] Y. Nishizawa and T. Kodama, *Vitamin* **33,** 589 (1966).

is filtered as described. The ethyl acetate layer containing PaA and HOPA is collected and evaporated at 40° under nitrogen gas to a volume of ca. 0.5 ml. The concentrate is transferred to a 3-ml reaction vial by washing with ethyl acetate (1.0 ml) and then mixed with ethyl acetate containing 100.0 ng of IS. The mixture is evaporated to dryness again, as described previously, before derivatization.

*Brain Samples*

Fresh brain samples of chicken (0.60 g) or soft-shelled turtle (0.66 g) are homogenized with 0.005 $M$ potassium hydroxide solution (2 ml) and centrifuged at 21,000 $g$ for 20 min at 4°. The supernatant (1.0 ml) is applied to 2.5 ml of ion-exchange resion (H type, MCI GEL CK08P, 17 cm × 10 mm i.d.) and taken through the same procedure as the plasma samples.

*Natural Products*

Distilled water (20.0 ml) is added to boiled rice (10.0 g), green tea (1.0 g), or dried yeast (0.1 g), and shaken vigorously at 70° for 30 min. After centrifugation at 1800 $g$ for 5 min at 4°, 2.0 ml of supernatant is applied to 2.5 ml of ion-exchange resin (H type, MCI GEL CK08P, 17 cm × 10 mm i.d.) and taken through the same procedure as described for the plasma samples.

Derivatization

The derivatization of hydroxy and carboxyl groups with several types of trimethylsilylation reagent has been investigated. It has been found that the reaction proceeds quantitatively at 80° for 60 min with BSTFA in a screw-capped reaction vial in the presence of pyridine. The best concentration of BSTFA in pyridine has been found to be 50% (v/v). Pyridine is an effective catalyst for trimethylsilylation of hydroxy and carboxyl groups in PaA, HOPA, and IS. The trimethylsilyl derivatives of PaA, HOPA, and IS show a high detection sensitivity for the MID method and are completely separated by a wide-bore fused-silica DB-17 column (15 m × 0.53 mm i.d.) (Fig. 1).

Mass Fragmentographic Analysis

The gas chromatograph–mass spectrometer is a Hitachi M-80A equipped with an M-003 computer system. The wide-bore fused-silica column is coated with DB-17 (15 m × 0.53 mm i.d.) (J & W Scientific, Folsom,

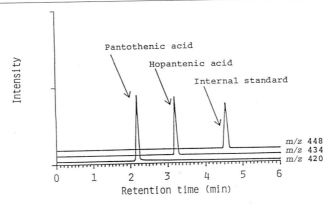

Fɪɢ. 1. Mass fragmentograms of the trimethylsilyl derivatives of pantothenic acid, hopan-tenic acid, and internal standard using a wide-bore fused-silica column coated with DB-17 for gas chromatography.

CA). The flow rate of helium as carrier gas is 15 ml/min, and the injection port, column oven, and separator temperature are 250, 200, and 250°, respectively. The mass spectrometer is operated in the electron-impact mode at 70 eV, the ionization current is 100 $\mu$A, and the temperature of the ion source is 200°.

The MID method for the simultaneous determination of trimethylsilyl-PaA and trimethylsilyl-HOPA has been investigated in a constant volume of the trimethylsilyl IS solution. The stable mass fragment ions are detected at $m/z$ 420, 434, and 448 as the characteristic $[M - CH_3]^+$ ions of the trimethylsilyl derivatives of PaA, HOPA, and IS, respectively (Fig. 2). These ions have been selected for the simultaneous determination of PaA and HOPA by MID because other ions do not interfere with quantitation. However, as may be seen from Fig. 2, the base peak ions of trimethylsilyl derivatives of PaA, HOPA, and IS are the $[M - 2TMS + 2H]^+$ ions at $m/z$ 291, 305, and 319, respectively. It has been found that base peak ions are not suitable for the quantitation because they cannot be separated completely from interfering substances in some samples.

The calibration curves are linear in the range of 5–100 ng/ml of plasma. The detection limits of PaA and HOPA using this method are ca. 1 ng/ml of plasma. The overall recoveries of PaA and HOPA are 92.9 ± 4.6% and 95.5 ± 5.1%, respectively.

Measurement in Natural Substances

The measurements of PaA and HOPA in a variety of biological samples and natural products were carried out as previously described. Samples

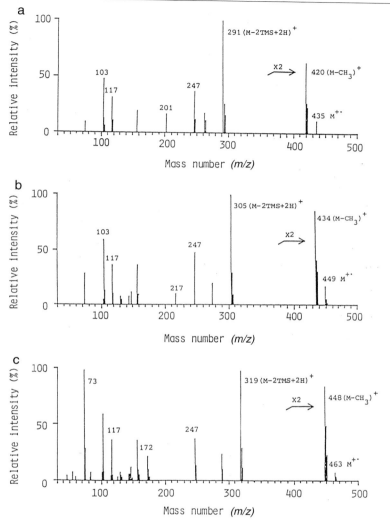

FIG. 2. Mass spectra of the trimethylsilyl derivatives of (a) panthothenic acid, (b) hopantenic acid, and (c) internal standard.

included the plasma from humans, fowl, and reptiles. Brain samples from fowl and reptiles were also analyzed, as were food products (boiled rice, green tea, and dried yeast).

Typical examples of mass fragmentograms obtained from the assay of PaA and HOPA in samples of human plasma, soft-shelled turtle brain, and

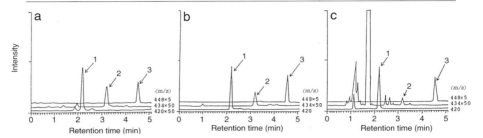

Fig. 3. Typical mass fragmentograms by the multiple ion detection method of (a) human plasma, (b) soft-shelled turtle brain, and (c) green tea. Peaks: 1, pantothenic acid; 2, hopantenic acid; 3, internal standard.

green tea are presented in Fig. 3. The contents of PaA and HOPA in samples were determined using previously obtained calibration curves with the peak intensity ratio of the trimethylsilyl derivatives of PaA ($m/z$ 420) and HOPA ($m/z$ 434) to that of the trimethylsilyl derivative of IS ($m/z$ 449). The retention times of the trimethylsilyl derivatives of PaA, HOPA, and IS in the samples were ca. 2.2, 3.2, and 4.5 min, respectively.

TABLE I

LEVELS OF PANTOTHENIC ACID AND HOPANTENIC ACID IN NATURAL SUBSTANCES

| Sample | PaA | HOPA |
|---|---|---|
| Plasma | | |
|    Human 1 | 23.2 ng/ml | 7.4 ng/ml |
|    Human 2 | 22.4 ng/ml | 8.0 ng/ml |
|    Human 3 | 25.1 ng/ml | 5.4 ng/ml |
|    Monkey | 32.5 ng/ml | 5.6 ng/ml |
|    Rabbit | 114.8 ng/ml | 8.4 ng/ml |
|    Mouse | 234.3 ng/ml | 15.7 ng/ml |
|    Pig | 74.8 ng/ml | 2.4 ng/ml |
|    Dog | 30.4 ng/ml | 11.6 ng/ml |
|    Rat | 403.3 ng/ml | 9.6 ng/ml |
|    Chicken | 510.3 ng/ml | 38.6 ng/ml |
|    Soft-shelled turtle | 267.9 ng/ml | 38.1 ng/ml |
| Brain | | |
|    Chicken | $3.46 \times 10^4$ ng/g | 123.3 ng/g |
|    Soft-shelled turtle | $1.09 \times 10^4$ ng/g | 121.3 ng/g |
| Natural products | | |
|    Dried yeast | $1.65 \times 10^5$ ng/g | $8.5 \times 10^3$ ng/g |
|    Boiled rice | $1.11 \times 10^3$ ng/g | 21.2 ng/g |
|    Green tea | $1.26 \times 10^4$ ng/g | 52.3 ng/g |

Both PaA and HOPA were identified in all samples studied at concentrations shown in Table I. It was found that an average of 23.6 ng of PaA, and 6.9 ng of HOPA, per milliliter is contained in plasma obtained from healthy men ($n = 3$). The resulting value for the PaA content in human plasma agrees with the microbiological values reported previously.[14,15] Plasma levels of PaA and HOPA in monkeys were almost the same as those in humans. Pantothenic acid and HOPA in plasma samples of other animals, such as fowl and reptiles, ranged between 30.4 and 510.3 ng/ml and between 2.4 and 38.6 ng/ml, respectively. Furthermore, the PaA content in chicken brain was 280 times higher than that in plasma, and that in brain of soft-shelled turtle was ca. 90 times higher than that in plasma. The HOPA content of brain was detected at a level only three times higher than that plasma. In contrast, the PaA content in natural products was determined to be much higher than that in biological samples (Table I). It was found that the HOPA content in dried yeast is more than 400 times greater than that in boiled rice, and ca. 160 times greater than that in green tea. It has been reported that PaA is widely distributed in every organ and that ingested PaA has a tendency to accumulate in the brain.[16] It has also been suggested that HOPA passes the blood–brain barrier.[13] The experimental results show that both PaA and HOPA are found in a wide range of concentrations in every animal studied, including microorganisms. Pantothenic acid is generally present in higher concentrations than HOPA in both biological samples and in natural products.

## Acknowledgments

The author thanks the coworkers in his laboratory whose research results are presented here. This chapter was adapted from *J. Chromatogr.* **525** (1990), K. Banno, M. Matsuoka, S. Horimoto, and J. Kato, Simultaneous determination of pantothenic acid and hopantenic acid in biological samples and natural products by gas chromatography–mass fragmentography, 255–264, Copyright 1990, with kind permission from Elsevier Science–NL, Sara Burgerhartstraat 25, 1055 KV Amsterdam, The Netherlands.

[14] M. Hatano, *J. Vitaminol.* **8,** 134 (1962).
[15] I. Masugi, *Vitamin* **46**(5), 261 (1972).
[16] T. Ariyama and S. Kimura, *Vitamin* **22,** 237 (1961).

## [24] Large-Scale Synthesis of Coenzyme A Esters

*By* Raghavakaimal Padmakumar, Rugmini Padmakumar, and Ruma Banerjee

Esters of coenzyme A (CoA) are prevalent metabolites involved in diverse pathways. It is estimated that approximately 4% of all known enzymes use CoA or CoA esters as substrates.[1] Many of the physiologically relevant CoA esters are either unavailable commercially or are relatively expensive. This has prompted the publication of several procedures for the synthesis of CoA esters of either mono- or diacids.[2-7] In this chapter, we describe the synthesis of methylmalonyl-CoA and succinyl-CoA (Scheme I), the substrate and product of the enzyme, methylmalonyl-CoA mutase.

### Synthesis of Methylmalonyl-CoA

#### Reagents

Methylmalonic acid, dicyclohexylcarbodiimide, thiophenol, and dimethylformamide are purchased from Aldrich (Milwaukee, WI). Coenzyme A (sodium salt) and sodium bicarbonate are purchased from Sigma (St. Louis, MO).

#### Step 1. Synthesis of Monothiophenylmethyl Malonate (2)

Dicyclohexylcarbodiimide (DCC, 5.04 g, 24.4 mmol) in dimethylformamide (DMF, 250 ml) is added dropwise with stirring over a period of 90 min to a solution of methylmalonic acid (**1**, 2.42 g, 20.5 mmol) and thiophenol (2 ml, 19.5 mmol) in dimethylformamide (250 ml) at 0°. Stirring is continued for an additional 5 hr at 0°. Reaction progress is monitored by the disappearance of thiophenol ($R_f$ 0.75) and the appearance of the monothiophenyl ester ($R_f$ 0.15) by thin-layer chromatography on a silica plate developed

[1] C. Lee and A. F. Chen, "The Pyridine Nucleotide Coenzymes." Academic Press, New York, 1982.
[2] E. Simon and D. Shemin, *J. Am. Chem. Soc.* **75**, 2520 (1953).
[3] E. G. Trams and R. O. Brady, *J. Am. Chem. Soc.* **82**, 2972 (1960).
[4] P. Overath, E. R. Stadtman, G. M. Kellerman, and F. Lynen, *Biochem. Z.* **336**, 77 (1962).
[5] K. Wölfe, M. Michenfelder, A. König, W. E. Hull, and J. Retey, *Eur. J. Biochem.* **156**, 545 (1986).
[6] R. Padmakumar, S. Gantla, and R. Banerjee, *Anal. Biochem.* **214**, 318 (1993).
[7] R. Padmakumar and R. V. Banerjee, *BioFactors* **6**, 1 (1995).

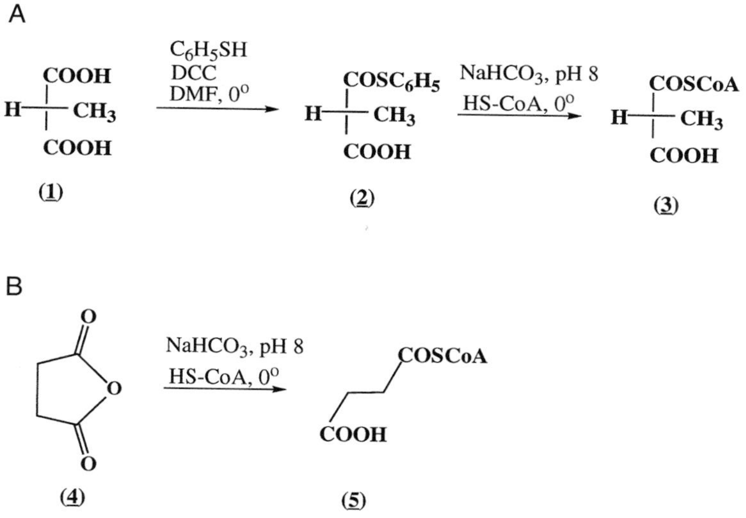

A

B

SCHEME I. Strategies for the synthesis of CoA esters.

with hexane–ethyl acetate (8:2, v/v). The reaction is stopped by adding 30 ml of cold water. The precipitate is removed by filtration and the filtrate is brought to pH 3 with 1 $N$ HCl. The product (2), present in the filtrate, is extracted with five 100-ml washes with ether. The ether extracts are combined, washed with 100 ml of 0.1 $N$ HCl, and concentrated to 75 ml *in vacuo*. The product is extracted from the organic phase with two 50-ml washes with 0.2 $N$ sodium bicarbonate. The aqueous layer containing the monothiophenol ester is washed with 100 ml of ether to remove traces of thiophenol and the side product, dithiophenyl ester. The solution is then brought to pH 2 with cold 1 $N$ HCl and the product (2) is reextracted into the organic phase with six 50-ml washes with ether. The combined organic extracts are washed with 50 ml of saturated sodium chloride and dried over 20 g of anhydrous sodium sulfate for 4–5 hr. Sodium sulfate is removed by filtration and the ether is evaporated off under reduced pressure to give 2.8 g of a colorless liquid (yield 65%). $^1$H NMR (CDCl$_3$, TMS): $\delta$ 1.54 (d, 3H), $\delta$ 3.7 (q, 1H), and $\delta$ 7.43 (s, 5H).

## Step 2. Synthesis of Methylmalonyl-Coenzyme A (3)

Monothiophenylmethyl malonate [2, 240 mg (1.142 mmol)] in 2 ml of 0.1 $M$ sodium bicarbonate, pH 8.0, is added to 200 mg of CoA (240 $\mu$mol) dissolved in 10 ml of the same buffer at 0°. Addition of 2 results in a drop in the pH, and cold 0.2 $N$ sodium hydroxide is added to maintain the pH

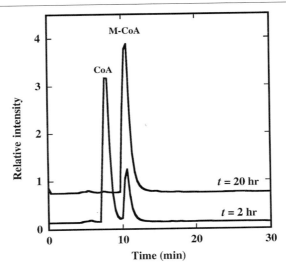

Fig. 1. Conversion of CoA to methylmalonyl-CoA. The reaction mixture from the transesterification step (step 2) is injected onto a reversed-phase HPLC column and eluted as described in text. The eluant is monitored by absorbance at 254 nm. Elution profiles of the reaction mixture after 2 hr (lower) and 20 hr (upper), showing almost complete conversion of CoA to the product, methylmalonyl-CoA.

at 8. Nitrogen is bubbled through the mixture and stirring is continued at 0°. Product formation is followed by injecting 10 $\mu$l of the reaction mixture onto a $C_{18}$ reversed-phase high-performance liquid chromatography (HPLC) column (Spherex-IP, 10 $\mu$m; Phenomenex, CA) eluted isocratically with buffer containing 220 m$M$ potassium phosphate (pH 4.5)–20% (v/v) methanol at a flow rate of 0.8 ml/min. When the reaction is complete (after 20–24 hr), as indicated by the disappearance of the CoA peak (retention time 7.87 min) and the presence of the methylmalonyl-CoA peak (**3**, retention time 10.53 min; Fig. 1), the mixture is acidified to pH 4 with 1 $N$ HCl. The reaction mixture is extracted eight times with 15 ml of ether to remove thiophenol and unreacted monothiophenylmethyl malonate. The aqueous phase is then washed five times with 15 ml of ethyl acetate to remove methylmalonic acid, and lyophilized to give 204 mg of methylmalonyl-CoA in 97% yield.

The purity of methylmalonyl-CoA is established by HPLC analysis as previously described, which is a modification of the procedure described by Clough *et al.*,[8] by fast atom bombardment (FAB)–mass spectroscopy,

[8] R. C. Clough, S. R. Barnum, and J. G. Jaqorski, *Anal. Biochem.* **176**, 82 (1989).

and by $^1$H nuclear magnetic resonance (NMR).[5,6,9] In addition, ascending paper chromatography (3MM; Whatman, Clifton, NJ) in ethanol–$H_2O$ (7:3, v/v), pH 5, separates methylmalonyl-CoA (UV positive, $R_f$ 0.36) from the starting material, coenzyme A ($R_f$ 0.21). The hydroxamic acid-to-adenine ratio is determined as described by Lipmann and Tuttle and should be 1.[10]

Synthesis of Succinyl-CoA (5)

*Reagents*

Succinic anhydride and coenzyme A (sodium salt) are purchased from Sigma. Anhydrous tetrahydrofuran is purchased from Aldrich and is used without further purification.

*Succinyl-CoA*

Succinyl-CoA is synthesized by thiolysis of succinic anhydride (4, 6.84 mg, 68.3 $\mu$mol) in 3 ml of anhydrous tetrahydrofuran under positive nitrogen pressure with CoA (30 mg, 36 $\mu$mol) dissolved in anaerobic distilled water (3 ml). The solution is maintained at pH 8 with 1 $N$ NaOH and the solution is stirred for 20 min. The reaction is monitored by chromatography on a $C_{18}$ HPLC column [220 m$M$ KP$_i$ (pH 5.3), 20% (v/v) $CH_3OH$; flow rate, 0.8 ml/min]. Succinyl-CoA (5) and CoA have retention times of 12.45 and 7.87 min, respectively, under these conditions. When thiolysis is complete (20 min), the reaction is stopped by adjusting the pH to 5 with 0.2 $N$ HCl. Tetrahydrofuran is evaporated off under reduced pressure and the remaining aqueous solution is extracted five times with 10 ml of ether to remove any residual organic soluble material. The aqueous solution is then lyophilized to give 32 mg of succinyl-CoA (96% yield). The purity of the product is judged by the methods described for methylmalonyl-CoA. This reaction can also be run on a larger scale as described for **3**.

Discussion

Two strategies for the synthesis of CoA esters are described. In the first, one of the carboxylates of a diacid precursor is activated by esterification, and subsequently converted to the CoA ester in a transthioesterification reaction (Scheme IA). Alternatively, use of a cyclic anhydride precur-

---

[9] S. Taoka, R. Padmakumar, M.-t. Lai, H.-w. Liu, and R. V. Banerjee, *J. Biol. Chem.* **269**, 31630 (1994).

[10] F. Lipmann and C. Tuttle, *Biochim. Biophys. Acta* **4**, 301 (1950).

sor of a diacid permits a convenient one-step thiolysis reaction, as in the synthesis of succinyl-CoA (Scheme IB) or glutaryl-CoA.[7] Both procedures give high yields of the desired products. CoA esters vary greatly in their lability toward hydrolysis, and attention needs to be paid to careful handling and storage procedures. While methylmalonyl-CoA is readily stored at $-80°$ under desiccation, succinyl-CoA is more labile and has a shorter shelf life when stored under similar conditions. The synthetic routes described here can be adopted for the preparation of other CoA esters of biochemical interest.

## Acknowledgments

This work was supported by a grant from the National Institutes of Health (NIDDK 45776). Mass spectral determinations were made at the Midwest Center for Mass Spectrometry (Lincoln, NE), with partial support from the National Science Foundation (DIR9017262).

# [25] Synthesis of Nonhydrolyzable Acyl-Coenzyme A Analogs

*By* ANDREAS ABEND and JÁNOS RÉTEY

Since the discovery of the thioester nature of acetylcoenzyme A (CoA),[1] acyl-CoA thioesters have been recognized as the activated forms of the corresponding acids. They are ubiquitous in living systems as common substrates and regulators in many biologically important reactions.

In 1952 Wilson[2] described the first synthesis of acetyl-CoA from the free coenzyme and acetic anhydride while the total synthesis of free coenzyme A was published only in the next decade by Moffatt and Khorana.[3] Taking advantage of the same convergent strategy, Stewart and Wieland[4] synthesized acetonyldethia-coenzyme A (acetyl-CH$_2$CoA), in which the sulfur of the normal thioester bond in acetyl-CoA is replaced by methylene. This was the first nonhydrolyzable acyl-CoA analog used for the mechanistic investigation of enzymatic reactions in which acetyl-CoA is either a substrate or a regulator. Among these are condensations, carboxylations, and acetylations. Several long-chain fatty acyl-CH$_2$CoA's, e.g., palmitoyl-

[1] F. Lynen and E. Reichert, *Angew. Chem.* **63,** 47 (1951).
[2] I. B. Wilson, *J. Am. Chem. Soc.* **74,** 3205 (1952).
[3] J. G. Moffatt and H. G. Khorana, *J. Am. Chem. Soc.* **83,** 663 (1961).
[4] C. J. Stewart and T. Wieland, *Liebigs Ann. Chem.* **1,** 57 (1978).

Propionyl-$CH_2CoA$ $\xrightleftharpoons{transcarboxylase}$ (S)-methylmalonyl-$CH_2CoA$

(S)-Methylmalonyl-$CH_2CoA$ $\xrightleftharpoons{epimerase}$ (R)-methylmalonyl-$CH_2CoA$

(R)-Methylmalonyl-$CH_2CoA$ $\xrightleftharpoons{mutase}$ succinyl-$CH_2CoA$

SCHEME I

$CH_2CoA$ and heptadecyl-$CH_2CoA$, were synthesized by Thorpe et al.[5] and used in mechanistic investigations of acyl-CoA dehydrogenases, which are key enzymes in the $\beta$-oxidation of fatty acids. As in the case of acetyl-$CH_2CoA$, these compounds turned out to be strong inhibitors in those enzymatic reactions in which the thioester bond of the natural substrate is broken.

In 1986 Michenfelder and Rétey[6] synthesized propionyl-$CH_2CoA$, which turned out to be an excellent substrate of the propionyl-CoA oxaloacetate transcarboxylase from *Propionibacterium shermanii*, ST33 being carboxylated stereospecifically to pure (S)-methylmalonyldethia-coenzyme A. The latter is epimerized by the corresponding epimerase to (R)-methylmalonyldethia-coenzyme A, which is finally rearranged by the coenzyme $B_{12}$-dependent methylmalonyl-CoA mutase to succinyl-$CH_2CoA$ (Scheme I). It is remarkable that all these acyl-$CH_2CoA$ derivatives have almost the same $K_m$ and $V_{max}$ values in the corresponding enzymatic reactions as their natural counterparts.

Kinetic $^1H$ nuclear magnetic resonance (NMR) studies of the methylmalonyl-CoA mutase reaction using the nonhydrolyzable (R)-methylmalonyl-$CH_2CoA$ in deuterium oxide uncovered details of the nonstereospecific incorporation of deuterium into the product. Using the dethia analog the reaction could be monitored for a much longer period, because the produced succinyl-$CH_2CoA$ was stable to hydrolysis.[7]

Brendelberger et al.[8] confirmed these observations for the coenzyme $B_{12}$-dependent isobutanoyl-CoA-n-butanoyl-CoA mutase by using isobutanoyl-$CH_2CoA$. This compound was also an excellent substrate of the enzyme

[5] C. Thorpe, T. L. Ciardelli, C. J. Stewart, and T. Wieland, *Eur. J. Biochem.* **118**, 279 (1981).
[6] M. Michenfelder and J. Rétey, *Angew. Chem. Int. Ed. Engl.* **25**, 366 (1986).
[7] K. Wölfle, M. Michenfelder, A. König, W. E. Hull, and J. Rétey, *Eur. J. Biochem.* **156**, 545 (1986).
[8] G. Brendelberger, J. Rétey, D. M. Ashworth, K. Reynolds, F. Willenbrock, and J. A. Robinson, *Angew. Chem. Int. Ed. Engl.* **27**, 1089 (1988).

TABLE I
COMPARISON OF $K_m$ AND $V_{max}$ VALUES OF PROPIONYL-CoA AND ITS
ANALOGS IN TRANSCARBOXYLASE REACTION[a]

| Substrate | $K_m$ (mM) | $V_{max}$ ($\mu$mol/min · U) |
|-----------|-----------|------------------------------|
| $CH_3CH_2COCH_2SCoA$ | 0.44 | 0.020 |
| $CH_3CH_2COCH_2CH_2CoA$ | 0.14 | 0.025 |
| $CH_3CH_2COSCoA$ | 0.047 | 1 |
| $CH_3CH_2COCH_2CoA$ | 0.15 | 0.5 |
| $CH_3CH_2CONHCoA$ | 0.24 | 0.04 |

[a] Adapted from Refs. 10 and 11, with permission.

and was rearranged to $n$-butanoyl-$CH_2$CoA. Isobutanoyl-$CH_2$CoA was also converted into a mixture of methacryl-$CH_2$CoA and 3-hydroxyisobutanoyl-$CH_2$CoA by a cell-free extract of *Pseudomonas putida* ATCC 21244.[9]

Further propionyl-CoA derivatives have been synthesized in which the sulfur atom is replaced by either ethylene ($-CH_2CH_2-$),[10] thiomethylene ($-CH_2-S-$),[10] or the NH group.[11] All three analogs, propionyldethia(dicarba)-CoA (propionyl-$CH_2CH_2$CoA), propionyl-$S$-(2-oxobutyl)-CoA (propionyl-$CH_2$SCoA),[6] and propionyldethia(aza)-CoA (propionyl-NH-CoA),[7] were reacted with propionyl-CoA oxaloacetate transcarboxylase. While the first two were carboxylated to the corresponding methylmalonyl derivatives, propionyldethia(aza)-CoA was not.

The kinetic constants ($K_m$ and $V_{max}$ values) are listed in Table I. When the above-mentioned propionyl-CoA derivatives (except propionyl-NH-CoA) were incubated with an enzyme "cocktail" containing transcarboxylase, epimerase, and the coenzyme $B_{12}$-dependent methylmalonyl-CoA mutase, the corresponding succinyl-$CH_2CH_2$CoA and succinyl-$CH_2$S-CoA were formed and could be isolated on a preparative scale. After spectroscopic characterization their kinetic constants with methylmalonyl-CoA mutase were determined (Table II).

Myristoyl-$CH_2$CoA, a good inhibitor of $N$-myristoyl-CoA : protein-$N$-myristoyltransferase, an enzyme important for signal transduction in several organisms, was also synthesized.[12]

Martin *et al.*[13] have described an elegant chemoenzymatic synthesis of CoA analogs, taking advantage of steps in the biosynthesis of the natural coenzyme A.

[9] G. Brendelberger and J. Rétey, *Isr. J. Chem.* **29,** 195 (1989).
[10] Y. Zhao, M. Michenfelder, and J. Rétey, *Can. J. Chem.* **72,** 164 (1994).
[11] H. Martini and J. Rétey, *Angew. Chem. Int. Ed. Engl.* **32,** 278 (1993).
[12] A. Popa-Wagner and J. Rétey, *Eur. J. Biochem.* **195,** 699 (1991).
[13] D. P. Martin, R. T. Bilbart, and D. G. Drueckhammer, *J. Am. Chem. Soc.* **116,** 4660 (1994).

TABLE II
COMPARISON OF $K_m$ AND $V_{max}$ VALUES OF SUCCINYL-CoA AND ITS
ANALOGS IN METHYLMALONYL-CoA MUTASE REACTION[a]

| Substrate | $K_m$ (mM) | $V_{max}$ ($\mu$mol/min · U) |
|---|---|---|
| HOOCCH$_2$CH$_2$COCH$_2$CoA | 0.136 | 1.1 |
| HOOCCH$_2$CH$_2$COSCoA | 0.025 | 1.0 |
| HOOCCH$_2$CH$_2$COCH$_2$CH$_2$CoA | 2.20 | 0.013 |
| HOOCCH$_2$CH$_2$COCH$_2$SCoA | 0.132 | 0.0047 |

[a] Adapted from Ref. 10, with permission.

The chemical synthesis of different precursors of the nonhydrolyzable acyl-CH$_2$CoA derivatives is described in the first part of the section on experimental methods, whereas the synthesis of precursors for the chemo-enzymatic strategy is outlined in the second part in the same section. The final synthesis of various nonhydrolyzable acyl-CH$_2$CoA derivatives is described in the last part of the section on experimental methods.

*Reagents*

Pyridine: Dried over CaH$_2$ under nitrogen
Triethanolamine: Dried over KOH
Ether, ethanol, methanol: Dried over molecular sieves
*tert*-Butanol: Dried over CaOH and freshly distilled before use
(*R*)(−)-Pantolactone, Pd/C (10%), cyanoethyl phosphate, acrylamide, *N*-acryloxysuccinimide, 2,2′-Azobisisobutyronitrile (AIBN), *p*-nitrophenylcarbobenzoxy-β-alanine, urea, dicyclohexylcarbodiimide (DCC), glacial acetic acid, glucose, dithiothreitol (DTT), and Dowex and Amberlite ion exchangers: From Fluka (Ronkonkoma, NY)
5′-AMP, ADP, ATP, phosphoenolpyruvate, dephospho-CoA: From Sigma (St. Louis, MO)
Raney nickel and dry THF: From Aldrich (Milwaukee, WI)
TSK-DEAE: From Merck
All commercial enzymes and coenzymes: From Boehringer Mannheim (Indianapolis, IN)
PAN 950[14]: Synthesized from 13.5 g of acrylamide (0.19 mmol) and 3 g of *N*-acryloxysuccinimide (17.8 mmol) dissolved in 150 ml of dry THF in a 200-ml centrifuge tube, capped with a serum stopper. The solution is kept under argon during the whole procedure. Eighty

[14] A. Pollak, H. Blumenfeld, M. Wax, R. L. Baughn, and G. M. Whitesides, *J. Am. Chem. Soc.* **102**, 6324 (1980).

milligrams of AIBN in 5 ml of dry THF is introduced slowly via a syringe and the reaction mixture is heated to 50° for 20 hr. PAN precipitates as a white solid and is removed by centrifugation at 5000 rpm for 5 min. The product is washed several times with dry THF under an argon atmosphere. Yields are usually about 16.3 g (99% of theory).

Enzyme Assays

See Ref. 15 for additional details.

*Pantetheine Phosphate Adenylyltransferase*

The activity of pantetheine phosphate adenylyltransferase is assayed in the reverse direction by monitoring the production of ATP from dephospho-CoA and PP$_i$.[16] The 1-ml assay system consists of 0.1 m$M$ dephospho-CoA, 2 m$M$ PP$_i$, 2 m$M$ MgCl$_2$, 1 m$M$ NADP$^+$, 5 m$M$ glucose, 2 units of hexokinase, and 3 units of glucose-6-phosphate dehydrogenase in 50 m$M$ Tris-HCl buffer, pH 8, containing 0.5 m$M$ dithiothreitol.

*Dephospho-CoA Kinase*

Dephospho-CoA kinase is assayed by monitoring the production of ADP from dephospho-CoA and ATP by using the standard pyruvate kinase/lactate dehydrogenase coupling enzyme system. The 1-ml assay incubation mixture consists of 0.1 m$M$ dephospho-CoA, 2 m$M$ ATP, 2 m$M$ MgCl$_2$, 0.2 m$M$ NADH, 2.5 m$M$ phosphoenolpyruvate, 3 units of pyruvate kinase, and 5 units of lactate dehydrogenase in 50 m$M$ Tris-HCl buffer, pH 8, containing 0.5 m$M$ dithiothreitol.

Experimental Methods

*Chemical Synthesis of Precursors of Nonhydrolyzable Acyl-CH$_2$CoA Derivatives*

The two key intermediates for the chemical synthesis of acyldethia-coenzyme A analogs are acyldethia(carba)pantetheine 4-phosphate (**7**) and adenosine 2′,3′-cyclophosphate 5′-phosphomorpholidate (**8**). The strategy of Moffatt and Khorana[3] is followed (see Scheme II).

---

[15] D. M. Worrall and P. K. Tubbs, *Biochem. J.* **215**, 153 (1983).
[16] W. Lambrecht and I. Trautschold, *Methods Enzymol. Anal.* (2nd Ed.), **4**, 2101 (1974).

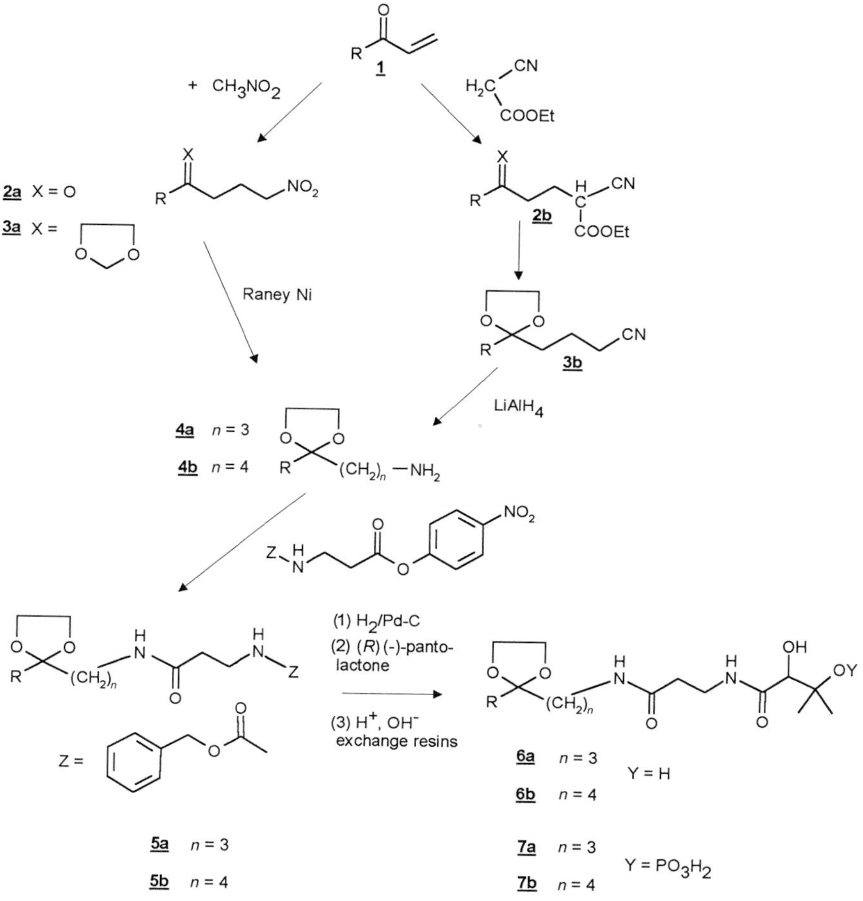

SCHEME II

## Synthesis of Acyldethia(carba)pantetheine 4-Phosphate

*Synthesis of 1-Nitro-3-alkanone (**2a**).* Sodium (0.2 g, 8.6 mmol) is dissolved in 8.5 ml of dry methanol. After addition of 5.3 g of nitromethane (86.8 mmol) dissolved in 10 ml of dry ether the solution is heated to boiling and 12 mmol of alkyl vinyl ketone (**1**; freshly distilled and dissolved in 10 ml of dry ether) is added dropwise. After 10 hr of reflux the reaction mixture is neutralized with glacial acetic acid and the solvent evaporated. The residue is dissolved in 30 ml of chloroform and the insoluble residue is filtered off. The solution is subsequently washed with 10 ml of HCl (5%

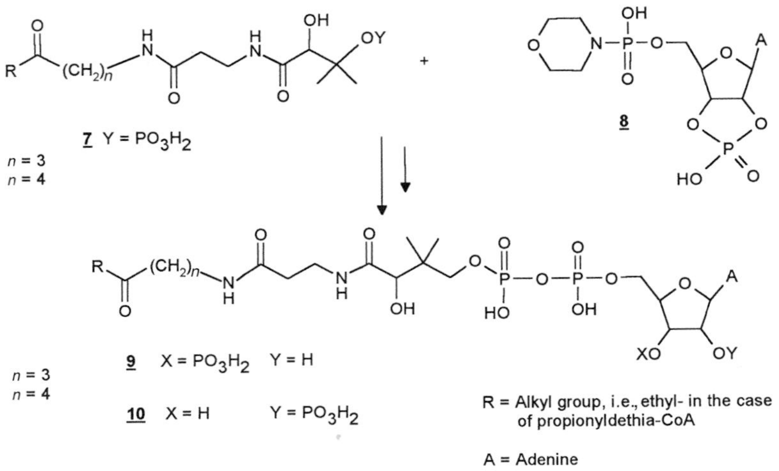

$7$  Y = PO$_3$H$_2$

n = 3
n = 4

$8$

$9$  X = PO$_3$H$_2$  Y = H

n = 3
n = 4

$10$  X = H          Y = PO$_3$H$_2$

R = Alkyl group, i.e., ethyl- in the case
    of propionyldethia-CoA

A = Adenine

SCHEME II (continued)

in water, v/v), 10 ml of urea (10% in water, w/w), and finally with water. Traces of water are removed by adding Na$_2$SO$_4$ to the chloroform phase. After filtration the chloroform is evaporated. The residue is distilled over a short Vigreux column. Yield: about 6 mmol (50%), depending on the alkyl vinyl ketone used.

*Synthesis of 2-(Nitropropyl)-2-ethyl-1,3-dioxolane (3a)*. Ethylene glycol (0.28 g) and catalytic amounts of *p*-toluenesulfonic acid are added to 4.5 mmol of 1-nitro-3-alkanone (**2a**) dissolved in toluene. The mixture is refluxed in a Dean–Stark extractor. After termination of the reaction the solvent is evaporated and the residue is distilled under reduced pressure. Yield: About 80% of theory.

*Synthesis of 2-(3-Cyano-3-carbethoxyethyl)-2-ethyldioxolane (2b)*. 2-Cyano-5-oxoheptanoic acid ethyl ester[17] (9.4 g, 48 mmol) in benzene (25 ml) in the presence of 4.2 ml (75 mmol) of ethylene glycol and 0.2 g of toluenesulfonic acid is refluxed for 6 hr. The reaction water is removed with a separatory funnel. After neutralization with 2% (w/v) NaOH (15 ml), the resulting solution is washed with water, then dried over anhydrous K$_2$CO$_3$. Distillation of the product yields 9.6 g (40 mmol, 81%) of **2b**, bp 135°/133 Pa.

*Synthesis of 2-(3-Cyanopropyl)-2-ethyl-1,3-dioxolane (3b)*. To 7.8 g (32 mmol) of **2b** in dimethyl sulfoxide (2.5 ml) are added 1.89 g of NaCl and

---

[17] T. A. Spencer, M. D. Newton, and S. W. Baldwin, *J. Org. Chem.* **29**, 787 (1962).

1.18 ml of water. The stirred mixture is heated to 180° for 7 hr. After cooling, 80 ml of water is added and the resulting solution is extracted five times with 40-ml portions of ether. The combined extracts are dried over anhydrous MgSO$_4$. Distillation of the product **3b** yields 3.7 g (22 mmol, 69%), 95°/100 Pa.

*Synthesis of 2-(3-Aminopropyl)-2-alkyl-1,3-dioxolane **4a** and **4b** from **3a** and **3b***. 2-(Nitropropyl)-2-ethyl-1,3-dioxolane (**3a**) (30.5 mmol) is dissolved in dry methanol. After addition of 0.6 g of activated Raney nickel the mixture is heated to 50° in an autoclave under a hydrogen pressure of 10 MPa for 2 hr. After an additional 12 hr at room temperature under the same pressure the Raney nickel is removed by filtration and the solvent is evaporated. The residue is distilled under reduced pressure. Yield: About 80% of theory.

To a suspension of 1 g of LiAlH$_4$ in 20 ml of dry ether a solution of 3.12 g (18 mmol) of **3b** in 25 ml of dry ether is introduced under stirring for 1 hr. After 4 hr of reflux ice is added. Work-up affords 2.4 g (13.8 mmol, 76%) of **4b**.

*Synthesis of 2-[3-(N-Benzyloxycarbonyl-β-alanyl)aminopropyl]-2-alkyl-1,3-dioxolane (**5**)*. Syntheses of the following acyl-CH$_2$CoA and acyl-CH$_2$CH$_2$CoA precursors are identical, and thus only the synthesis of the next acyl-CH$_2$CoA precursor is described in detail: 6.3 mmol of the aminoketal 2-(3-aminopropyl)-2-alkyl-1,3-dioxolane **4a** and 6.3 mmol of triethylamine and 2.18 g (6.3 mmol) of *p*-nitrophenylcarbobenzoxy-β-alanine are dissolved in 20 ml of dichloromethane. After stirring for 3 hr at room temperature, 80 ml of benzene is added. The solution is washed with 2% (w/v) aqueous sodium hydrogen carbonate solution until the aqueous phase appears colorless. The benzene phase is washed with water and dried over Na$_2$SO$_4$, the salt is removed, and the solvent is then evaporated. Ether is added to the residue until a homogeneous solution is obtained. *n*-Pentane is added until the solution becomes turbid. Compound **5** crystallizes after a few hours at 4°. The crystals are collected by suction, washed with pentane, and dried. Yield: About 65%.

*Synthesis of Acyldethiapantetheine (**6**)*. A total of 0.09 g of Pd/C (10%) is added to a solution of 2.4 mmol of 2-[3-(N-benzyloxycarbonyl-β-alanyl)aminopropyl]-2-alkyl-1,3-dioxolane (**5**) in 20 ml of dry methanol. Hydrogenation is performed under vigorous stirring. The catalyst is removed by centrifugation and the methanol evaporated. A total of 0.34 g of D-2-hydroxy-3,3-dimethyl-4-butyrolactone  [(*R*)(−)-pantolactone] is added to the residue and heated to 90° until the lactone melts. The flask is kept at 60° overnight. The viscous residue is dissolved in a small amount of water and centrifuged. The supernatant is applied to a Dowex 50W-X4 (H$^+$, 1.5 × 5 cm) column and then to a Dowex 2-X8 (OH$^-$, 1.5 × 5 cm) column. Both columns are washed with 200 ml of water, the eluate (with the unabsorbed product) is collected, and the water is evaporated under reduced pressure. The residue is dissolved in 5 ml of dry ethanol/1 ml of dry benzene. To remove traces of water, the solution is several times evaporated and redissolved in ethanol/benzene (5:1, v/v). The highly viscous residue is dried *in vacuo*. Yield: About 50% of theory.

*Synthesis of Acyldethiapantetheine 4-Phosphate (7).* One gram of cyanoethyl phosphate (barium salt, 2.63 mmol) is dissolved in 3 ml of water in the presence of Dowex 50W-X2 (H[+]) and then charged on a Dowex 50W-X2 (H[+]) column (1 × 12 cm). Water (400 ml) is charged on the column and the water is removed under reduced pressure at 30°. The colorless oil is dissolved in 10 ml of dry pyridine (dried over calcium hydride) and the solvent is then evaporated. This procedure is repeated three times to remove traces of water.

Acyldethiapantetheine (**6**, 2 mmol) is dissolved in 10 ml of dry pyridine and traces of water are removed in the same way as described above. Both components are dissolved in 10 ml of dry pyridine, each in a flask under nitrogen atmosphere. Dicyclohexylcarbodiimide (0.6 g, 2.7 mmol), dissolved in 2 ml of dry pyridine, is added to the reactants via a syringe through a serum cap. The mixture is stirred overnight, after which 15 ml of water is added. After stirring for 1 hr the precipitated urea is removed by filtration. The filtrate is diluted with 20 ml of water and the solvent is removed under reduced pressure. The slightly yellow oil is twice dissolved in water; the water is then evaporated to remove traces of pyridine. The residue is then dissolved in 12.5 ml of water, the pH is adjusted to pH 7.5 with a saturated barium hydroxide solution, and 200 ml of ethanol is added to precipitate the excess cyanoethyl phosphate. The precipitate is removed and the solvent is evaporated. The residue is redissolved in 12.5 ml of water, cooled on ice, and 12.5 ml of cold 0.2 *M* NaOH is added dropwise. After stirring for 40 min on ice the pH is adjusted to pH 1 by addition of Amberlite IR100 (H[+]) to remove the dioxolane group. The whole suspension is charged on an Amberlite column (2 × 12 cm) and the product is eluted with 500 ml of water. The pH is adjusted to pH 7.5 with a barium hydroxide solution and the water is removed by evaporation, avoiding temperatures above 40°. The residue is redissolved in 1 ml of absolute ethanol and evaporated to dryness. This procedure is repeated several times to remove traces of water. The oil is redissolved in 2 ml of ethanol and the product, **7**, is precipitated by addition of 30 ml of acetone/ether (1:1, v/v). The precipitate is collected by centrifugation and dried *in vacuo*. Yield: About 40% of theory.

*Synthesis of 2'-(3'-)Cyclophosphate 5'-Phosphomorpholidate of Adenosine (8).* A total of 890 mg of 5'-AMP (2.26 mmol, sodium salt) and 13 g of tritetramethylammonium trimetaphosphate are dissolved in 21.7 ml of 1 $N$ NaOH. The mixture is stirred for 4 days and the pH is controlled and readjusted to pH 14 if necessary. The precipitate is removed by filtration and 10 g of degassed activated charcoal is added to the filtrate. The suspension is applied to an activated charcoal (8 g) column. After washing of the column with water, the product is eluted with 500 ml of pyridine/water (1:1, v/v). The eluate is evaporated under reduced pressure and the residue is dissolved in 3 ml of water and applied to a Dowex 1-X8 column (Cl⁻, 3.5 × 33 cm). The column is washed with 500 ml of water, the educt is eluted with 0.04 $M$ LiCl in 0.01 $M$ HCl (1 liter), and the bisphosphate is eluted with 0.1 $M$ LiCl in 0.01 $M$ HCl. The pH is adjusted to pH 4.5 with a saturated LiOH solution and the solution is concentrated to a final volume of 2 ml. The bisphosphate [2'-(3'-),5'-ADP] can be precipitated with 300 ml of acetone/ethanol (1:1, v/v) and collected by centrifugation. The yield is 514 mg (1.13 mmol), 50% of theory.

To the solution of 162.6 mg of 2'-(3'-),5'-ADP (lithium salt) in 10 ml of water, 2 ml of Dowex 50W-X2 (H⁺) is added. After 30 min the mixture is filtered through a Dowex 50W-X2 (H⁺, 1 × 12 cm) column, eluted with 300 ml of water, and 0.7 ml of morpholine is added. The water is evaporated and the residue is dissolved in 4.5 ml of water/*tert*-butanol (1:2, v/v) and 124 µl of morpholine (1.35 mmol) is added. While boiling the mixture under a nitrogen atmosphere, 309 mg of DCC (1.5 mmol) dissolved in 6 ml of *tert*-butanol is added. Refluxing is continued for 4.5 hr, after which the excess DCC is hydrolyzed with 10 ml of water and the resulting urea is removed by filtration. The filtrate is evaporated and redissolved in 25 ml of water. The aqueous solution is washed three times with ether and then evaporated. The residue is dissolved in 2 ml of dry methanol and the product precipitated on addition of 60 ml of dry ether. The white precipitate **8** is dried *in vacuo*. Yield: 255 mg (0.239 mmol), 79.3% of theory.

*Chemoenzymatic Strategy*

Scheme III outlines the chemoenzymatic strategy.

*Synthesis of Pantothenic Acid Methyl Ester (12).* The sodium salt of pantothenic acid **11** (20 g, 83 mmol) is suspended in 150 ml of methanol. Methyl *p*-toluene sulfonate (15.4 g, 83.7 mmol) is added to yield a homogeneous solution.[18] The solution is heated to reflux overnight, concentrated to 50 ml *in vacuo*, and filtered to remove the precipitate. The filtrate is concentrated *in vacuo* to yield a viscous liquid. The liquid is dissolved in

---

[18] A. Kleeman, W. Leuchtenberger, J. Martens, and H. Weigel, *Angew. Chem. Int. Ed. Engl.* **19**, 627 (1980).

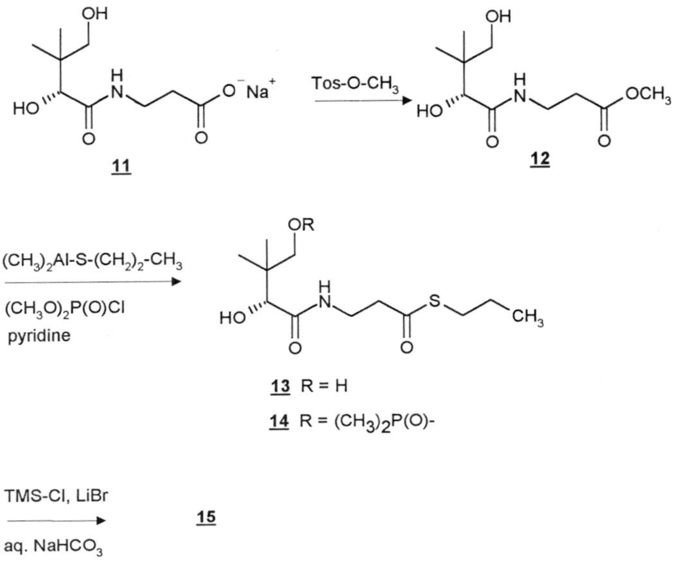

Scheme III

100 ml of chloroform and 15 ml of ether, refiltered, and concentrated *in vacuo* to yield 20 g of crude material as a yellow viscous liquid. An analytical sample is purified by chromatography on a silica gel column, using ethyl acetate/hexane (2:1, v/v), to yield a colorless viscous oil.

*Synthesis of Pantothenic Acid Propyl Thioester (13).* In a dry, round-bottom flask under a nitrogen atmosphere, dimethyl(propylthio)aluminum is generated by the slow addition of propanethiol (2.74 g, 35.9 mmol) to a cooled (0°) solution of trimethylaluminum (2 *M* in hexane, 18 ml, 36 mmol) in 55 ml of dry $CH_2Cl_2$.[19] The dimethyl(propylthio)aluminum solution is stirred at 0° for 5 min and **12** (3 g, 13 mmol) is dissolved in about 10 ml of dry toluene. The toluene is evaporated *in vacuo* to remove residual water, and compound **12** is redissolved in 20 ml of dry $CH_2Cl_2$. The dimethyl(propyldethio)aluminum solution is cannulated into the solution of **12**, and the reaction mixture is stirred for 5 min and then extracted three times with ethyl acetate (75 ml each time). The combined organic extracts are washed sequentially with 1 *M* NaOH (50 ml) and brine (25 ml), dried over $MgSO_4$, and concentrated *in vacuo* to yield a viscous oil. The product is

[19] M. A. J. Palmer, E. Differding, R. Gamboni, S. F. Williams, O. P. Peoples, C. T. Walsh, A. J. Sinskey, and S. Masamune, *J. Biol. Chem.* **266**, 3369 (1991).

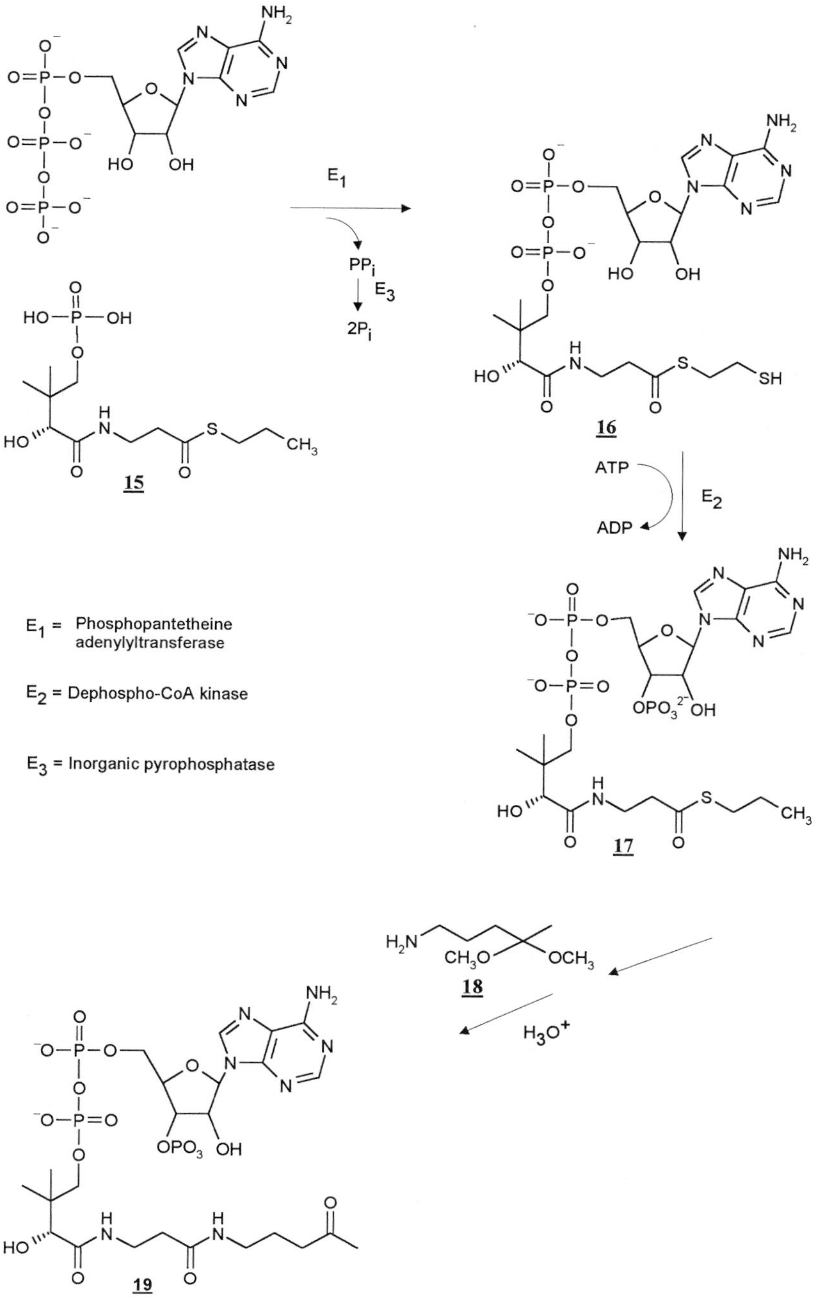

E1
PPi
E3
2Pi
15
16
ATP
ADP
E2
17
18
H3O+
19
E1 = Phosphopantetheine adenylyltransferase
E2 = Dephospho-CoA kinase
E3 = Inorganic pyrophosphatase


SCHEME III (*continued*)

purified by silica gel chromatography, using ethyl acetate/hexane (2:1, v/v) to give a colorless oil. Yield: 1.23 g (35%).

*Synthesis of S-Propylthiopantothenate 4'-(Dimethyl Phosphate) (14).* Dimethyl phosphochloridate is generated from dimethyl phosphite (2.5 ml, 27 mmol) and N-chlorosuccinimide (3.63 g, 27.3 mmol) in 35 ml of dry toluene.[20] The mixture is stirred under nitrogen for about 2 hr. During the exothermic reaction succinimide precipitates from solution, which is cooled in an ice bath to prevent refluxing. The resulting solution is used immediately in the next step. Compound 14 (2.52 g, 9.10 mmol) is dissolved in dry pyridine (10 ml). The pyridine is evaporated *in vacuo* to remove residual water. The residue is dissolved in 40 ml of dry pyridine and cooled in a −40° bath (CH₃CN–dry ice). The dimethyl phosphochloridate solution is cannulated into the reaction vessel over 10 min. The reaction mixture is allowed to warm slowly to room temperature and is stirred overnight. Water (15 ml) is added, and the mixture is concentrated under reduced pressure to yield a brown oil, which is dissolved in 100 ml of ethyl acetate and washed sequentially with 1 $M$ sulfuric acid (twice, 25 ml each), 1 $M$ sodium hydrogen carbonate (twice, 25 ml each), and saturated sodium sulfate (once, 25 ml). The organic layer is dried over $MgSO_4$ and concentrated *in vacuo*. The resulting oil is purified by flash chromatography on silica gel, using ethyl acetate, yielding 1.78 g of a colorless oil in 51% yield.

*Synthesis of S-Propylthiopantothenate 4'-Phosphate (15).* Compound 14 (1.2 g, 3.12 mmol) and LiBr (0.56 g, 6.5 mmol) are added to 4 ml of dry CH₃CN. The CH₃CN is evaporated under reduced pressure to remove residual water. Dry CH₃CN (6.0 ml) and trimethylsilyl chloride (4.0 ml, 31 mmol) are added to the residue to yield a clear solution. The reaction mixture is heated overnight with stirring at 50° under nitrogen. White crystals form and pervade the solution. ¹H NMR analysis of the reaction mixture shows complete disappearance of the starting material. The reaction mixture is filtered, evaporated, treated with methanol, and reevaporated. The residue is dissolved in water, brought to pH 4.5 with lithium hydroxide (0.5 $M$), and filtered through cotton. Lyophilization yields a tan foam containing inorganic salts. The crude material is dissolved in water, centrifuged to remove insolubles, and purified by preparative $C_{18}$ reversed-phase high-performance liquid chromatography (HPLC) using a gradient of methanol in aqueous phosphate (50 m$M$, pH 4.5). While being monitored at 215 and 275 nm, compound 15 is eluted at a flow rate of 3 ml/min with 10% (v/v) methanol for 5 min, followed by a linear gradient of methanol increasing to 70% (v/v) over 40 min. Fractions containing 15 are pooled and lyophilized. Compound 15 is assayed with 5,5'-dithiobis(2-nitrobenzoic

---

[20] B. M. Trost and D. P. Curran, *Tetrahedron Lett.* **14**, 1287 (1981).

acid) [DTNB, $\varepsilon_{412} = 13.6 \times 10^3$ mol$^{-1}$ cm$^{-1}$, in 0.1 $M$ Tris buffer (pH 8), 0.1 $M$ hydroxylamine hydrochloride, 30 m$M$ NaHCO$_3$], and the yield is 0.11 mol (34% yield). The material containing inorganic salts is used for the enzymatic synthesis of compound **16**. For analysis, residual phosphates are removed by solid-phase extraction on a C$_{18}$ reversed-phase SPICE cartridge from Rainin (Woburn, MA) to yield the free acid. An aqueous solution (2 ml) of the HPLC residue of compound **15** is acidified with 1 $M$ H$_3$PO$_4$ to pH 2 and is loaded onto the SPICE pack (prewashed with 2 ml of 3 m$M$ HCl), and **15** is eluted with 25% (v/v) methanol in water and lyophilized.

*Phosphopantetheine Adenylyltransferase and Dephospho-CoA Kinase: Preparation and Immobilization.* A 12-liter culture of *Brevibacterium ammoniagenes* is grown and the crude extract prepared and chromatographed on DEAE-Sepharose as described previously.[21] The fractions containing phosphopantotheine adenylyltransferase (PPAT) activity are pooled (0.40–0.43 $M$ NaCl, 63 ml, 0.026 unit/ml, 0.019 unit/mg), and the Tris buffer is replaced with HEPES buffer [50 m$M$ (pH 7.5), 5 m$M$ DTT] by two serial dilutions via ultrafiltration (PM10 Diaflo membrane; Amicon, Danvers, MA). The fractions containing dephospho-CoA kinase (DPCK) activity (0.36–0.39 $M$ NaCl, 0.08 unit/ml, 0.028 unit/mg) are pooled and dialyzed vs HEPES buffer [20 m$M$ (pH 8), 5 m$M$ DTT, 0.1 m$M$ phenylmethylsulfonyl fluoride (PMSF), 10 m$M$ MgCl$_2$, 1 liter]. Each enzyme solution is concentrated to 1 ml (ultrafiltration by PM10 membrane, followed by Centriprep 10 from Amicon). The PPAT and DCPK are then incubated separately in 0.1 $M$ HEPES buffer (pH 8), 0.07 m$M$ dephospho-CoA, 0.7 m$M$ MgCl$_2$, 2 m$M$ DTT, 0.07 m$M$ ATP (DPCK only), and 0.07 m$M$ inorganic pyrophosphate (PPAT only) to protect the active sites, and the enzymes are coimmobilized in 1.5 g of PAN 950[22] (yielding 36 ml of gel suspension; PPAT, 0.012 unit/ml; DPCK, 0.021 unit/ml; 17%).

*Synthesis of Adenosine 5'-(Trihydrogen Diphosphate) 3'-(Dihydrogen Phosphate) 5'-[(R)-3-Hydroxy-4-[3-(propylthio)-3-oxopropyl]amino]-2,2-dimethyl-4-oxobutyl] Ester (16).* Compound **15** (0.45 mmol) is dissolved in HEPES buffer (0.1 $M$, 16 ml, pH 7.5). Phosphopantetheine adenylyltransferase and dephospho-CoA kinase (0.13 and 0.23 unit, respectively), coimmobilized in polyacrylamide gel, are added to the solution, which is then sparged with nitrogen for 15 min. ATP (0.54 mmol, dissolved in 2.7 ml of water, pH adjusted to pH 7 with KOH), MgCl$_2$ (0.54 mmol), and inorganic pyrophosphatase (10 units) are added, and the reaction mixture is swirled at 30°. After 5 hr, pyruvate kinase (22 units) and phosphoenol pyruvate

---

[21] S. Shimizu, T. Yoshiki, and K. Ogata, *Methods Enzymol.* **62,** 236 (1979).
[22] G. Wiegand and S. J. Remington, *Annu. Rev. Biophys. Biophys. Chem.* **15,** 97 (1986).

(0.45 mmol as the potassium salt, pH adjusted to pH 7 with KOH) are added. The progress of the reaction is monitored via HPLC and is judged complete by the disappearance of compound **15** (about 2 days). The reaction mixture is centrifuged and the supernatant decanted. The immobilized enzymes are washed twice with HEPES buffer (0.1 $M$, 5 ml). The combined washes and supernatant are filtered through a 0.45-$\mu$m (pore size) nylon filter and lyophilized. The crude material (2.6 g) is dissolved in 3 m$M$ HCl (80 ml), and the solution is adjusted to pH 3.5 with phosphoric acid. At 4°, the solution is loaded onto a DEAE-cellulose column (2.5 × 17 cm$^2$) that has been equilibrated with 3 m$M$ HCl. The column is washed with 100 ml of 3 m$M$ HCl and then with a linear gradient of LiCl (0–0.2 $M$) in 3 m$M$ HCl (total volume of gradient, 1 liter). Individual fractions of 18 ml each are collected and analyzed by HPLC. Compound **16** elutes between 0.11 and 0.14 $M$ LiCl. Fractions containing compound **16** are adjusted to pH 4 with 1 $M$ K$_2$HPO$_4$ and are lyophilized (0.11 g of **16**, 31% yield). Compound **16** is further purified by preparative C$_8$ reversed-phase HPLC using a gradient of methanol in aqueous phosphate (50 m$M$, pH 4.5). Compound **16** is eluted at a flow rate of 3 ml/min with 5% (v/v) methanol for 10 min, followed by a linear gradient of methanol increasing to 60% (v/v) over 60 min. Fractions containing **16** are pooled and lyophilized. For analysis, residual phosphates are removed by solid-phase extraction on a SPICE cartridge (as outlined for compound **15**) to yield the free acid.

## Synthesis of Acyl-CH$_2$CoA

The final steps in the synthesis of acyl-CH$_2$CoA are different, depending on whether its precursors are obtained by the chemical or the chemoenzymatic strategy.

*Synthesis of Acyl-CH$_2$CoA (**9**) Using Precursors from Chemical Strategy.* Acyldethiapantetheine 4'-phosphate (**7**) (0.25 mmol) is dissolved in 2 ml of water and passed through a Dowex 50W-X4 (pyridine form, 1.2 × 12 cm) column. The compound is eluted with 300 ml of water (which is subsequently evaporated) and carefully dried *in vacuo*. The water-free residue is dissolved in 5 ml of dry pyridine under an argon atmosphere. Adenosine 2', 3'-cyclophosphate 5'-morpholidate (**8**) dissolved in 5 ml of dry pyridine is added. To remove traces of water the mixture is evaporated and redissolved in 10 ml of dry pyridine. After stirring for 40 hr at room temperature the solvent is evaporated and the residue is redissolved in 10 ml of water. Ring opening of the cyclophosphate is achieved by addition of 5 ml of 0.1 $M$ HCl. The solution is stirred for 1 hr at room temperature. After evaporation of the solvent the residue is dissolved in 5 ml of 0.001 $M$ trifluoroacetic acid. The separation of the isomers is achieved by preparative

HPLC on a TSK-DEAE column (30 × 1.5 cm, preequilibrated with 0.001 $M$ TFA; detection at 254 nm; flow, 3 ml/min), using a linear gradient from 0.001 $M$ TFA to 0.1 $M$ LiTFA/0.001 $M$ TFA within 100 min. Fractions are collected and evaporated after the pH is adjusted to pH 6. The residue is dried for several minutes *in vacuo* and dissolved in 2 ml of methanol. The product is precipitated after addition of 50 ml of dry acetone at 4°. After standing overnight, the white precipitate is collected by centrifugation and washed three times with cold, dry acetone. The precipitate is redissolved in 1 ml of water. This solution is lyophilized and the product **9**, is stored at −25°. Characterization of the two isomers, **9** ($R_t$ = 55 min) and **10** ($R_t$ = 67 min), is achieved by $^1$H NMR and $^{31}$P NMR. The yields are usually about 40% of theory.

### Synthesis from Precursors Obtained in Chemoenzymatic Synthesis

*Synthesis of Acetyl-CH$_2$CoA* (**19**). 5-Amino-2,2-dimethoxypentane[12] (1 ml, 1.9 $M$, pH adjusted to pH 10.5 with 1 $M$ H$_3$PO$_4$) is added to compound **16** (5 $\mu$mol, 4 mg as an HPLC residue containing KH$_2$PO$_4$). The pH is found to be pH 10.2. The solution is allowed to stand for 20 hr. The HPLC analysis shows the disappearance of **16** and the appearance of three new peaks corresponding to product **19** (11 min), its dimethyl acetal (13.2 min), and the carboxylic acid (9.5 min) resulting from hydrolysis of the thioester bond of **16**. The reaction mixture is neutralized to pH 4.5 with 1 $M$ H$_3$PO$_4$ and lyophilized. The material is purified by reversed-phase C$_{18}$ HPLC as outlined for compound **16** to yield 2 mg of compound **19** (2.5 $\mu$mol, 50%). For analysis, residual phosphates are removed using a SPICE cartridge as outlined for compound **15** to yield the free acid.

## Final Comments

Nonhydrolyzable acyl-CoA analogs are useful as alternative substrates, inhibitors, and regulators of a large number of enzymatic reactions. Their enhanced stability may be advantageous not only in mechanistic investigations but also in drug development. Here we have described some conventional chemical syntheses of nonhydrolyzable acyl-CoA analogs and also a more recent chemoenzymatic strategy for their preparation. Continuing activity both in the synthesis and use of nonhydrolyzable acyl-CoA analogs is expected in future.

# [26] Synthesis, Purification, and Characterization of Dicarboxylylmono-coenzyme A Esters

*By* Morteza Pourfarzam and Kim Bartlett

## Introduction

The excretion of dicarboxylic acids of chain length $C_6$–$C_{10}$ has been documented in diabetic ketoacidosis,[1] riboflavin deficiency,[2–4] and prolonged fasting.[5] In addition, inherited disorders of mitochondrial $\beta$-oxidation[6–8] and acquired disorders, such as Jamaican vomiting sickness caused by the ingestion of hypoglycine[9] and valproate intoxication, are characterized by the excretion of dicarboxylic acids.[10] The biogenesis of these compounds is thought to involve $\omega$- and $\beta$-oxidation pathways although the precise sequence of events remains uncertain. Studies of the mitochondrial and peroxisomal $\beta$-oxidation of [U-$^{14}$C]hexadecanedioylmono-CoA required methods for the analysis of the probable coenzyme A ester intermediates. We describe the synthesis, characterization, and analysis of these compounds by reversed-phase high-performance liquid chromatography (HPLC) and illustrate the application of these methods by reference to studies of the intermediates of [U-$^{14}$C]hexadecanedioylmono-CoA $\beta$-oxidation.

## Methods

### Materials

Dicarboxylic acids are obtained from Aldrich Chemical Co. (Gillingham, Dorset, U.K.). Acetyl-CoA, succinyl-CoA, CoA (lithium salt), ATP,

[1] J. E. Pettersen, *Clin. Chim. Acta* **38,** 17 (1972).
[2] S. I. Goodman, *Am. J. Clin. Nutr.* **34,** 2434 (1981).
[3] N. Gregersen and S. Kolvraa, *J. Inherit. Metab. Dis.* **5,** 17 (1982).
[4] K. Vietch, J.-P. Draye, J. Vamecq, A. G. Causey, K. Bartlett, H. S. A. Sherratt, and F. Van Hoof, *Biochim. Biophys. Acta* **1006,** 335 (1989).
[5] P. B. Mortensen and N. Gregersen, *Biochim. Biophys. Acta* **666,** 394 (1981).
[6] H. Przyrembel, U. Wendel, K. Becker, H. J. Bremer, L. Bruinvis, D. Ketting, and S. Wadman, *Clin. Chim. Acta* **66,** 227 (1976).
[7] N. Gregersen, S. Kolvraa, K. Rasmussussen, P. B. Mortensen, P. Divry, M. David, and N. Hobolth, *Clin. Chim. Acta* **132,** 181 (1983).
[8] C. Vianey-Liaud, P. Divry, N. Gregersen, and M. Matthieu, *J. Inherit. Metab. Dis.* **10**(1), 159 (1987).
[9] K. Tanaka, *J. Biol. Chem.* **247,** 7465 (1972).
[10] P. B. Mortensen, N. Gregersen, S. Kolvraa, and E. Christensen, *Biochem. Med.* **24,** 153 (1980).

NAD, acyl-CoA synthetase (long-chain-fatty-acid–CoA ligase), acyl-CoA oxidase, and crotonase (enoyl-CoA hydratase) are purchased from Sigma Chemical Co. (Poole, Dorset, U.K.). Acetonitrile (S grade) is purchased from Rathburn Chemicals (Walkerburn, Scotland, U.K.). HPLC-grade water, Analar-grade $CHCl_3$, $CH_3OH$, Scintran-grade xylene, Triton X-100, and 2,5-diphenyloxazole are supplied by BDH, Ltd. (Poole, Dorset, U.K.). [U-$^{14}$C]Hexadecanoic acid (specific radioactivity, 30.6 GBq/mmol) is obtained from Amersham International (Amersham, Bucks, U.K.). DEAE-Sephacel is purchased from Pharmacia-LKB (Uppsala, Sweden). $C_{18}$ solid-phase extraction cartridges (Sep-Pak) are obtained from Millipore, Waters Chromatography (Milford, MA). All other reagents are of the highest available purity unless otherwise specified.

## Chemical Synthesis of Dicarboxylylmono-CoA Esters: Applicable to All Chain Lengths

Dicarboxylylmono-CoA esters are prepared from their corresponding acylmonochlorides, which are synthesized by treating dicarboxylic acid (10 mmol) with an equimolar amount of thionyl chloride (10 mmol) in 15 ml of anhydrous boiling dioxane for 5–7 hr. Dioxane and unreacted thionyl chloride are removed using a rotary evaporator under reduced pressure. The resultant acyl chloride is used without further purification and 0.3 mmol (with respect to the starting acid) is dissolved in 7 ml of anhydrous dioxane and added to a solution of 0.05 mmol of CoA (lithium salt) in 4.5 ml of degassed water (pH 8 with triethylamine) dropwise over a period of 20 min. The pH is maintained at approximately pH 8 by the addition of triethylamine. The reaction is carried out under a gentle stream of $N_2$ and a single phase maintained by addition of small amounts of either dioxane or water as necessary. Once all the acyl chloride is added, the reaction is stirred at room temperature for an additional 10 min under $N_2$. After completion of the reaction, the pH is adjusted to pH 3–4 with HCl, and dioxane is removed with a rotary evaporator. The remaining solution is extracted with 20 ml of ethyl acetate and 20 ml of diethyl ether and, after removal of dissolved ether under $N_2$, the aqueous layer containing the product is purified as described. The yields are typically 50–60% with respect to CoA.

## Enzymatic Synthesis of Dicarboxylylmono-CoA Esters: For $C_{16}$–$C_{12}$

The free acid (0.1 mmol) is incubated with 1 U of acyl-CoA synthase in a medium containing 10 m$M$ $MgCl_2$, 10 m$M$ ATP, 5 m$M$ CoA, 0.1% (v/v) Triton X-100, and 50 m$M$ $K_2HPO_4$, pH 7.4, in a volume of 5 ml for 2 hr at 35° in a shaking water bath. The reaction is terminated by the addition of HCl to bring the pH to 3–4 and the product is purified by means of a $C_{18}$ solid-phase extraction cartridge as described in this chapter.

## Synthesis of 2,3-Enoyldicarboxylylmono-CoA Esters

Dicarboxyl-2-enoyl-CoA esters are synthesized enzymatically from the respective saturated acyl-CoA esters as follows: Dicarboxylyl-CoA (4 $\mu$mol) is incubated with 5 U of acyl-CoA oxidase in 25 m$M$ Tris buffer, pH 7.8, in a total volume of 2 ml at 30° in a shaking water bath. The formation of 2-enoyl-CoA is monitored by measuring the increase in absorbance at 225 nm owing to the conjugated 2,3-double bond. The reaction goes to completion typically after 30 min. The yields of conversion are 55–85% (increasing with increasing chain length).

## Synthesis of 3-Hydroxydicarboxylylmono-CoA Esters

3-Hydroxydicarboxylylmono-CoA esters are prepared from the corresponding 2-enoyl-CoA esters. Enoyl-CoA (4 $\mu$mol) is treated with 10 U of crotonase in 2 ml of 25 m$M$ Tris buffer, pH 7.8, at 35°. The production of 3-hydroxyacyl-CoA is monitored by following the decrease in the absorbance at 225 nm owing to the disappearance of the double bond.

## Purification of Synthetic Acyl-CoA Esters

The following method has been adopted to purify a wide range of acyl-CoA esters. $C_{18}$ solid-phase extraction (Sep-Pak, 500 mg) cartridges are conditioned prior to use by rinsing successively with two column volumes (5 ml) each of (1) acetonitrile, (2) water, and (3) 50 m$M$ potassium phosphate, pH 5.3, under vacuum (10 kPa). The composition of acetonitrile–phosphate buffer mixture required to elute the product of interest is previously determined by HPLC analysis on a $C_{18}$ analytical column. Residual organic solvent, if any, is removed from the crude product, the phosphate concentration is adjusted to about 50 m$M$, and the pH is adjusted to pH 5–6. The sample is then applied to the conditioned cartridge under gravity, and the eluant is retained and reapplied to the cartridge to promote maximum binding. The cartridge is washed under vacuum (10 kPa) with 5 ml of 10% (v/v) acetonitrile–50 m$M$ phosphate buffer to elute CoA and CoA-disulfide. Following this the concentration of acetonitrile is increased by 5% (v/v) in successive 2-ml fractions until the concentration of acetonitrile required to elute the product is reached. The content of each eluant is monitored by HPLC and the fraction(s) containing the product of interest are collected and acetonitrile removed under $N_2$ at 40°. The recoveries are typically 90–95%.

*Synthesis and Purification of [U-$^{14}$C]Hexadecanedioylmono-CoA*

[U-$^{14}$C]Hexadecanedioic acid is synthesized from [U-$^{14}$C]hexadecanoic acid by incubation with a washed microsomal fraction prepared from the livers of rats treated with ciprofibrate.[11] Microsomal fractions are prepared by differential centrifugation of liver homogenate. Following successive centrifugations at 10,000 $g_{av}$ for 10 min at 4° and 30,000 $g_{av}$ for 15 min at 4°, the final supernatant is centrifuged at 126,000 $g_{av}$ for 60 min at 4° to generate a microsomal pellet, which is resuspended and recentrifuged twice. [U-$^{14}$C]Hexadecanoic acid (5 $\mu$mol; specific radioactivity, 1.85 GBq/mmol) is incubated with washed microsomal fraction (5 mg of microsomal protein per milliliter) in the presence of 5 m$M$ MgCl$_2$, 5 $\mu M$ MnCl$_2$, 1 m$M$ NADP, 0.5 m$M$ NADH, 5 m$M$ isocitrate, 10 U of isocitrate dehydrogenase, 50 m$M$ Tris buffer (pH 7.5), and bovine serum albumin (BSA; 1 mg ml$^{-1}$) in a total volume of 20 ml for 14 hr at 30°. The reaction is terminated by the addition of KOH to pH 12 and the reaction mixture freeze-dried. The residue is resuspended in 2 ml of water and the pH taken to less than pH 1 with HCl. Organic acids are extracted three times with 3 vol of ethyl acetate, the extract is dried over anhydrous sodium sulfate, and the solvent is removed by rotary evaporation.

Preliminary experiments with unlabeled hexadecanoate, and analysis by gas chromatography/mass spectrometry of the products methylated by the addition of an excess of diazomethane, show that there is about 80% conversion of hexadecanoate to hexadecanedioate. The mass spectra of the biosynthesized hexadecanedioate and standard compound are identical (results not shown). The radioactive preparation is used without further purification and converted to the corresponding mono-CoA ester by incubation with 1 U of acyl-CoA synthetase in medium containing 10 m$M$ MgCl$_2$, 10 m$M$ ATP, 3 m$M$ CoA, 0.1% (v/v) Triton, and 50 m$M$ KH$_2$PO$_4$, pH 7.4, in a total volume of 5 ml for 4 hr at 30°. The reaction is terminated by the addition of HCl to bring the pH to 3–4. The product is purified by preparative HPLC using a 250 × 10 mm, 10-$\mu$m C$_{18}$ reversed-phase LiChrosorb column (E. Merck AG, Darmstadt, Germany) developed with a gradient of acetonitrile in 50 m$M$ potassium phosphate, pH 5.3, as follows: isocratic 10% (v/v) acetonitrile (5 min); linear gradient to 20% (v/v) acetonitrile (5 min); linear gradient to 50% (v/v) acetonitrile (25 min). The flow rate is 10 ml/min. Radioactivity and ultraviolet (UV) absorbance are monitored continuously and the major radioactive peak, which cochromatographs with standard hexadecanedioylmono-CoA, is collected. This fraction is concen-

[11] K. Bartlett, R. Hovik, S. Eaton, N. J. Watmough, and H. Osmundsen, *Biochem. J.* **270**, 175 (1990).

trated and the phosphate removed using a 500-mg $C_{18}$ solid-phase extraction cartridge, which is eluted with acetonitrile–water (50 : 50, v/v), freeze-dried, and redissolved in water. Reanalysis of the product by radio-HPLC using an analytical (250 × 4.6 mm) 5-$\mu$m $C_{18}$ LiChrosorb column reveals a single radioactive peak, which cochromatographs with standard hexadecanedioyl-mono-CoA and has a UV absorbance spectrum characteristic of saturated thioesters (Fig. 1). A portion of the product was saponified and the resultant free acid derivatized with 4-nitrobenzoyldi-isopropylisourea and analyzed by radio-HPLC as the 4-nitrobenzoyl ester, again a single radioactive peak was observed which cochromatographed with standard di(4-nitrobenz-oyl)hexadecanedioate.

## High-Performance Liquid Chromatography

*Instrumentation.* A Waters 600 solvent delivery system fitted with a 5-$\mu$m LiChrosorb $C_{18}$ column (250 × 4.6 mm i.d.; E. Merck AG) is used for HPLC. The column is maintained at 30°. Samples are introduced with a Waters U6K injector. All solvents are deaerated with helium.

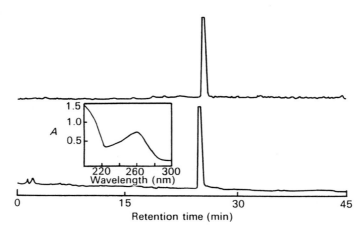

FIG. 1. Radio-HPLC analysis of synthetic [U-$^{14}$C]hexadecanedioylmono-CoA. Upper trace, radioactivity; lower trace, $A_{260}$; inset in lower trace, UV absorption spectrum. The product was analyzed using a 5-$\mu$m Lichrosorb $C_{18}$ column (250 × 4.6 mm) developed with the following binary gradient of acetonitrile in 50 m$M$ potassium phosphate, pH 5.3: isocratic 5% (v/v) acetonitrile (5 min), linear gradient to 10% (v/v) acetonitrile (0.1 min), linear gradient to 30% (v/v) acetonitrile (9.9 min), linear gradient to 50% (v/v) acetonitrile (30 min). The flow rate was 1.7 ml/min. This modified gradient was used to ensure that the product was free of the CoA ester of the starting material ([U-$^{14}$C]hexadecanoyl-CoA) or its derivatives. [Reprinted from M. Pourfarzam and K. Bartlett, *Biochem. J.* **273**, 205 (1991), with permission.]

Analytes are detected with a Waters 990 photodiode array detector ($8\text{-}\mu\text{l}$ flow cell; 10-mm path length). Spectra are acquired at 0.35-sec intervals over the range of 200 to 300 nm, with a bandwidth of 1.4 nm. Spectral data are acquired on a microcomputer using dedicated software. Radioactivity associated with eluted compounds is detected on-line with a LKB Betacord radioactivity monitor fitted with a 0.4-ml flow cell as described previously.[12,13] Because the flow rate of effluent from the photodiode array detector is 1.5 ml/min the flow rate of the scintillation fluid [10 g of 2,5-diphenyloxazole (PPO), 330 ml of Triton X-100, 150 ml of $CH_3OH$, and 670 ml of xylene] is 6 ml/min. The photodiode array detector and radioactivity monitor are connected in series and analog signals from each detector are acquired by a Waters chromatography data station (model 840). This allows the generation of superimposable radiochemical and UV traces. The time lag between the detectors is determined using [U-$^{14}$C]hexadecanoyl-CoA.

*High-Performance Liquid Chromatography of Dicarboxylylmono-CoA Esters.* Initial attempts to analyze intact dicarboxylylmono-CoA esters were made on the basis of the method of Causey *et al.*[12] This method and its modifications,[13] however, did not provide sufficient resolution. The method has been modified by changing the gradient profile and reducing the flow rate. Adequate separation of the saturated acyl- and 2,3-enoyl-CoA esters of chain length $C_6$ to $C_{16}$ is achieved by this means, using a 10-$\mu$m LiChrosorb $C_{18}$ column. However, the system still fails to resolve some of the dicarboxylyl $(DC)_n$-2,3-enoyl-CoA from $DC_{n+2}$-3-hydroxyacyl-CoA esters. In an attempt to further improve the resolution of these pairs, four alternative columns have been evaluated: 5-$\mu$m Hypersil $C_{18}$ (250 × 4.6 mm i.d.), 3-$\mu$m Hypersil $C_{18}$ (250 × 4.6 mm i.d.), 10-$\mu$m LiChrosorb $C_{18}$ (250 × 4.6 mm i.d.,), and 5-$\mu$m LiChrosorb $C_{18}$ (250 × 4.0 mm i.d.). The best resolution is obtained using the 5-$\mu$m LiChrosorb column with the following gradient of acetonitrile and 50 m$M$ phosphate buffer, pH 5.3, isocratic 5% (v/v) acetonitrile (5 min); linear gradient to 10% (v/v) acetonitrile (0.1 min); linear gradient to 40% (v/v) acetonitrile (29.9 min); linear gradient to 50% (v/v) acetonitrile (5 min); linear gradient to 5% (v/v) acetonitrile (5 min). The flow rate is 1.5 ml min$^{-1}$ and the total run time is 45 min. The column is reequilibrated for at least 10 min under the starting conditions between analyses to maintain reproducibility of retention times. A typical chromatogram is shown in Fig. 2 and the retention times are listed in Table I. This system is able to resolve all of the CoA esters, with the exception of tetradecane-2,3-dioyl-CoA and 3-hydroxyhexadecanedioyl-CoA, which co-

[12] A. G. Causey, B. Middleton, and K. Bartlett, *Biochem. J.* **235**, 343 (1986).
[13] N. J. Watmough, D. M. Turnbull, H. S. A. Sherratt, and K. Bartlett, *Biochem. J.* **262**, 261 (1989).

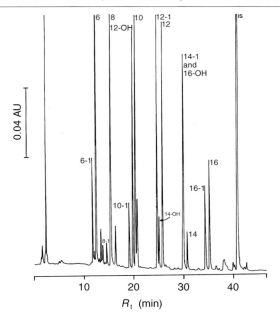

FIG. 2. HPLC separation of a standard mixture of dicarboxylylmono-CoA esters. Chromatographic conditions were as described in text. The carbon numbers are as indicated; 2,3-enoyl-CoA esters are indicated by the suffix -1; 3-hydroxyacyl-CoA esters are indicated by the suffix -OH. The chromatographic internal standard (IS) was undecanoyl-CoA. (Reprinted from *J. Chromatogr.* **570**, M. Pourfarzam and K. Bartlett, 253, Copyright 1991, with kind permission of Elsevier Science–NL, Sara Burgerhartstraat 25, 1055 KV Amsterdam, The Netherlands.)

eluted. Acyl-CoA esters of a given chain length are eluted in the order 3-hydroxyacyl, enoyl, acyl, as has been reported previously for monocarboxylate acyl-CoA esters.[13]

*Fast Atom Bombardment Mass Spectrometry*

A Kratos MS 80RF mass spectrometer in fast atom bombardment (FAB) ionization mode is used. Aqueous solutions of samples (2–3 $\mu$l) are mixed with 10 $\mu$l of glycerol matrix and introduced via a vacuum lock direct insertion probe. A charge-transfer gun is used to generate a xenon atom beam of 7-eV kinetic energy and mass spectra are acquired over the range of 60–2000 amu (atomic mass units). Good-quality spectra are obtained from 5 to 10 nmol of acyl-CoA esters.

The CoA used for the synthesis of CoA esters is the $Li_3^+$ salt. During preparation $Na^+$ is also added to the reaction mixture; therefore, the products contain both these ions. The presence of alkali metals can complicate

TABLE I
RETENTION TIMES OF DICARBOXYLYLMONO-CoA ESTERS[a,b]

| CoA ester | Retention time (min) | Relative retention time |
|---|---|---|
| Acetyl | 9.76 | 0.428 |
| $DC_4$ | 4.70 | 0.206 |
| $DC_6$ | 12.76 | 0.559 |
| $DC_{6:1}$ | 12.23 | 0.536 |
| $DC_8$ | 15.73 | 0.689 |
| $DC_{8:1}$ | 14.88 | 0.652 |
| $DC_{10}$ | 20.36 | 0.892 |
| $DC_{10:1}$ | 19.23 | 0.840 |
| $DC_{10:OH}$ | 15.50 | 0.679 |
| $DC_{11}$ | 22.81 | 1.000 |
| $DC_{12}$ | 25.27 | 1.107 |
| $DC_{12:1}$ | 24.23 | 1.062 |
| $DC_{12:OH}$ | 19.79 | 0.867 |
| $DC_{14}$ | 29.73 | 1.303 |
| $DC_{14:1}$ | 28.95 | 1.269 |
| $DC_{14:OH}$ | 24.57 | 1.077 |
| $DC_{16}$ | 33.76 | 1.480 |
| $DC_{16:1}$ | 33.00 | 1.446 |
| $DC_{16:OH}$ | 28.99 | 1.271 |

[a] Reprinted from *J. Chromatogr.* **570,** M. Pourfarzam and K. Bartlett, 253, Copyright 1991, with kind permission of Elsevier Science–NL, Sara Burgerhartstraat 25, 1055 KV Amsterdam, The Netherlands.
[b] CoA esters were analyzed using a 5-$\mu$m Lichrosorb $C_{18}$ (250 × 4.6 mm i.d.) column eluted with a binary gradient of acetonitrile in 50 m$M$ $KH_2PO_4$ (pH 5.3) as described in Methods.

FAB–mass spectrometry (MS) analysis by adduct formation with the analyte and glycerol matrix, resulting in dilution of the analyte signal with reduction of sensitivity. A variety of methods are used to overcome this problem, including addition of $Ag^+$ to the glycerol matrix[14] and desalting the sample with ion-exchange resins.[15] Figure 3 shows the FAB mass spectrum of octanedioylmono-CoA in the region of the molecular ion. The upper spectrum (Fig. 3A) is obtained without removal of the alkali metals from the sample and multiple adducts are clearly present. The lower spectrum (Fig. 3B) is obtained after removal of alkali metals with Dowex 50

[14] K. W. S. Chan and K. D. Cook, *Macromolecule* **16,** 1736 (1983).
[15] B. D. Musselman, J. T. Watson, and J. Allison, *in* "Extended Abstracts of the 31st Annual Meeting on Mass Spectrometry and Allied Topics," p. 728. 1983.

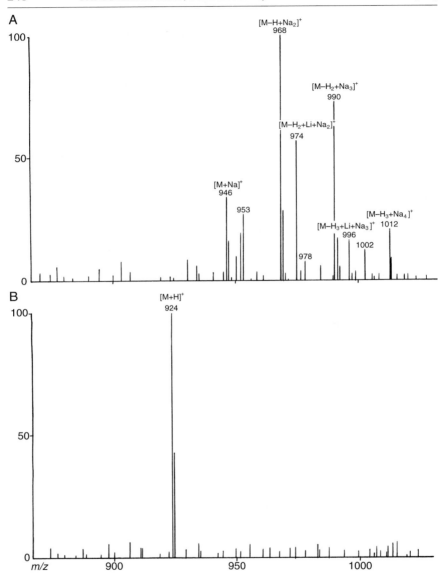

FIG. 3. FAB mass spectra of octanedioylmono-CoA. (A) Without removal of alkali metals; (B) with removal of alkali metals. The assignations of the ions are as indicated. (Reprinted from *J. Chromatogr.* **570,** M. Pourfarzam and K. Bartlett, 253, Copyright 1991, with kind permission of Elsevier Science–NL, Sara Burgerhartstraat 25, 1055 KV Amsterdam, The Netherlands.)

(hydrogen form; 250 mg/ml) and only the protonated molecular ion is present resulting in considerably enhanced sensitivity. Accordingly, alkali metals are removed from all samples prior to FAB–MS analysis.

Figure 4 shows the FAB mass spectrum of dodecanedioylmono-CoA, derived after subtracting an averaged background spectrum to remove signals from the glycerol matrix. The spectrum illustrates the characteristic major fragmentation process of this class of compounds. The ion at $m/z$ 980 corresponds to the molecular ion of the protonated free acid $[M + H]^+$. Cleavage of the carbon–oxygen and phosphorus–oxygen bonds resulted in prominent ions at $m/z$ 428, $m/z$ 473, $m/z$ 508, and $m/z$ 553. The $m/z$ 473 and $m/z$ 553 ions are of particular importance because they preserve the identity of the dicarboxylyl group, and therefore shift in mass as the acyl group changes. These ions have been observed in all of the CoA esters examined. Other fragments derived from cleavages at adenine–ribose and ribose–phosphate bonds of the CoA moiety ($m/z$ 136 and $m/z$ 330) are present in all spectra and are not shown. Table II summarizes the major characteristic ions with their relative intensities, normalized to the $[MH_2 - 508]^+$ ion, observed in the FAB mass spectra of the dicarboxylyl-

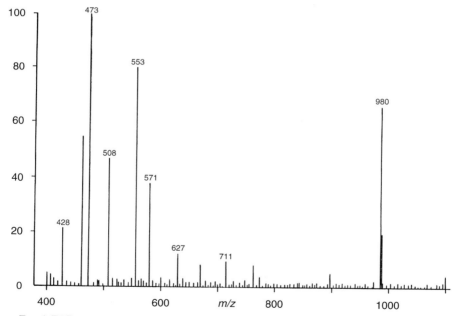

FIG. 4. FAB mass spectrum of dodecanedioylmono-CoA. (Reprinted from *J. Chromatogr.* **570**, M. Pourfarzam and K. Bartlett, 253, Copyright 1991, with kind permission of Elsevier Science–NL, Sara Burgerhartstraat 25, 1055 KV Amsterdam, The Netherlands.)

TABLE II

MAJOR CHARACTERISTIC FRAGMENTS AND RELATIVE INTENSITIES[a] OBSERVED IN FAST
ATOM BOMBARDMENT MASS SPECTRA OF DICARBOXYLYLMONO-CoA ESTERS[b]

| Fragment (relative intensities) | $[M + H]^+$ (20–65) | $[MH_2 - 508]^+$ (100) | $[MH_2 - 428]^+$ (35–60) | $[M - 386]^+$ (15–30) | $[M - 408]^+$ (20–40) |
|---|---|---|---|---|---|
| $DC_{16}$-CoA | 1036 | 529 | 609 | nd[c] | nd |
| $DC_{14}$-CoA | 1008 | 501 | 581 | nd | nd |
| $DC_{12}$-CoA | 980 | 473 | 553 | nd | 571 |
| $DC_{10}$-CoA | 952 | 445 | 525 | 565 | 543 |
| $DC_8$-CoA | 924 | 417 | 497 | 537 | 515 |
| $DC_6$-CoA | 896 | 389 | 469 | 509 | 487 |

[a] Normalized to $[MH_2 - 508]^+$.

[b] Reprinted from *J. Chromatogr.* **570**, M. Pourfarzam and K. Bartlett, 253, Copyright 1991, with kind permission of Elsevier Science–NL, Sara Burgerhartstraat 25, 1055 KV Amsterdam, The Netherlands.

[c] nd, Not determined.

CoA esters studied. In no case is there any evidence for the presence of dicarboxylyldi-CoA esters.

## Incubation of Peroxisomal Fractions

Peroxisomal fractions are prepared from ciprofibrate-treated rats by isopycnic centrifugation on self-generated Percoll gradients as described previously.[11] [U-14C]Hexadecanedioylmono-CoA is incubated with peroxisomes, using nonsolubilizing (isosmotic) conditions in the presence or absence of an $NAD^+$-regenerating system as previously described.[16]

## Preparation of Samples for Radio-HPLC Analysis of [14C]Acyl-CoA Esters

Incubations are terminated by the addition of 50 μl of 0.5 M $H_2SO_4$ to 1 ml of the reaction mixture. Internal standard (20 nmol of undecanedioyl-CoA) is added, followed by 100 μl of saturated ammonium sulfate. Each sample is extracted twice with 10 vol of ethyl acetate–diethyl ether (1:1, v/v) to remove free carboxylic acids and then extracted with 10 ml of methanol–chloroform (2:1, v/v) for 1 hr with continuous agitation. After centrifugation (5000 $g_{av}$; 5 min at 4°) the supernatant is retained and the pellet is reextracted with 5 ml of methanol–chloroform (2:1, v/v) and

[16] M. Pourfarzam and K. Bartlett, *Biochem. J.* **273**, 205 (1991).

recentrifuged. The combined supernatants are evaporated to dryness under a stream of $N_2$ at 30°. In the case of peroxisomal incubations this preparation is dissolved in 500 $\mu$l of 5% (v/v) acetonitrile in phosphate buffer (50 m$M$, pH 5.3) and analyzed directly as described previously for acyl-CoA esters. The recovery of acyl-CoA esters by this procedure is shown in Table III. In the case of mitochondrial incubations the residue is dissolved in 4 ml of methanol–water (4:1, v/v) and insoluble material is removed by centrifugation (5000 $g_{av}$; 5 min at 4°). The supernatant is applied to a DEAE-Sephacel column (60 × 6 mm i.d., acetate form, prepared by treatment of the ion-exchanger with 5 $M$ acetic acid overnight and then rinsing with water to pH 5–6). The eluant is retained and reapplied to the column to promote maximal binding. The column is washed with 4 ml of methanol–water (4:1, v/v) and the eluant retained for analysis of acylcarnitines. The acyl-CoA esters are eluted with 4 ml of methanol–water (4:1, v/v) containing 0.5 $M$ ammonium acetate and 10 m$M$ acetic acid. The methanol is removed under $N_2$ and after the addition of 2 ml of water the sample is freeze-dried. The residue is dissolved in 500 $\mu$l of 5% (v/v) acetonitrile in phosphate buffer (50 m$M$, pH 5.3) and analyzed by radio-HPLC as described. The recovery by this procedure is shown in Table III.

## Radio-HPLC Analysis of [$^{14}$C]Dicarboxylyl-CoA Esters

The objective of the studies described in the previous section was to develop a comprehensive range of analytical methods to study the metabolism of hexadecanedioic acid by isolated mitochondria and peroxisomes. The results presented in Fig. 5 show the acyl-CoA intermediates generated by incubation of purified peroxisomes with [U-$^{14}$C]hexadecanedioylmono-CoA in the presence (Fig. 5A) and absence (Fig. 5B) of an NAD$^+$-regenerating system for 2 min. In the presence of NAD$^+$ the $C_6$–$C_{16}$ saturated acyl-CoA intermediates and acetyl-CoA were detected. However, in the absence of NAD$^+$, 3-hydroxyacyl- and 2,3-enoyl-CoA intermediates were seen with

TABLE III
RECOVERY OF ACYL-CoA ESTERS FROM QUENCHED MITOCHONDRIAL AND PEROXISOMAL INCUBATIONS

| Sample | Amount (nmol) | Recovery[a] | | | | |
|---|---|---|---|---|---|---|
| | | Acetyl-CoA | DC$_6$-CoA | DC$_{11}$-CoA | DC$_{12}$-CoA | DC$_{16}$-CoA |
| Peroxisomes | 5 | 85.0 ± 2.2 | 89.6 ± 4.5 | 91.7 ± 3.8 | 94.5 ± 2.7 | 93.2 ± 4.5 |
| Mitochondria | 2 | 68.8 ± 6.9 | 70.2 ± 4.2 | 71.8 ± 4.0 | 72.4 ± 3.8 | 72.8 ± 3.6 |
| | 10 | 70.5 ± 4.9 | 72.9 ± 3.6 | 74.3 ± 3.3 | 73.7 ± 3.1 | 74.8 ± 4.7 |

[a] Mean ± SD; $n$ = 5.

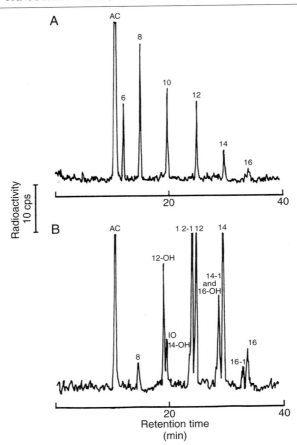

FIG. 5. Intermediates of β-oxidation of [U-$^{14}$C]hexadecanedioylmono-CoA in the presence (A) and absence (B) of an NAD$^+$-regenerating system by purified rat liver peroxisomes. Peaks are as designated in Fig. 2. AC, Acetyl-CoA. (Reprinted from *J. Chromatogr.* **570**, M. Pourfarzam and K. Bartlett, 253, Copyright 1991, with kind permission of Elsevier Science–NL, Sara Burgerhartstraat 25, 1055 KV Amsterdam, The Netherlands.)

a concomitant lowering of flux (12.9 ± 2.9 vs 8.0 ± 3.3 nmol of acetate/min/mg protein; mean ± SD, $n = 3$).

Conclusion

The chemical synthesis of saturated dicarboxylylmono-CoA esters using acyl chlorides as acylating reagent provided reasonable yields, typically in excess of 60% (with respect to CoA). Two factors are important to obtain reproducible yields. First, the pH of the reaction mixture, which was found to be optimal at about pH 8, and second, the acyl chloride/CoA ratio. When using acyl chlorides for the preparation of CoA esters an excess molar ratio of acylchloride to free CoA is required. We found a molar ratio of 3–4 to be optimal for preparation of a wide range of acyl-CoA esters. Increasing the acyl chloride/CoA ratio not only did not improve yield any further but decreased it significantly. Presumably this was due to the reaction of acyl chlorides at other functional groups of the CoA molecule. When the starting fatty acid is present in limited supply, for example, if it is isotopically labeled, chemical methods therefore are not suitable. The enzymatic method described in Methods provides a simple and efficient method for the synthesis of a wide range of acyl-CoA esters in high yield. [U-$^{14}$C]Hexadecanedioylmono-CoA and [U-$^{14}$C]hexadecanoyl-CoA were synthesized by this method with a specific radioactivity of 2 GBq/mmol. The yields obtained for these syntheses were greater than 70 and 85%, respectively, with respect to the recovery of the label.

The choice of purification strategy for synthetic acyl-CoA esters depends on the purity requirement, the chain length of the acyl-CoA, and the instrumentation available. Long-chain acyl-CoA esters are insoluble in strong acid and diethyl ether, and can be purified from CoASH, CoA-disulfide, and free fatty acid. The reaction mixture is acidified to pH 1–2, the remaining organic solvent is removed, and the residual acidic solution is extracted with ether to remove any fatty acid or unreacted acylated intermediates. Because long-chain acyl-CoAs are insoluble in both phases they form a precipitate at the interface. After removing the aqueous and etheral layers the precipitated acyl-CoA is recovered and washed with ether. This method is suitable only for large-scale preparation of acyl-CoA esters. For small-scale preparations such as labeled acyl-CoA esters, for acid-soluble acyl-CoA esters, or for applications that require very high purity products alternative methods were adopted on the basis of semipreparative HPLC or the use of reversed-phase $C_{18}$ solid-phase extraction cartridges. Acyl-CoA esters of high purity can be obtained using semipreparative HPLC. This method, however, can be time consuming. The use of $C_{18}$ solid-phase extraction cartridges allows rapid purification of acyl-CoA esters of any chain-length acyl group, using small volumes of solvent. Because the composition of the solvent required to elute the product of interest can be determined by HPLC, using an analytical column with a similar stationary phase, the product can be recovered efficiently and in high purity.

The results shown in Fig. 5 demonstrate the application of radio-HPLC analysis of dicarboxylylmono-CoA esters to the study of the intermediates generated by incubation of [U-$^{14}$C]hexadecanedioylmono-CoA with rat liver peroxisomes. If the NAD$^+$-regenerating system is omitted from the incubation (Fig. 5B), flux is slowed with concomitant lower production of acetyl-CoA (cf. Fig. 5A). Furthermore, under these conditions 2-enoyl-CoA and 3-hydroxyacyl-CoA esters accumulate. Similar effects have been observed when the intermediates of peroxisomal [U-$^{14}$C]hexadecanoate oxidation were examined,[11] suggesting that the peroxisomal β-oxidation of dicarboxylates and monocarboxylates may be modulated by the prevailing intracellular redox state.

Detailed studies of the peroxisomal and mitochondrial β-oxidation of [U-$^{14}$C]hexadecanedioylmono-CoA, utilizing the dicarboxylylmono-CoA esters whose synthesis and analysis are described here, have been reported elsewhere.[16–19]

## Acknowledgment

Action Research and the Muscular Dystrophy Group of Great Britain are thanked for the provision of HPLC equipment.

[17] M. Pourfarzam and K. Bartlett, *J. Chromatogr.* **570**, 253 (1991).
[18] M. Pourfarzam and K. Bartlett, *Eur. J. Biochem.* **208**, 301 (1992).
[19] M. Pourfarzam and K. Bartlett, *Biochim. Biophys. Acta* **1141**, 81 (1993).

# [27] Holo-[Acyl-Carrier-Protein] Synthase of *Escherichia coli*

*By* Ralph H. Lambalot and Christopher T. Walsh

Holo-[acyl-carrier-protein] synthase (ACP synthase; EC 2.7.8.7) of *Escherichia coli* transfers the 4'-phosphopantetheinyl moiety of coenzyme A (CoA) to Ser-36 of apo-[acyl-carrier-protein] (apo-ACP) to produce holo-ACP and 3',5'-ADP in a Mg$^{2+}$-dependent reaction[1]:

$$CoA + apo\text{-}ACP \rightarrow holo\text{-}ACP + 3',5'\text{-}ADP$$

In their holo form, ACPs play central roles in a broad range of biosynthetic pathways that depend on iterative acyl transfer steps, including po-

[1] J. Elovson and P. R. Vagelos, *J. Biol. Chem.* **243**, 3603 (1968).

lyketide,[2] nonribosomal peptide,[3] and depsipeptide biosynthesis,[4] as well as in the transacylation of oligosaccharides[5] and proteins.[6] Overexpression of the *E. coli* ACP[7,8] and several homologs, including the polyketide synthase ACPs from *Streptomyces*[9] and the D-alanyl carrier protein (Dcp) from *Lactobacillus casei,*[10] in *E. coli* yields variable mixtures of the apo and holo forms of the proteins. Because structure–function studies of ACPs require pure preparations of completely phosphopantetheinylated protein, a method for converting milligram quantities of apo-ACPs or apo–holo mixtures quantitatively to the holo form is needed. We have reported the cloning and overexpression of ACP synthase from *E. coli*[11] and have found that ACP synthase is capable of recognizing and modifying several ACP homologs from other species, namely Dcp from *L. casei,*[10] NodF from *Rhizobium leguminosarum,* and several polyketide synthase ACPs from *Streptomyces.*[12] This broad specificity of ACP synthase has allowed the development of methods for the quantitative conversion of recombinant apo-ACPs to their holo form, thereby making structure–function studies of these proteins possible.

## Overproduction and Purification of Recombinant ACP Synthase

### Materials

*Escherichia coli* strain BL21(DE3)pACPS1: Available from the author
Ampicillin: 50 mg/ml, sterile filtered
Luria–Bertani (LB) ampicillin agar: Prepare by autoclaving 10 g of tryptone, 5 g of yeast extract, 5 g of NaCl, 15 g of agar, and 1 ml of 1 *N* NaOH per liter of water, cooling to 65°, and adding ampicillin to 50 $\mu$g/ml before pouring 20-ml plates

[2] B. Shen, R. G. Summers, H. Gramajo, M. J. Bibb, and C. R. Hutchinson, *J. Bacteriol.* **174,** 3818 (1992).

[3] M. A. Marahiel, *FEBS Lett.* **307,** 40 (1992).

[4] F. Rusnak, M. Sakaitani, D. Drueckhammer, J. Reichart, and C. T. Walsh, *Biochemistry* **30,** 2916 (1991).

[5] T. Ritsema, O. Geiger, P. van Dillewijn, B. J. J. Lugtenberg, and H. P. Spaink, *J. Bacteriol.* **176,** 7740 (1994).

[6] J. P. Issartel, V. Koronakis, and C. Hughes, *Nature (London)* **351,** 759 (1991).

[7] R. B. Hill, K. R. MacKenzie, J. M. Flanagan, J. E. Cronan, Jr., and J. H. Prestegard, *Protein Express. Purif.* **6,** 394 (1995).

[8] D. H. Keating, M. Rawlings-Carey, and J. E. Cronan, Jr., *J. Biol. Chem.* **270,** 22229 (1995).

[9] J. Crosby, D. H. Sherman, M. J. Bibb, W. P. Revill, D. J. Hopwood, and T. J. Simpson, *Biochim. Biophys. Acta* **1251,** 32 (1995).

[10] D. V. Debabov, M. P. Heaton, Q. Zhang, K. D. Stewart, R. H. Lambalot, and F. C. Neuhaus, *J. Bacteriol.* **178,** 3869 (1996).

[11] R. H. Lambalot and C. T. Walsh, *J. Biol. Chem.* **270,** 24658 (1995).

[12] A. M. Gehring, R. H. Lambalot, and C. T. Walsh, unpublished results (1996).

LB ampicillin medium: Prepare as described previously, without agar

Isopropyl-$\beta$-D-thiogalactopyranoside (IPTG): 100 m$M$, sterile filtered

Dithiothreitol (DTT): 0.5 $M$, freshly prepared

Tris(hydroxymethyl)aminomethane (Tris)

2-($N$-Morpholino)ethanesulfonic acid (MES): 1 $M$, free acid

MgCl$_2$: 1 $M$

Glycerol

Tris buffer: Contains 50 m$M$ Tris, 10 m$M$ MgCl$_2$, 2 m$M$ DTT, 5%
    (w/v) glycerol, titrated to pH 8.1 with 1 $M$ MES

MES buffer: Contains 50 m$M$ MES, 10 m$M$ MgCl$_2$, 2 m$M$ DTT, 5%
    (w/v) glycerol, titrated to pH 6.5 with 10 $N$ NaOH

DE-52 slurry (Whatman, Clifton, NJ): 50% (w/v) in Tris buffer

SP-Sepharose (Pharmacia, Piscataway, NJ): Equilibrated with MES
    buffer

## Overproduction of ACP Synthase

*Escherichia coli* strain BL21(DE3)pACPS1 is streaked onto LB ampicillin agar and grown in a 37° incubator overnight. LB ampicillin media (3 ×
5 ml) are then inoculated with single colonies and grown at 37° with agitation
to an optical density at 600 nm (OD$_{600}$) of 1.0. The cultures are centrifuged
at 3000 $g$ for 10 min at 4° and the pellets are resuspended in fresh LB
ampicillin media. These inocula can be stored at 4° overnight. LB ampicillin
media (3 × 1 liter) are inoculated with the 5-ml cultures and grown at 37°
and 250 rpm in a shaker incubator until the OD$_{600}$ reaches 0.6. At this time
the cultures are each induced with 10 ml of 100 m$M$ IPTG and grown at
30° and 250 rpm for an additional 3 hr. The induced culture is harvested
in a 4° centrifuge at 5000 $g$ for 15 min. The pellet is then resuspended in
3 ml of Tris buffer per gram of wet cell paste. The cell suspension is lysed
by two passages through a French pressure cell at 16,000 psi. Cellular debris
is removed by centrifugation at 20,000 $g$ for 30 min at 4° and the supernatant
is decanted into a clean centrifuge tube chilled on ice.

## Batch DE-52

All purification procedures should be carried out in a 4° cold room
using chilled buffers and chromatography media. DE-52 slurry (1 ml/ml
extract) is added to the clarified extract and mixed gently for 15 min. The
anion-exchange material and adsorbed protein is removed by centrifugation
at 5000 $g$ for 15 min and the supernatant is decanted to a clean centrifuge
tube. DE-52 slurry is again added (1 ml/ml of original extract volume) and
the previous procedure is repeated. Following removal of DE-52, the extract

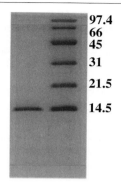

| | 97.4 |
| | 66 |
| | 45 |
| | 31 |
| | 21.5 |
| | 14.5 |

FIG. 1. Tris–Tricine SDS–PAGE gel (16% T, 6% C) of DE-52/SP-Sepharose-purified recombinant ACP synthase. (Adapted from Ref. 11, with permission.)

is adjusted to pH 6.5 by the slow addition of 1 *M* MES. The extract should then be clarified again by centrifugation at 20,000 *g* for 15 min.

## SP-Sepharose Chromatography

The DE-52-treated supernatant is applied to a 3 × 30 cm SP-Sepharose column preequilibrated with MES buffer at a flow rate of 2 ml/min. The column is then washed with 250 ml of MES buffer and the protein is eluted with a 500-ml linear gradient of 0 to 1 *M* NaCl in MES buffer while collecting 8-ml fractions. Fractions are monitored for ACP synthase using either Tris–Tricine sodium dodecyl sulfate-polyacrylamide gel electrophoresis (SDS–PAGE)[11,13] (Fig. 1) or one of the activity assays described in the next section. This procedure yields ~20 mg of pure ACP synthase per liter of culture, with a specific activity of 320 mU/mg (Table I). One unit of activity produces 1 $\mu$mol of holo-ACP per minute at 37°, using the conditions described for the radioassay.

## Radioassay of ACP Synthase Activity

### Principle

ACP synthase activity is assayed by monitoring the rate of transfer of tritium label from [$^3$H]pantetheinyl-CoA to holo-ACP. The reaction is stopped and the protein precipitated by the addition of trichloroacetic acid (TCA). The precipitated protein is washed with TCA, resolubilized in

[13] H. Schagger and G. von Jagow, *Anal. Biochem.* **166,** 368 (1987).

TABLE I
PURIFICATION OF RECOMBINANT ACP SYNTHASE FROM *Escherichia coli*
BL21(DE3)pACPS1[a]

| Fraction | Protein (mg) | Activity (U)[b] | Specific activity (mU mg$^{-1}$) | Purification (-fold) | Overall yield (%) |
|---|---|---|---|---|---|
| Crude extract | 600 | 120 | 200 | — | — |
| DE-52 | 160 | 40 | 250 | 1.3 | 27 |
| SP-Sepharose | 50 | 16 | 320 | 1.6 | 13 |

[a] R. H. Lambalot and C. T. Walsh, *J. Biol. Chem.* **270**, 24658 (1995).
[b] One unit of activity produces 1 $\mu$mol of holo-ACP per minute.

1 *M* Tris base, and the amount of tritium transferred to ACP is quantified
by liquid scintillation counting.

*Materials*

Tris-HCl: 1 *M*, pH 8.8
Tris base: 1 *M*
DTT: 0.5 *M*, freshly prepared
MgCl$_2$: 0.2 *M*
[$^3$H]Pantetheinyl-CoA: 1 m*M*, 100 Ci/mol, prepared by tritium gas
  exposure by NEN-Du Pont (Boston, MA)[1,11]
Apo-ACP: 1 m*M*, purified from *E. coli* strain DK554[8] following the
  procedure of Rock and Cronan[14]
Trichloroacetic acid (TCA): 10% (w/v)
Bovine serum albumin (BSA): 25 mg/ml
Ultima-Gold LSC cocktail (Packard, Downers Grove, IL)

*Procedure*

The incubation mixture contains 100 $\mu$*M* [$^3$H]CoA, 50 $\mu$*M* apo-ACP,
50 m*M* Tris-HCl (pH 8.8), 10 m*M* MgCl$_2$, 2 m*M* DTT, and ACP synthase
in a final volume of 100 $\mu$l. The reactions are incubated at 37° for 30 min
in 1.5-ml microcentrifuge tubes and are quenched by the addition of 800
$\mu$l of TCA followed by 20 $\mu$l of BSA solution. Following centrifugation at
12,000 *g* for 5 min at 4°, the supernatants are removed by pipette and the
pellets are rinsed three times with 900 $\mu$l of TCA. Residual TCA is collected
by brief centrifugation and removed by pipette. Pellets are then resuspended in 150 $\mu$l of 1 *M* Tris base and transferred to scintillation vials

[14] C. O. Rock and J. E. Cronan, Jr., *Methods Enzymol.* **71**, 341 (1981).

containing 2.5 ml of LSC cocktail and the amount of [3]H-labeled holo-ACP formed is quantified by liquid scintillation counting.

## Autoradiography

### Principle

Incubation products may be resolved by native PAGE following the procedure described by Rock and Cronan.[14] Specific incorporation of 4'-[[3]H]phosphopantetheine into ACP may then be confirmed by autoradiography.

### Materials

Native polyacrylamide gel: 20% (w/v) resolving gel with 4% (w/v) stacking gel
Coomassie stain: 0.5% (v/v) Coomassie Brilliant Blue in methanol–water–acetic acid (9 : 9 : 2, v/v)
Destain solution: Methanol–water–acetic acid (9 : 9 : 2, v/v)
Amplify (Amersham, Arlington Heights, IL)
Reflection NEF autoradiography film (NEN Research Products)
Gel dryer
Film exposure cassette

### Procedure

The ACP synthase radioassay is performed as previously described, except that no BSA is added and the protein is not TCA precipitated. The incubation mixture is resolved by 20% (w/v) native PAGE[14] and the protein bands are visualized by Coomassie stain followed by destaining. The stained gel is then soaked for 15 min in Amplify solution. The gel is removed from the Amplify solution and dried under vacuum at 60–80°. All electrophoresis buffers and staining solutions should be collected in a radioactive waste container after use. The dried gel is placed in close contact with autoradiography film in a film exposure cassette and exposed for 1–3 days in a −80° freezer. After exposure, the film is developed according to manufacturer instructions (Fig. 2).

## HPLC Assay

### Principle

The conversion of apo-ACP to holo-ACP may be monitored by HPLC following the procedure described by Hill *et al.*[7]

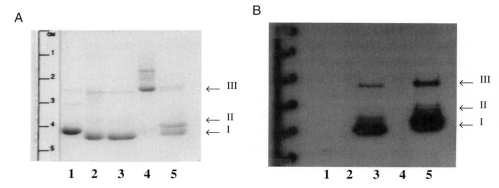

Fig. 2. (A) Coomassie-stained 20% native PAGE gel; lane 1, apo-ACP; lane 2, holo-ACP standard (Sigma), reduced with DTT; lane 3, $^{3}$H-labeled holo-ACP formed *in vitro* using recombinant ACP synthase, reduced with DTT; lane 4, holo-ACP standard, not reduced with DTT; lane 5, $^{3}$H-labeled holo-ACP formed *in vitro* using recombinant ACP synthase, not reduced with DTT. (B) Autoradiogram of gel A confirms introduction of [$^{3}$H]phosphopantetheine into holo-ACP (**I**), holo-ACP-CoA mixed disulfide (**II**), and (holo-ACP)$_2$ disulfide (**III**). [Reproduced from R. H. Lambalot and C. T. Walsh, *J. Biol. Chem.* **270**, 24658 (1995), with permission.]

*Materials*

Buffer A: 15% (v/v) 2-propanol, 0.1% (w/v) trifluoroacetic acid (TFA) in water

Buffer B: 75% (v/v) 2-propanol, 0.1% (w/v) TFA in water

Analytical $C_{18}$ reversed-phase HPLC column, 4.6 × 250 mm, 5-$\mu$m particle size (Vydac, The Separations Group, Hesperia, CA)

*Procedure*

The ACP synthase assay is performed as described for the radioassay except that 1 m$M$ unlabeled CoA is used as substrate and the 100-$\mu$l reaction is quenched with 900 $\mu$l of 0.1% (v/v) TFA without the addition of BSA. Alternatively, a 100-$\mu$l aliquot of a preparative-scale incubation may be added to 900 $\mu$l of 0.1% (v/v) TFA. A sample (200 $\mu$l) of this solution is injected onto the column, which has been equilibrated with buffer A at 0.5 ml/min. Absorbance at 220 nm is monitored. The column is then eluted with 2.5 ml of 100% buffer A, a 10-ml linear gradient to 40% (v/v) buffer B, followed by a 7.5-ml linear gradient to 100% (v/v) buffer B at a constant flow rate of 0.5 ml/min. Under these conditions holo-ACP elutes before apo-ACP (Fig. 3). This procedure has been adapted for several ACP homologs from various organisms as well as hexahistidine-tagged derivatives.

A

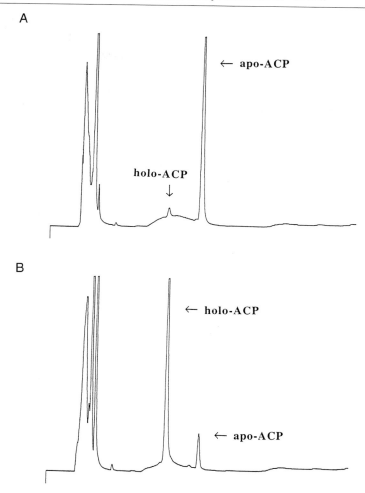

holo-ACP
↓

← apo-ACP

B

← holo-ACP

← apo-ACP

FIG. 3. HPLC chromatograms of ACP synthase incubation mixtures. (A) No ACP synthase added to the incubation mixture. (B) Complete incubation mixture.

## Preparation of Holo-ACPs from Recombinant Apo-ACPs

### Materials

Purified apo-ACP or apo–holo mixture
ACP synthase
CoA: 10 mM
Tris-HCl: 1 M, pH 8.8

DTT: 0.5 $M$, freshly prepared
$MgCl_2$: 0.2 $M$

*Procedure*

Apo-ACP (15 mg) is dissolved to 1 mg/ml in 50 m$M$ Tris-HCl, pH 8.8, and dialyzed against the same buffer using a 3000 molecular weight cutoff (MWCO) dialysis membrane to remove any salts or low molecular weight impurities. The final incubation mixture contains 1 m$M$ CoA, 10 m$M$ $MgCl_2$, 5 m$M$ DTT, and 5 mU of ACP synthase per milligram of apo-ACP. The mixture is incubated at 37° and the progress of the reaction is monitored by HPLC. The enzymatic product may also be confirmed by MALDI-TOF mass spectrometry. The change in molecular weight following attachment of 4′-phosphopantetheine is 339 amu. This procedure has been employed for the conversion of 10–15 mg of NodF and $^{15}$N-enriched ACP.

### Physical Properties of ACP Synthase

ACP synthase behaves as a homodimer on Sephadex-75 gel-filtration chromatography, eluting at a molecular weight of 28,000. The calculated molecular weight per monomer is 14,053. The molar extinction coefficient as determined by the method of Gill and von Hippel[15] is in agreement with the calculated value of 18,470 $M^{-1}$. The calculated isoelectric point is 9.25. Kinetic constants for ACP synthase as measured by the radioassay are $K_m$(CoA) 50 $\mu M$, $K_m$(apo-ACP) 1 $\mu M$, and $k_{cat}$ 70 min$^{-1}$.

### Acknowledgments

This work was supported by National Institutes of Health Grant GM20011. R.H.L. was supported by National Institutes of Health Post-Doctoral Fellowship GM16583.

---

[15] S. C. Gill and P. H. von Hippel, *Anal. Biochem.* **182**, 319 (1989).

# Section V

# Biotin and Derivatives

# [28] Determinations of Biotin in Biological Fluids

*By* DONALD M. MOCK

## Introduction

Interest in accurate measurement of biotin and biotin metabolites in human plasma and urine has been stimulated by advances in the understanding of biotin-responsive inborn errors of metabolism, by several clinical studies suggesting that mild to moderate biotin deficiency may not be a rare event, and by wider application of avidin–biotin techniques to tumor and organ imaging and to assay of many molecules of biological importance.[1]

This chapter describes a method for determination of biotin and biotin metabolites. The analytic rationale is similar to that of a biotin assay described in a previous volume of this series,[2] but the assay described here differs in three important respects: (1) The detector molecule is avidin-linked to horseradish peroxidase rather than avidin-linked to [125]I. This change eliminates radioactive waste and provides a detector molecule that is stable at $-70°$ for at least 12 months. Moreover, $\gamma$ counting in a conventional $\gamma$ counter after separating individual "snap" wells is supplanted by spectrophotometry in a 96-well format; (2) this assay also quantitates the two principal biotin metabolites: bisnorbiotin (BNB) and biotin sulfoxide (BSO); and (3) the high-performance liquid chromatography (HPLC) separation of biotin and its metabolites has been optimized for discrete separation of metabolites.

This HPLC/avidin-binding assay has increased sensitivity compared to most published methods based on avidin binding. The limit of detection is about 5 fmol and about 10 parts per trillion.

## Principles of Assay Method

Biotin and biotin metabolites in a sample to be assayed must first be separated by HPLC. In Incubation 1 (Fig. 1), avidin linked to horseradish peroxidase (HRP–avidin) is incubated with the HPLC fractions containing

---

[1] D. M. Mock, *in* "Present Knowledge in Nutrition" (E. E. Ziegler and L. J. Filer, Jr., eds.), 7th Ed., p. 220. International Life Sciences Institutes–Nutrition Foundation, Washington, DC, 1996.

[2] D. M. Mock, *Methods Enzymol.* **184,** 224 (1990).

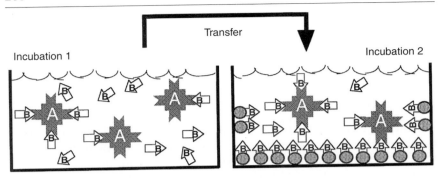

FIG. 1. Incubation 1: Incubate the HPLC eluate fractions containing biotin or biotin metabolites (B) in a microtiter well with avidin linked to horseradish peroxidase (A). Incubation 2: Transfer a portion of Incubation 1 to microtiter wells that have been coated with BSA covalently linked to biotin. HRP–avidin with unoccupied biotin-binding sites binds to the biotinyl-BSA.

biotin or a biotin metabolite. The biotin or biotin metabolite binds to HRP–avidin occupying a portion of the total biotin-binding sites. An aliquot of the mixture (Fig. 1, Incubation 2) is then transferred to a well previously coated with biotin covalently linked to bovine serum albumin (biotinyl-BSA). During Incubation 2, those HRP–avidin molecules that have unoccupied biotin-binding sites will bind to the biotinyl-BSA. After Incubation 2, the free HRP–avidin molecules (i.e., those with no available biotin-binding sites) are washed away. In Incubation 3, the amount of HRP–avidin bound to the plate is quantitated by measuring the rate of peroxidation of the indicator dye $o$-phenylenediamine (OPD). The peroxidation is stopped with the addition of acid ($H_2SO_4$), which also intensifies the absorbance of the peroxidized OPD. Optical density (OD) at 490 nm is then quantitated in a multiwell format spectrophotometer.

As increasing amounts of biotin in the standards or unknowns occupy an increasing proportion of the biotin-binding sites on HRP–avidin in Incubation 1, progressively fewer molecules of HRP–avidin bind to the well in Incubation 2, and color development in Incubation 3 is progressively less. The concentration of biotin or a metabolite is determined by comparing change in OD to the authentic standard curve. To construct the needed standard curves, varying concentrations of biotin or the biotin metabolite are added to separate wells in Incubation 1.

Biotin and each metabolite must be quantitated against a standard curve of the same compound because the biotin-binding affinity varies among biotin and its metabolites. For example, Fig. 2 depicts the OD at 490 nm for a series of standard concentrations of biotin, biotin sulfoxide, and bisnor-

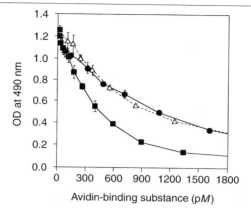

Fig. 2. Detect the amount of HRP–avidin bound to the biotinyl-BSA of Incubation 2 by peroxidation of a dye, *o*-phenylenediamine. Increasing OD at 490 nm indicates increasing HRP–avidin bound and, thus, decreasing biotin or biotin metabolite in Incubation 1. Note that differences in binding affinity of biotin sulfoxide ($\triangle$) and bisnorbiotin ($\bullet$) cause a shift of their respective standard curves to the right of the biotin ($\blacksquare$) standard curve. Symbols represent mean $\pm$ SD for a minimum of triplicate determinations. Where error bars are not shown, they are less than the symbol size.

biotin assayed using the HRP–avidin assay. For the two metabolites, the point of half-maximal OD is shifted toward greater concentrations of the metabolites; the true concentration of each metabolite would have been underestimated by a factor of 2–3 if the unknown were read against the biotin standard curve.

Limitations of Assay

A direct assay of total avidin-binding substances in serum or urine without chromatographic separation overestimates biotin and underestimates the total of biotin plus biotin metabolites. The potential for artifacts arising from the lack of chemical specificity of avidin-binding assays and bioassays has been a concern since Lee *et al.* demonstrated the presence of substantial amounts of biotin metabolites in urine using paper chromatography and relative migration with radiolabeled standards.[3] We have confirmed that substantial amounts of bisnorbiotin and biotin sulfoxide (relative to biotin) are present in normal urine and have reported that substantial amounts are also present in normal serum. For this avidin-binding assay without prior HPLC, the measured total avidin-binding substances do not

[3] H. M. Lee, L. D. Wright, and D. B. McCormick, *J. Nutr.* **102**, 1453 (1972).

equal either the stoichiometric sum of biotin and biotin metabolites or the concentration of biotin alone when applied to complex biological mixtures of biotin and biotin metabolites such as serum and urine. We have observed a clinical situation in which conclusions about biotin status would be erroneous if judged by total avidin-binding substances in urine.

This assay detects only free biotin and biotin metabolites in serum, milk, and urine. More than 99% of biotin in urine is free in aqueous solution. More than 95% of biotin in human milk is free in aqueous solution. However, controversy exists concerning the amounts of biotin that are reversibly and covalently bound to macromolecules (presumably proteins) in serum.

## Materials and Methods

### Equipment

The following list identifies essential equipment and potential suppliers:
HPLC system capable of at least a binary gradient and manual injector with 1-ml sample loop
$C_{18}$ reversed-phase column, 2 mm × 25 cm (Spherisorb microbore column; Phase Sep, Norwalk, CT)
Fraction collector capable of timed advancement
Concentrator (SC210A Speed-Vac Plus vacuum centrifuge; Savant, Hicksville, NY)
12-Well plate washer (mini-washer; Skatron, Lier, Norway) or 96-multiwell plate washer (EL 403 automated microplate washer; Bio-Tek, Winooski, VT)
Multiwell format spectrophotometer with 490- and 630-nm wavelength filters
Scintillation counter
Optional: Detection of radiolabeled metabolites in HPLC eluates in fraction collector vials or via flow detection of HPLC eluate (A-500 Radio-Flo detector; Radiomatic, Packard, Meriden, CT)

### Materials and Reagents

Multiwell plates for Incubation 1: U-bottomed multiwell plates to allow recovery of the greatest volume for pipetting
Multiwell plates for Incubation 2 and Incubation 3: Flat-bottomed multiwell plates for use in multiwell format spectrophotometer. We have tested plates from several manufacturers, using the methods described here. Immulon II plates (Dynatech Laboratories, McLean,

VA), enzyme immunoassay/radioimmunoassay (EIA/RIA) plates by Costar (Cambridge, MA), and MaxiSorb plates (Nunc, Roskilde, Denmark) have the best binding capacity

d-Biotin, BSA (RIA grade), o-phenylenediamine (Sigma, St. Louis, MO)

HRP–avidin (Pierce, Rockford, IL)

N-2-Hydroxyethylpiperazine-N'-2-ethanesulfonic acid (HEPES; Calbiochem, La Jolla, CA)

[$^3$H]Biotin: Specific activity about 50 Ci/mmol (Du Pont–New England Nuclear, Boston, MA)

[$^{14}$C]Biotin: Specific activity about 50 mCi/mmol (Amersham, Arlington Heights, IL)

All other chemicals: Reagent grade or better; ultrapure deionized water is used throughout

## Buffers and Solutions

HEPES buffer (0.2 $M$ HEPES, 2 $M$ NaCl at pH 7.0): Dissolve 190.64 g of HEPES, 467.52 g of NaCl in about 3 liters of water. Adjust the pH to 7.0 with 10 $M$ NaOH and bring to 4.0 liters

Coating buffer (50 m$M$ bicarbonate at pH 9.0): Dissolve 1.59 g of sodium carbonate and 2.93 g of sodium bicarbonate in 900 ml of water. Bring to 1.0 liter with water. This buffer can be stored at 4° for at least 2 weeks

Blocking buffer [0.1 $M$ HEPES, 1 $M$ NaCl at pH 7.0 with 0.01% (w/v) BSA and 0.05% (v/v) Tween 20]: Dissolve 0.1 g of BSA in 1 liter of HEPES buffer diluted 1:1 (v/v) in deionized water. Add 0.5 ml of Tween 20. This buffer can be stored at 4° for at least 2 weeks

Buffer for HRP–avidin [0.1 $M$ HEPES, 1 $M$ NaCl at pH 7.0 with 0.1% (w/v) BSA]: Dissolve 1.0 g of BSA in HEPES buffer that has been diluted 1:1 (v/v) in deionized water

Buffer for OPD solution (0.1 $M$ citric acid, 0.2 $M$ sodium phosphate, dibasic at pH 5.0): Mix 243 ml of 0.1 $M$ citric acid, 257 ml of 0.2 $M$ Na$_2$HPO$_4$, and 500 ml of deionized water to make 1 liter of substrate buffer. Adjust the pH to 5.0 with either the acid or the base. This buffer is stable for approximately 1 month at 4°

OPD solution (7.4 m$M$ o-phenylenediamine, 350 m$M$ H$_2$O$_2$): Dissolve 0.08 g of OPD and 40 $\mu$l of 30% (v/v) H$_2$O$_2$ in 100 ml of substrate buffer. Make fresh for each assay run

Sulfuric acid solution (1 $M$ sulfuric acid): Add 55 ml of 18 $M$ H$_2$SO$_4$ to 945 ml of deionized water

*Biotin and Biotin Metabolites for Standards*

*Preparation of Biotin Standard*

Biotin stock (100 n$M$): Dissolve 24.43 mg of *d*-biotin in 1 liter of distilled water. When aliquoted into 1-ml fractions and stored at −70°, this reference biotin stock is stable for at least 2 years

Biotin standard [3000 p$M$ (3000 fmol/ml)]: Aliquot 0.5 ml of the 100 n$M$ biotin stock into a 50-ml volumetric flask and bring to 50 ml with HEPES buffer. After mixing, dilute 3.0 ml into 997 ml of HEPES buffer. Store 1-ml aliquots at −20°

Cross-check the concentration of the 3000 p$M$ biotin standard by comparison to radiolabeled biotin of known specific activity (e.g., [³H]biotin or [¹⁴C]biotin as described in a previous volume of this series).[2] The concentration of the 300 p$M$ biotin standard can be independently verified against a standard curve from this "gold standard" radiolabeled biotin.

*Synthesis of [³H]Biotin Sulfoxide.* [³H]Biotin sulfoxide serves as both a chromatography retention time standard and as an independent quantitation standard for unlabeled biotin sulfoxide in a fashion analogous to [³H]biotin as previously described. Synthesize [³H]biotin sulfoxide ([³H]BSO) from [³H]biotin by oxidizing the sulfur atom with hydrogen peroxide in an acidic environment.[4] To 100 $\mu$l of a 1-$\mu$Ci/$\mu$l solution of [³H]biotin, add 100 $\mu$l of the acidic peroxide solution. [The acid peroxide is made by mixing 10 ml of 0.1 $M$ HCl with 0.34 ml of 30% (v/v) $H_2O_2$.] Allow the peroxidation to proceed overnight (16–20 hr) at room temperature. Stop the reaction by addition of 130 $\mu$l of 0.1 $M$ NaOH. Evaporate to dryness with $N_2$ gas to remove the hydrogen peroxide. Dissolve the resulting white powder in 200 $\mu$l of 10 m$M$ potassium phosphate buffer, pH 7. Use HPLC (as described below) to confirm that >99% of radioactivity is BSO. Yields are typically better than 90%. Store aliquots at −70°.

Because the conversion of biotin to BSO is stoichiometric, the specific activity of the [³H]BSO is equal to that of [³H]biotin. Thus, concentrations of unlabeled BSO can be independently standardized by radioactivity.

*Synthesis of [¹⁴C]Bisnorbiotin Standard.* [¹⁴C]Bisnorbiotin serves as both a chromatography retention time standard and as an independent quantitation standard for unlabeled bisnorbiotin in a fashion analogous to [¹⁴C]biotin as previously described. Biosynthesize [¹⁴C]bisnorbiotin (BNB) from [¹⁴C]biotin using a fungal strain such as *Rhodotorula rubra*.[5] $\beta$-Oxida-

---

[4] J. L. Chastain, D. M. Bowers-Komro, and D. B. McCormick, *J. Chromatogr.* **330**, 153 (1985).

[5] H.-C. Yang, M. Kusomoto, S. Iwahara, T. Tochikura, and K. Ogata, *Agric. Biol. Chem.* **32**, 399 (1968).

tion of the valeric acid side chain of biotin causes the loss of the $^3$H label on carbons 8 and 9 of the side chain and thus precludes the use of [$^3$H]biotin.

Grow a subculture of *R. rubra* (NRRL Y-1592; Northern Regional Research Center, Peoria, IL) in a carbon-containing medium [8 g of glucose, 5 g of neopeptone, 10 g of yeast extract, 9 g of KH$_2$PO$_4$, 0.3 g of CaCl$_2$, 0.5 mg of MgSO$_4 \cdot$ 7H$_2$O, 6 g of (NH$_4$)$_2$SO$_4$ in 1000 ml of H$_2$O adjusted to pH 5.8]. Wash the cells four times in 0.9% (w/v) saline by centrifugation (10 min at 1500 $g$) to remove the growth medium. Resuspend the cells in 10 ml of 0.25 $M$ phosphate-buffered saline (PBS), pH 7.2. Dilute [$^{14}$C]biotin to 50 $\mu$Ci/ml in 2 ml of 0.25 $M$ PBS. Add 1.75 ml (87.5 $\mu$Ci) of the [$^{14}$C]biotin to 1 ml of cells. Culture the fungus at 24–25° for 48 hr with shaking at 190–200 rpm. After 48 hr, remove the cells by centrifugation (20 min, 1500 $g$, 4°), and retain the supernatant, which contains the [$^{14}$C]BNB and any residual [$^{14}$C]biotin. Centrifuge the supernatant at 1500 $g$ for 10 min at 4° to remove any remaining cells. Use HPLC as described below to confirm that >99% of radioactivity is BNB. Yields are typically better than 95%. Store aliquots at −70° in 0.2% (v/v) 2-mercaptoethanol.

Because the conversion of biotin to BNB is stoichiometric, the specific activity of the [$^{14}$C]BNB is equal to that of [$^{14}$C]biotin. Thus, concentrations of unlabeled BNB can be standardized by radioactivity.

*Synthesis of Biotinyl-BSA.* Bovine serum albumin is biotinylated according to the method of Heitzmann and Richards.[6] Dissolve 500 mg of BSA in 50 ml of ice-cold 0.1 $M$ NaHCO$_3$ at pH 7.5. Dissolve 60 mg of biotin *N*-hydroxysuccinimide ester (NHS–biotin; Pierce) in 5 ml of dimethyl sulfoxide to produce a final concentration of 12 mg/ml. Add 5 ml of the NHS–biotin solution to 50 ml of BSA solution and incubate with gentle stirring overnight at 4°. Dialyze this mixture exhaustively against deionized water to remove unreacted NHS–biotin. Then dialyze the biotinyl-BSA against HEPES buffer. Aliquot in 1-ml fractions and store at −20°; biotinyl-BSA is stable for at least 1 year.

Assess the degree of biotinylation of BSA by measuring the amount of HRP–avidin that binds to a well coated with biotinyl-BSA as follows. Use flat-bottomed plates coated with either biotinyl-BSA or native BSA as described. Incubate serial dilutions of HRP–avidin in the wells overnight at 4°; wash, develop, and read the wells as described. Wells coated with biotinyl-BSA reproducibly bind 50–100 times more HRP–avidin than wells coated with native BSA.

*Preparation of Biotinyl-BSA-Coated Plates.* The biotinyl-BSA-coated plates must be prepared at least 1 day before the assay. Dilute 2 ml of the biotinyl-BSA stock previously described in 1000 ml of coating buffer. Add

---

[6] H. Heitzmann and F. M. Richards, *Proc. Natl. Acad. Sci. U.S.A.* **71**, 3537 (1974).

75 $\mu$l of this coating solution to each well, cover the plate, and incubate at 4° for at least 16 hr but not more than 4 days. After incubation, aspirate the coating solution. Wash the coated plates twice as follows: Fill each well to the brim with water and then empty by aspiration. Add 400 $\mu$l of blocking buffer to each well and incubate for at least 2 hr at room temperature before using. Plates can be stored at 4° in blocking buffer for at least 1 month. At the time of assay, aspirate the blocking buffer, and wash the plates twice immediately before transferring the contents from Incubation 1. Do not allow drying of the biotinyl-BSA-coated plates for more than 1 min.

## Preparation of Biological Samples for Analysis

### Serum Preparation

Thaw serum samples in a 37° water bath until completely free of ice.

*Ultrafiltration of Serum.* This assay detects only free biotin and biotin metabolites because only protein-free (really, macromolecule-free) ultrafiltrates of serum are used. For quantitation of biotin metabolites by HPLC, the ultrafiltrate must be concentrated; at least 5 ml of serum is required. For this volume of serum, Amicon (Beverly, MA) Centriprep units are preferred because the ultrafiltrate is pushed up through the bottom of the membrane and larger aggregates of protein are sedimented, thus avoiding clogging the membrane. An initial 1:1 dilution with water prevents an artifactual decrease in the concentration of free biotin passing through the membrane, presumably by preventing a build-up of high molecular weight complexes on the bottom of the Centriprep membrane. The units may be centrifuged in either fixed-angle or swinging bucket rotors at 1500 $g$ for 60 min at 4°. Remove the ultrafiltrate. Centrifuge for an additional 30 min. Immediately remove the ultrafiltrate from the concentrator to prevent back diffusion into the original sample. The combined ultrafiltrate may be stored overnight at −20°, long term at −70°, or immediately used in the next step.

For limited serum volumes (e.g., <5 ml), only biotin can be detected reliably by HPLC. Ultrafilter in one (or more) Amicon Centrifree units that hold 1 ml. One milliliter of serum produces about 0.5 ml of ultrafiltrate. Centrifuge at 1500 $g$ for 60 min at 4° in a fixed-angle rotor. Stir the retentate in the top of the unit with a Pasteur pipette. Centrifuge for an additional 30 min.

*Serum Concentration.* Because of the low concentration and weak binding affinity of biotin metabolites and analogs for avidin, most serum ultrafiltrate samples must be concentrated about fivefold for accurate measurement. For example, concentrate 5 ml of ultrafiltrate in a centrifugal vacuum concentrator (Savant Instruments) until the volume reaches approximately

200 $\mu$l. Transfer the ultrafiltrate to a new vial by Pasteur pipette. Wash the original tube three times with 200 $\mu$l of deionized water and add the wash to the new tube. By weight, bring the sample with pooled washings to a total volume of 1 ml. The final concentration factor for this example is 5.0.

*HPLC Injection.* Just before HPLC injection, adjust the pH of the ultrafiltrate concentrate to pH 2.5, using about 50 $\mu$l of 6 $M$ HCl. Filter through a 0.45-mm (pore size) filter and inject as described as follows.

### Urine Preparation

Urinary excretion rates of biotin and biotin metabolites may be expressed per unit time if timed samples are available or per milligram of urinary creatinine if untimed samples are available.

Centrifuge at 5000 $g$ for 10 min at 4° to remove cells and debris before storage at −20°. Biotin and biotin metabolites are stable for as long as 1 year. For analysis, thaw the supernatant in a water bath at 65° for 30 min to resolubilize any particulates formed by freezing. (If samples were not centrifuged before freezing, centrifuge after thawing as previously described.)

*HPLC Injection.* The amount of urine to be injected on HPLC will depend on the expected biotin concentration. For urine from subjects who are receiving no biotin supplementation, inject the maximum volume— 1 ml. Just before injection, adjust the pH of the urine to pH 2.5, using about 10 $\mu$l of 6 $M$ HCl. Filter through a 0.45-mm (pore size) filter and inject as described. For urine from subjects who are receiving biotin supplementation, as little as 200 $\mu$l may be injected. Because this volume will not change the pH of the column or alter retention times, adjustment of the pH to 2.5 is not necessary. Filter through a 0.45-mm (pore size) filter and inject as described. For urine from subjects whose biotin status is not known, estimates of biotin concentration can be made before HPLC by measurement of total avidin-binding substances using the HRP–avidin assay.

### HPLC Method

$C_{18}$ reversed-phase chromatography separates biotin and analogs on the basis of polarity. The HPLC method described here is a modification of the method of Chastain *et al.*[5] In the binary gradient system, the reversed-phase $C_{18}$ column is initially equilibrated in 0.05% (v/v) trifluoroacetic acid (TFA) adjusted to pH 2.5 with ammonium acetate (solution A). The constant flow rate is 1 ml/min. The second mobile phase is acetonitrile– 0.05% (v/v) TFA (1:1, v/v) (solution B). The linear gradient begins at 0% (v/v) solution B and reaches 40% (v/v) by 35 min. Nonpolar constituents of the injectate are removed after the analytical gradient by increasing to

100% solution B over 5 min and holding at 100% solution B for 5 min. The column is then equilibrated to initial conditions by returning to 0% solution B for 10 in. The total run time is 60 min/sample.

Discrete resolution of BNB and BSO requires accurate retention times and closely spaced fractions. Retention times of biotin and biotin metabolites can be determined by prior injection of radiolabeled standards of [$^3$H]biotin, [$^3$H]biotin sulfoxide, and [$^{14}$C]bisnorbiotin. On the basis of these retention times, collect fractions as small as 0.25 ml (= 0.25 min). Neutralize the fractions with a 1.5:1 mixture of 0.125 $M$ NaOH:0.02 $M$ HEPES, 2 $M$ NaCl buffer, using 120 $\mu$l per 1-ml fraction. Dry the fractions to remove the chromatography solvents. Dissolve the dried sample of 0.5 ml of deionized water and add 0.5 ml of HEPES buffer.

### Performing Assay

*Incubation 1.* To each well of the U-bottom multiwell plate, add 100 $\mu$l of either a dilution of a biotin or biotin metabolite standard, a dilution of the unknown sample to be assayed, or HEPES buffer ("zero" value). Run replicates of each standard; we run quadruplicates. A range of concentrations of standards in the standard curve and the unknowns are prepared by serial dilution. The dilutions can be made in separate tubes or by sequential dilution and transfer in the Incubation 1 plate, but the total volume in each well must be equal prior to addition of HRP–avidin.

After all standards and unknowns have been added to the wells in the multiwell plate, add 50 $\mu$l of HRP–avidin solution to each well. The HRP–avidin solution is made by diluting the stock HRP–avidin (1 mg/ml) with buffer for HRP–avidin (see above) to a dilution of approximately 125,000. Generally, 1 $\mu$l of the stock is diluted to 125 ml. Mix the contents of Incubation 1 by alternately drawing up and expelling the contents of each well with a pipette. Incubate for 1 hr at room temperature. The length of this incubation is not critical.

*Incubation 2.* Transfer 100 $\mu$l from each well to a corresponding well of the biotinyl-BSA-coated plate. Incubate the samples in the coated wells at room temperature for at least 4 hr but at 4° for up to 3 days. At the end of Incubation 2, wash the plates three times with water to remove all unbound HRP–avidin.

*Incubation 3.* Add 200 $\mu$l of substrate solution to each well and incubate for 45 min at room temperature. Because the OD depends directly on the incubation time, timing of Incubation 3 for all plates in the same assay and each well within the same plate must be the same. At the end of Incubation 3, quench the reaction by adding 100 $\mu$l of 1.0 $M$ H$_2$SO$_4$. Read the OD at 490 nm with background correction at 630 nm.

*Data Analysis and Interpretation.* Figure 2 depicts the standard curves of absorbance versus the concentration of biotin or biotin metabolite. Note that the decrease in binding affinity for avidin by biotin sulfoxide and bisnorbiotin causes a shift to the right. We use a "point-to-point" line fit to calculate unknowns. This requires a point series with minimal deviation from the natural curve progression and occasional exclusion of a "bad" point. A fourth or fifth degree polynomial curve-fitting program also yields acceptable results. The fit to the curve should allow $r^2 > 0.99$.

The assay sensitivity is typically about 50 p$M$ 5 fmol/assay well) but is occasionally as good as 15 p$M$ (>2 fmol/assay well) with an HRP–avidin preparation that is fresh and fully active. Typically, the intraassay variability in a single assay is ≤5%. The long-term precision (interassay variability) was estimated from results of triplicate determinations from assays of the same sample by three different technicians on six different days; the coefficient of variation (calculated from the standard deviation of the means of the triplicates) was 6%, and the intraclass correlation coefficient for that data was −0.77.

# [29] Fluorophore-Linked Assays for High-Performance Liquid Chromatography Postcolumn Reaction Detection of Biotin and Biocytin

*By* NATHANIEL G. HENTZ and LEONIDAS G. BACHAS[1]

## Introduction

Immunoassays, and more generally speaking, protein-binding assays have become an important analytical tool, in part because of their high sensitivity. These techniques also have excellent selectivity among dissimilar compounds that is determined by the molecular recognition provided by the protein binder (antibody, binding protein, receptor, lectin, etc.). A limitation, however, of protein-binding assays is the relatively low selectivity among compounds that belong to the same class. In other words, binding assays alone may have difficulty distinguishing between a compound and its metabolites or structurally related derivatives. The term *cross-reactivity* has been used to describe this property.

[1] Author to whom correspondence should be addressed: Department of Chemistry, University of Kentucky, Lexington, KY 40506-0055.

High-performance liquid chromatography (HPLC), however, offers excellent selectivity by being able to separate structurally related metabolites or derivatives from each other. Thus, by coupling HPLC and protein-binding assays, the excellent separation ability of HPLC and the sensitivity of binding assays can be achieved within a single technique. The resultant technique combines two methods that provide orthogonal selectivity properties, namely, the class selectivity offered by protein-binding assays and the resolution of chemical separations.

This chapter describes HPLC postcolumn reaction detection systems in which two types of protein-binding assay principles were used. In a competitive-binding approach, the fluorescent probe 2-anilinonaphthalene-6-sulfonic acid (2,6-ANS) competes with the analyte (biotin or biocytin) for the same binding sites on avidin. The two competing equilibria are shown as follows:

$$\text{Biotin + avidin} \rightleftharpoons \text{avidin–biotin}$$
$$\text{2,6-ANS + avidin} \rightleftharpoons \text{avidin–2,6-ANS}$$

The excitation wavelength, $\lambda_{ex}$, for both free and avidin-bound 2,6-ANS is 328 nm, while the emission wavelength, $\lambda_{em}$, is at 462 and 438 nm for free and bound 2,6-ANS, respectively. Therefore, the fluorescence emission of 2,6-ANS is shifted toward the red region on release from avidin. Given the difference in the values of the dissociation constant ($K_d = 2 \times 10^{-4}$ $M$ for 2,6-ANS and $K_d = 10^{-15}$ $M$ for biotin),[2,3] when biotin is added to a solution containing the avidin–2,6-ANS complex biotin displaces the 2,6-ANS probe from the binding sites of avidin, yielding a decrease in fluorescence intensity at 438 nm, which is proportional to the analyte concentration.

Two noncompetitive binding approaches are also described, in which the interaction that takes place is between the analyte (biotin or biocytin) and avidin or streptavidin. In these cases (strept)avidin is labeled with fluorescein isothiocyanate (FITC). On binding with the analyte, the fluorescence intensity of (strept)avidin–FITC increases in a manner proportional to the concentration of the analyte. The corresponding equilibria are illustrated as follows:

$$\text{Biotin + avidin}^{F} \rightleftharpoons \text{avidin–biotin}^{F}$$
$$\text{Biotin + streptavidin}^{F} \rightleftharpoons \text{streptavidin–biotin}^{F}$$

[2] D. M. Mock, G. Langford, and P. Horowitz, *Biochim. Biophys. Acta* **956,** 23 (1988).
[3] N. M. Green, *Adv. Protein Chem.* **29,** 85 (1975).

where the smaller Fs denote FITC with background fluorescence and the larger Fs denote FITC with enhanced fluorescence ($\lambda_{ex}$ = 495 nm and $\lambda_{ex}$ = 518 nm).

The coupling of these binding assays with HPLC results in a highly selective detection of biotin and its derivatives owing to their selective interaction with (strept)avidin. These types of approaches also lead to enhanced sensitivity because the HPLC detection is fluorescence based rather than the conventional ultraviolet (UV) absorbance-based detection. In addition to the three biotin detection systems outlined in this chapter, an absorbance-based postcolumn reaction detection system has also been developed in this laboratory. In this system, biotin competes with and displaces the dye 2-(4'-hydroxyphenylazo)benzoic acid (HABA), where the free and bound HABA have different absorbance characteristics.[4] In a different mode, biotin has been detected by allowing unlabeled biotin to compete with fluorescein-labeled biotin for binding sites on avidin.[5]

## Methods

### Apparatus

The experimental setup (Fig. 1) consists of a Rainin Instruments (Woburn, MA) HPLC system. The separations are performed on a Rainin Microsorb 5-$\mu$m $C_{18}$ analytical column (250 × 4.6 mm), preceded by a guard column (15 × 4.6 mm) of the same material. The mobile phase is delivered by a Rainin Rabbit solvent delivery system. Samples are introduced through a Rheodyne (Berkeley, CA) model 7125 injector equipped with a 20-$\mu$l sample loop. Detection is carried out with a Knauer (Berlin, Germany) model 87 variable-wavelength UV–Vis detector set at 220 nm. The HPLC system is interfaced with a Macintosh computer (Apple Computer, Cupertino, CA). The mobile phase for the separation of biotin and biocytin consists of a mixture of 0.100 $M$ phosphate buffer (pH 7.0) and methanol that is pumped at a flow rate of 0.40 ml/min under isocratic conditions.

The postcolumn reagent solution is merged with the HPLC effluent through a tee-connector preceding a 10.0-m knitted open-tubular (KOT) reactor made from polytetrafluoroethylene (PTFE) tubing (0.5-mm i.d., 14-mm helix diameter).[6] Owing to the residence time in the KOT reactor, the retention time for biotin using direct UV absorbance detection is lower

[4] A. Przyjazny, T. L. Kjellström, and L. G. Bachas, *Anal. Chem.* **62,** 2536 (1990).
[5] A. J. Oosterkamp, H. Irth, U. R. Tjaden, and J. van der Greef, *Anal. Chem.* **66,** 4295 (1994).
[6] C. M. Selavka, K.-S. Jiao, and I. S. Krull, *Anal. Chem.* **59,** 2221 (1987).

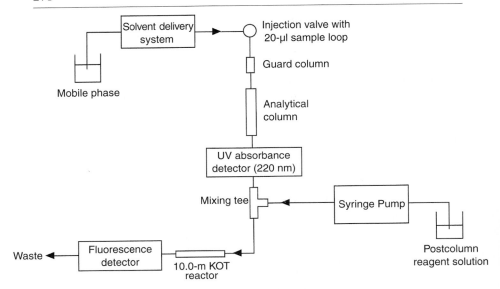

FIG. 1. Postcolumn reaction detection system.

than that obtained by using the postcolumn reaction detection. The flow of the postcolumn reagent is controlled by an ISCO (Lincoln, NE) model LC-2600 syringe pump.

A SPEX Fluorolog-2 spectrofluorometer (SPEX Industries, Edison, NJ), operated in a photon-counting mode, is used with the (strept)avidin–FITC postcolumn reaction detection systems. The excitation and emission wavelengths are set at 495 and 518 nm, respectively. The excitation and emission slit widths are each set at 2 mm. When using 2,6-ANS for postcolumn reaction detection, a Perkin-Elmer (Norwalk, CT) model LS 50 luminescence spectrometer is employed. The excitation wavelength is set at 328 nm, while the emission wavelength is monitored at 438 nm; the excitation and emission slits are set at 5 and 20 nm, respectively. Finally, in all systems the spectrofluorometers are equipped with a 20-$\mu$l $\mu$-fluorescence flow cell (NSG Precision Cells, Farmingdale, NY), and a back-pressure regulator is placed after the fluorescence detector to prevent outgassing problems.

*Real Sample Analyses*

Four different classes of samples are analyzed for their biotin content: liquid vitamin preparations, vitamin tablets/horse-feed supplement, liquid cell culture media, and liquid infant formula. The liquid vitamin preparations and the liquid cell culture media require no sample preparation other than preinjection filtration. The solid tablet samples (i.e., vitamin tablets and a horse-feed supplement) are powdered and dissolved in 0.10 $M$ NaOH. The samples are then centrifuged[7] and the supernatant is collected and adjusted to pH 6–7 with 1.0 $M$ HCl. The infant formula is prepared by first denaturing the proteins with concentrated HCl. After filtration and pH adjustment to pH 7.0, the lipids/oils are extracted from the supernatant with $n$-hexane.[8] Once the samples are centrifuged, the aqueous portion is diluted to a fixed volume.

## Results and Discussion

### Postcolumn Reaction Detection System Using 2-Anilinonaphthalene-6-sulfonic Acid

#### Conditions

Postcolumn reagent: Avidin (24.0 mg/liter) and $4.0 \times 10^{-6}$ $M$ 2,6-ANS
Postcolumn reagent flow rate: 1.00 ml/min
Fluorescence detection: $\lambda_{ex} = 328$ nm, $\lambda_{em} = 438$ nm
Mobile phase: 0.100 $M$ phosphate buffer (pH 6.0)–methanol (77:23, v/v)
Mobile phase flow rate: 0.40 ml/min

The analytical characteristics of the 2,6-ANS postcolumn reaction detection system were determined by using a concentration range from $1.0 \times 10^{-6}$ to $2.0 \times 10^{-4}$ $M$ biotin and biocytin. Under optimum conditions, the calibration plots were linear from $1.0 \times 10^{-6}$ to $2.0 \times 10^{-5}$ $M$ biotin and biocytin. The detection limits ($5.0 \times 10^{-7}$ $M$ for biotin and biocytin) were calculated by diluting a biotin/biocytin standard until a signal-to-noise ratio ($S/N$) of 3 was achieved.

Three typical organic solvents [i.e., $N,N$-dimethylformamide (DMF), acetone, and methyl ethyl ketone (MEK)] were chosen for an interference study because they are commonly used in extractions or as solvents in protein modification reactions and contain functionalities that absorb at 220 nm (the wavelength used in conventional absorbance-based HPLC determination of biotin). The three organic solvents severely interfered with the determination of biotin and biocytin when detected by UV absorbance at 220 nm. Using the postcolumn reaction detection system, concentrations up to 70 m$M$ DMF, acetone, or MEK did not interfere with the determination of biotin. Finally, the stability of the postcolumn reagent

[7] A. Przyjazny and L. G. Bachas, *Anal. Chim. Acta* **246**, 103 (1991).
[8] E. C. Nicholas and K. A. Pfender, *J. Assoc. Off. Anal. Chem.* **73**, 792 (1990).

was determined by preparing a batch and using portions of it with a biotin/ biocytin test standard over a 3-day period. During these analyses, the detector response did not change significantly, which indicates the stability of the postcolumn reagent for at least this time period.

*Analysis of Real Samples.* The analysis of biotin in commercial vitamin tablets (W. T. Thompson, Carson, CA; or L. Perrigo, Allegan, MI) and a horse-feed supplement (GEN-A-HOOF; Nickers International, Staten Island, NY) demonstrated the analytical utility of this postcolumn reaction detection system. The results of the analyses agreed well with those claimed by the manufacturer[7] (i.e., 3–6% difference). A typical chromatogram (Fig. 2) demonstrates the excellent selectivity of the described system. Under the conditions reported previously, correct determination of biotin would not be possible using direct UV absorbance at 220 nm owing to severe matrix interferences (Fig. 2A). In contrast, when the postcolumn reaction detection system is employed, no interferences are observed (Fig. 2B). This is due to the highly selective interaction between biotin/biocytin and avidin. Note that when biotin displaces 2,6-ANS from the binding sites of avidin, a decrease in fluorescence intensity is observed as result of the change in fluorescence properties of free vs bound 2,6-ANS.

*Postcolumn Reaction Detection System Using Avidin–Fluorescein Isothiocyanate*

*Conditions*

    Postcolumn reagent: Avidin–FITC (2 mg/liter) in 0.100 $M$ phosphate buffer, pH 7.0

    Postcolumn reagent flow rate: 1.00 ml/min

    Fluorescence detection: $\lambda_{ex}$ = 495 nm, $\lambda_{em}$ = 518 nm

    Mobile phase: 0.100 $M$ phosphate buffer (pH 7.0)–methanol (80:20, v/v)

    Mobile phase flow rate: 0.40 ml/min

The range of linearity for this postcolumn reaction detection system extended from $4.0 \times 10^{-8}$ to $6.0 \times 10^{-7}$ $M$ biotin and biocytin. Concentrations higher than $6.0 \times 10^{-7}$ $M$ biotin or biocytin resulted in appreciable band broadening and deteriorated resolution. The detection limits ($1.8 \times 10^{-8}$ $M$) for biotin and biocytin were estimated by decreasing the concentration of the analytes until an $S/N$ ratio of 3 was obtained.

The avidin–FITC reagent was also evaluated with respect to its stability by preparing a 2.0-mg/liter solution and storing it in the syringe pump. This reagent solution was used repeatedly to analyze a biotin/biocytin test standard. The reagent produced a stable fluorescence signal for up to 8 hr. However, the avidin–FITC stock solution (50 mg/liter) was stable for at

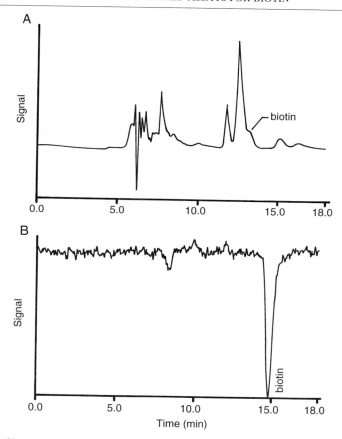

FIG. 2. Chromatograms of a 20-$\mu$l injection of a horse-feed supplement extract containing biotin (231 mg/kg). (A) UV absorbance detection at 220 nm; (B) fluorescence detection by postcolumn reaction using the avidin–2,6-ANS system ($\lambda_{ex}$ = 328 nm, $\lambda_{em}$ = 438 nm). The amount of biotin found was 238 mg/kg. (Reprinted from *Anal. Chim. Acta,* **246,** A. Przyjazny and L. G. Bachas, pp. 103–112, Copyright 1991 with kind permission of Elsevier Science–NL, Sara Burgerhartstraat 25, 1055 KV Amsterdam, The Netherlands.)

least 2 weeks when refrigerated and used to prepare fresh working reagent (2 mg/liter) on a daily basis.

*Analysis of Real Samples.* The avidin–FITC-based postcolumn reaction detection system was evaluated in terms of accuracy and selectivity by determining biotin in several real samples, including ABDEC liquid vitamin (Parke-Davis, Morris Plains, NJ), Centrum liquid vitamin (Lederle Laboratories, Pearl River, NY), Basic's Stress vitamin tablet (Basic Pharmaceutical, Vandalia, OH), and a horse-feed supplement (Nickers International). The

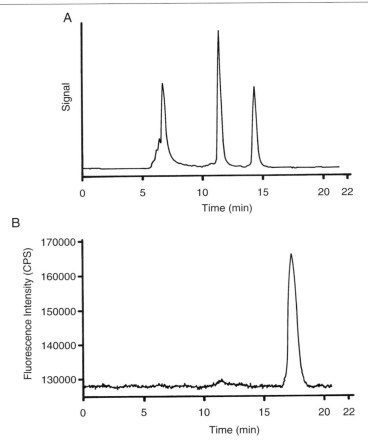

FIG. 3. Chromatograms of 20-$\mu$l of Centrum liquid vitamin preparation containing biotin (36 $\mu$g/ml). Detection was carried out by (A) UV absorbance at 220 nm and (B) fluorescence by the avidin–FITC postcolumn reaction detection system ($\lambda_{ex}$ = 495 nm, $\lambda_{em}$ = 518 nm). The amount of biotin found was 33 $\mu$g/ml.

content of biotin found by this method agreed with the amount of biotin on the package label (differing by only 3–6%).[9]

The selectivity of the described system is illustrated by comparing the UV absorbance chromatogram of Centrum liquid vitamin preparation (Fig. 3A) with that obtained using the postcolumn reaction detection system (Fig. 3B). Because the concentration of biotin is well below the UV absorbance

[9] A. Przyjazny, N. G. Hentz, and L. G. Bachas, *J. Chromatogr.* **654,** 79 (1993).

detection limit, quantification of biotin would not be possible by using direct UV absorbance detection alone. However, because the avidin–FITC provides a highly selective and sensitive detection, only a peak due to biotin is observed when using the postcolumn reaction detection system (Fig. 3B).

### Postcolumn Reaction Detection System Using Streptavidin–Fluorescein Isothiocyanate

#### Conditions

Postcolumn reagent: Streptavidin–FITC (2 mg/liter) in 0.100 $M$ phosphate buffer, pH 8.2

Postcolumn reagent flow rate: 0.10 ml/min

Fluorescence detection: $\lambda_{ex}$ = 495 nm, $\lambda_{em}$ = 518 nm

Mobile phase: 0.100 $M$ phosphate buffer (pH 7.0)–methanol (80:20, v/v)

Mobile phase flow rate: 0.40 ml/min

Calibration plots yielded a linear range extending from $8 \times 10^{-8}$ to $1 \times 10^{-6}$ $M$ biotin and biocytin. The detection limits were found to be $2 \times 10^{-8}$ $M$ biotin and biocytin, assuming an $S/N$ ratio of 3. The stability of the streptavidin–FITC system, investigated by preparing a 2-mg/liter solution and storing it in the syringe pump, was found to be greater than 96 hr. In contrast, the avidin–FITC system was stable for only 8 hr. This has been attributed to avidin being a glycoprotein, while streptavidin is not. The glycosylation allows for the avidin to be readily adsorbed to the wall of the syringe pump.

The streptavidin—FITC-based system was also evaluated in terms of its selectivity (i.e., cross-reactivity) toward compounds that are structurally similar to biotin. It was shown that the described system does indeed respond to a class of compounds that contain the biotin moiety.[10] Biotin, biocytin, biotin ethylenediamine, 6-(biotinoylamino)caproic acid, and 6-(biotinoylamino)caproic acid hydrazide could all be separated under the previously described conditions, as well as induce fluorescence enhancement.[11]

*Analysis of Real Samples.* A variety of real samples was analyzed for their biotin content in order to verify the accuracy and selectivity of the streptavidin–FITC postcolumn reaction detection system. The biotin content in four cell culture media and Isomil infant formula agreed well with that claimed by the manufacturer.[11] A typical chromatogram is depicted

[10] T. Smith-Palmer, M. S. Barbarakis, T. Cynkowski, and L. G. Bachas, *Anal. Chim. Acta* **279**, 287 (1993).

[11] N. G. Hentz and L. G. Bachas, *Anal. Chem.* **67**, 1014 (1995).

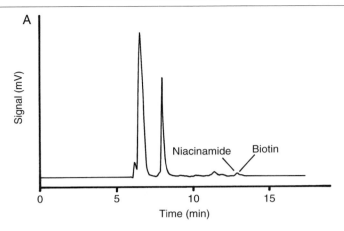

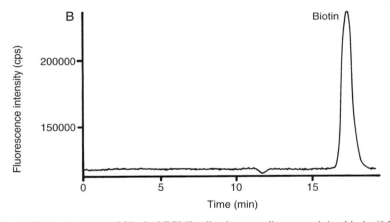

FIG. 4. Chromatograms of 20-$\mu$l of RPMI cell culture medium containing biotin (0.2 $\mu$g/ml). Detection was carried out by (A) UV absorbance at 220 nm and (B) fluorescence by the streptavidin–FITC postcolumn reaction detection system ($\lambda_{ex}$ = 495 nm, $\lambda_{em}$ = 518 nm). The amount of biotin found was 0.19 $\mu$g/ml.

in Fig. 4, where the RPMI cell culture medium was analyzed. Of particular importance is the niacinamide peak (at approximately 12.8 min in the chromatogram using UV detection; Fig. 4A) that coelutes with biotin. Also noteworthy is that the concentration of biotin present in the sample is well below the detection limit for biotin using UV absorbance detection. Therefore, if UV absorbance were to be used for the detection of biotin in this sample, preconcentration would be required, as well as altering the chromatographic conditions to resolve the niacinamide from biotin. In contrast, because streptavidin–FITC is selective, only biotin produces a fluorescence signal (Fig. 4B).

## Concluding Remarks

The use of the described postcolumn reaction detection systems (based on competitive and noncompetitive binding principles) greatly enhances the detection capability of biotin with respect to sensitivity and selectivity. The detection limits were at least an order of magnitude better than that for biotin detection by conventional UV absorbance (Table I). For further comparison, the quantity of reagent and cost per analysis (1996 prices) have also been presented (Table I).

The three described systems exhibited typical chromatographic behavior with respect to repeatability. The relative standard deviation for the retention times ranged from 0.25 to 0.9%, while that for peak areas ranged from 2.0 to 3.3% ($n = 8$). Typically, the band broadening due to the introduction of the postcolumn reactor was 15–20%. The subsequent loss in resolution did not affect quantification of the analyte because the baseline separation was not compromised.

The systems were also found to be selective. Organic solvents such as acetone, DMF, and MEK did not interfere with the postcolumn reaction detection. It was also demonstrated that compounds structurally related to biotin did not interfere with the analyses as long as their separation on the HPLC column was possible. In addition, because of the inherent selectivity of (strept)avidin, coelution of biotin with structurally dissimilar compounds had no effect on the quantification of biotin. If UV absorbance detection were to be used under these conditions, biotin could not be properly quanti-

TABLE I

COMPARISON OF POSTCOLUMN REACTION DETECTION SYSTEMS FOR BIOTIN

| System | Amount of reagent[a] (mg) | Cost in 1996[b] | Detection limit (ng) |
|---|---|---|---|
| UV absorbance | NA[c] | NA | 20 |
| 2,6-ANS–avidin | 0.48 | $1.94 | 2.4 |
| Avidin–FITC | 0.04 | $0.50 | 0.089 |
| Streptavidin–FITC | 0.004 | $0.24 | 0.097 |

[a] The amount of reagent per analysis refers only to the (strept)avidin reagent.

[b] The cost per analysis includes all reagents and assumes a 20-min analysis; the (strept)avidin reagent contributes >91% of the total cost.

[c] NA, Not applicable.

fied. The systems described in this chapter were found to be accurate in the determination of biotin in real samples.

Finally, sample preparation was minimal. For example, most of the liquid samples needed only preinjection filtration. It is important to note that no preconcentration step was necessary for the samples analyzed, because the detection limits were low enough to accommodate the low analyte concentration in the original sample.

### Acknowledgments

The research described in this chapter was supported by grants from the National Institutes of Health (GM40510), National Science Foundation (CTS-9307518), and American Cyanamid.

## [30] High-Performance Liquid Chromatographic Determination of Biotin in Biological Materials after Crown Ether-Catalyzed Fluorescence Derivatization with Panacyl Bromide

By Gertrud I. Rehner and Jürgen Stein

Biotin (hexahydro-2-oxo-1H-thienol[3,4-d]imidazole-4-pentanoic acid) is a coenzyme of carboxylating enzymes, such as pyruvate carboxylase (EC 6.4.1.1) and acetyl-CoA carboxylase (EC 6.4.1.2), catalyzing key reactions in main metabolic pathways. Exact quantification of this water-soluble vitamin in complex biological matrices in its free as well as its protein-bound form is of interest for some questions of cell metabolism. The amounts to be determined in biological material are mostly in the picomolar range.

Various techniques have been applied for quantification of biotin: microbiological assays, biotin–avidin reaction combined with radiometric tests, gas chromatography of trimethylsilyl derivatives, and others. Different attempts were made to detect biotin after conversion to ultraviolet (UV)-absorbing or fluorescing compounds by thin-layer chromatography or high-performance liquid chromatography (HPLC) without being able to attain the sensitivity needed. Reviews of these methods, and as far as possible, of their applicability to biological problems have been published by Bowers-Komro et al.[1] and Stein et al.[2]

[1] D. M. Bowers-Komro, J. L. Chastain, and D. B. McCormick, Methods Enzymol. **122**, 63 (1986).
[2] J. Stein, A. Hahn, B. Lembcke, and G. Rehner, Anal. Biochem. **200**, 89 (1992).

Improvements in HPLC techniques gave rise to the development of assays applying different modifications of this method. Ultraviolet detection of the carbonyl function of the biotin molecule has been used by Chastain *et al.*[3] Electrochemical detection,[4] or fluorimetric detection of biotin-9-anthryldiazomethane esters,[5] has also been used without being able to solve the difficulty of applicability to biological samples with their complex matrix.

The main problem in biotin analysis arises from the lack of a suitable UV absorbance or other characteristic of the molecule that would make selective and sensitive detection possible. Thus, adequate derivatization is an essential prerequisite for separation and quantification of this molecule, its analogs, and metabolites. Using reversed-phase or anion-exchange HPLC, McCormick's group[1] could differentiate biotin from its analogs, which possess variations in the side chain, the thiophane ring, or the ureido ring.

As could be shown by labeling different organic acids, a carboxylic function is well suited for derivatization. The only attempt to derivatize the carboxylic group of biotin using 9-anthryldiazomethane as a derivatization reagent[5] and applying fluorescence detection yielded unsatisfactory results, owing to numerous impurities.

The aim of our study was to quantify biotin by HPLC after precolumn derivatization of the carboxylic group with fluorescent panacyl bromide [*p*-(9-anthroyloxy)phenacyl bromide] in the presence of crown ether. The analytical procedure should be applicable to biological samples containing biotin in picomolar amounts and in free as well as protein-bound form.

## Analytical Procedures

### Chemicals

Biotin and dethiobiotin obtained from Sigma (Deisenhofen, Germany) are diluted in distilled water to a final concentration of 100 $\mu M$. [*carboxy-*$^{14}C$]Biotin (specific activity, 22.8 mCi/mmol) is a gift from T. Weber (Hoffmann-La Roche, Basel, Switzerland). It is checked by thin-layer chromatography on silica 60 layers (E. Merck AG, Darmstadt, Germany) using chloroform–methanol (95/5, v/v) as an eluent and found to be 98% pure. Dibenzo-18-crown-6 (2,3,11,12-dibenzo-1,4,7,10,13,16-hexaoxacyclo-octadeca-2,11-diene) and panacyl bromide are obtained from Aldrich

[3] J. L. Chastain, D. M. Bowers-Komro, and D. B. McCormick, *J. Chromatogr.* **330,** 153 (1985).
[4] K. Kamata, T. Hagiwara, M. Takahashi, S. Uehara, K. Nakayama, and K. Akiyama, *J. Chromatogr.* **356,** 326 (1986).
[5] K. Hayakawa and J. Oizumi, *J. Chromatogr.* **413,** 247 (1987).

(Steinheim, Germany). All substances are of the highest purity available. Water is purified with a Millipore Q system (Waters, Eschborn, Germany).

### Precolumn Derivatization Procedure

Standards as well as sample extracts are dried under a nitrogen stream and dissolved in 100 $\mu$l of acetone containing 250 nmol of panacyl bromide and 50 nmol of dibenzo-18-crown-6. In all cases 2.5 nmol of dethiobiotin is added as an internal standard. After adding 20–30 mg of $K_2CO_3$, the reaction tubes are closed and transferred to a water bath at either 37 or 57°. At certain intervals 20 $\mu$l is withdrawn from the tube and analyzed by thin-layer chromatography as mentioned previously.

A derivatization reagent for carboxylic functions should have two properties: The derivatives formed should permit sensitive detection and the reaction should be relatively specific for carboxylic functions. These requirements are fulfilled by several fluorescent labels including panacyl bromide. When standard solutions of biotin are analyzed after carrying out the whole cleanup procedure the detection limit is about 10 pmol, using normal-phase, and about 100 pmol when applying reversed-phase HPLC.

A third important factor responsible for adequate derivatization is the reactivity of the labeling reagent in the picomolar range, which is necessary for the analysis of biological samples. This is sufficient with panacyl bromide only when small amounts of biotin (10 to 20 pmol) are to be derivatized.

Formation of biotin esters such as panacyl derivatives may be impeded by the poor reactivity of the carboxylic function. Reactivity can be enhanced by adding a crown ether as a catalyst as described by Durst et al.[6] Crown ethers complex metal salts, i.e., they are also able to form the anionic form of biotin. Furthermore, they contribute to the dissolution of the anionic molecule into an aprotic eluent. Thus the nucleophilic properties of the biotinate anion are enhanced and nucleophilic attack of biotin on panacyl bromide is facilitated. Furthermore, in contrast to other chemicals used for initiating the reaction, such as triethylamine, crown ether-catalyzed derivatization is not affected by traces of water remaining in the sample.

Nevertheless, to correct possible deviations arising from the derivatization procedure as well as from the sample pretreatment, we propose using dethiobiotin as an internal standard. We assume that higher precision and recovery revealed by the method presented are at least partially due to the use of an internal standard.

[6] H. D. Durst, M. Milano, E. J. Kikta, S. A. Connelly, and E. Grushka, *Anal. Chem.* **47**, 1797 (1975).

As can be seen from the radiochromatograms in Fig. 1a, conversion of biotin into the corresponding panacyl ester is a function of incubation time. The derivatization kinetics shown in Fig. 1b emphasize that about 90% of the biotin is converted within 3 hr at 57° but efficiency varied from 73 to 98%. Nevertheless, when correcting these variations using dethiobiotin as an internal standard the technique is highly reproducible. Results of derivatizing the same sample 10 times indicate a coefficient of variation of 7.8%.

## *Nuclear Magnetic Resonance Spectra of Derivatization Products*

To characterize the panacyl esters of biotin and dethiobiotin, we recorded [1]H nuclear magnetic resonance (NMR) spectra of the derivatization product. The substances were dried under a nitrogen stream and subsequently dissolved in [2H]chloroform. Readings were taken with a Varian (Palo Alto, CA) HA-100 spectrometer at 400 MHz. Chemical shifts were expressed in parts per million relating to tetramethylsilane as an internal standard.

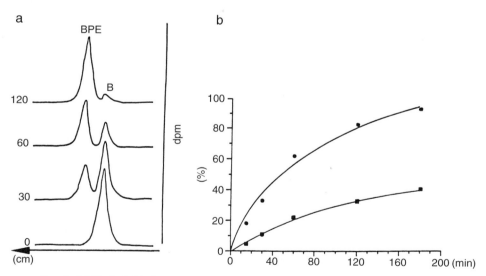

Fig. 1. (a) Typical thin-layer radiochromatograms of biotin (B) and biotin panacyl ester (BPE) obtained at distinct intervals, demonstrating the conversion of biotin in its panacyl ester (ca. 22 pmol of biotin). (b) Time-dependent formation of biotin panacyl ester at 37° (■) and 57° (●) (ca. 22 pmol of biotin). [Reprinted from J. Stein, A. Hahn, B. Lembcke, and G. Rehner, *Anal. Biochem.* **200,** 89 (1992), with permission.]

From the spectra of the panacyl esters of biotin and dethiobiotin, it could be concluded that only the carboxylic function of biotin participates in the formation of the panacyl ester whereas the amide functions of the molecule could still be detected at 5.8 and 6.3 ppm. Thus the reaction sequence shown in Fig. 2 is established.

## Chromatography and Detection

Separation of the panacyl esters of biotin and dethiobiotin is carried out on a Merck–Hitachi HPLC system consisting of a gradient former L-5000, a solvent metering pump 655A-11, a fluorescence detector F-1000, and a loop injector (model 7125; Rheodyne, Berkeley, CA) with a 100-$\mu$l syringe. All analyses are performed at room temperature. Two chromatographic systems are tested on the basis of either normal-phase or reversed-phase chromatography, respectively. In both cases we use a 4.6 × 150 mm column that has been filled by the upward slurry technique, using 2-propanol for preparing the slurry.

For reversed-phase analyses Hypersil ODS 3 $\mu$m (Shandon, Frankfurt, Germany) is used as a stationary phase. Gradient elution is performed with water and methanol. Composition of the eluent is linearly changed from 60:40 (water–methanol, v/v) to 30:70 within 15 min. The flow rate is 1 ml/min.

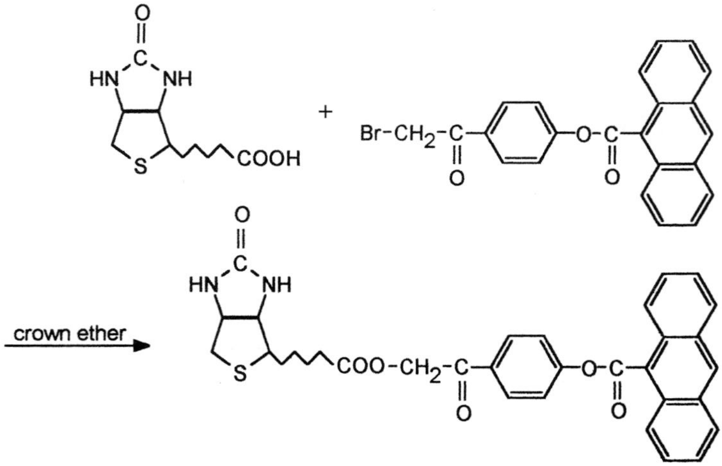

FIG. 2. Reaction sequence for the formation of biotin panacyl ester. [Reprinted from J. Stein, A. Hahn, B. Lembcke, and G. Rehner, *Anal. Biochem.* **200,** 89 (1992), with permission.]

Normal-phase chromatography is done on Shandon Hypersil 3 $\mu$m. Elution is obtained isocratically with 5% methanol/95% dichloromethane (v/v) at a flow rate of 1.4 ml/min. Fluorescence maxima of the biotin panacyl esters in different mobile phases are determined using a Hitachi F-2000 fluorescence spectrometer. Peaks are recorded at excitation and emission wavelengths of 380 and 470 nm, respectively. Calibration curves are calculated on the basis of peak area using least-squares regression analysis.

As Fig. 3a and b indicates, both reversed-phase and normal-phase HPLC are well suited to achieve complete separation of the panacyl esters of biotin and dethiobiotin within 7 min. The fluorescence maxima of both derivatives are found to be 380 and 470 nm for excitation and emission

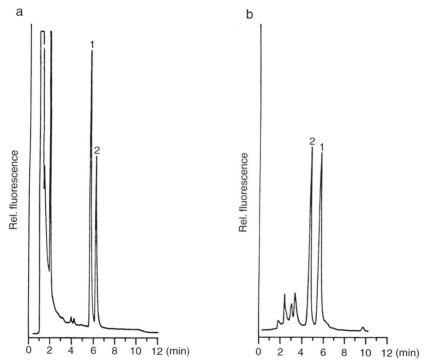

FIG. 3. Chromatographic separation of the panacyl esters of dethiobiotin (1) and biotin (2) applying normal-phase (a) and reversed-phase (b) HPLC. Retention times: (a) normal-phase HPLC: biotin, 6.46 min; dethiobiotin, 5.56 min. (b) Reversed-phase HPLC: biotin, 4.67 min; dethiobiotin, 5.71 min. [Reprinted from J. Stein, A. Hahn, B. Lembcke, and G. Rehner, *Anal. Biochem.* **200,** 89 (1992), with permission.]

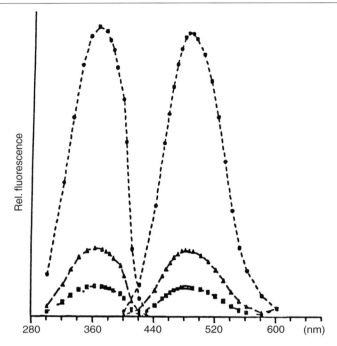

FIG. 4. Fluorescence spectra of biotin panacyl ester in different mobile phases, indicating that relative fluorescence is higher in the presence of an aprotic eluent. (●) Dichloromethane; (▲) methanol; (■) methanol–water (1 : 1, v/v). [Reprinted from J. Stein, A. Hahn, B. Lembcke, and G. Rehner, *Anal. Biochem.* **200,** 89 (1992), with permission.]

wavelengths, respectively, in both mobile phases (Fig. 4). As can be seen from Fig. 5, normal-phase chromatography (i.e., an aprotic eluent) provides higher sensitivity. Furthermore, additional tests reveal that reversed-phase chromatography is disturbed by side products of the derivatization reaction and by free panacyl bromide. These impurities must be removed by an additional thin-layer chromatographic procedure before HPLC analysis is performed.

## Validation of Applicability of Method to Biological Samples

### *Extraction of Biotin from Biological Materials, Sample Cleanup Procedure, and Analysis*

The method described has been validated using intestinal tissue as a model matrix. Rat gut is obtained from male Wistar rats (body weight,

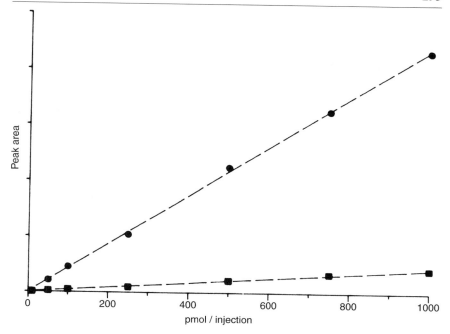

FIG. 5. Calibration curves for the HPLC determination of biotin panacyl ester as obtained after carrying out the whole sample cleanup with standard solutions of biotin. (●) Normal-phase HPLC ($r = 0.999$); (■) reversed-phase HPLC ($r = 0.993$). [Reprinted from J. Stein, A. Hahn, B. Lembcke, and G. Rehner, *Anal. Biochem.* **200**, 89 (1992), with permission.]

230–250 g). The animals are killed by a blow on the neck; the whole small intestine is removed and rinsed with ice-cold physiological saline. All further steps are carried out at 4°. Two to 3 g of gut tissue is cut, using a pair of scissors, and subsequently homogenized in 5 ml of a 5% (w/v) trichloroacetic acid (TCA) solution after dethiobiotin is added (5 nmol). The homogenate is centrifuged for 15 min at 10,000 g and the pellet is reextracted with 5 ml of TCA two more times.

The combined supernatants are applied to a $C_{18}$ solid-phase extraction column (Sep-Pak; Waters) that has been pretreated with 10 ml of methanol and 10 ml of water. Polar compounds are removed by washing the column with 10 ml of 2% (v/v) acetonitrile and biotin is eluted with 10 ml of 15% (v/v) acetonitrile.

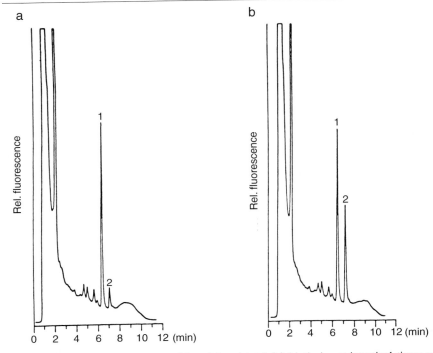

FIG. 6. Typical chromatograms of free (a) and total (b) biotin in rat intestinal tissue as determined by normal-phase chromatography of biotin and dethiobiotin panacyl ester. Peak 1, dethiobiotin; peak 2, biotin. [Reprinted from J. Stein, A. Hahn, B. Lembcke, and G. Rehner, *Anal. Biochem.* **200,** 89 (1992), with permission.]

TABLE I
BIOTIN CONTENT IN DIFFERENT PARTS OF RAT SMALL INTESTINE[a]

| | Content | | | | | |
|---|---|---|---|---|---|---|
| | Duodenum | | Jejunum | | Ileum | |
| Biotin | nmol/g | % | nmol/g | % | nmol/g | % |
| Free | 0.43 ± 0.07 | 13.9 | 0.44 ± 0.08 | 13.3 | 0.41 ± 0.07 | 12.9 |
| Protein bound | 2.66 ± 0.45 | 86.1 | 2.87 ± 0.52 | 86.7 | 2.77 ± 0.49 | 87.1 |
| Total | 3.09 ± 0.56 | 100 | 3.31 ± 0.58 | 100 | 3.18 ± 0.57 | 100 |

[a] Nanomoles per gram ± SD; percentage of total biotin, $N = 12$.

The eluted fraction is transferred to a 0.8 × 7 cm Dowex 1-X8 formate column (200–400 mesh; Serva, Heidelberg, Germany), which is then washed with 10 ml of water and 10 ml of 0.1 $N$ potassium formate. Biotin and dethiobiotin are eluted with an additional 30 ml of 0.1 $N$ potassium formate. The eluate is again applied to a preconditioned Sep-Pak $C_{18}$ cartridge, washed with 10 ml of water, and eluted with 10 ml of methyl formate. The methyl formate is evaporated under a nitrogen stream and the extract is dissolved in 50 $\mu$l of chloroform.

For further purification this solution is chromatographed on silica 60 layers using chloroform–methanol–potassium formate (90 : 9 : 1, v/v) as an eluent. To simplify the identification of the substance spots, standard solutions of biotin and dethiobiotin are chromatographed on every layer. When chromatography is terminated, the edges of the layer containing the standards are cut off with a pair of scissors and colored with $p$-dimethylaminocinnamaldehyde ($p$-DACA[7]). The corresponding sample spots are scraped off, transferred to reaction tubes, lyophilized, and derivatized as described previously.

The loss of substance within the whole procedure has been calculated after adding 0.5 $\mu$Ci of [$^{14}$C]biotin to the homogenate. Recovery of added unlabeled biotin has been determined by carrying out the whole analysis with homogenates from rat gut that is divided in half and adding 0.5 nmol of biotin to one half.

Samples of rat intestinal tissue (duodenum, proximal jejunum, distal jejunum) are analyzed as previously described. Free biotin is analyzed directly whereas protein-bound biotin is measured after hydrolytic treatment of the homogenate with 6 $N$ sulfuric acid.[8]

As judged by the radioactivity of [$^{14}$C]biotin, the final HPLC peak contains 61.2 ± 4.9% ($n = 8$) of the substrate added to the gut homogenates. Recovery of biotin is 93.8 ± 3.9% ($n = 10$) as calculated by means of the internal standard. Fluorescence spectra recorded from the peaks during the analysis of biological samples correspond to those of standard solutions. Figure 6 shows chromatograms of free biotin and total biotin (i.e., after hydrolysis of protein-bound biotin), emphasizing that only minor impurities are still present in the sample. Table I summarizes the biotin content of different parts of the rat small intestine, indicating that the major amount of biotin (i.e., ca. 85%) exists as bound biotin. Significant differences concerning the biotin content and the biotin pattern of the different intestinal segments have not been found.

[7] D. B. McCormick and J. A. Roth, *Anal. Biochem.* **34**, 226 (1970).
[8] J. M. Schreiner, *Ann. N.Y. Acad. Sci.* **447**, 420 (1985).

# [31] Bioluminescence Competitive Binding Assays for Biotin Based on Photoprotein Aequorin

*By* SERGIO LIZANO, SRIDHAR RAMANATHAN, AGATHA FELTUS, ALLAN WITKOWSKI, and SYLVIA DAUNERT[1]

## Introduction

Competitive binding assays have been particularly useful in bioanalytical chemistry and biomedical analysis owing to their high selectivity and sensitivity, both of which are dependent on the strength of the interaction between the analyte (ligand) and the corresponding biological binder molecule (i.e., binding protein, antibody, receptor, lectin, etc.).[2] The interaction of the vitamin biotin with the binding protein avidin is characterized by a strong affinity ($K_d = 10^{-15} M$),[3] and this interaction can be used in the development of sensitive assays for the detection of biotin in a variety of samples. Labels such as bioluminescent proteins can be employed to enhance the sensitivity of competitive binding assays compared to more conventional assays that use enzymes as labels.[4]

In this chapter, competitive binding assays for biotin are described that use the bioluminescent protein aequorin as the label. Aequorin is a 21.4-kDa calcium-activated photoprotein found in the jellyfish *Aequorea victoria*.[4] Aequorin associates noncovalently with its chromophore, coelenterazine, which in the presence of $O_2$ and $Ca^{2+}$ is oxidized into a metastable excited state that results in the emission of photons at 469 nm (quantum yield of 11–17.5%, depending on the isoform of aequorin used).[5] This calcium-triggered emission of photons exhibits flash kinetics in which most of the bioluminescence signal is generated within a few seconds of the addition of calcium (Fig. 1).

Heterogeneous and homogeneous competitive binding assays using aequorin bioluminescence have been developed in our laboratory for detection of biotin.[6] In the heterogeneous system, an aequorin–biotin conjugate

---

[1] Address correspondence to this author at the Department of Chemistry, University of Kentucky, Lexington, Kentucky 40506-0055.

[2] R. M. Nakamura, Y. Kasahara, and G. A. Rechnitz, "Immunochemical Assays and Biosensor Technology for the 1990's." American Society of Microbiology, Washington, DC, 1992.

[3] M. Wilchek and E. A. Bayer (eds.), *Methods Enzymol.* **184,** 14 (1990).

[4] K. Campbell, "Chemiluminescence." Ellis Horwood, Chichester, England, 1988.

[5] O. Shimomura and F. H. Johnson, *in* "Detection and Measurement of Free $Ca^{2+}$ in Cells" (C. C. Ashley and A. K. Campbell, eds.), p. 73. Elsevier, Amsterdam, 1979.

[6] A. Witkowski, S. Ramanathan, and S. Daunert, *Anal. Chem.* **66,** 1837 (1994).

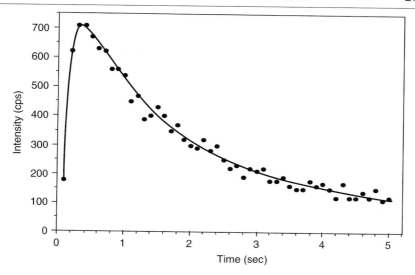

FIG. 1. Typical flash kinetics plot of the bioluminescence generated by 20 amol of biotinylated aequorin. The intensity of the bioluminescence signal, reported in counts per second (cps), was collected over a 5-sec time period at 0.1-sec intervals. [Reprinted with permission from A. Witkowski, S. Ramanathan, and S. Daunert, *Anal. Chem.* **66,** 1839 (1994). Copyright 1994 American Chemical Society.]

competes with unlabeled biotin for binding to immobilized avidin, followed by a separation step to remove the unbound conjugate. The bioluminescence of the avidin-bound conjugate is measured, and the signal is related to the concentration of biotin present in a sample. However, the homogeneous assay is a one-step procedure in which the aequorin–biotin conjugate competes with free biotin for binding to avidin in solution (Fig. 2). In this case, binding of the conjugate to avidin results in partial quenching of the bioluminescence signal. The decrease in bioluminescence can thus be related to the concentration of biotin in a sample. Both the homogeneous and heterogeneous systems are sensitive, with detection limits in the attomole range.

### Requirements for Instrumentation and Special Reagents

Bioluminescence assays are best performed on luminometers operated in a photon-counting mode. A variety of such instruments is commercially available. In addition, modern scintillation counters may be used to perform bioluminescence assays. For a review on bioluminescence instrumentation

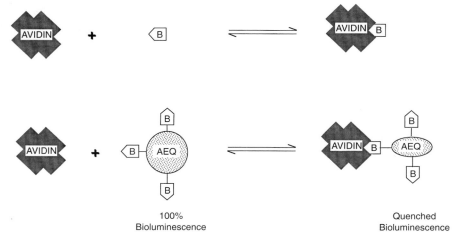

FIG. 2. Schematic of the homogeneous bioluminescence competitive binding assay for biotin, using avidin as the binder. B, Biotin; AEQ, aequorin.

see Ref. 7. In our studies, bioluminescence measurements are performed on an Optocomp I luminometer from GEM Biomedical (Carrboro, NC), using a 100-$\mu$l fixed injector.

The aequorin label, a photoprotein, can be purchased from SeaLite Sciences (Atlanta, GA) under the trade name AquaLite or from Molecular Probes (Eugene, OR). It is commercially available both as recombinant aequorin and as biotinylated recombinant aequorin (2.6 mol of biotin per mole of aequorin). Laboratories with molecular biology expertise could express the apoaequorin in *Escherichia coli.* At least one of the isoforms of the photoprotein encoded by the *AQ440* gene is available on plasmids sold commercially by Molecular Probes. If the route of expressing the protein is selected, apoaequorin can be converted to aequorin (the active photoprotein) by incubating with coelenterazine.[8] In addition to the native coelenterazine, several synthetic coelenterazine analogs are commercially available (Molecular Probes) that confer different spectral properties to aequorin (e.g., different bioluminescence emission maxima and enhanced quantum yields). Detailed biotinylation instructions are given by Stults *et al.*[9] However, it should be noted that special precautions should be taken while biotinylating and storing the protein to avoid accidental discharge

[7] P. E. Stanley, *J. Biolumin. Chemilumin.* **7,** 77 (1990).

[8] O. Shimomura, *Cell Calcium* **12,** 635 (1991).

[9] N. L. Stults, N. F. Stocks, H. Rivers, J. Gray, R. O. McCann, D. O'Kanne, R. D. Cummings, M. J. Cormier, and D. F. Smith, *Biochemistry* **31,** 1433 (1992).

of bioluminescence,[8] and thus, purchase of the stabilized commercially available form may be preferable.

For the assays described in this study, all reagents are obtained from Sigma (St. Louis, MO) unless otherwise stated. Tris(hydroxymethyl)amino-methane (Tris) is from Research Organics (Cleveland, OH). Biotinylated recombinant aequorin (biotinylated AquaLite) is purchased from SeaLite Sciences (Atlanta, GA) or from Molecular Probes (Eugene, OR). All exper-iments are performed in triplicate at room temperature and have been corrected for the contribution of the blank.

### Bioluminescence of Biotinylated Aequorin

The relative amounts of avidin and biotinylated aequorin (conjugate) influence the response characteristics (e.g., detection limits and sensitivity) of the assay. Therefore, the amounts of avidin and biotinylated aequorin must be optimized in order to enhance assay performance. To determine the optimal concentration of biotinylated aequorin for use in these assays, serial dilutions of biotinylated aequorin are prepared in buffer A [2.0 m$M$ EDTA, 1.0 $M$ KCl, bovine serum albumin (BSA, 1.0 mg/ml), and 0.10% (w/v) sodium azide in 10 m$M$ Tris-HCl, pH 7.5]. Then, 100 $\mu$l of each dilution of the biotinylated aequorin is mixed with 200 $\mu$l of buffer B (0.15 $M$ NaCl and 2.0 m$M$ EDTA in 10 m$M$ Tris-HCl, pH 8.0) in a 12 × 75 mm glass test tube and placed in the Optocomp I luminometer. The presence of EDTA in buffers A and B is necessary to complex any $Ca^{2+}$ present, which can cause a premature discharge of aequorin. To induce bioluminescence emission, a Tris-HCl buffer containing calcium at concentrations higher than that of EDTA in buffers A and B is added. The composition of this Tris/$Ca^{2+}$ luminescence-triggering solution is 100 m$M$ $CaCl_2$ in 100 m$M$ Tris-HCl, pH 7.5. The bioluminescence of the biotinylated aequorin is integrated over a 3-sec period after injection of 100 $\mu$l of the Tris/$Ca^{2+}$ luminescence-triggering solution. The corresponding calibration plot is shown in Fig. 3. An amount of biotinylated aequorin that produces a light intensity at least 10-fold higher than the background signal is chosen for the assays, as this signal provides sufficient sensitivity. It should be noted that similar results in terms of bioluminescence emission kinetics and signal intensity are obtained with nonbiotinylated aequorin, indicating that bio-tinylation of the protein with 2.6 mol of biotin per mole of aequorin has a minimum effect on the bioluminescence emission.

### Effect of Avidin on Biotinylated Aequorin

The development of a homogeneous assay for biotin using aequorin is based on the ability of avidin in solution to inhibit the bioluminescence of

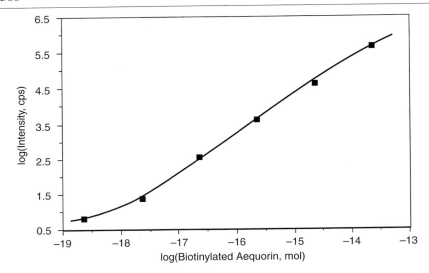

FIG. 3. Calibration plot for the biotinylated aequorin. The intensity of the bioluminescence signal was measured by a luminometer over a period of 3 sec. [Reprinted with permission from A. Witkowski, S. Ramanathan, and S. Daunert, *Anal. Chem.* **66,** 1839 (1994). Copyright 1994 American Chemical Society.]

biotinylated aequorin. To establish the optimal amount of avidin to use in the assay, a binder dilution curve is constructed by varying the amount of avidin in the presence of a fixed amount of biotinylated aequorin, and measuring the degree of signal inhibition. The avidin solutions are prepared using buffer B, also containing BSA (1.0 mg/ml). A volume of 10 $\mu$l of each avidin solution is mixed with 200 $\mu$l of assay buffer C [500 m$M$ NaCl, 2.0 m$M$ EDTA, and BSA (1.0 mg/ml) in 20 m$M$ $Na_2HPO_4/NaH_2PO_4$, pH 7.5] and 100 $\mu$l of a $1.0 \times 10^{-13}$ $M$ solution of biotinylated aequorin, and incubated under constant shaking for 30 min at room temperature. The bioluminescence of each reaction mixture is then determined on injection of 100 $\mu$l of the Tris/$Ca^{2+}$ luminescence-triggering solution as previously described. A blank (400 $\mu$l of assay buffer C) and a positive control sample containing no avidin (300 $\mu$l of assay buffer C plus 100 $\mu$l of $1.0 \times 10^{-13}$ $M$ biotinylated aequorin) are also run to determine background and maximum bioluminescence signals, respectively. Maximum inhibition is typically reached at levels of avidin higher than 5 pmol (Fig. 4).

## Homogeneous Assay for Biotin

Standard solutions of biotin are prepared in buffer A. A volume of 100 $\mu$l of each biotin standard solution is mixed with 100 $\mu$l of the avidin

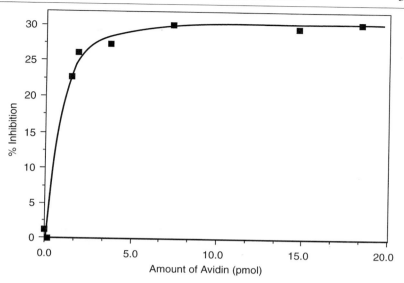

Fig. 4. Binder dilution curve showing the effect of varying the amount of avidin in the presence of 10 amol of biotinylated aequorin. The percent inhibition indicates the percent decrease in bioluminescence intensity owing to the interaction between avidin and biotinylated aequorin relative to the intensity in the absence of avidin. [Reprinted with permission from A. Witkowski, S. Ramanathan, and S. Daunert, *Anal. Chem.* **66**, 1839 (1994). Copyright 1994 American Chemical Society.]

solution ($2.0 \times 10^{-8}$ *M* avidin) and 100 $\mu$l of buffer C, and incubated with shaking for 30 min at room temperature. After this incubation step, 100 $\mu$l of the biotinylated aequorin solution ($1.0 \times 10^{-13}$ *M* biotinylated aequorin) is added and incubated for an additional 30 min, after which the bioluminescence is measured on addition of 100 $\mu$l of the Tris/Ca$^{2+}$ luminescence-triggering solution. The following controls are also measured as reference signals: 400 $\mu$l of assay buffer C (blank); 300 $\mu$l of assay buffer C plus 100 $\mu$l of biotinylated aequorin (full bioluminescence control); and 200 $\mu$l of assay buffer C plus 100 $\mu$l of avidin and 100 $\mu$l of biotinylated aequorin (full inhibition control). The corresponding dose–response plot is shown in Fig. 5, and is symbolized by open circles. To determine the concentration of biotin in a real sample, an appropriate dilution of that sample is processed in the same way as described for the biotin standard solutions and the concentration of biotin can be obtained from its corresponding bioluminescence signal by extrapolation from this dose–response curve.

Two additional dose–response curves are prepared using higher concentrations of avidin and/or biotinylated aequorin. These curves are also shown

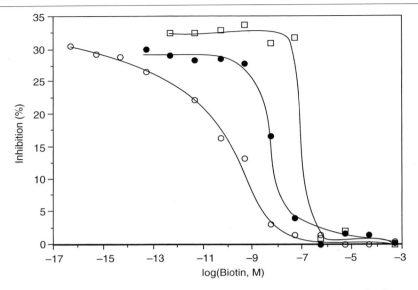

Fig. 5. Family of dose–response curves for biotin prepared by incubating a fixed amount of biotinylated aequorin and avidin with varying concentrations of biotin: (○) 2.0 pmol of avidin and 10 amol of biotinylated aequorin; (□) 20 pmol of avidin and 10 amol of biotinylated aequorin; (●) 4.0 pmol of avidin and 20 amol of biotinylated aequorin. [Reprinted with permission from A. Witkowski, S. Ramanathan, and S. Daunert, *Anal. Chem.* **66,** 1840 (1994). Copyright 1994 American Chemical Society.]

in Fig. 5, and indicate that when the avidin concentration is increased, the limits of detection of biotin are worsened. This effect can be explained by the fact that more biotin-binding sites are present at higher amounts of avidin, and thus more biotin is required to prevent the inhibition of aequorin bioluminescence.

Heterogeneous Competitive Binding Assay

Because the biotinylated aequorin maintains part of its bioluminescence activity when bound to avidin, it is possible to use this conjugate to develop a heterogeneous binding assay for biotin. In this assay, biotin competes with the biotinylated aequorin for binding to solid phase-bound avidin. After separation of the unbound species, the luminescence of the solid phase-bound biotinylated aequorin is measured and correlated to the concentration of biotin. Assays based on this principle, using avidin immobilized on agarose, polyacrylamide, or superparamagnetic beads, are currently being optimized in our laboratory. These assays yield detection limits that are at least as low as those that correspond to the described bioluminescence homogeneous assays for biotin.

## Concluding Remarks

Recombinant aequorin can be detected at subattomole levels, and therefore, can provide an ideal label for the development of highly sensitive competitive binding assays. This photoprotein can be modified through biotinylation to yield conjugates that maintain their bioluminescence properties, while at the same time preserving the ability of the attached biotin to bind to avidin. It was found that avidin causes up to 30% inhibition of the bioluminescence of biotinylated aequorin, a property that can be used to determine low concentrations of avidin. It was also demonstrated that this inhibition of the bioluminescence signal can form the basis of a sensitive homogeneous competitive binding assay for biotin. This assay has a detection limit of $1 \times 10^{-14}$ $M$ biotin, which to the best of our knowledge is the most sensitive reported thus far for homogeneous assays. It is important to mention that dose–response curves with relative standard deviations of less than 3% were routinely obtained, which demonstrates the high reproducibility of the method. Our preliminary data using immobilized avidin and a heterogeneous competitive binding assay format demonstrate at least as good detection limits for biotin.

## Acknowledgments

This research was supported by grants from the National Institutes of Health (GM47915), and the Department of Energy (DE-TG05-95ER62010). A.F. is an Otis Singletary Scholar at the University of Kentucky and gratefully acknowledges support from a fellowship from the Barry Goldwater Foundation (Springfield, VA).

# [32] Competitive Enzymatic Assay of Biotin

*By* Eric Z. Huang and Yu-Hui Rogers

## Introduction

Biotin is a small molecule that has been widely found as a vitamin in living organisms and used as a tag for detection and isolation in biomedical research.[1-3] A simple and sensitive assay of biotin is often needed in the study of metabolism and in the determination of the degree of conjugation of biotin to biomolecules. Although the bioassay of biotin on the basis of growth of biotin-dependent bacteria is sensitive, it is rarely used in routine laboratory practice because of the long culture time, low reproducibility, and interference caused by many growth-promoting factors other than biotin in a sample.[1] However, with proper design biotin can be assayed with relative ease in a reagent system consisting of its specific protein binder, avidin or streptavidin. One of the numerous sensitive and nonradioactive methods derived from the avidin–biotin interaction involves coating of biotinylated bovine serum albumin (B-BSA) on a microtiter plate and several subsequent detection steps.[4] In this chapter we substantially simplify this assay method while enhancing its reproducibility and sensitivity.

## Materials and Methods

*d*-Biotin, BSA, B-BSA (8–12 mol of biotin per mole of albumin), and *o*-phenylenediamine dihydrochloride (OPD) are purchased from Sigma Chemical Co. (St. Louis, MO); streptavidin–horseradish peroxidase conjugate (SA–HRP, 1.0 mg/ml) is from Vector Laboratories (Burlingame, CA); polystyrene 96-well microtiter plates (Immulon 4, flat bottom) and automatic washer Ultrawash Plus are from Dynatech Laboratories (Chantilly, VA); a microtiter plate reader $V_{max}$ is from Molecular Devices (Menlo Park, CA).

The competitive assays for free biotin are performed at room temperature as described below. To wells of a microtiter plate are added 100 $\mu$l of

[1] R. P. Bhullar, S. H. Lie, and K. Dakshinamurti, *in* "Biotin" (K. Dakshinamurti and H. N. Bhagavan, eds.), p. 122. The New York Academy of Science, New York, 1985.
[2] G. Bers and D. Garfin, *BioTechniques* **3,** 276 (1985).
[3] M. Wilchek and E. A. Bayer, *in* "Protein Recognition of Immobilized Ligands" (T. W. Hutchins, ed.), p. 83. Alan R. Liss, New York, 1989.
[4] E. A. Bayer, H. Ben-Hur, and M. Wilchek, *Anal. Biochem.* **154,** 367 (1986).

B-BSA:BSA mixes made in phosphate-buffered saline (PBS) at various molarity ratios while keeping a constant total protein concentration of 10 mg/ml. After a 1.0-hr incubation, the plate is washed three times in the automatic washer with 300 $\mu$l of 10 m$M$ Tris-HCl (pH 7.5), 150 m$M$ NaCl, 0.050% (v/v) Tween 20 (TNTw). Immediately afterward, the following is added in sequence to the wells: 25 $\mu$l of biotin standard solutions at various concentrations and biotin samples to be assayed, and 50 $\mu$l of SA–HRP prepared at various dilutions in PBS containing BSA (1.0 mg/ml). After a 0.50-hr incubation, the plate is washed three times in the automatic washer with 300 $\mu$l of TNTw. To the wells are added 100 $\mu$l of 1.0 mg/ml OPD dissolved in 0.10 $M$ citric acid (pH 4.5) with 0.012% (v/v) $H_2O_2$. After a 0.50-hr reaction time, the plate is read at 450 nm in the microtiter plate reader. The absorbance reading of the assay system background from a well filled simply with 100 $\mu$l of the OPD solution is subtracted from all the absorbance readings of the wells processed as described.

## Results

Figure 1 shows the results of SA–HRP binding to the biotin immobilized on the surfaces of the microtiter plate wells through coating of the B-BSA:BSA mixes. This reproducible binding experiment is a special scenario of the competition experiment described in Fig. 2 (with no free biotin) that allows analysis and optimization of the following real competitive assay system. The wells with no coating of B-BSA (zero spiking of B-BSA) serve as controls to observe the nonspecific binding of SA–HRP conjugate. It is apparent that the optimal conditions for the binding or competition experiment in terms of low nonspecific binding and high and dynamic specific binding should correspond to an SA–HRP dilution of 1:1000 and a B-BSA:BSA spiking ratio of 1:100 (Fig. 1). When the B-BSA spiking ratio was fixed at 1:100, we found that the SA–HRP dilution can be varied between 1:500 and 1:1500 to reproduce essentially the same result patterns in the binding and competition experiments (data not shown).

Analysis of the data in Fig. 1 indicates that the SA–HRP binding with the immobilized biotin is roughly proportional to the product of the concentrations of these two binding components. Furthermore, all of the free biotin concentrations involved in this work are significantly higher than its dissociation constant with streptavidin.[5] Therefore, the binding between SA–HRP and the surface biotin should be linearly reduced by the quantitative binding of the free competing biotin and make an excellent basis for

[5] N. M. Green, *Adv. Protein Chem.* **29**, 85 (1975).

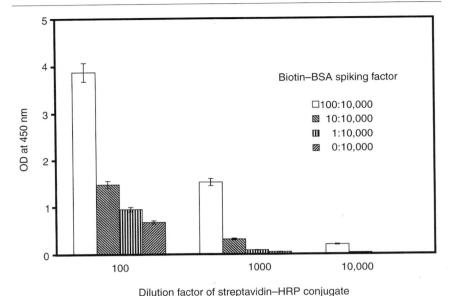

FIG. 1. Absorbance readings related to the enzymatic activities immobilized on the microtiter plate wells as a result of binding of SA–HRP at the indicated dilutions to the B-BSA : BSA
coated (at the indicated spiking ratios) on the wells. Each datum point is a result of four
replicate experiments and is presented as an average ± standard deviation.

a competitive assay of the free biotin. Indeed, Fig. 2 demonstrates the
quantitative nature of the competition of free biotin against the immobilized
biotin in binding to SA–HRP. Such competition curves are also the standard
curves for quantitation of the free biotin. The two standard curves obtained
with the two B-BSA spiking ratios are essentially identical in shape, in
agreement with the notion that the formation of the SA–HRP complex
with the immobilized biotin varies only with the B-BSA spiking ratio as a
proportionality factor. However, the standard curve associated with the
larger B-BSA spiking ratio is more reliable for use because of the greater
absorbance readings. The competitive assay is also characterized by the
high reproducibility (coefficient variation <5.0%) and the clean nonspecific
binding of SA–HRP to the microtiter plate well (Fig. 1 and the points of
excess free biotin in Fig. 2). The high reproducibility and low background
are further combined with the HRP enzymatic amplification to produce an
assay sensitivity (lower detection limit) of 0.10 pmol or 4.0 n$M$ free biotin
in the 25-$\mu$l sample volume (Fig. 2). We also used the standard curve
constructed with the larger B-BSA spiking ratio to determine the biotin

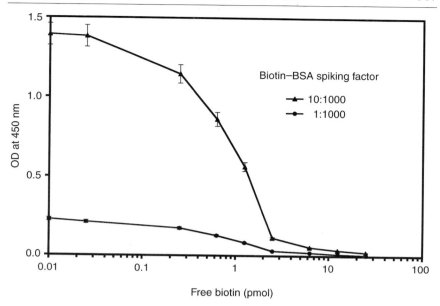

FIG. 2. Standard curves correlating the free biotin at the indicated amounts to the absorbance readings produced by the residual activities of SA–HRP bound at 1:1200 dilution to the microtiter plate wells coated with the indicated B-BSA : BSA spiking ratios on competing with the free biotin. Each datum point is a result of four replicate experiments and is presented as an average ± standard deviation.

moiety conjugated to protein and oligonucleotide and obtained satisfactory results with ease (data not shown).

Discussion

One of the desirable aspects of a separation-based assay is the freedom from the interference of the spectroscopic impurities or turbidity of the sample, as these interfering factors are washed out or removed before commencing the final spectroscopic measurement. Therefore, our competitive assay method as described in this work can be used for specific quantitation of biotin in a crude sample, the absorbance or turbidity of which is otherwise a problem in a homogeneous assay such as the one based on biotin disruption of an avidin–dye complex.[6] A separation-based assay in general has a lower analyte-independent background, which can be translated into a higher assay sensitivity.

[6] N. M. Green, *Biochem. J.* **94**, 23c (1965).

The competitive assay described in this work represents a significant improvement in procedural ease, reproducibility, and sensitivity over the one developed by Bayer et al.[4] In their method, B-BSA at a low concentration is incubated with a microtiter plate at 4° overnight followed by BSA blocking, streptavidin binding, and final competition between free sample biotin and biotin alkaline phosphatase. This multistep method is cumbersome, less reproducible, and less sensitive because of the variability. In our method, immobilization of biotin on the microtiter plate well with spiking of B-BSA in a high total BSA concentration (10 mg/ml) is fast (1 hr) and reproducible (coefficient variation <5.0%). When the enzyme amplification is used with the reproducible coating of a small amount of B-BSA (such as the one done with the 1 : 100 B-BSA spiking ratio), a 50-fold enhancement in assay sensitivity is achieved.[4,7] Therefore, our method is better suited for routine use in sensitive biotin quantitation of a biological or manned sample.

It is also interesting to explore the possibility of manipulating our competitive assay system to accommodate different assay sensitivities and dynamic ranges. One approach is to dilute further the streptavidin enzyme conjugate, because binding of the enzyme conjugate at the larger dilution to the immobilized biotin can be disrupted by a lower amount of free biotin. Unfortunately, the HRP and OPD detection system in this work is not sensitive enough to produce a reliable signal with the much diluted conjugate. A solution to this problem would rely on switching to a more potent enzyme label or more sensitive substrate such as the fluorogenic version.[8]

---

[7] Z. Huang, R. P. Haugland, D. Szalecka, and R. P. Haugland, *BioTechniques* **13,** 543 (1992).
[8] Z. Huang, N. A. Olson, W. You, and R. P. Haugland, *J. Immunol. Methods* **149,** 261 (1992).

# [33] Competitive Agglutination Assay of Biotin

*By* Eric Z. Huang

## Introduction

A homogeneous assay in which the reaction between analyte and its detecting agent is directly measured without physical separation is preferred in clinical laboratories because of the operational ease and compatibility with automation. The latex particle is an excellent vehicle for such a homogeneous assay, because it is often possible to modify the particle surface in such a manner that the analyte-related reaction leads to visually or instrumentally observable agglutination of the particle.[1] A latex-particle homogeneous assay with a spectrophotometer suitable for measuring turbidity can be sensitive, as a single molecular event of binding between analyte and its detecting agent on the particle surface is amplified by numerous photons scattered from the particle. In fact, various types of latex-particle agglutination systems have been developed for sensitive nonradioactive immunoassays.[2–5]

An agglutination immunoassay for a large analyte is usually achieved on the cross-linking of antibody-coated latex particles by the determinants of the analyte. This method cannot be adapted to a small analyte that has only one determinant. One approach to determining such a small analyte is to utilize the ability of the free analyte to competitively disrupt agglutination of the analyte-coated latex particles caused by the multiple (usually two) binding sites of an antibody. In this work, we demonstrate the use of such an analyte-coated latex particle for assay of a small analyte, biotin. Furthermore, heterogeneity in affinities of analytes and detecting agent is often encountered in biological assays and adversely affects the accuracy of a competitive assay system.[6] The biotin competitive assay system in this work is therefore constructed under a so-called *saturating condition* so that a uniform quantitation of a mixture of sample biotin analogs having different affinities for its protein binder, avidin or streptavidin, is possible.

## Materials and Methods

### Reagents and Instruments

The chemicals and instruments used in this work are obtained from the following suppliers:
   Biotin-coated latex particles (BCLPs; 1.0 $\mu$m, carboxylate modified), *d*-biotin (biotin standard), avidin ($M_r$ 68,000), streptavidin ($M_r$ 60,000), and streptavidin–agarose gel (85 nmol of biotin–fluorescein binding activity per milliliter gel): Molecular Probes (Eugene, OR)
   4-Hydroxyazobenzene-2'-carboxylic acid (HABA), plane agarose: Sigma (St. Louis, MO)

[1] J. Camey, *Anal. Proc.* **27,** 99 (1990).

[2] J. R. Delanghe, J. P. Chapelle, and S. C. Vanderschueren, *Clin. Chem.* **36,** 1675 (1990).

[3] E. A. Medcalf, D. J. Newman, A. Gilboa, E. G. Gorman, and C. P. Price, *J. Immunol. Methods* **129,** 97 (1990).

[4] E. A. Medcalf, D. J. Newman, E. G. Gorman, and C. P. Price, *Clin. Chem.* **36,** 446 (1990).

[5] D. Collet-Cassart, J. N. Limet, L. Van Krieken, and R. De Hertogh, *Clin. Chem.* **35,** 141 (1989).

[6] H. Markowitz and N.-S. Jiang, *in* "Fundamentals of Clinical Chemistry" (N. W. Tietz, ed.), p. 274. W. B. Saunders, Philadelphia, 1982.

Carnation nonfat dry milk (CNDM): Carnation Co. (Los Angeles, CA)

Bovine serum albumin [BSA, radioimmunoassay (RIA) grade]: United States Biochemical (Cleveland, OH)

Syringe filter Anotop 25 Plus (0.20 $\mu$m): Alltech Association (Deerfield, IL)

Round-bottom, 96-well microtiter plates: Corning Glass Works (Corning, NY)

Microtiter plate reader MR 600: Dynatech (Boston, MA)

Microtiter plate shaker 4G25: Lab-Line Instruments (Melrose Park, IL)

## Preparation of Carnation Nonfat Dry Milk Assay Samples

Because of the relatively low biotin content, a raw CNDM dispersion at the concentration required for a minimally detectable amount of biotin is too turbid to be directly applied to the agglutination assay system described in this work. To eliminate the turbidity, the raw CNDM dispersion should be processed by acid precipitation or filter filtration. Because biotin in CNDM appears to be water soluble and free from macromolecules such as proteins, removal of these macromolecules by precipitation or filtration procedures does not cause a loss of biotin content.[7] However, acid precipitation is easier to perform than filtration, although both methods give a sufficiently clear liquid sample for running the agglutination assay and for obtaining essentially the same results.

For acid precipitation, a homogeneous dispersion of CNDM at a concentration of about 0.50 g/ml in distilled water is prepared by brief (1.0-min) sonication in a water-bathed sonicator. To the CNDM dispersion is added 6.0 $M$ HCl to a final concentration of 0.80 $M$. The CNDM sample is manually shaken for 15 sec and the precipitate formed is removed by a brief centrifugation (2000 rpm, 2 min, 26°). The supernatant is collected, 1.0 $M$ K$_2$HPO$_4$ is added to a final concentration of 10 m$M$, and 6.0 $M$ NaOH is used to adjust the pH to 7.2. The CNDM sample is again briefly centrifuged (2000 rpm, 2 min, 26°) to remove the newly formed precipitate. The supernatant is collected, solid NaCl is added to a final concentration of 150 m$M$, 10% (v/v) Tween 20 is added to a final concentration of 0.50% (v/v), and solid BSA is added to a final concentration of 0.20% (w/v). The CNDM sample is filtered through a syringe filter to yield a clear solution. This solution should have a low absorbance reading at 340 nm as read in a microtiter plate reader (<0.0010/$\mu$l). The final volume of this CNDM sample should be recorded for calculating the CNDM content per unit volume of the

[7] W. L. Hoffman and A. A. Jump, *Anal. Biochem.* **181,** 318 (1989).

sample preparation. To minimize the loss of the liquid CNDM sample during the preparation process, CNDM should be used at a total amount of more than 5.0 g in each preparation.

A clear solution of CNDM may also be prepared by direct filtration of a raw CNDM dispersion through a syringe filter. The raw CNDM dispersion is prepared in an assay buffer [10 m$M$ potassium phosphate (pH 7.2) containing 150 m$M$ NaCl, 0.50% (v/v) Tween 20, and 0.20% (w/v) BSA] on brief sonication as previously described. The CNDM concentration should not exceed 0.25 g/ml to avoid plugging the filter.

To assess the binding specificities of the biotin in CNDM, the clear CNDM sample prepared as described above is split into two parts to be treated, respectively, with the plain and streptavidin–agarose gels. To two Eppendorf tubes are added 0.50 ml of each of the agarose gels. After a brief centrifugation (2000 rpm, 2 min, 26°) to remove the free aqueous content of the gels, 0.25 ml of the CNDM sample is added to the tubes. Following a 30-min incubation at room temperature under constant shaking in a shaker, the tubes are briefly centrifuged (2000 rpm, 2 min, 26°). The supernatants are collected as the final two types of CNDM sample for biotin assay in the competitive agglutination system described in this work. Essentially no volume dilution occurs in the CNDM samples during this treatment step.

## Spectrophotometric Assay of Biotin

This assay is based on the ability of biotin to disrupt the HABA–avidin complex as monitored by absorbance around 500 nm.[8] At first, a stock solution of HABA–avidin complex is prepared by adding 10 m$M$ HABA solution (in 10 m$M$ NaOH) to a 0.50-mg/ml avidin solution [in 50 m$M$ phosphate buffer (pH 6.0) containing 0.90% (w/v) NaCl] for a final HABA concentration of 0.24 m$M$. The assay procedure is modified for the microtiter plate setting. To the microtiter plate is added 0–20 $\mu$l of biotin standard at a concentration of 0.50 m$M$ in distilled water. After adjusting the volume to 20 $\mu$l with distilled water, 180 $\mu$l of the HABA–avidin complex solution is added to the plate. After a 10-min incubation at room temperature with constant shaking in a shaker, the plate is read in the microtiter plate reader at 490 nm. This absorbance reading versus the biotin amount forms a standard curve.

## Competitive Agglutination Assay

As both avidin and streptavidin have four binding sites for biotin, they can cross-link the BCLPs. To the microtiter plate are added 100 $\mu$l of the

[8] N. M. Green, *Biochem. J.* **94**, 23c (1965).

BCLPs at a final concentration of about 0.20 mg/ml (or absorbance reading at 340 nm of about 0.96) and 100 $\mu$l of avidin or streptavidin solutions at various concentrations in the assay buffer. The agglutination reaction is allowed to proceed for 2.0 hr at room temperature under constant shaking in the shaker. The agglutination of the particles is read in the microtiter plate reader as a reduction in turbidity or absorbance at 340 nm relative to the reading of a pure BCLP dispersion at the same concentration.

The concentration of avidin or streptavidin that induces a maximal agglutination of the fixed BCLP concentration is chosen as an optimum for the competitive assay that is based on disruption of the agglutination by the biotin to be assayed. To the microtiter plate wells are added biotin standard at various amounts in the assay buffer and the CNDM samples. After adjusting the volume to 75 $\mu$l with assay buffer, 75 $\mu$l of theBCLPs in the assay buffer is added to the plate for a fixed final concentration of 0.20 mg/ml. The BCLP agglutination in the presence of biotin is initiated by adding 50 $\mu$l of the avidin or streptavidin solution to the optimal final concentration in the assay buffer. After a 2.0-hr incubation at room temperature with constant shaking in the shaker, the plate is read in the microtiter plate reader for absorbance at 340 nm to measure the agglutination reaction.

## Theoretical Aspect of the Saturating Competition Assay

In this work we deal with a saturating competition assay system. This system consists of each binding component with a concentration being much greater than its dissociation constant, and the labeled and competing (assayed) analytes with a total amount being much greater than the number of the total binding sites of their detecting agent. Under such condition, the mass action law states that the amount of the complex of the competing analyte and detecting agent, $C$, should be proportional to its fractional binding activity:

$$C = D(K_c C_0)/(K_c C_0 + K_1 L_0) \tag{1}$$

where $D$, $C_0$, $L_0$, $K_c$, and $K_1$ are concentrations of the detecting agent, competing and labeled analytes, and affinities of the competing and labeled analytes for the detecting agent, respectively. It should be mentioned that Eq. (1) is also applicable to a mixture of analyte analogs with different affinities for the detecting agent, if the binding activity of the competing analyte, $K_c C_0$, is replaced by the summation of the binding activity of each competing analyte. Thus, a measurable signal that is monotonically related to the extent of the competition, $S$, should be determined by a function of

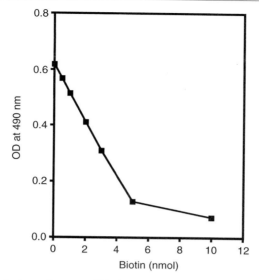

FIG. 1. Spectrophotometric assay of biotin with a fixed amount of 1.3 nmol of avidin. All of the data are an averaged result of four replicated experiments with coefficient variations less than 5.0%.

the concentration of the competing complex, or the binding activity of the competing analyte relative to that of the labeled analyte:

$$S = f(C) = f[D/(1 + K_1 L_0/K_c C_0)]  \qquad (2)$$

Results

Figure 1 gives the results of the HABA-based assay for biotin. The 490-nm absorbance measurements show a 4:1 HABA-displacing or binding stoichiometry of biotin to avidin. The lower detection limit of biotin is about 0.50 nmol, or 2.5 $\mu M$ in concentration, with the microtiter plate format (Fig. 1). This detection sensitivity (2.5 $\mu M$ in concentration) agrees with that obtained in a spectrophotometer, if one considers the extinction coefficient of the HABA–avidin complex of about 20,000 $M^{-1}$ cm$^{-1}$ at 490 nm and a minimally detectable absorbance change of 0.05.[8,9]

The presence of avidin indeed changes the turbidity of the BCLP disper-

[9] N. M. Green, *Methods Enzymol.* **18**, 418 (1970).

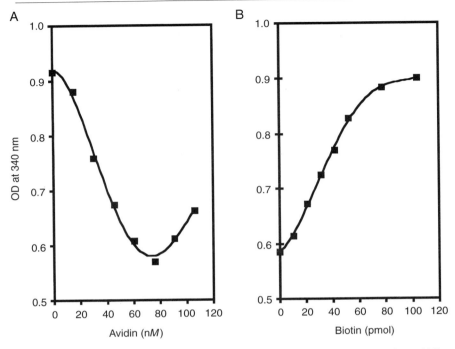

Fig. 2. (A) Agglutination of BCLP (0.20 mg/ml) by avidin as measured by the turbidity change at 340 nm. (B) Disruption of biotin standard on the maximal agglutination of BCLP (0.20 mg/ml) caused by 74 nM avidin. All of the data are an averaged result of four replicated experiments with coefficient variations less than 10%.

sion, indicating a particle aggregation process (Fig. 2A). The possibility that the particle aggregation caused by avidin results from the nonspecific interactions between avidin and BCLPs is ruled out by the following observations. First of all, the BSA concentration in the assay buffer is sufficiently high to prevent nonspecific binding of avidin to BCLPs.[10] Second, addition of biotin blocks the turbidity decrease of BCLPs induced by avidin in a stoichiometric manner (Fig. 2B). It must be emphasized that the constant shaking is important for completing the agglutination reaction, because the relatively large particles need to collide to cross-link with each other.

A maximal degree of cross-linking of BCLPs by avidin corresponds to a condition under which the avidin binds to a certain fraction of the biotin coated on BCLPs. If the amount of avidin added is too small, the avidin-

[10] M. A. Cohen-Stuart, J. M. Scheutjens, and G. J. Fleer, *J. Polymer Sci. Polymer Phys. Ed.* **18,** 559 (1980).

bound density on the BCLP surface is insufficient for effective cross-linking with the other BCLPs. However, when too much of the biotin on the BCLP surface is bound by avidin in large amounts, the cross-linking reaction between the particles is also reduced because of the decrease in unbound biotin on BCLPs that is required for the cross-linking. The previously described agglutination mechanism is confirmed in Fig. 2A. Thus the optimal avidin concentration for a maximal agglutination of 0.20 mg of BCLP per milliliter is determined to be 74 n$M$. This concentration implies that an average coating molecular density of avidin of $2.2 \times 10^5$ for each particle is required for maximal agglutination, if a weight density of 1.0 g/cm$^3$ is assumed for BCLPs.

The disruption of the optimized BCLP agglutination system (with 74 n$M$ avidin) by the competitive binding of free biotin to the avidin was further observed as a recovery in turbidity of the agglutinated particles (Fig. 2B). As the disruption curve of biotin standard shown in Fig. 2B is quantitatively measured by the absorbance at 340 nm, it can serve as a standard curve to assay an unknown biotin sample. This avidin-based agglutination and disruption system responds to biotin standard ranging from 10 to 120 pmol with a lower detection limit 50-fold less than that based on the HABA method (Figs. 1 and 2). It should be noted that this biotin disruption derives from a saturating competition condition, because the competition between the free biotin and coated biotin occurs with the concentrations of biotin and avidin being much larger than the avidin–biotin dissociation constant of 1.0 f$M$.[11]

Similarly, streptavidin causes the agglutination of the same BCLP dispersion with an optimal concentration identical to that of avidin, 74 n$M$ (Fig. 3A). The streptavidin-induced BCLP agglutination can also be disrupted by free biotin (Fig. 3B). However, biotin standard disrupts the BCLP agglutination by streptavidin more effectively than that by avidin (Figs. 2B and 3B). According to Eq. (2), the competitive disruption of the agglutinated BCLPs is determined by the binding activity of the competing biotin relative to the particle-coated biotin (i.e., labeled analyte). Therefore, the relative binding activity of biotin standard to avidin corresponding to half-disruption of the BCLP agglutination should be equal to that to streptavidin. Comparing the amount of biotin standard for the half-disruption of the agglutination induced by avidin (32 pmol) to that by streptavidin (8.0 pmol) leads one to conclude that the affinity of biotin standard for streptavidin relative to the particle-coated biotin is threefold higher than for avidin. This relatively higher affinity of free biotin for streptavidin is consistent with reports

[11] N. M. Green, *Biochem. J.* **89**, 585 (1963).

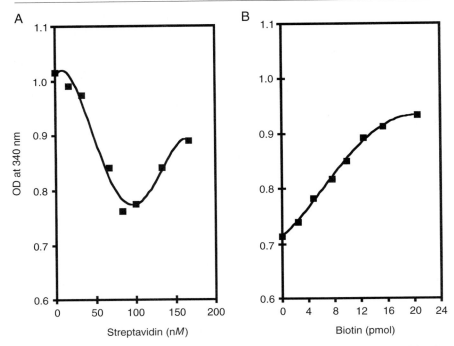

F<small>IG</small>. 3. (A) Agglutination of BCLP (0.20 mg/ml) by streptavidin as measured by the turbidity change at 340 nm. (B) Disruption of biotin standard on the maximal agglutination of BCLP (0.20 mg/ml) caused by 74 n$M$ streptavidin. All of the data are an averaged result of four replicated experiments with coefficient variations less than 10%.

that reveal a largely reduced affinity of stereo-hindered biotin analogs for streptavidin, likely owing to the depth of its biotin-binding sites.[12–14]

Table I summarizes the results of the assays of the biotin content in CNDM in the saturating competitive agglutination systems. The biotin in CNDM consists of at least two types of analogs according to their ability and inability to bind to streptavidin, since the CNDM sample treated with the streptavidin–agarose still disrupts the avidin-induced agglutination (Table I). As the properties of each biotin analog in CNDM are unknown, an arbitrary biotin standard is necessary to quantitate all the analogs as a whole. In this work, an equivalent of biotin standard that causes a binding

[12] G. Gitlin, I. Khait, E. A. Bayer, M. Wilchek, and K. A. Muszkat, *Biochem. J.* **259,** 493 (1989).
[13] W. A. Hendrickson, A. Pähler, J. L. Smith, Y. Satow, E. A. Merritt, and R. P. Phizackerley, *Proc. Natl. Acad. Sci. U.S.A.* **86,** 2190 (1989).
[14] K. Fujita and J. Silver, *BioTechniques* **14,** 608 (1993).

TABLE I
ANALYSIS OF CARNATION NONFAT DRY MILK BIOTIN CONTENT IN COMPETITIVE BIOTIN-
COATED LATEX PARTICLE AGGLUTINATION SYSTEM[a]

| Binder[b] | Sample ($\mu$l)[c] | Absorbance reading at 340 nm[d] | Uncorrected biotin content (nmol/g)[e] | Recovery (%)[f] | Corrected biotin content (nmol/g)[g] |
|---|---|---|---|---|---|
| Avidin | 20 | 0.686 | 3.6 | 106 | 3.4 |
| | 50 | 0.753 | 2.3 | 73 | 3.2 |
| | 60 | 0.767 | 2.0 | 60 | 3.3 |
| | 60[h] | 0.648 | 1.0 | 60 | 1.7 |
| Streptavidin | 75 | 0.863 | 0.42 | 100 | 0.42 |
| | 75[h] | 0.717 | 0.00 | 100 | 0.00 |

[a] Results of at least four replicated experiments with coefficient variations less than 10%.
[b] Refers to agglutinating agent used in the assay system, avidin or streptavidin.
[c] CNDM sample volume used. The solution sample is prepared by the acid precipitation and plain agarose treatment unless specified otherwise, and contains 0.33 mg dry weight of CNDM per microliter.
[d] Reading of the competitive agglutination assay system in the presence of CNDM sample after subtracting the sample background absorbance at 340 nm ($<0.0010/\mu$l of sample volume).
[e] Equivalent of biotin standard per gram CNDM obtained by referring the absorbance reading of the sample agglutination system to its corresponding standard curve.
[f] Ratio of the difference in assay results between CNDM samples $\pm$ 15 pmol of biotin standard by the same corresponding standard curve to 15 pmol.
[g] Equivalent of biotin standard per gram CNDM after correction by the corresponding recovery rate.
[h] CNDM biotin sample treated with streptavidin–agarose gel.

activity to avidin or streptavidin equal to a CNDM sample is used as a measure of the total unknown biotin content. This binding activity is reflected by disruption of the BCLP agglutination that is in turn read as absorbance at 340 nm. In essence, the biotin standard equivalent of a CNDM sample is determined by referring the absorbance reading related to the disrupting action of a sample to its corresponding biotin standard curve. According to Eq. (2), the biotin standard equivalent, $C_{0,\text{standard}}$, may be expressed as

$$C_{0,\text{standard}} = [(K'_{c,\text{sample}}/K'_1)/(K_{c,\text{standard}}/K_1)]C_{0,\text{sample}} \qquad (3)$$

where $C_{0,\text{sample}}$ and $(K'_{c,\text{sample}}/K'_1)/(K_{c,\text{standard}}/K_1)$ are the true biotin content in CNDM and the ratio of the binding affinities (relative to the coated biotin) of the sample biotin in the sample medium and biotin standard in the standard medium. Equation (3) shows the linear relationship between the equivalent of the standard and the quantity of the sample analytes in a given saturating competitive assay system, which forms the basis of the standard referral method.

The results of biotin assay with several CNDM sample volumes by the avidin-based biotin standard curve of Fig. 2B are presented in column 4 of Table I. Owing to the CNDM sample medium effect that varies with the sample amount, the uncorrected biotin content determined by the direct referral to the standard curve is dependent on the sample volume used. A calibration of such sample medium effect by measuring a CNDM sample free of the interfering factors is impossible. However, it should be reasonable to assume that the interfering factors in the CNDM sample affect the biotin analytes and biotin standard to the same extent, or that

$$K'_{c,\text{sample}}/K_{c,\text{sample}} = K'_{c,\text{standard}}/K_{c,\text{standard}} \tag{4}$$

Combining Eqs. (3) and (4) yields

$$C_{0,\text{standard}}/[(K'_{c,\text{standard}}/K'_1)/(K_{c,\text{standard}}/K_1)] = (K_{c,\text{sample}}/K_{c,\text{standard}})C_{0,\text{sample}} \tag{5}$$

where $(K'_{c,\text{standard}}/K'_1)/(K_{c,\text{standard}}/K_1)$ is the ratio of the relative affinities of biotin standard in the sample and standard media, respectively. This ratio can be determined by a recovery test in which the assayed value of biotin standard spiked in a CNDM sample as determined by referring to one of the biotin standard curves compares with the true value. According to Eq. (5), division of the uncorrected result of CNDM biotin assay by the corresponding recovery rate gives a corrected value of the true biotin content that is related only to an affinity ratio under the standard condition, $K_{c,\text{sample}}/K_{c,\text{standard}}$. The recovery rates of 15 pmol of biotin standard added to 20-, 50-, and 60-$\mu$l CNDM samples in the avidin-based agglutination assay system are determined to be 106, 73, and 60%, respectively (column 5 of Table I). These recovery rates suggest that the unknown molecules in CNDM could nonspecifically interact with avidin and thus reduce the binding of avidin to free biotin.[15,16] The corrected results about the avidin-bindable biotin analogs give a consistent number averaged at 3.3 nmol of biotin per gram of CNDM (column 6 of Table I). The consistency of these corrected results also confirms the linear additive nature of the saturating competition assay in relation to the affinity and concentration heterogeneities of the competing analytes, as follows from Eqs. (2) and (3).

The biotin per gram of CNDM is, however, determined as an equivalent of 0.42 nmol of biotin standard in the streptavidin-based agglutination

[15] R. C. Bruch and H. B. White III, *Biochemistry* **21**, 5334 (1982).
[16] N. M. Green, *Adv. Protein Chem.* **29**, 85 (1975).

competitive assay system (Table I). Again, this is done by referring the disrupting action of the sample to the corresponding biotin standard curve through the absorbance measurement at 340 nm (Table I, Fig. 3B). As seen in Table I, the recovery rate of 15 pmol of biotin standard spiked in 75 $\mu$l of CNDM sample is about 100%, indicating essentially no effect of the sample medium on the biotin-binding activity of this nonglycoprotein, streptavidin, in contrast to avidin.[16]

Natural biotin usually occurs in several metabolic forms that may differ in affinity for avidin and streptavidin.[16,17] The CNDM sample treated with the streptavidin–agarose is still able to disrupt the avidin-induced agglutination, with an avidin-binding activity equivalent to 1.7 nmol of biotin standard per gram CNDM (Table I). This means that the avidin-bindable biotin analogs in CNDM can be divided into two types, one (52%) able to bind to streptavidin and the other (48%) unable to do so. As a control, the streptavidin–agarose-treated CNDM sample shows no binding activity in the streptavidin agglutination assay system (Table I). Applying Eq. (5) concludes that the total biotin content and its streptavidin-bindable portion in CNDM bind to avidin with respective activities seven- and threefold higher than streptavidin. This could again be explained by the reduced accessibility of the CNDM biotin to the deep binding sites of streptavidin.[12-14] Table I highlights the fundamental role of a detecting agent in determining the absolute assay value in a saturating competitive assay system.

Discussion

The good sensitivity of the latex-based agglutination assay results from the amplification of the specific assay reaction by the light scattering of the particles. This is supported by the signal efficiencies of the conventional absorbance method and the agglutination method described in this work. Figure 1 shows that the HABA assay system has an absorbance signal coefficient to the biotin–avidin binding of 23,000 $M^{-1}$, as the saturation of 26 $\mu M$ biotin-binding sites in avidin causes an absorbance change of about 0.6. However, the saturation of 0.29 $\mu M$ biotin-binding sites in avidin produces an absorbance response of 0.35 through particle scattering, yielding a signal coefficient of 1,200,000 $M^{-1}$ (Fig. 2A). This 52-fold higher signal coefficient of the latex system explains the 50-fold higher assay sensitivity of biotin by the competitive agglutination assay than that based on the HABA assay (Figs. 1 and 2B). The HABA method cannot be used to determine the relatively low CNDM biotin content at all. The agglutination

[17] D. M. Mock, *Methods Enzymol.* **184**, 224 (1990).

assay of biotin described in this work also holds promise of being integrated into an automatic system because of its homogeneous nature.

On the basis of the mass action law, we have proposed the use of a saturating competitive assay to overcome the potential quantitation inconsistency caused by the different affinities of analyte analogs for their detecting agent. If a competitive assay of these heterogeneous analytes is done under nonsaturating conditions, in which the concentrations of the analytes and detecting agent are less than their dissociation constants, the result will vary depending the fraction of each analyte.[6] As shown in Eq. (2), the saturating condition produces an assay result that is a linear summation of the heterogeneous binding or competing activity of each analyte regardless of its relative abundance.

It should also be noted that a different detecting agent used in a competition assay yields a different value for a given analyte because the result depends on the affinity of the analyte for the detecting agent relative to the labeled analyte. This accounts for the different values of the CNDM biotin determined in the avidin and streptavidin agglutination systems in this work (Table I). Therefore, caution must be taken when one is to compare assay results obtained with different detecting agents. For example, the biotin content per gram of whole dry milk is determined to be 0.22 $\mu$g of biotin standard equivalent by the bioassay method, in opposition to 0.51 $\mu$g by the avidin-based agglutination assay in this work (on calibration for the fat content).[18] A higher value of biotin in human serum, obtained by an avidin-based competition method, than that obtained by a bioassay method is also reported.[19] It is possible that the bacterial reactor utilizes biotin standard and milk and serum biotin analogs with relative affinities or activities different from those of avidin. Unfortunately, it is impossible to resolve and compare the affinity spectra of the milk and serum biotin analogs to the bioreactor and avidin based on the assay results alone.

[18] R. Jenness, in "Lactation: A Comprehensive Treatise" (B. L. Larson and V. R. Smith, eds.), Vol. III, p. 3. Academic Press, New York, 1974.
[19] R. P. Bhullar, S. H. Lie, and K. Dakshinamurti, in "Biotin" (K. Dakshinamurti and H. N. Bhagavan, eds.), p. 122. The New York Academy of Sciences, New York, 1985.

# [34] Competitive Enzyme-Linked Immunosorbent Assay for Biotin

*By* DAVID SHIUAN, CHWEN-HUEY WU, YUO-SHENG CHANG, and RE-JIIN CHANG

Several methods for the determination of biotin concentration exist for research and clinical diagnosis. Typically, biotin concentration has been determined by the microbiological assay methods based on biotin-requiring strains such as *Lactobacillus arabinosus* and *Amphidinium carterae*[1,2] with a sensitivity of 5–10 pg/ml. More advanced techniques such as the reciprocal enzyme assay,[3] spectrophotometric assay,[4] fluorescent polarization method,[5] and the isotope dilution method[6] have been employed for the determinations of biotin concentrations. However, these methods are either rapid but less sensitive or have high sensitivity but require more complicated and time-consuming protocols. Therefore, we have developed a competitive enzyme-linked immunosorbent assay (ELISA) method for the rapid and sensitive determination of biotin concentrations. Our method is based on the measurement of residual horseradish peroxidase [conjugated to streptavidin, streptavidin–horseradish peroxidase (HRP)] activities after streptavidin–HRP has reacted with biotin in sample solutions. The detection limit is 1 pg/ml in simple aqueous media and 5 pg/ml in bacterial growth media.

## Principle

The strategy of the competitive ELISA method for the determination of biotin concentration is based on the competition between streptavidin-conjugated horseradish peroxidase (streptavidin–HRP, in excess amount) in solutions of variable biotin concentrations. The residual free form of streptavidin–HRP was immobilized with biotinylated goat anti-rabbit immunoglobulin (IgG) precoated on the wells of microtiter plates, and finally, the peroxidase activities were assayed and correlated with the biotin concen-

[1] L. D. Wright and H. R. Skeggs, *Proc. Soc. Exp. Biol. Med.* **56,** 95 (1994).
[2] A. F. Carucci, *Methods Enzymol.* **18A,** 379 (1970).
[3] E. A. Bayer, H. Ben-Hur, and M. Wilchek, *Anal. Biochem.* **154,** 364 (1986).
[4] R. S. Neidbala, F. Gergits, and K. J. Schray, *J. Biochem. Biophys. Methods* **13,** 205 (1986).
[5] K. J. Schray and P. G. Artz, *Anal. Biochem.* **60,** 853 (1988).
[6] S. A. Yankofsky, R. Gurevitch, A. Niv, G. Cohen, and L. Goldstein, *Anal. Biochem.* **118,** 307 (1982).

trations. When more biotin molecules were present initially in the competition reactions, less free streptavidin–HRP would be available to react with the biotinylated IgG. Biotin concentrations could be estimated from the calibration curves relating standard biotin concentrations and the horseradish peroxidase activities.

## Materials and Methods

Biotin and thimerosal (ethylmercurithiosalicylic acid, sodium salt) are purchased from Sigma (St. Louis, MO). Biotinylated goat anti-rabbit IgG, streptavidin–HRP (streptavidin-conjugated horseradish peroxidase), and ABT [2,2-azinodi(3-ethylbenzthiazoline)sulfonic acid] are obtained from Zymed Laboratories (San Francisco, CA). Biotin-free casein, Tween 20, and most other chemicals are obtained from Serva (Heidelberg, Germany). The 0.5% (w/v) boiled casein solution is prepared by dissolving 5 g of casein in 100 ml of 0.1 $N$ NaOH and 900 ml of phosphate-buffered saline (PBS) (see below) and adjusting to pH 7.4 after cooling to room temperature. The solution is filtered (0.45-$\mu$m pore size) after the addition of 0.1 g of thimerosal. Phosphate-buffered saline contains 8 g of NaCl, 1.15 g of $Na_2HPO_4$, 0.2 g of $KH_2PO_4$, and 0.2 g of KCl per liter at pH 7.4. The 1% tryptone (w/v) broth is prepared by the addition of 5.0 g of NaCl and tryptone (Difco, Detroit, MI) to 10 g/liter. After autoclaving for 20 min, 10% (w/v) thiamin hydrochloride is added to a final concentration of 0.1% (v/v).

*Mycoplasm hyopneumoniae* total protein in prepared from *Mycoplasma hyopneumoniae* 232[7] cultured in Friis broth medium supplemented with 20% (v/v) porcine serum, to stationary phase, concentrated 100 times, and resuspended in Hanks' balanced salt solution (GIBCO, Grand Island, NY) with 0.15% (v/v) formalin. Rabbit anti-*M. hyopneumoniae* antiserum is prepared and purified as described previously.[8]

### Assay of Biotin Concentration in Simple Aqueous Medium

Biotinylated goat anti-rabbit IgG is diluted 1:5000 in PBS, and 100-$\mu$l aliquots are pipetted into the wells of Nunc (Rochester, NY) Immuno microtiter plates. After incubation at room temperature for 2 hr (the coated plate can be kept at 4° for up to 1 week), the unbound IgG is removed by washing (with PBS solution), and 0.5% (v/v) boiled biotin-free casein (150 $\mu$l) is added to each well to block the remaining available sites. After a

[7] T. F. Young and R. F. Ross, *Am. J. Vet. Res.* **48**, 651 (1987).

[8] L. H. Ro, R. J. Chen, and D. Shiuan, *J. Biochem. Biophys. Methods* **28**, 155 (1994).

1-hr incubation at room temperature the excess casein solution is removed and the wells are washed twice with PBS (150 $\mu$l/well). Streptavidin–HRP is diluted with 0.025% (v/v) Tween 20 (1:20,000 or 1:40,000 dilution), mixed with equal volumes of standard biotin solutions (2 ng/ml, and 500, 100, 25, 5, and 1 pg/ml, prepared in aqueous medium by serial dilutions of the stock solution, 100 $\mu$g/ml), loaded onto the microtiter plates, and incubated at 37° for 30 min to perform the competitive ELISA experiment. After the wells are washed twice with PBS, the substrate ABTS [0.5 mg/ml in 0.1 $M$ citrate buffer (pH 4.2) and 0.03% (v/v) $H_2O_2$] is added (100 $\mu$l/well) and the plates are incubated for 30 min at room temperature. Color development is read at $OD_{405\,nm}$ in an ELISA reader (Titertek Multiskan Plus MKII; Labsystem & Flow Laboratories, Finland). Each experiment is performed at least three times and the standard deviations of these measurements are estimated to be ±10%.

As shown in Fig. 1, the 1:40,000 dilution of streptavidin–HRP is more effective for determinations of lower biotin concentrations (with a detection limit of approximately 1 pg/ml) while the 1:20,000 dilution is slightly better for measurements of biotin concentration near the nanogram per milliliter range. It is likely that with more refinements the detection limits should improve even further.

### Modified Competitive ELISA Method for Determination of Biotin Concentration in Complex Medium

Nonspecific binding of various proteins may interfere with the ELISA method and result in unstable readings in complex media such as bacterial growth media. Applying a principle similar to the sandwich assay,[9] this interference may be circumvented by the incorporation of an extra antibody–antigen interaction step into the protocol. We use the rabbit anti-*M. hyopneumoniae* antiserum and the *M. hyopneumoniae* protein currently available in this laboratory to examine their effects on stabilizing the ELISA readings. After the initial reactions of the sample solutions, the streptavidin–HRP is reacted with the biotinylated goat anti-rabbit IgG, which is linked to rabbit anti-*M. hyopneumoniae* antiserum complexed with the precoated *Mycoplasma* protein coated on the wells of the microtiter plates.

Experimentally, the total protein of *M. hyopneumoniae* is diluted with PBS (1:20) to approximately 10$\mu$g/ml and incubated at room temperature in the microtiter plates (100 $\mu$l/well) for 2 hr. After rinsing the excess unbound protein solution, the plate wells are blocked with 0.5% (v/v) boiled casein solution (150 $\mu$l/well) at room temperature for 1 hr.

[9] E. Harlo and D. Lane, (eds.), "Antibodies: A Laboratory Manual," Ch. 14. Cold Spring Harbor Laboratory Press, Cold Spring Harbor, New York, 1988.

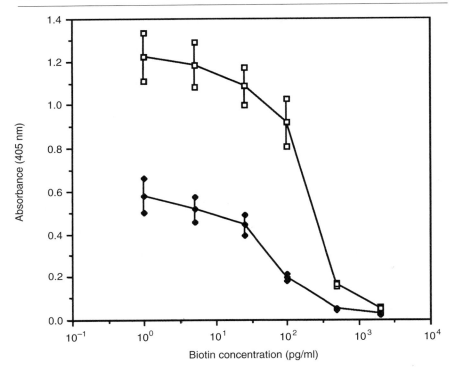

FIG. 1. The optimized curve for biotin concentration determinations. The streptavidin–HRP was incubated with the sample solution at 37° for 30 min. The curves with open and closed squares represent the dilutions of streptavidin–HRP with 0.025% (v/v) Tween 20 at 1:20,000 and 1:40,000 ratios, respectively. [Reprinted from *J. Biochem. Biophys. Methods* **29**, Y.-S. Chang, C.-H. Wu, R.-J. Chang, and D. Shiuan, Determination of biotin concentration by a competitive enzyme-linked immunosorbent assay (ELISA) method, 321, Copyright 1994 with kind permission of Elsevier Science–NL, Sara Burgerhartstraat 25, 1055 KV Amsterdam, The Netherlands.]

Rabbit anti-*M. hyopneumoniae* antiserum[8] diluted 1:6000 with a solution containing 0.025% (v/v) Tween 20 and 0.5% (v/v) boiled casein is added (150 μl/well) and incubated for 1 hr at 37°. The excess solution is removed and the wells are washed with 0.5% (v/v) boiled casein solution (150 μl/well) three times before the addition to 100 μl/well, of biotinylated goat anti-rabbit IgG diluted with PBS (1:10,000). After incubation at 37° for 1 hr, the solutions are removed and the wells washed three times with 0.5% (v/v) boiled casein (150 μl/well). The competitive ELISA procedure is then followed as described for the assays of biotin in simple aqueous media.

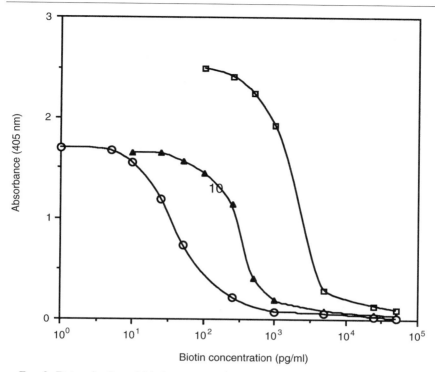

FIG. 2. Determination of biotin concentrations with the modified competitive ELISA method. Curves with open squares, closed triangles, and open circles represent the range of biotin concentration determination corresponding to the dilution factor of streptavidin–HRP: 1 : 50,000, 1 : 100,000, and 1 : 150,000. Samples with high absorbance (with a streptavidin–HRP dilution of 1 : 50,000) were measured after appropriate dilution. [Reprinted from *J. Biochem. Biophys. Methods* **29,** Y.-S. Chang, C.-H. Wu, R.-J. Chang, and D. Shiuan, Determination of biotin concentration by a competitive enzyme-linked immunosorbent assay (ELISA) method, 321, Copyright 1994 with kind permission of Elsevier Science–NL, Sara Burgerhartstraat 25, 1055 KV Amsterdam, The Netherlands.]

As shown in Fig. 2, this additional antibody–antigen step stabilizes the biotin concentration assay in 1% (v/v) tryptone broth: in the range of 50 ng/ml to 500 pg/ml, 5 ng/ml to 25 pg/ml, and 1 ng/ml to 5 pg/ml with the dilution factors of streptavidin–HRP of 1 : 50,000, 1 : 100,000, and 1 : 150,000, respectively. This improvement could be due to the specific interaction between the biotinylated goat anti-rabbit IgG and the rabbit anti-*M. hyopneumoniae* antiserum, as well as the thorough blocking step using the 0.5% (v/v) boiled casein solution.

Discussion

Accurate measurements of biotin concentrations by the competitive ELISA method depend partly on the optimization of the dilution factors of the steptavidin–HRP solution, incubation time and temperature for the competition reaction, coating and specificity of the biotin-labeled goat anti-rabbit IgG, and the blocking step to minimize nonspecific binding of streptavidin–HRP after coating the biotinylated goat anti-rabbit IgG onto the wells of the microtiter plates. Slightly more streptavidin–HRP molecules remained in the wells after blocking with the 0.5% (v/v) boiled casein. The results suggested that the biotinylated goat anti-rabbit IgG may cross-react with the 0.5% (v/v) boiled casein (a mixture of related phosphoproteins existing in milk) and may be more stable after washing. Similarly, the effects of incubation temperature on the competition interaction were examined at 37° and 25° (room temperature) for 30 min (or 1 hr). It appears that blocking and incubation at 37° for 30 min provide slightly better conditions for the current method.

In conclusion, the competitive ELISA method measures biotin concentrations accurately and efficiently. The biotin-containing solution was reacted with excess streptavidin–HRP. The remaining free streptavidin–HRP was immobilized by biotin-labeled antibodies precoated on the microtiter plates. The peroxidase activities were measured and related to the initial biotin concentrations present in solution. The incorporation of the extra antibody–antigen interaction step minimized interference by various proteins when biotin was assayed in more complex media. The major advantages of this method are its speed and high sensitivity compared to other biotin assay procedures.[3–6]

The entire assay can be completed within 6 hr, with detection limits of approximately 1 pg/ml for biotin in simple aqueous solutions and of 5 pg/ml in complex media. Furthermore, the enzyme-conjugated streptavidin, biotin-labeled antibodies, and the antibody–antigen pairs used in the present study can be substituted by any antibody–antigen pair available in the laboratory. Enzymes such as alkaline phosphatase and $\beta$-galactosidase can also be conjugated to streptavidin. The present method should be applicable to the efficient determination of biotin concentrations in simple aqueous solutions, bacterial growth media, and, in principle, to clinical samples.

Acknowledgments

This work was supported in part by Grant NSC82-0203-B110-024 from the National Science Council (ROC). We thank Dr. Larry Steinrauf of the Department of Biology, National Sun Yat-Sen University, for valuable suggestions.

# [35] Biotin Synthesis in Higher Plants

*By* Pierre Baldet, Claude Alban, and Roland Douce

Biotin (vitamin H) has been known for about 30 years to be the prosthetic group of four major carboxylases in higher organisms.[1] The importance of biotin in plants has been illustrated by the discovery of a mutation that causes defective embryo development in *Arabidopsis thaliana* and requires biotin at a critical stage of embryogenesis.[2] Investigations have established that plant cells contain the four biotin-dependent activities found in mammals,[3] among which acetyl-CoA carboxylase is the most studied, because it constitutes the target of powerful herbicides.[4]

Biotin is synthesized by many microorganisms[5] and plants.[6] Its biosynthesis has been widely investigated in bacteria[5,7–9] especially in a biotin mutant of *Escherichia coli* K12 and *Bacillus sphaericus*. In contrast, our knowledge of the metabolism of biotin in higher plants is rather meagre. In this chapter, we describe the methods used to determine the localization of free and protein-bound biotin in cells from green pea leaves, using isolated intact protoplasts. The characterization of intermediates of biotin biosynthesis from a stable biotin-overproducing cell line of lavender cells is also described.

## Isolation and Gentle Rupture of Intact Protoplasts

### Plant Material

Improvements in the methods available for protoplast preparation have made possible the use of a large number of plant species. However, a high

[1] F. Lynen, *Biochem. J.* **102,** 380 (1967).

[2] J. Shellhammer and D. Meinke, *Plant Physiol.* **93,** 1161 (1990).

[3] E. S. Wurtele and B. J. Nikolau, *Arch. Biochem. Biophys.* **278,** 179 (1990).

[4] J. L. Harwood, *Annu. Rev. Plant Physiol. Plant Mol. Biol.* **39,** 101 (1988).

[5] M. A. Eisenberg, *in* "*Escherichia coli* and *Salmonella typhimurium:* Cellular and Molecular Biology" (F. C. Neidhardt, J. L. Ingraham, K. B. Low, B. Magasanik, M. Schaechter, and M. E. Umbarger, eds.), p. 544. American Society for Microbiology, Washington, DC, 1987.

[6] P. Baldet, H. Gerbling, S. Axiotis, and R. Douce, *Eur. J. Biochem.* **217,** 479 (1994).

[7] Y. Izumi, Y. Kano, K. Inagaki, N. Kawase, Y. Tani, and H. Yamada, *Agric. Biol. Chem.* **45,** 1983 (1981).

[8] I. Ohsawa, D. Speck, T. Kisou, K. Hayakawa, M. Zinsius, R. Gloeckler, Y. Lemoine, and K. Kamogawa, *Gene* **80,** 39 (1989).

[9] O. Ifuku, N. Koga, S. Haze, J. Kishimoto, and Y. Wachi, *Eur. J. Biochem.* **224,** 173 (1994).

yield of protoplasts is more easily obtained from peas. Pea (*Pisum sativum* L. 'Douce Provence') plants are grown from seeds in soil for 8 to 12 days under a 12-hr photoperiod of white light from fluorescent tubes [10–40 $\mu$E (Einstein)/m$^2$] at 18°.

## Preparation of Intact and Purified Mesophyll Protoplasts

Intact protoplasts are isolated as previously described by this laboratory.[10] The use of a combination of isosmotic layers is highly suitable for the rapid and effective purification of intact mesophyll protoplasts. Under these conditions, the protoplast membrane can be disrupted with little damage to subcellular structures, allowing the isolation and purification of the cytosolic compartment, chloroplasts, mitochondria, and vacuoles.

### Solutions

Isolation medium: Sorbitol, 0.5 *M*; CaCl$_2$, 1 m*M*; polyvinylpyrrolidone (PVP 25), 0.05% (w/v); morpholineethanesulfonic acid (MES)– NaOH 10 m*M*, pH 5.5

Digestion medium: Same as isolation medium plus cellulase Onozuka R10, 2% (w/v); macerozyme R10 (Yakult Honsha Co., Shingikancho, Nishinomiya, Japan), 0.5% (w/v); pectolyase Y23 (Seishin Pharmaceutical Co., Tokyo, Japan), 0.2% (w/v)

Suspension medium: Sucrose, 0.5 *M*; CaCl$_2$, 1 m*M*; PVP 25, 0.05% (w/v); morpholinepropanesulfonic acid (MOPS)–KOH, 10 m*M*, pH 7.0

Isosmotic medium 1: Sucrose, 0.4 *M*; sorbitol, 0.1 *M*; CaCl$_2$, 1 m*M*; PVP 25, 0.05% (w/v); MOPS–KOH, 10 m*M*, pH 7.0

Isosmotic medium 2: Sorbitol, 0.5 *M*; CaCl$_2$, 1 m*M*; PVP 25, 0.05% (w/v); MOPS–KOH, 10 m*M*, pH 7.0

Washing medium: Same as isosmotic medium 2

Hypotonic medium: EDTA, 1 m*M*; dithiothreitol (DTT), 5 m*M*; phenylmethylsulfonyl fluoride (PMSF), 1 m*M*; $\varepsilon$-aminocaproic acid, 5 m*M*; benzamidine hydrochloride, 1 m*M*; leupeptin, 0.5 $\mu$g/ml; pepstatin, 0.7 $\mu$g/ml; HEPES–NaOH, 50 m*M*, pH 8.0

### Materials

Nylon mesh, 100-$\mu$m aperture (Züricher Beuteltuchfabrik AG, Rüschlikon, Switzerland)

[10] J. E. Lunn, M. Droux, J. Martin, and R. Douce, *Plant Physiol.* **94**, 1345 (1990).

Glass centrifuge tubes (Corex), 15 and 30 ml

Superspeed centrifuge, refrigerated [J2-21 M/E (Beckman, Palo Alto, CA) or equivalent] and the following rotors:

Swinging bucket rotor, 4 × 50 ml (JS 13; Beckman)

*Procedure.* Derib 6- to 8-day-old pea leaves and cut them into fine strips (0.5–1 mm) immersed in isolation medium (6 to 8 g of leaves for 100 ml of isolation medium). Remove this medium and replace it with 100 ml of fresh digestion medium. To improve the infiltration of the digestion medium in the intercellular spaces several vacuum rounds are necessary until sedimentation of the strips is complete. Incubate the suspension at 25° for 2.5 hr. All the subsequent procedures are carried out at 4°. Release protoplasts from the digested tissue by gentle shaking and then separate the protoplasts from the leaf debris by filtering through a 100-$\mu$m nylon mesh. Centrifuge the filtrate at 100 $g$ for 5 min (Beckman J2-21, JS 13 rotor) and discard the supernatant fluid. Resuspend the pelleted protoplasts in a total volume of 20 ml of suspension medium and divide the suspension into four 15-ml glass centrifuge tubes. To each tube, add 2 ml of isosmotic medium 1, and 1 ml of isosmotic medium 2. Centrifuge the three-step gradients at 250 $g$ for 5 min (Beckman J2-21, JS 13 rotor) using low acceleration (value 1 on the selector) and without braking (value 0 on the selector). Collect intact mesophyll protoplasts at the interface between the top two layers, using a Pasteur pipette, and resuspend them in 40 ml of washing medium. Centrifuge at 100 $g$ for 5 min (Beckman J2-21, JS 13 rotor) and repeat the washing. Resuspend the pelleted protoplasts in the washing medium to a chlorophyll concentration of 0.2 mg/ml.

To obtain a total extract, resuspend the pellet of washed protoplasts in hypotonic medium and submit the suspension to three cycles of freeze-thawing to ensure complete lysis. Under these conditions, all the subcellular structures are ruptured. The lysate suspension comprises the total extract of protoplasts.

### Gentle Rupture of Protoplasts and Separation of Organelles from Cytosolic Fraction

Because pea leaf protoplasts from the mesophyll have an average diameter of 20 to 30 $\mu$m, a rapid and effective procedure for the gentle rupture of intact protoplasts (i.e., for stripping the cell membrane) is to pass protoplasts through a 20-$\mu$m fine nylon mesh affixed to the cut end of a 1-ml disposable syringe. Four passes through allow the rupture of more than 90% of protoplasts, which can be checked with a microscope.

*Solutions*

Rupture medium: Same as hypotonic medium (see above)
Washing medium: Same as isosmotic medium 2 (see above)

*Materials*

Nylon mesh, 20-$\mu$m aperture (Züricher Beuteltuchfabrik AG)
Superspeed centrifuge, refrigerated [Heræus Sepatech Biofuge 17S/
     RS (Hanau, Germany), or equivalent] and the following rotor:
Drum rotor, 80 × 1.5 ml (Heræus Sepatech)

*Procedure.* Centrifuge lysed protoplasts in 1-ml fractions at 100 *g* in Eppendorf tubes (Heræus Sepatech Biofuge 17S/RS) for 5 min. Keep the pellet containing most of the chloroplasts on ice, and centrifuge the supernatant at 300 *g* for 5 min. For this second centrifugation, invert the rack containing the tubes in the drum (i.e., top of the tubes opposite to the axis). This allows the pellet of chloroplasts to collect in the lid of the Eppendorf tubes, thus providing a chloroplast-free supernatant on opening the cap. Keep the second pellet, which also consists of chloroplasts on ice, and centrifuge the supernatant at 14,000 *g* for 45 min. The last pellet contains the mitochondria and the supernatant corresponds to cytosol contaminated by the vacuolar compartment. Resuspend the mitochondria pellet (14,000 *g*) and the two chloroplast pellets (100 *g* and 300 *g*) in the rupture medium and mix them together.

*Comments.* To prevent starch grains from tearing through the plastid envelope, the leaves must be harvested 12 hr prior to protoplast extraction and stored in the dark at 4°, and the centrifugal force must be increased gradually.

## Gentle Rupture of Protoplasts for Preparation of Purified Vacuoles

In pea leaves, vacuoles constitute about 90% of the cell and are very fragile organelles. For that reason, it is necessary to develop a suitable method to isolate intact vacuoles. The method used for purification of vacuoles is adapted from that described by Wagner.[11] The use of a combination of density layers has been adapted for the rapid and effective purification of intact vacuoles.

*Solutions*

Suspension medium: Sorbitol, 0.5 *M*; CaCl$_2$, 1 m*M*; PVP 25, 0.05%
     (w/v); MOPS–KOH, 10 m*M*, pH 7.0

---

[11] G. J. Wagner, *Methods Enzymol.* **148,** 55 (1987).

Rupture medium: $KH_2PO_4/K_2HOP_4$, 0.1 $M$, pH 8.0; DTT, 1 m$M$; $MgCl_2$, 1 m$M$; PVP 25, 0.05% (w/v)
Ficoll solutions: Ficoll, 5, 2, or 1% (w/v) in rupture medium

*Materials*

Gyratory shaker (Rota-test 74403, Bioblock, or equivalent)
Glass centrifuge tubes (Corex), 15 ml
Superspeed centrifuge, refrigerated [J2-21 M/E (Beckman) or equivalent], and the following rotor:
Swinging bucket rotor, 4 × 50 ml (JS 13; Beckman)

*Procedure.* Pellet intact protoplasts and resuspend them in 0.4 ml of washing medium equivalent to 20 × $10^6$ cells, and dilute rapidly (2 to 5 sec) in a round-bottom glass tube with 3.6 ml of rupture medium. Stir the suspension gently for 4 min, using the gyratory shaker at 30 rpm, to effect complete emergence of vacuoles from lysing protoplasts. To obtain pure intact vacuoles, prepare four 15-ml glass centrifuge tubes containing a three step-gradient consisting (bottom to top) of 5 ml of 5% Ficoll, 2 ml of 2% Ficoll, and 1.5 ml of 1% Ficoll each in the rupture medium. Layer 1 ml of suspension of lysed protoplasts into each tube. Centrifuge the resulting four-step gradients at 1000 $g$ (swinging bucket rotor) using low acceleration (value 1 on the selector) and without braking (value 0 on the selector) for 30 min at 17°. Collect intact vacuoles at the interface between the 2 and 1% Ficoll layers using a Pasteur pipette and check with a microscope the integrity of vacuoles.

## Distribution of Marker Enzymes

To assess the effectiveness of the gentle rupture of protoplasts, the following markers may be used to estimate possible contamination in the fractions. For cytosol, pyrophosphate–fructose-6-phosphate 1-phosphotransferase[12] (EC 2.7.1.90); for chloroplasts, glyceraldehyde-3-phosphate dehydrogenase $(NADP^+)$[12] (EC 1.2.1.13); for mitochondria, fumarase[13] (EC 4.2.1.2, fumarate hydratase).

## Characterization of Biotin

Among the numerous methods to assay biotin, the most widely used are the biological assay and the assay using the avidin–biotin complex based on isotopic dilution, spectrophotometric, fluorimetric, and immunological

[12] H. Weiner, M. Stitt, and H. W. Heldt, *Biochim. Biophys. Acta* **893,** 13 (1983).
[13] R. L. Hill and R. A. Bradshaw, *Methods Enzymol.* **13,** 91 (1969).

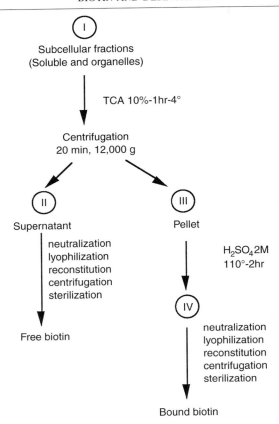

Fig. 1. Procedure for extraction of free and bound biotin from subcellular fractions.

methods (for a review see Wilchek and Bayer[14]). Each system has its merits; however, we use the biological assay (using biotin-auxotroph mutants of *E. coli* K12) for its higher sensitivity.

*Extraction of Biotin*

The biotin extraction protocol is described in Fig. 1. In all cell fractions prepared from pea leaf protoplasts, biotin is extracted according to the following method.

[14] M. Wilchek and E. A. Bayer, *Methods Enzymol.* **184,** 1 (1990).

*Materials*

Set of glass tubes with ground glass join end (acting as primitive condensor, 50-cm length)

Set of Pyrex test tubes with ground glass joints (20 ml)

Electrical heating jacket

Antibumping granules or glass beads, glass wool

Sterile syringe filter unit, 0.2 $\mu$m (DynaGard; Microgon, Inc., Laguna Hills, CA)

Freeze-dryer Flexi-Dry FDX-3-55-MP (FTS Systems, Inc., Stone Ridge, NY)

*Procedure.* Treat each extract (stage I) (Fig. 1) with trichloroacetic acid (10%, w/v) for 1 hr at 4° and centrifuge for 20 min at 12,000 g. All of the free biotin remains soluble in the supernatant fluid (stage II), whereas biotinylated proteins are recovered in the pellet (stage III). Neutralize the supernatant with 10 $M$ NaOH and lyophilize. This fraction is then resuspended in a minimum volume of sterile water (<100 $\mu$l) and contains free biotin. To detach biotin bound to the biotinylated proteins, resuspend the pellet (containing all of the biotinylated proteins) with 2 ml of 2 $M$ $H_2SO_4$. Put the suspension in a 20-ml Pyrex tube containing four or five antibumping granules, attach the condensor tube, and heat for 2 hr at 110° (stage IV). Let the suspension cool down, filter through a 0.2-$\mu$m filter unit, neutralize with 10 $M$ NaOH, and then lyophilize. Resuspend the lyophilized powder in a minimum volume (<400 $\mu$l) of water and centrifuge to discard the bulk of insoluble salts.

*Comments.* Extraction yields are determined by incorporating an internal standard of [$^{14}$C]biotin (85 × 10$^{-12}$ mol, 53 Ci/mol) prior to extraction of free biotin and bound biotin (stage II). The yield of free biotin extraction is 98% whereas that for bound biotin is 80%. The lowest yield obtained for bound biotin results from the trapping of 15–20% of biotin in the bulk of insoluble salts. Furthermore, we must take into account for the calculation that 10–15% of biotin is degradated during the course of drastic acid hydrolysis.[15] Thus, free biotin concentration values determined by the biological assay are unchanged whereas those for bound biotin must be corrected accordingly.

## Biological Assay of Biotin

*Bacterial Strain.* Strain *bioB105*, a single mutant of *E. coli* K12 strain Y-10, is the test organism for this study. The mutation, which blocks the

[15] E. P. Lau, B. C. Cochran, and R. R. Fall, *Arch. Biochem. Biophys.* **205,** 352 (1980).

synthesis of biotin from dethiobiotin, lies on the r strand most proximal to the operator region.[16] Furthermore, a plasmid conferring resistance to carbenicillin has been introduced in this strain.

*Solutions*

LB broth (per liter):[17] Bacto-tryptone, 10 g; yeast extract, 5 g; NaCl, 10 g; supplemented with carbenicillin, 0.1 mg/ml

Minimum medium M9: $Na_2HPO_4$, 49 m$M$; $KH_2PO_4$, 22 m$M$; NaCl, 17 m$M$; $NH_4Cl$, 18 m$M$; glucose, 22 m$M$; $MgSO_4$, 1 m$M$; $CaCl_2$, 0.1 m$M$; supplemented with casein hydrolysate (1 mg/ml; Difco, Detroit, MI) and carbenicillin, 0.1 mg/ml

Phosphate-buffered saline (PBS) buffer: $KH_2PO_4$, 15 m$M$; $Na_2HPO_4$, 8.6 m$M$; KCl, 2.7 m$M$; NaCl, 138 m$M$; supplemented with 7% (v/v) dimethyl sulfoxide (DMSO)

*Procedure.* Grow *E. coli* strain *bioB105* for 5 to 6 hr at 37° in 5 ml of LB until the turbidity reaches 0.4–0.6 at 600 nm. Take 1 ml from the culture, centrifuge at 3000 g for 5 min, wash the bacteria twice with 2 ml of M9 (free of biotin), and resuspend the pellet in 1 ml of M9. Take 250 $\mu$l of the suspension to inoculate 25 ml of M9 in a 150-ml Erlenmeyer flask and incubate at 37°. After 18 hr, centrifuge the culture ($OD_{600 \text{ nm}}$ is about 0.5) for 5 min at 3000 g. Resuspend the exhausted pelleted bacteria in 1 ml of PBS buffer and dilute 250 $\mu$l of this suspension in 250 ml of M9–agar (0.6%, w/v; ultrapure agarose, electrophoresis grade; GIBCO-BRL, Gaithersburg, MD) maintained at 35° and supplemented with 0.3 m$M$ 2,3,5-triphenyltetra-zolium chloride (freshly prepared), and pour the whole solution in a petri dish (24.5 cm × 24.5 cm). Make wells (2-mm diameter over 5-mm depth) using a hole punch connected to an aspirator. For the detection of biotin in various extracts, spot the samples (20 $\mu$l), previously sterilized (using sterile syringe filter units), into wells. After incubation at 37° for 18 to 20 hr, biotin is detected as growth of bacteria, which results in the appearance of a reddish color. Biotin concentration is determined by reporting the square diameter of the corresponding halo on the standard curve established with standard solutions of 0.05 to 2 ng of biotin and plotted as the logarithm of biotin concentration versus the squared diameter of the halo.

*Comments.* We have used these methods in our laboratory to study the presence of bound biotin in organelles, and have demonstrated, for the first time, the presence of a soluble pool of biotin in the cytosol of plant

[16] M. A. Eisenberg, *J. Bacteriol.* **123**, 248 (1975).
[17] J. Sambrook, E. F. Fritsch, and T. Maniatis, *in* "Molecular Cloning: A Laboratory Manual." Cold Spring Harbor Laboratory Press, Cold Spring Harbor, New York, 1989.

TABLE I

SUBCELLULAR LOCALIZATION OF FREE AND BOUND BIOTIN IN
PEA LEAF PROTOPLASTS[a]

| Assay | Total extract | Supernatant | Pellet |
|---|---|---|---|
| | | Distribution | |
| mg protein/$10^6$ cells[b] | 1.1 | 0.28 | 0.88 |
| | mU/mg protein | Percentage of total activity | |
| Enzyme activity[b] | | | |
| PFP | 23.5 | 89 | 11 |
| NADP-GAPDH | 302 | 10 | 90 |
| Fumarase | 5.8 | 12 | 88 |
| | | Picomoles | |
| Biotin[c] | | | |
| Total | 4.98 ± 0.72 | 4.31 ± 0.45 | 0.90 ± 0.22 |
| Free | 4.07 ± 0.59 | 4.31 ± 0.45 | nd[d] |
| Bound | 0.81 ± 0.18 | nd | 0.90 ± 0.22 |

[a] Preparation of intact protoplasts ($20 \times 10^6$ cells) and broken protoplasts (total extract), as well as fractionation of protoplasts leading to supernatant (cytosol) and pellet (organelles), were carried out as described in text. All fractions were assayed for marker enzymes or chloroplasts[12] (NADP-GAPDH), mitochondrial[13] (fumarase), and cytosol[12] (PFP). Extraction and biological testing of free and bound biotin were realized as described in text.

[b] These data are from a representative experiment and have been reproduced five times.

[c] Values of biotin content are given as means ± SD from five experiments.

[d] nd, Not detectable (<0.05 ng).

cells (Tables I and II).[18] In pea leaf cells, chloroplast from the mesophyll contain a prokaryotic form of acetyl-CoA carboxylase associated with an $M_r$ 34,000 biotinylated polypeptide, whereas the eukaryotic form associated with an $M_r$ 220,000 polypeptide has been proposed to be localized in the cytosol of epidermal cells.[19] Mitochondria contain a unique biotinylated $M_r$ 76,000 polypeptide associated with 3-methylcrotonyl-CoA carboxylase.[20] Using this procedure, we have not been able to detect free biotin in highly purified chloroplasts and mitochondria.

[18] P. Baldet, C. Alban, S. Axiotis, and R. Douce, *Arch. Biochem. Biophys.* **303,** 67 (1993).
[19] C. Alban, P. Baldet, and R. Douce, *Biochem. J.* **300,** 557 (1994).
[20] P. Baldet, C. Alban, S. Axiotis, and R. Douce, *Plant Physiol.* **99,** 450 (1992).

TABLE II
LOCALIZATION OF BIOTIN IN TOTAL EXTRACT AND
VACUOLES FROM PEA LEAF PROTOPLASTS[a]

| Assay | Protoplasts | Vacuoles |
|---|---|---|
| mg protein/10⁶ cells[b] | 0.88 | 0.07 |

| | mU/mg protein | Percentage of total activity |
|---|---|---|
| Enzyme activity[b] | | |
| PFP | 25.4 | 0.5 |
| NADP-GAPDH | 378 | 0.4 |
| Fumarase | 6.3 | 2.3 |

| | Picomoles | |
|---|---|---|
| Biotin[c] | | |
| Total | 4 ± 0.72 | nd[d] |
| Free | 3.51 ± 0.36 | nd |
| Bound | 0.54 ± 0.07 | nd |

[a] Pea leaf protoplasts (20 × 10⁶ cells) were prepared as described in text. Vacuoles were obtained by gentle rupture of intact protoplasts and purified by centrifugation using 5, 2, and 1% Ficoll layers. With a yield of 20%, about 20 × 10⁶ protoplasts are needed to obtain 4 × 10⁶ vacuoles. All fractions were assayed for marker enzymes for chloroplasts[12] (NADP-GAPDH), mitochondrial[13] (fumarase), and cytosol[12] (PFP). Extraction and biological testing of free and bound biotin were realized as described in text.

[b] These data are from a representative experiment and have been reproduced four times.

[c] Values of biotin content are given as means ± SD from four experiments.

[d] nd, Not detectable (<0.05 ng).

## Identification of Intermediates of Biotin Biosynthesis

### Solutions

Lavender medium LS (Linsmaier and Skoog[21]): This medium consists of Murashige and Skoog basal salt mixture (Sigma, St. Louis, MO) supplemented with thiamin hydrochloride (400 $\mu$g ml$^{-1}$), inositol

[21] E. M. Linsmaier and F. Skoog, *Physiol. Plant.* **18**, 100 (1965).

(100 $\mu$g ml$^{-1}$), sucrose (30 mg ml$^{-1}$), indole-3-butyric acid (10$^{-5}$ $M$), and benzylaminopurine (10$^{-6}$ $M$)

Yeast medium (MY, per liter): 20 g of malt extract, 2 g of yeast extract
Minimum medium: Sucrose, 60 m$M$: $(NH_4)_2SO_4$, 23 m$M$; $KH_2PO_4$, 14 m$M$; $MgSO_4$, 1 m$M$; $CaCl_2$, 1.7 m$M$; $H_3BO_3$, 16 $\mu$$M$; $ZnSO_4$, 3.5 $\mu$$M$; $MgCl_2$, 5 $\mu$$M$; inositol, 28 $\mu$$M$; $FeCl_3$, 3 $\mu$$M$; $CuSO_4$, 0.4 $\mu$$M$; KI, 0.6 $\mu$$M$; aspartic acid, 0.7 m$M$; $\beta$-alanine, 6 $\mu$$M$; thiamin, 60 n$M$; vitamin B$_6$, 10 n$M$ (Snell et al.[22])

## Plant Material

A stable biotin-overproducing cell line of lavender cell cultures (*Lavandula vera* L.) was a gift from Y. Yamada (Department of Agricultural Chemistry, Kyoto University; Watanabe et al.[23]). The cells (1 g fresh weight) are cultivated in 250-ml Erlenmeyer flasks containing 50 ml of LS medium for 21 days under a 16-hr photoperiod of white light from fluorescent tubes (15 mol of photons m$^{-2}$) at 27° and a horizontal agitation of 85 rpm (Rota-Test 74403; Bioblock).

## Extraction of Intermediates

Filter lavender cells (8 g) corresponding to 10 days of growth through a nylon mesh (100 $\mu$m) and wash with 400 ml of sterile water. Freeze the cells in liquid nitrogen, homogenize to a very fine powder, and resuspend in 15 ml of 5 m$M$ EDTA plus 50 m$M$ HEPES/NaOH, pH 8.0. Centrifuge at 48,000 g for 30 min, discard the pellet, and dilute the supernatant with cold acetone (−20°) to a final acetone concentration of 80% (v/v). Leave for 15 min at −80°, and eliminate precipitated proteins by centrifugation for 30 min at 48,000 g. Discard the pellet and concentrate the supernatant first under a nitrogen flux to evaporate acetone and then lyophilize. Dissolve the residue in 800 $\mu$l of sterile water.

## Thin-Layer Chromatography

Using a 1-$\mu$l capillary, apply 1- to 10-$\mu$l aliquots of the soluble cell extract to thin-layer chromatography (TLC) plates (20 × 20 cm) coated with 0.2 mm silica gel 60 F254 (E. Merck AG, Darmstadt, Germany) and air dry between each spot. For better separation, develop the plates twice at room temperature: first, with ethyl acetate–heptane–acetic acid–water (45:45:1:9, by volume) for 1 hr, then air dry the plate for 10 min; second,

[22] E. E. Snell, R. E. Eakin, and R. J. Williams, *J. Am. Chem. Soc.* **62**, 175 (1940).
[23] K. Y. Watanabe, S. C. Yano, and Y. Yamada, *Phytochemistry* **21**, 513 (1982).

with butanol–acetic acid–water (60:15:25, by volume) for 4.5 hr. Dry the plate for 12 hr in a sterile hood.

*Biological Assay*

*Saccharomyces cerevisiae* L. [American Type Culture Collection (Rockville, MD); ATCC 7754], a mutant that requires the presence of one of four biotin intermediates (7-oxo-8-aminopelargonic acid, 7,8-diaminopelargonic acid, dethiobiotin, and biotin) in the growth medium, has been used to detect biotin intermediates. However, this mutant does not recognize pimeloyl-CoA or pimelic acid, two of the first vitamers of the pathway described

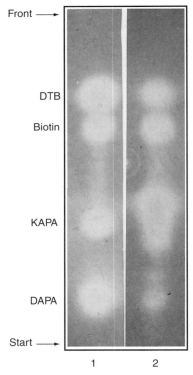

FIG. 2. Biological characterization of vitamers in lavender cells. Standard solutions of vitamers and lavender cell extract are separated on TLC plates and plates are biologically revealed with a mutant of *S. cerevisiae* as described in the procedure. Lane 1: Two microliters of standard containing biotin (5 pg, $R_f$ 0.56), dethiobiotin (DTB) (5 pg, $R_f$ 0.66), 7-oxo-8-aminopelargonic acid (KAPA) (2 ng, $R_f$ 0.28), and 7,8-diaminopelargonic acid (DAPA) (0.5 ng, $R_f$ 0.1). Lane 2: Soluble (defatted and deproteinated) lavender extract (2 μl spotted).

in bacteria. Furthermore, the yeast is 10 times more sensitive than the *E. coli bioF* mutant, which recognizes the same vitamers. Cultivate the yeast strain in petri dishes containing MY medium agarose (15 g/liter) at 25°. For the biological test, transfer one colony to 25 ml of minimal medium and incubate for 16 hr at 30° until a concentration of $10^8$ cells/ml is reached. Centrifuge the culture at 3000 *g* for 15 min and wash with 25 ml of minimal medium. Repeat the centrifugation, suspend the pellet in 250 ml of minimal medium containing agarose (8 g/liter), pour into a petri dish (24.5 × 24.5 cm), and store at 4°. Carefully adpress the dried TLC plates, after separation of biotin intermediates, for 3 hr at 4° under sterile conditions. Remove the TLC plates gently in one movement to avoid silica remaining on the agarose, and incubate for 48 hr at 30° for analysis. Biotin intermediates are detected as growth of yeast cells in the agar, which results in the appearance of white spots (Fig. 2). $R_f$ values for 7-oxo-8-aminopelargonic acid, 7,8-diaminopelargonic acid, dethiobiotin, and biotin on the plates are determined and compared with values obtained from standard compounds.

## Comments

Biological analysis of the lavender extract reveals five halos, among which four have migration characteristics identical to those of the standards mixture. In the extract around the spot of 7-oxo-8-aminopelargonic acid, two comigrating spots are detected at $R_f$ 0.32 and $R_f$ 0.28. Comparison with standards vitamers has shown that the spot at $R_f$ 0.32 represents the oxidized form of biotin as *d*-biotin sulfoxide.[5] Results obtained by TLC analysis are confirmed by high-performance liquid chromatography (HPLC) after the vitamer spots were scraped into vials containing water.[5]

# [36] Analysis of Biotin Biosynthesis Pathway in Coryneform Bacteria: *Brevibacterium flavum*

*By* Kazuhisa Hatakeyama, Miki Kobayashi, and Hideaki Yukawa

## Introduction

Coryneform bacteria are gram-positive bacteria that have a high GC content and constitute the Actinomycetes subdivision of the eubacteria. Among these, the glutamic acid-producing bacteria that have been used for the industrial production of numerous metabolites are exacting in their nutritional requirements and are all dependent on the presence of biotin

METHODS IN ENZYMOLOGY, VOL. 279                                    0076-6879/97 $25

in the growth medium.[1] Biotin concentration in the growth medium has been successfully used as a parameter to optimize coryneform bacteria-mediated amino acid production. For instance, biotin limitation in the medium could improve glutamic acid secretion.[2,3] On the contrary, addition of excess biotin could improve L-lysine production through the activation of pyruvate carboxylase.[4]

We have studied the coryneform bacterium, *Brevibacterium flavum* MJ233, for use as a bioconverter to produce amino acids and organic acids,[5–9] and we have developed a new bioprocess by using biotin removal from a minimal medium as a means to repress cellular division.[10,11]

The biosynthetic pathway of biotin has been studied mainly by using strains of *Escherichia coli* (Fig. 1). The biotin biosynthetic genes are organized in an operon in *E. coli*,[12] but in two different clusters in *Bacillus sphaericus*.[13] Among the enzymes of the pathway, all that are involved in dethiobiotin synthesis from pimelic acid have been documented: pimelyl-CoA synthetase[14] (*bioW*), 7-keto-8-aminopelargonic acid synthetase[15] (*bioF*), 7,8-diaminopelargonic acid aminotransferase[16] (*bioA*), and dethiobiotin synthase[17,18] (*bioD*). However, the final step in biotin synthesis, the

[1] W. Liebl, *in* "The Prokaryotes" (A. Balows, H. G. Trüper, M. Dworkins, W. Harder, and K. H. Schleifer, eds.), 2nd Ed., p. 1157. Springer-Verlag, New York, 1992.

[2] I. Shiio, S. I. Otsuka, and M. Takahashi, *J. Biochem.* **51**, 56 (1962).

[3] C. Hoischen and R. Krämer, *J. Bacteriol.* **172**, 3409 (1990).

[4] O. Tosaka, H. Morioka, and K. Takinami, *Agric. Biol. Chem.* **43**, 1513 (1979).

[5] M. Terasawa, H. Yukawa, and Y. Takayama, *Process Biochem.* **20**, 124 (1985).

[6] H. Yukawa, H. Yamagata, and M. Terasawa, *Process Biochem.* **21**, 164 (1986).

[7] H. Yukawa and M. Terasawa, *Process Biochem.* **21**, 196 (1986).

[8] M. Terasawa, M. Inui, M. Goto, K. Shikata, M. Imanari, and H. Yukawa, *J. Ind. Microbiol.* **5**, 289 (1990).

[9] H. Yamagata, M. Terasawa, and H. Yukawa, *in* "Biochemical Engineering for 2001" (S. Furusaki, I. Endo, and R. Matsuo, eds.), p. 475. Springer-Verlag, New York, 1992.

[10] M. Terasawa, M. Inui, M. Goto, Y. Kurusu, and H. Yukawa, *Appl. Microbiol. Biotechnol.* **35**, 348 (1991).

[11] M. Terasawa and H. Yukawa, *in* "Industrial Application of Immobilized Biocatalysts" (A. Tanaka, T. Tosa, and T. Kobayashi, eds.), p. 37. Marcel Dekker, New York, 1992.

[12] A. J. Otsuka, M. R. Buoncristiani, P. K. Howard, J. Flamm, C. Johnson, R. Yamamoto, K. Uchida, C. Cook, J. Ruppert, and J. Matsuzaki, *J. Biol. Chem.* **263**, 19577 (1988).

[13] R. Gloeckler, I. Ohsawa, D. Speck, C. Ledoux, S. Bernard, M. Zinsius, D. Villeval, T. Kisou, K. Kamogawa, and Y. Lemoine, *Gene* **87**, 63 (1990).

[14] O. Ploux, P. Soularue, A. Marquet, R. Gloeckler, and Y. Lemoine, *Biochem. J.* **287**, 685 (1992).

[15] O. Ploux and A. Marquet, *Biochem. J.* **287**, 327 (1992).

[16] G. L. Stoner and M. A. Eisenberg, *J. Biol. Chem.* **250**, 4037 (1975).

[17] K. Krell and M. A. Eisenberg, *J. Biol. Chem.* **245**, 6558 (1970).

[18] K. Ogata, Y. Izumi, K. Aoike, and Y. Tani, *Agric. Biol. Chem.* **37**, 1093 (1973).

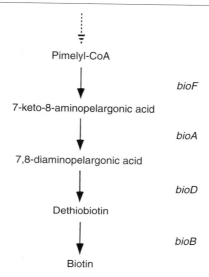

FIG. 1. The biotin biosynthetic pathway of *E. coli* and *B. sphaericus*.

conversion of dethiobiotin to biotin, remains to be elucidated and might involve a flavodoxin, ferredoxin (flavodoxin–NADP$^+$ reductase, and unidentified enzyme(s) in addition to the *bioB* gene product.[19–21]

The lack of biotin biosynthesis is a taxonomical characteristic of the glutamic acid-producing group of coryneform bacteria. However, there have been no practical data regarding the deleted steps of biotin biosynthesis. We have revealed the deleted steps of the biotin biosynthetic pathway in several coryneform bacteria by using cross-feeding experiments with *E. coli bio* mutants. In addition, we have cloned and sequenced biotin biosynthetic genes from a strain of coryneform bacterium, *Brevibacterium flavum*, MJ233. We have also obtained *bio* mutants by *in vivo* disruption of the *bio* gene.[22,23]

[19] O. Ifuku, J. Kishimoto, S. Haze, M. Yanagi, and S. Fukushima, *Biosci. Biotech. Biochem.* **56,** 1780 (1992).

[20] O. M. Birsh, M. Fuhrmann, and N. M. Shaw, *J. Biol. Chem.* **270,** 19158 (1995).

[21] I. Sanyal, K. J. Gibson, and D. H. Flint, *Arch. Biochem. Biophys.* **326,** 48 (1996).

[22] K. Hatakeyama, K. Kohama, A. A. Vertès, M. Kobayashi, Y. Kurusu, and H. Yukawa, *DNA Sequence* **4,** 87 (1993).

[23] K. Hatakeyama, K. Kohama, A. A. Vertès, M. Kobayashi, Y. Kurusu, and H. Yukawa, *DNA Sequence* **4,** 177 (1993).

TABLE I

CROSS-FEEDING STUDIES BETWEEN *Escherichia coli bio* MUTANTS
AND CORYNEFORM BACTERIA

| E. coli mutant | | Coryneform bacteria | | |
|---|---|---|---|---|
| Strain[a] | Genotype | B. flavum MJ233 | B. lactofermentum ATCC 13869 | C. glutamicum ATCC 31831 |
| BM360 | bioH | −[b] | − | − |
| R876 | bioC | − | − | − |
| R874 | bioF | − | − | − |
| R873 | bioA | + | + | + |
| R877 | bioD | + | + | + |
| R875 | bioB | + | + | + |

[a] Data from Refs. 27 and 28.
[b] −, Cross-feeding did not occur; +, cross-feeding occurred.

## Cross-Feeding Methods to Identify Deleted Steps in Biotin Biosynthetic Pathway of Coryneform Bacteria

Cross-feeding experiments between *E. coli bio* mutants and coryneform bacteria provide a precise means for understanding the deleted steps of biotin biosynthetic pathway in coryneform bacteria.[24] In *E. coli*, four genes, *bioF, bioA, bioD,* and *bioB,* have been shown to be involved in the biosynthetic pathway of biotin from pimelyl-CoA. The functions of *bioC* and *bioH* have not yet been determined, but these genes may be involved in pimelyl-CoA biosynthesis.

Cross-feeding studies are carried out as described by Rolfe and Eisenberg.[25] To make M9-based medium, 100 mg of vitamin assay Casamino Acids (Difco Laboratories, Detroit, MI) and 10 $\mu$g of thiamin hydrochloride are added to 100 ml of M9 medium[26]; this medium is used for *E. coli bio* mutants. Six diferent *E. coli bio* mutants[27,28] (Table I) are separately streaked onto M9-based medium agar plates in which washed cells of a test strain of coryneform bacteria are suspended to a concentration of $5.0 \times 10^5$ cells/ml. Cross-feeding interactions are scored after a 72-hr incuba-

[24] S. Okumura, R. Tsugawa, T. Tsunoda, and S. Morisaki, *J. Agric. Chem. Soc. (Japan)* **36,** 204 (1962).
[25] B. Rolfe and M. A. Eisenberg, *J. Bacteriol.* **96,** 515 (1968).
[26] J. Sambrook, E. F. Fritsch, and T. Maniatis (eds.), "Molecular Cloning: A Laboratory Manual," 2nd Ed. Cold Spring Harbor Laboratory Press, Cold Spring Harbor, New York, 1989.
[27] P. P. Cleary and A. Campbell, *J. Bacteriol.* **112,** 830 (1972).
[28] D. F. Baker and A. M. Campbell, *J. Bacteriol.* **143,** 789 (1980).

tion at 33°. Cross-feeding under these conditions results from diffusion of biotin precursors excreted between each *E. coli* mutant and test strain.

We have tested three strains of coryneform bacteria, *B. flavum* MJ233, *Brevibacterium lactofermentum* ATCC 13869, and *Corynebacterium glutamicum* ATCC 31831, with *E. coli* mutants R873 (*bioA*), R877 (*bioD*), and R875 (*bioB*) (Table I). The results demonstrate that these coryneform bacteria have the ability to convert 7-keto-8-aminopelargonic acid into biotin. The findings also correspond well to the biotin precursor experiments to promote L-glutamic acid fermentation.[24]

### Cloning and Sequencing of *Brevibacterium flavum bio* Genes

Biotin biosynthesis genes have been cloned from *B. flavum* MJ233 by the complementation of *E. coli bio* mutants.[27,28] Propagation and transformation of *E. coli* and *Brevibacterium* strains have been described previously.[5] Chromosomal DNA is isolated from *B. flavum* MJ233 by the following protocol described by Saito and Miura,[29] modified by using 4 mg of lysozyme per milliliter at 37° for 30 min. *Brevibacterium flavum* MJ233 chromosomal DNA is partially digested with *Sau*3AI so as to yield high molecular weight DNA fragments. The digested DNA fragments are ligated to the *Bam*HI site of the cosmid vector pWE15.[30] The ligation mixture is packaged using GigapackII GOLD packaging extracts (Stratagene, La Jolla, CA) as per the manufacturer instructions. The M9-based medium (mentioned previously) is used as the medium for complementation.

### Cloning and Sequencing of *Brevibacterium flavum bioB* Gene

Ampicillin-resistant transformants are selected and screened for the complementation of the biotin-requiring auxotroph *E. coli* R875 (*bioB*). Out of $2.0 \times 10^4$ transformants screened, 6 have been able to grow without biotin supplementation. Each cosmid contains chromosomal DNA inserts of approximately 20 to 40 kb. The cosmids do not complement the other *E. coli* biotin auxotrophs [BM360 (*bioH*), R873 (*bioA*), R874 (*bioF*), R876 (*bioC*), and R877 (*bioD*)].

One cosmid containing chromosomal DNA inserts of approximately 30 kb has been digested by *Hin*dIII and ligated to *Hin*dIII-digested pUC118 plasmid DNA. The ligation mixture is used to transform *E. coli* R875. A clone bearing a 5.5-kb *Hin*dIII fragment containing *B. flavum bioB* gene has been obtained. Deletion and complementation analysis of this fragment

[29] H. Saito and K. I. Miura, *Biochim. Biophys. Acta* **72,** 619 (1963).
[30] G. M. Wahl, K. A. Lewis, J. C. Ruiz, B. Rothenberg, J. Zhao, and G. A. Evans, *Proc. Natl. Acad. Sci. U.S.A.* **84,** 2160 (1987).

indicate that the *bioB* gene is located on a 1.7-kb *Hind*III–*Sac*I fragment (Fig. 2).

The nucleotide sequence of the 1.7-kb *Hind*III–*Sac*I fragment has been determined on both strands. It includes an open reading frame comprising 1005 nucleotides corresponding to 334 amino acids. The deduced polypeptide is 67.6, 36.2, 35.5, 34.6, 32.1, and 28.0% homologous to the *bioB* genes of *Mycobacterium leprae*,[31] *E. coli*,[7] *Serratia marcescens*,[32] *Haemophilus influenzae*,[33] *B. sphaericus*,[34] and *Saccharomyces cerevisiae*,[35] respectively.

## Cloning and Sequencing of Brevibacterium flavum bioA and bioD Genes

*Brevibacterium flavum bioA* and *bioD* genes are cloned by the complementation of the biotin-requiring auxotrophs *E. coli* R873 (*bioA*) and R877 (*bioD*). Out of $1.0 \times 10^4$ R873 and R877 ampicillin-resistant transformants screened, 20 transformants of R873 (*bioA*) and 13 transformants of R877 (*bioD*) have been able to grow without biotin supplementation. Each cosmid contains a chromosomal DNA insert of approximately 20 to 40 kb. The *E. coli* strains R873 (*bioA*) and R877 (*bioD*) are complemented by the same cosmids. However, these cosmids cannot complement the other *E. coli* biotin auxotrophs BM360 (*bioH*), R874 (*bioF*), R875 (*bioB*), and R876 (*bioC*).

One of the isolated cosmids containing a chromosomal DNA insert of approximately 30 kb has been digested by *Sal*I and ligated to *Sal*I-digested pUC118 plasmid DNA. The ligation mixture is used to transform *E. coli* R873 and R877. A clone bearing a 4.0-kb *Sal*I fragment containing the *B. flavum bioA* and *bioD* genes has been obtained. The nucleotide sequence of the 4.0-kb *Sal*I fragment has been determined on both strands. It includes two open reading frames comprising 1272 (*bioA*) and 675 (*bioD*) nucleotides corresponding to 423 and 224 amino acids, respectively (Fig. 2).

The deduced amino acid sequence of the *B. flavum bioA*-encoded 7,8-diaminopelargonic acid aminotransferase is 53.6, 51.0, 49.0, 47.8, and 31.9% identical to those of *M. leprae*,[31] *E. coli*,[7] *S. marcescens*,[32] *H. influenzae*,[33] and *B. sphaericus*,[34] respectively. A potential SAM (*S*-adenosyl-L-methionine)-binding site (GFGRTG),[7] which resembles an ATP-binding

[31] D. R. Smith and K. Robinson, data from GenBank (1994).
[32] N. Sakurai, Y. Imai, H. Akatsuka, E. Kawai, S. Komatsubara, and T. Tosa, data from DDBJ (1993).
[33] R. D. Fleischmann *et al., Science* **269**, 496 (1995).
[34] I. Osawa, D. Speck, T. Kisou, K. Hayakawa, M. Zinsius, R. Gloeckler, Y. Lemoine, and K. Kamogawa, *Gene* **80**, 39 (1989).
[35] S. Zahng, I. Sanyal, G. H. Bulboaca, A. Rich, and D. H. Flint, *Arch. Biochem. Biophys.* **309**, 29 (1994).

*Escherichia coli*

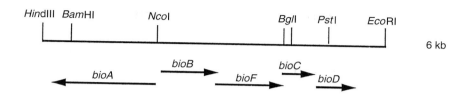

*Bacillus sphaericus*

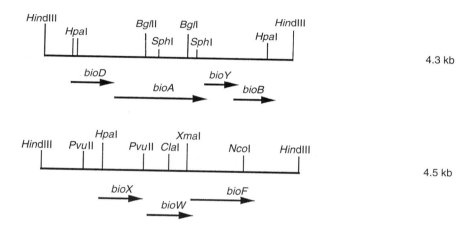

*Brevibacterium flavum*

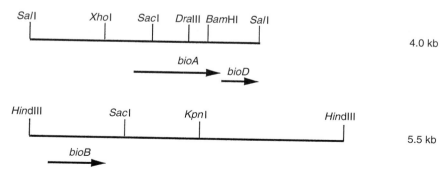

FIG. 2. Comparison of the genomic organization of the biotin biosynthetic genes in *E. coli*, *B. sphaericus*, and *B. flavum*.

site $(GXGXXG)$,[36] is conserved in each organism. A putative PLP-binding site, which can be inferred by strong homology to the proposed PLP-binding site in ornithine aminotransferase, is also conserved in each organism.[7] A computer search of the National Biomedical Research Foundation (NBRF) database has revealed a strong homology with several other aminotransferases as follows: *E. coli* and yeast acetylornithine aminotransferase (28.6 and 27.7%)[37]; *E. coli* and *Emericella* 4-aminobutyrate aminotransferase (29.1 and 25.5%)[38]; rat, human, and yeast ornithine aminotransferase (26.0, 25.8, and 27.5%)[39]; and *Streptomyces* L-lysine 6-aminotransferase (23.9%),[40] respectively.

The deduced amino acid sequence of the *B. flavum bioD*-encoded dethiobiotin synthetase is 35.8, 34.1, 33.2, 26.7, and 26.6% identical to those of *M. leprae*,[30] *H. influenzae*,[33] *B. sphaericus*,[13] *E. coli*,[7] and *S. marcescens*[32] dethiobiotin synthetases.

*Genomic Organizations of Biotin Biosynthetic Genes*

The genomic organizations of biotin biosynthetic genes seem to vary widely in the microorganisms analyzed to date, as shown in Fig. 2. In *E. coli* and *S. marcescens*, the biotin biosynthetic genes are organized in a divergent operon (*bioABFCD*).[7,32] However, these of *B. sphaericus* are separated into two DNA clusters, *bioXWF* and *bioDAYB*.[13] In the case of *B. flavum*, the *bioA–bioD* and *bioB* genes seem to be scattered on the chromosome, but the *bioA*, *bioB*, and *bioD* genes of *B. sphaericus* are located in one cluster.

Disruption of Biotin Biosynthetic Gene on *Brevibacterium flavum* MJ233 Chromosome

Gene disruption and replacement methods are convenient for analyzing a function of a certain gene *in vivo*. We have developed these methods in coryneform bacteria.[41] In one procedure, the disruption of the *bioB* gene is shown using the integrative plasmid. The expected modes of integration of the plasmid pKH1 in the *B. flavum* chromosome are represented in

[36] R. K. Wierenga and W. G. J. Hol, *Nature (London)* **302**, 842 (1983).
[37] H. Heimberg, A. Boyen, M. Crabeel, and N. Glansdorff, *Gene* **90**, 69 (1990).
[38] K. Bartsch, A. John-Marteville, and A. Schulz, *J. Bacteriol.* **172**, 7035 (1990).
[39] M. Mueckler and H. C. Pitot, *J. Biol. Chem.* **260**, 12993 (1985).
[40] M. B. Tobin, S. Kovacevic, K. Madduri, J. A. Hoskins, P. Skatrud, L. C. Vining, C. Stuttard, and J. R. Miller, *J. Bacteriol.* **173**, 6223 (1991).
[41] A. V. Vertès, K. Hatakeyama, M. Inui, M. Kobayashi, Y. Kurusu, and H. Yukawa, *Biosci. Biotech. Biochem.* **57**, 2036 (1993).

A

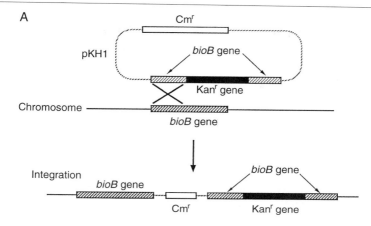

B

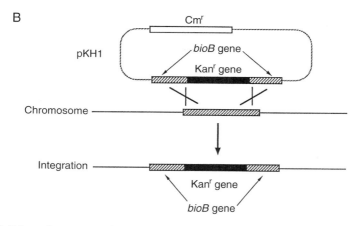

Fig. 3. Schematic representation of plasmid pKH1 and its integration into the chromosome of *B. flavum* MJ233. The integrative plasmid pKH1, which is unable to replicate in *B. flavum,* carries a kanamycin resistance gene (Kanamycin Resistance GenBlock; Pharmacia, Uppsala, Sweden) that was inserted into the coding region of the *bioB* gene. (A) Integration via a Campbell-like mechanism; (B) integration via a double cross-over mechanism. Filled box, kanamycin resistance gene (Kan$^r$); open box, chloramphenicol resistance gene (Cm$^r$); hatched box, *bioB* gene; solid line, chromosomal DNA; dashed line, plasmid pHSG398.

Fig. 3. *B. flavum* is cultured at 33° in BT medium [glucose, 20 g; (NH$_4$)$_2$SO$_4$, 7 g; urea, 2 g; K$_2$HPO$_4$, 0.5 g; KH$_2$PO$_4$, 0.5 g; MgSO$_4$ · 7H$_2$O, 0.5 g; FeSO$_4$ · 7H$_2$O, 6 mg; MnSO$_4$ · 6H$_2$O, 6 mg; biotin, 200 $\mu$g; thiamin hydrochloride, 200 $\mu$g; and deionized water, 1 liter (pH adjusted to 7.5 with NaOH)] with the addition of 1 g of yeast extract and 1 g of Casamino Acids (per liter). *B. flavum* transformants are selected by resistance to kanamycin (50 $\mu$g/ml), or chloramphenicol (3 $\mu$g/ml) on the medium.[42–45]

CD medium, which is BT medium with the addition of 200 $\mu$g of dethiobiotin (Sigma, St. Louis, MO) per liter instead of biotin, is used for the growth test of *B. flavum bioB* mutants. Integration via a Campbell-like mechanism results in a Kan$^r$, Cm$^r$ phenotype. And the integrants of this type represent the majority number of integrants and are able to grow on CD medium, like the wild-type strain. Integration via a double cross-over event (gene replacement) results in a *bioB,* Kan$^r$, Cm$^s$ phenotype, as demonstrated by the inability of these mutants to grow on CD medium. These results demonstrate that the *bioB* gene product from *B. flavum* MJ233 is involved in the conversion of dethiobiotin to biotin.

### Nucleotide Sequence Accession Number

The nucleotide sequence presented in this chapter has been deposited in GenBank/EMBL under the accession number D14084 for *bioB* and D14083 for *bioA–bioD* genes.

### Acknowledgment

We are greatly indebted to Professor Allan Campbell (Stanford University, Stanford, CA) for generously providing us with *E. coli bio* mutants.

[42] Y. Satoh, K. Hatakeyama, K. Kohama, M. Kobayashi, Y. Kurusu, and H. Yukawa, *J. Ind. Microbiol.* **5,** 159 (1989).
[43] Y. Kurusu, M. Kainuma, M. Inui, Y. Satoh, and H. Yukawa, *Agric. Biol. Chem.* **54,** 443 (1990).
[44] Y. Kurusu, Y. Satoh, M. Inui, K. Kohama, M. Kobayashi, M. Terasawa, and H. Yukawa, *Appl. Environ. Microbiol.* **57,** 759 (1991).
[45] A. V. Vertès, M. Inui, M. Kobayashi, Y. Kurusu, and H. Yukawa, *Res. Microbiol.* **144,** 181 (1993).

# [37] Purification and Characterization of Biotin Synthases

## By Dennis H. Flint and Ronda M. Allen

### Introduction

The final step in the biotin biosynthetic pathway consists of the addition of a sulfur atom between the methyl and methylene carbon atoms adjacent to the imidazolinone ring of dethiobiotin to form the vitamin biotin as in reaction (1).[1]

$$\text{Dethiobiotin} \rightarrow \text{Biotin} \tag{1}$$

Chemically this is a difficult reaction and little is known about the mechanism. This reaction requires the removal of one hydrogen atom from each of the two carbon atoms to which the sulfur is added. The other hydrogen atoms on these two carbons and the carbon atoms in the imidazolinone ring are all preserved.[2] The direct source of the sulfur atom is not known.[3-5] Only recently has this reaction been carried out in cell-free extracts.[6-11]

[1] M. Eisenberg, in "*Escherichia coli* and *Salmonella typhimurium*—Cellular and Molecular Biology" (F. Neidhardt, ed.), p. 544. American Society for Microbiology, Washington, DC, 1987.

[2] R. J. Parry, *Tetrahedron* **39,** 1215 (1983).

[3] E. Demoll and W. Shive, *Biochem. Biophys. Res. Commun.* **110,** 243 (1983).

[4] R. H. White, *Biochemistry* **21,** 4271 (1982).

[5] T. Nimura, T. Suzuki, and Y. Sahashi, *Vitamin* **29,** 86 (1964).

[6] O. Ifuku, J. Kishimoto, S. Haze, M. Yanagi, and S. Fukushima, *Biosci. Biotech. Biochem.* **56,** 1780 (1992).

[7] I. Sanyal, G. Cohen, and D. H. Flint, *Biochemistry* **33,** 3625 (1994).

[8] I. Sanyal, K. J. Gibson, and D. H. Flint, *Arch. Biochem. Biophys.* **326,** 48 (1996).

[9] O. M. Birch, M. Fuhrmann, and N. M. Shaw, *J. Biol. Chem.* **270,** 19158 (1995).

[10] A. Fujisawa, T. Abe, I. Ohsawa, K. Kamogawa, and Y. Izumi, *FEMS Microbiol. Lett.* **110,** 1 (1993).

Purification

The *Escherichia coli bioB* gene encodes a protein that catalyzes this reaction.[1,6] This gene has been cloned and sequenced.[12] Highly homologous genes have been identified, cloned, and sequenced from *Bacillus sphaericus*,[13] *Saccharomyces cerevisiae*,[14] and *Arabidopsis thaliana*.[15] The *Bacillus* and yeast genes are known to encode proteins that catalyze the biotin synthase reaction.[11,14]

From the sequence of the *E. coli bioB* gene, its gene product was expected to be a 38.7-kDa peptide. We have used this information along with the predicted $NH_2$-terminal amino acid sequence to purify this protein prior to the establishment of a cell-free assay.[7]

An *E. coli* strain that overexpresses biotin synthase has been constructed starting with the plasmid pAOB7A (obtained from A. J. Otsuka), which contains the *bio* operon. A 2.2-kb *Nco*I–*Bgl*II fragment from this plasmid containing the *bioB* and *bioF* genes (encodes 7-keto-8-aminopelargonic acid synthase) has been cloned into the plasmid pET-11d.[16] The new plasmid, named pBioBF2, has been used to transform *E. coli* strain HMS174. HMS174/pBioBF2 cells are grown at 37° in a fermentor in minimal medium with the addition of 2% (w/v) glucose, 2% (w/v) Casamino acids, 1% (v/v) glycerol, and ampicillin (100 $\mu$g/ml). When the $A_{600}$ of the culture reaches 5, the cells are induced by isopropyl-$\beta$-D-thiogalactopyranoside, and harvested 3 hr later.

An extract from these cells is loaded onto a DEAE-Sepharose column, the sample is eluted with an increasing concentration of NaCl, the fractions are collected. Sodium dodecyl sulfate (SDS) electrophoresis gels are run on a sample from each fraction, using a Pharmacia (Piscataway, NJ) Phast-System. Fractions containing a significant quantity of peptide with a molecular mass of approximately 39 kDa are identified. Samples from each of these fractions are electrophoresed on SDS Daiichi gels, blotted, and the $NH_2$-terminal sequences of the peptides whose molecular masses are near

[11] A. Méjean, B. T. S. Bue, D. Florentin, O. Ploux, Y. Izumi, and A. Marquet, *Biochem. Biophys. Res. Commun.* **217**, 1231 (1995).

[12] A. J. Otsuka, M. R. Buoncristiani, P. K. Howard, J. Flamm, C. Johnson, R. Yamamoto, K. Uchida, C. Cook, J. Ruppert, and J. Matsuzaki, *J. Biol. Chem.* **263**, 19577 (1988).

[13] I. Ohsawa, D. Speck, T. Kisou, K. Hayakawa, M. Sinsius, R. Gloeckler, Y. Lemoine, and K. Kamogawa, *Gene* **80**, 39 (1989).

[14] S. Zhang, I. Sanyal, G. H. Bulboaca, A. Rich, and D. H. Flint, *Arch. Biochem. Biophys.* **309**, 29 (1994).

[15] L. M. Weaver, F. Yu, E. S. Wurtele, and B. J. Nikolau, *Plant Physiol.* **110**, 1021 (1996).

[16] J. Sambrook, E. F. Fritsch, and T. Maniatis, "Molecular Cloning: A Laboratory Manual," 2nd Ed. Cold Spring Harbor Laboratory, Cold Spring Harbor, New York, 1989.

39 kDa are determined. A set of fractions that contain a 39-kDa protein, the $NH_2$-terminal sequence from which matches that predicted from the gene sequence of the *bioB* gene, is identified.

Fractions containing the *bioB* gene product are combined, brought to 10% (w/v) $(NH_4)_2SO_4$, and loaded onto a phenyl-Sepharose column. The column is eluted with a decreasing gradient of $(NH_4)_2SO_4$, and fractions are collected. The fractions containing the *bioB* gene product as determined by the method described previously are pooled.

The combined fractions from the phenyl-Sepharose column are chromatographed on a Superdex 35/600 column and fractions are collected. The fractions containing the *bioB* gene product as determined by the method described are pooled. The results from a typical purification batch are shown in Table I. Starting from 1 kg of *E. coli* cell paste, approximately 500 mg of deeply colored, wine-red protein is obtained.[7]

The crude extract from the HMS174/pBioBF2 cells has an unusual reddish color, and during the purification of biotin synthase it becomes clear that the color is due to the presence of large amounts of this protein. More specifically, the color is due to the [2Fe-2S] cluster that biotin synthase contains. Judging by the red color in the various column fractions, it appears that there are at least three forms of biotin synthase. One has a native molecular mass of 82 kDa, one has a native molecular mass of 104 kDa, and one has a molecular mass of approximately 160 kDa. The two lower molecular mass species have been purified and characterized in some detail.

After the protein was purified, an *in vitro* assay for biotin synthase activity was established.[7,8] This assay has been used to detect biotin synthase in chromatographic fractions during subsequent purifications. The enzymatic activity in the various fractions is proportional to the amount of red color ($A_{453}$) present. Thus there appears to be close correspondence between the amount of Fe-S cluster in biotin synthase and the amount of enzymatic activity.

TABLE I
PURIFICATION OF *Escherichia coli bioB* GENE PRODUCT[a]

| Purification step | Amount of protein (mg) | Estimated purity (%) |
|---|---|---|
| Crude extract | 12,800 | 4 |
| Protamine sulfate precipitation | 9,200 | 6 |
| DEAE-Sepharose | 2,200 | 23 |
| Phenyl-Sepharose | 810 | 56 |
| Superdex | 520 | 90 |

[a] Reprinted with permission from I. Sanyal, G. Cohen, and D. H. Flint, *Biochemistry* **33**, 3625 (1994). Copyright 1994 American Chemical Society.

## Characterization of Biotin Synthase

Elemental analysis, ultraviolate (UV)–visible spectroscopy, and electron paramagnetic resonance (EPR) studies reveal that the 82-kDa form of *E. coli* biotin synthase is a homodimer containing one [2Fe-2S] cluster per monomer. Its UV–visible spectrum exhibits peaks at 275 nm ($\varepsilon$ = 3.3 × $10^4$ $M^{-1}$ cm$^{-1}$, based on monomer molecular mass), 330 nm ($\varepsilon$ = 1.4 × $10^4$ $M^{-1}$ cm$^{-1}$), 420 nm (shoulder, $\varepsilon$ = 6.0 × $10^3$ $M^{-1}$ cm$^{-1}$), 453 nm ($\varepsilon$ = 7.1 × $10^3$ $M^{-1}$ cm$^{-1}$), and 540 nm (shoulder, $\varepsilon$ = 3.5 × $10^3$ $M^{-1}$ cm$^{-1}$).[7] The cluster is unstable to reduction by dithionite.[7] The presence of a [2Fe-2S] cluster in biotin synthase from *B. sphaericus* has also been reported.[11]

The 104-kDa form of biotin synthase that has been isolated from *E. coli* cells in which the *bioB* gene is overexpressed appears to be a homodimer that contains a single [2Fe-2S] cluster per dimer. This species, which lacks a complete complement of cluster, can be converted to the 82-kDa molecular mass form (one [2Fe-2S] cluster/monomer) by incubation in the presence of $Fe^{3+}$, $S^{2-}$, and 2-mercaptoethanol.[7]

## Assay of Biotin Synthase

A defined mixture of components that supports biotin synthase activity has been described; however, optimization of the assay clearly remains to be established.[8] The defined *in vitro* reaction mixture that we have developed is adapted from a system described by Ifuku *et al.*[6] that utilizes cell-free extract of *E. coli*.

Although biotin synthase is not oxygen labile, the activity assay is usually performed anaerobically to avoid oxidative chemical reactions that can occur in the presence of some of the reaction components. The assay is carried out in 250 m$M$ morpholinepropanesulfonic acid (MOPS, pH 7.5). A typical assay mixture contains 2 $\mu M$ biotin synthase monomer, 50 $\mu M$ dethiobiotin, 10 m$M$ dithiothreitol (DTT), 10 m$M$ KCl, 1 m$M$ NADPH, 2.5 m$M$ Fe(NH$_4$)$_2$(SO$_4$)$_2$, 5 m$M$ fructose 1,6-diphosphate, 6.7 $\mu M$ flavodoxin reductase, 12.5 $\mu M$ flavodoxin, 500 $\mu M$ L-cysteine, and 150 $\mu M$ S-adenosylmethionine (AdoMet) in a volume of 0.1 to 1 ml. The mixture is incubated for up to 24 hr and the reaction is terminated by heating at 95° for 5 min. After centrifugation of the reaction mixtures, the biotin formed is quantitated using the microbiological assay described in the next section.

Photoreduced deazaflavin can substitute for flavodoxin reductase, flavodoxin, and NADPH as the source of reductant in the biotin synthase assay.[11]

Microbiological Assay for Biotin

A microbiological assay is used to quantitate the amount of biotin formed *in vitro*. *Lactobacillus plantarum,* a biotin auxotroph, and *E. coli* strain KS302Δ*bio,* which lacks the biotin operon and therefore requires biotin for growth, are both suitable organisms for this assay.[7] Our laboratory utilizes the *E. coli* deletion strain in the following way. A 25-ml culture of *E. coli* is grown in tryptone broth [1% (w/v) Difco (Detroit, MI) tryptone, 0.5% (w/v) NaCl] overnight at 37°. The cells are harvested by centrifugation and the cell pellet is washed three times with 25 ml of sterile saline [0.9% (w/v) NaCl] to remove any extracellular biotin. The pellet is resuspended in sterile saline such that the $A_{650}$ is approximately 1 unit. One hundred microliters of the washed cells are added to the top agar of bilayer assay plates that are prepared as follows.

*Bottom Layer.* The bottom layer is composed of M9 basal salts (per 1 liter: $Na_2HPO \cdot 7H_2O$, 13.2 g; $KH_2PO_4$, 3 g; $NH_4Cl$, 1 g; NaCl, 0.5 g) containing the following (autoclaved separately): 1 m$M$ $MgCl_2$, 0.02 m$M$ $CaCl_2$, 0.2% (w/v) glucose, 0.1% (w/v) vitamin-free Casamino Acids (Difco), thiamine (10 $\mu$g/ml), and 1.5% (w/v) Bacto-grade agar (Difco). Twenty-five milliliters of the constituted agar are poured into 9-cm petri dishes at >55°. The plates are allowed to cool before being stored in plastic bags at 4°. Plates containing only the bottom agar can be stored for up to 2 months.

*Top Layer.* The top agar is prepared by transferring 10 ml of melted 1.5% (w/v) agar into sterile 18 × 150 mm culture tubes. The agar is equilibrated at 47° in a water bath and 100 $\mu$l of 2% (w/v) Tetrazolium Red dye (2,3,5-triphenyltetrazolium chloride) and 100 $\mu$l of washed cells (final $A_{650}$ of approximately 0.01 units; see above) are added. The tube is vortexed and the bottom agar is immediately overlaid with the top agar. The plates are allowed to cool and can be used immediately or can be stored in plastic bags at 4° for up to 5 days.

*Quantitation of Biotin.* Three 6-mm diameter sterile paper disks are placed on the top agar of the petri plates prepared as described above. Ten microliters of the appropriate sample (standard D-biotin solutions or *in vitro* reaction mixtures) are applied per disk within a few minutes of placing the disks on the plate. The plates are incubated at 37° for approximately 16 hr. The diameter of the growth ring surrounding each disk (which appears red) is measured and compared to a standard curve prepared using biotin standards in which 10 $\mu$l contains 15, 10, 1, 0.5, 0.4, 0.3, 0.2, and 0.1 ng of biotin. The standard curve is conveniently prepared as a log–log plot, which yields a straight line.

Comments on *in Vitro* Biotin Synthase Reaction Mixture

Although the reaction mixture described previously supports biotin synthesis, only 1 mol of biotin is typically found per biotin synthase monomer, which suggests that the system is not functioning catalytically. One or more of the following hypotheses could explain this observation: (1) the components of the reaction mixture have not been optimized (i.e., a component is missing or limiting), (2) the biotin synthase is somehow modified to an inactive form after one turnover, or (3) only a small percentage of the biotin synthase added to the reaction is functional. A survey of the literature reveals that much remains to be established regarding the assay of biotin synthase *in vitro*.[7–9,11]

The physiological relevance of fructose 1,6-diphosphate in biotin biosynthesis is questionable. We observed that a number of phosphorylated intermediates of the glycolytic pathway, including glucose 6-phosphate, fructose 6-phosphate, dihydroxyacetone phosphate, and phosphoenolpyruvate, were able to substitute (to varying degrees) for fructose 1,6-diphosphate.[8] Fructose 1,6-diphosphate is not a component of the biotin synthase reaction mixture described by Birch *et al.*[9] Interestingly, these authors report that thiamine pyrophosphate (TPP) and one of the amino acids asparagine, aspartate, glutamine, or serine are required for activity. In our hands, however, TPP has no effect on the assay. We have found that fructose 1,6-diphosphate can be replaced by asparagine. Perhaps all of these reagents provide some nonspecific ionic effect, or are contaminated with a small amount of an agent required by the reaction.

A number of other issues regarding the *in vitro* assay of biotin synthase also remain to be solved. A requirement for a thiamine pyrophosphate-dependent protein for activity has been proposed.[9] Clearly, however, there is not an absolute requirement for such a protein because one is not included in our defined mixture. Stimulation of activity by a low molecular weight product of the 7,8-diaminopelargonic acid aminotransferase reaction (which precedes biotin synthase in the biotin biosynthetic pathway) has also been reported.[8] In the defined reaction mixture the formation of biotin is slow for an enzyme-catalyzed reaction. Typically it takes about 8 hr for 1 mol of biotin to be formed per mole of biotin synthase added to the assay. Although it is tempting to continue searching for a component of the cell-free extract that will increase the turnover number of biotin synthase, it should be noted that a single turnover is also typical of biotin synthase in the presence of cell-free extract.

## Source of Sulfur for Reaction

Biotin synthase is involved in the insertion of a sulfur atom into dethiobiotin to form the biotin molecule. As with most aspects of this system, attempts to identify the source of the sulfur atom for the *in vitro* reaction have yielded complex findings. In the presence of cell-free extract, the incorporation of $^{35}$S from [$^{35}$S]cysteine and [$^{35}$S]cystine has been re-ported.[9,17] In defined reaction mixtures that lack cell-free extract, however, no incorporation of $^{35}$S from either [$^{35}$S]cysteine or [$^{35}$S]AdoMet has been observed.[8] It appears that cell-free extract contains a component that is capable of converting the sulfur atom from cysteine to a form that can ultimately be incorporated into biotin. The three potential sources of sulfur in the defined reaction mixture that remain include (1) the [2Fe-2S] cluster of biotin synthase, (2) an unidentified sulfur-containing species associated with the biotin synthase, or (3) DTT. Although source 2 is an intriguing possibility, especially because it is consistent with the observed upper limit of one turnover per biotin synthase monomer, we have been unable to observe such a molecule. If DTT is serving as the source of sulfur *in vitro*, the lack of a physiological sulfur donor in our defined mixture might explain the low turnover number.

Mechanism

By analogy to other enzyme systems that require a similar set of cofactors for activity, much has been inferred about the possible mechanism of biotin synthase. Biotin synthase shares a number of similarities with anaerobic ribonucleotide reductase, lysine 2,3-aminomutase, and pyruvate formate-lyase. Most notably, AdoMet is required by all of these systems, and it has been demonstrated to be involved in the formation of a radical in all but the biotin synthase system.[18–20] Other features that biotin synthase shares with one or more of the aforementioned enzymes include (1) the presence of an Fe-S cluster, (2) the requirement for a flavodoxin reducing system, (3) the requirement for iron, and (4) the reactions catalyzed require abstraction of a hydrogen atom from an unactivated carbon atom. These observations are all consistent with the hypothesis that radical chemistry is involved in the mechanism of biotin synthase. No definitive evidence, however, for the existence of a radical in the biotin synthase system has been documented. Formation of the radicals in the pyruvate formate-lyase and anaerobic ribonucleotide reductase systems requires specific activating

[17] D. Florentin, B. Tse Sum Bui, A. Marquet, T. Oshiro, and Y. Izumi, *C.R. Acad. Sci. (Paris)* **317,** 485 (1994).

[18] E. Mulliez, M. Fontecave, J. Gaillard, and P. Reichard, *J. Biol. Chem.* **268,** 2296 (1993).

[19] J. Knappe, F. A. Neugebauer, H. P. Blaschkowski, and M. Gänzler, *Proc. Natl. Acad. Sci. U.S.A.* **81,** 1332 (1984).

[20] M. L. Moss and P. A. Frey, *J. Biol. Chem.* **265,** 18112 (1990).

enzymes.[19,21] It is possible that an unidentified activating enzyme is also involved in the biotin synthase system and that the absence of such an enzyme might contribute to the sluggishness and/or single turnover of the *in vitro* assay system.

[21] X. Sun, R. Eliasson, E. Pontis, J. Andersson, G. Buist, B.-M. Sjoeberg, and P. Reichard, *J. Biol. Chem.* **270**, 2443 (1995).

## [38] Biotin Synthase of *Bacillus sphaericus*

*By* B. TSE SUM BUI and A. MARQUET

**Dethiobiotin**  
**(DTB)**     **Biotin**

Biotin synthase catalyzes a unique transformation, shared only by lipoate synthase, namely the insertion of sulfur at nonactivated positions of the substrate. *S*-Adenosylmethionine (AdoMet) and an electron source are absolutely required for activity,[1,2] indicating that biotin synthase belongs to the family of deoxyadenosyl radical-dependent enzymes[3] (pyruvate formate-lyase,[4,5] lysine 2,3-aminomutase,[6] and anaerobic ribonucleotide reductase[7]).

[1] O. Ifuku, J. Kishimoto, S. Haze, M. Yanagi, and S. Fukushima, *Biosci. Biotech. Biochem.* **56**, 1780 (1992).
[2] T. Ohshiro, M. Yamamoto, Y. Izumi, B. Tse Sum Bui, D. Florentin, and A. Marquet, *Biosci. Biotech. Biochem.* **58**, 1738 (1994).
[3] D. Florentin, B. Tse Sum Bui, A. Marquet, T. Ohshiro, and Y. Izumi, *C.R. Acad. Sci.* (*Paris*) **317**, 485 (1994).
[4] M. Frey, M. Rothe, A. F. V. Wagner, and J. Knappe, *J. Biol. Chem.* **269**, 12432 (1994).
[5] C. V. Parast, K. K. Wong, J. W. Kozarich, J. Peisach, and R. S. Magliozzo, *Biochemistry* **34**, 5712 (1995).
[6] M. D. Ballinger, P. A. Frey, and G. H. Reed, *Biochemistry* **31**, 10782 (1992).
[7] E. Mulliez, M. Fontecave, J. Gaillard, and P. Reichard, *J. Biol. Chem.* **268**, 2296 (1993).

Although the sulfur of cysteine is incorporated to some extent in the presence of crude cell-free extracts,[3,8] the nature of the ultimate sulfur donor is still unknown.

## Assay Method

### Principle

Biotin synthase catalyzes the synthesis of biotin from dethiobiotin (DTB). Biotin is determined by a microbiological assay (paper disk method) with *Lactobacillus plantarum*.[9] The enzyme assays are carried out using procedures A and B described as follows.

### Reagents

Tris-HCl buffer, 1 *M*, pH 8
(+)-Dethiobiotin, 2.5 m*M*
Biotin standards, 0.25 to 25 μ*M*
NADPH, 25 m*M*
AdoMet (*p*-toluenesulfonate salt), 25 m*M*
Dithiothreitol (DTT), 100 m*M*
L-Cysteine, 25 m*M*
5-Deazaflavin (0.6 m*M*), prepared in 100 m*M* argon-saturated Tris-HCl, pH 8
Protein solution
Crude extracts from *Bacillus sphaericus* IFO 3525

### Procedure A

The standard assay system contains biotin synthase (1 mg/ml), crude extract (5 mg/ml), 50 μ*M* DTB, 1 m*M* NADPH, 0.5 m*M* AdoMet, 0.5 m*M* L-Cys, 10 m*M* DTT, 40 m*M* Tris-HCl buffer (pH 8), and distilled water to bring the volume to 100 μl. The reaction mixture is incubated at 37° for 1 hr and stopped by the addition of 10 μl of 12% (w/v) trichloroacetic acid. Precipitated proteins are removed by centrifugation and 5 μl of the supernatant is spotted onto 5-mm paper assay disks (Sanofi Diagnostics Pasteur, Garches, France), which are placed onto the agar surface of a biotin assay medium (Difco, Detroit, MI) plate seeded with *L. plantarum*. Disks containing 5 μl of appropriate standards of biotin are included on the plate. The plates are incubated overnight and the concentration of

---

[8] A. Méjean, B. Tse Sum Bui, D. Florentin, O. Ploux, Y. Izumi, and A. Marquet, *Biochem. Biophys. Res. Commun.* **217**, 1231 (1995).
[9] L. D. Wright and H. R. Skeggs, *Proc. Soc. Exp. Biol. Med.* **56**, 95 (1944).

biotin is estimated from the diameter of growth. The amount of biotin formed under these conditions is about 2 $\mu M$.

### Procedure B

The standard assay system is the same as that described in procedure A except that biotin synthase (2 mg/ml) is used and a 30 $\mu M$ concentration of photoreduced 5-deazaflavin replaces NADPH and the crude extract.[8] The amount of biotin formed under these conditions is about 0.7 $\mu M$.

*Definition of Specific Activity.* Specific activity is defined as nanomoles of biotin formed per milligram of protein per hour. Protein content is determined by the method of Bradford[10] with bovine serum albumin as standard.

### Purification Procedure

All chromatographic steps are performed at 4°. All buffers are degassed and saturated with argon to limit the amount of oxygen present.

### Strains

*Bacillus sphaericus* BT(250C [bioR⁻, actithiazic acid$^r$, 5-(2-thienyl)-valeric acid$^r$, 1-(2′-thenoyl)-3,3,3-trifluoroacetone$^r$], derived from *B. sphaericus* IFO3525[11] (wild type) transformed with plasmid pBHB5022,[12] which contains the *bioB* gene is employed.

*Step 1. Bacterial Cultivation.* BT(250)C[pBHB5022] cells are grown in the presence of kanamycin (10 $\mu$g/ml) in 800 ml of GP medium in 2-liter flasks. The GP broth contains per liter: 20 g of glycerol, 30 g of polypeptone (Organotechnie, La Courneuve, France), 5 g of vitamin-free Casamino Acids (Difco), 1 g of $K_2HPO_4$, 0.5 g of KCl, 0.5g of $MgSO_4 \cdot 7H_2O$, 0.01 g of $FeSO_4 \cdot 7H_2O$, 0.01 g of $MnSO_4$, and 0.02 g of thiamin hydrochloride, pH 7. The inoculum is 4 ml of an overnight culture grown at 30° in the same medium. The cells are grown for 40 hr with reciprocal shaking (140 strokes min⁻¹). When the $A_{600}$ of the culture reaches 6, the cells are harvested from the culture broth by centrifugation at 16,000 g for 30 min, suspended in 30 m$M$ Tris-HCl buffer (pH 8) to $A_{600}$ 300, and kept at −80° until use.

[10] M. M. Bradford, *Anal. Biochem.* **72**, 248 (1976).
[11] Y. Izumi, Y. Kano, K. Inagaki, N. Kawase, Y. Tani, and M. Yamada, *Agric. Biol. Chem.* **45**, 1983 (1981).
[12] R. Gloeckler, I. Ohsawa, D. Speck, C. Ledoux, S. Bernard, M. Zinsius, D. Villeval, T. Kisou, K. Kamogawa, and Y. Lemoine, *Gene* **87**, 63 (1990).

*Bacillus sphaericus* IFO 3525 is cultivated in the same way except that no kanamycin is added to the medium.

*Step 2. Preparation of Cell-Free Extracts.* The cells are thawed and without further dilution are disrupted by intermittent sonication with a Vibracell sonicator (Bioblock Scientific, Strasbourg, France). The cell debris is removed by centrifugation at 30,000 g for 30 min. The supernatant solution is referred to as the cell-free extract.

*Step 3. Immobilized Metal Chelate Affinity Chromatography (IMAC) on Co(II)-Chelating Sepharose.* The affinity matrix is prepared by charging a column packed with 60 ml (bed volume) of chelating Sepharose Fast Flow (Pharmacia, Uppsala, Sweden) with 200 ml of 50 m$M$ CoSO$_4$ followed by washing with water to remove the excess Co$^{2+}$. Then, 200 ml of each of the eluting buffers is passed through the column to remove any loosely bound Co$^{2+}$ that might be carried along during the elution of the proteins. After that, the column is equilibrated with 50 m$M$ Tris-HCl buffer plus 0.5 $M$ KCl, pH 7.5, and the cell-free extract is loaded on the column at a flow rate of 3 ml/min. Under the injection conditions used, most of the contaminating proteins do not bind and are eliminated in the nonretained fraction. The retained fractions are eluted stepwise at a rate of 5 ml/min with 200 m$M$ Tris-HCl buffer plus 0.5 $M$ KCl, pH 8, followed by 200 m$M$ Tris-HCl plus 0.5 $M$ KCl plus 20 m$M$ imidazole, pH 8. The latter fraction, which is the most active in the production of biotin, is concentrated to a final volume of about 50 ml in an Amicon (Beverly, MA) ultrafiltration cell using a YM10 membrane under argon atmosphere. The concentrated protein is diluted with 200 ml of 30 m$M$ Tris-HCl buffer containing 5 m$M$ DTT, pH 8, and reconcentrated to give a final protein concentration of about 10 mg/ml. The enzyme is kept frozen at $-80°$ until use. This fraction is about 80% pure as determined by gel densitometry performed on a Shimadzu (Tokyo, Japan) dual-wavelength flying spot scanning densitometer.

*Step 4. Fast Protein Liquid Chromatography on Mono Q.* The enzyme solution is injected by 2-ml fractions (20 mg of protein) into a Mono Q HR 10/10 column and a gradient elution is performed using 30 m$M$ Tris-HCl, pH 8, containing 5 m$M$ DTT in channel A and the same buffer plus 0.5 $M$ KCl in channel B. The gradient is 0–100% B in 20 min and the flow rate is 2.5 ml/min. The first peak coming out at about 0.4 $M$ KCl contains pure biotin synthase and is concentrated in Centriprep-30 concentrators (Amicon) to 1 ml, diluted 15-fold with 30 m$M$ Tris-HCl buffer containing 5 m$M$ DTT, pH 8, and reconcentrated to 10 mg of protein per milliliter. This enzyme preparation is homogeneously pure [single band on sodium dodecyl sulfate (SDS)-12% polyacrylamide gel]. The overall purification

TABLE I
PURIFICATION OF BIOTIN SYNTHASE

| Step | Protein (mg) | Specific[a] activity (nmol/hr · mg) | Total activity[a] (nmol/hr) | Activity[a] yield (%) |
|---|---|---|---|---|
| Crude extract | 3124 | 0.5 | 1562 | 100 |
| IMAC | 240 | 1.9 | 456 | 29 |
| Mono Q | 48 | 1.6 | 76.8 | 4.9 |

[a] Although biotin synthase has been purified to homogeneity, the specific activity has increased by only threefold, which indicates that the enzyme has either been inactivated owing to loss of its [Fe-S] cluster or more likely that essential cofactors or proteins have been lost during purification.

results are summarized in Table I for a typical enzyme purification from 4 liters of culture.

Properties of Purified Enzyme

*Molecular Weight.* The molecular weight of the enzyme as predicted by cDNA sequence is 36,954.[13] The molecular weight as determined by SDS-polyacrylamide gel electrophoresis is about 37,000. In the native state, the molecular weight based on gel filtration with Superdex 200 HR 10/30 is approximately 75,000, indicating that biotin synthase is a homodimer. Pure biotin synthase gives the expected following $NH_2$ sequence: M-N-W-L-Q-L-A-D-E-V-I-A-G-K-.

*Spectroscopic Characterization and [Fe-S] Cluster Content.* The absorption spectrum of pure biotin synthase is shown in Fig. 1 and presents absorption bands compatible with the presence of a [2Fe-2S] cluster.[14] Protein-bound iron and inorganic sulfur as assayed by the method of Fish[15] and Beinert[16] are, respectively, 1 and 0.8 mol/monomer, indicating that the enzyme is cluster deficient. Biotin synthase is oxygen sensitive and loses all its cluster if the purification is carried out with buffers that have not been saturated with argon.

*Cofactor Requirements.* Optimum concentrations of the cofactors are used in the assay system; of these, AdoMet and crude extracts of the wild strain are the essential components without which no activity can be detected. KCl (0.5 to 20 m$M$) and FAD (0.1 to 1 m$M$), which stimulated

[13] I. Ohsawa, D. Speck, T. Kisou, K. Hayakawa, M. Zinsius, R. Gloeckler, Y. Lemoine, and K. Kamogawa, *Gene* **80,** 39 (1989).

[14] W. H. Orme-Johnson and N. R. Orme-Johnson, *Methods Enzymol.* **53,** 259 (1978).

[15] W. W. Fish, *Methods Enzymol.* **158,** 357 (1988).

[16] H. Beinert, *Anal. Biochem.* **131,** 373 (1983).

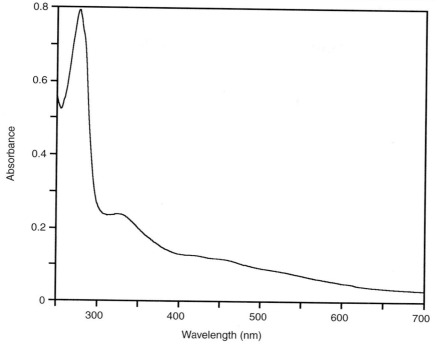

FIG. 1. Absorption spectrum of pure biotin synthase (1 mg/ml) in Tris-HCl buffer, pH 8.

activity with the crude extracts of the transformant,[2] have no effect. Fructose 1,6-bisphosphate (5 m$M$) has no effect and Fe$^{2+}$ (>1 m$M$) is inhibitory. These cofactors were reported to have stimulating effects.[1]

*Sulfur Donor.* [$^{35}$S]Cysteine is found to give $^{35}$S to biotin to some extent under the conditions of procedure A but not under the conditions of procedure B. The ultimate sulfur donor has not yet been identified.

*Kinetic Parameters.* Under the assay conditions described previously, biotin synthase is not catalytic as only 2 nmol of biotin is produced by 25 nmol of enzyme. The double-reciprocal plot of 1/v of the biotin synthase reaction plotted against 1/[DTB], however, gives a straight line from which the *apparent* $K_m$ and $V_{max}$ values were determined to be 1.6 $\mu M$ and 3.6 nmol of biotin/mg · hr, respectively.

*Substrate Specificity.* A postulated intermediate, 9-mercaptodethiobiotin)[3] is also a substrate of the enzyme and is transformed into biotin. The optimum concentrations of cofactors NADPH, AdoMet, L-Cys, DTT, and of crude extracts of the wild strain, are identical to those needed for DTB. The *apparent* $K_m$ and $V_{max}$ values for 9-mercaptodethiobiotin as deduced from a double-reciprocal plot were found to be 5 $\mu M$ and 1.2 nmol of biotin/mg · hr, respectively.

# [39] *Escherichia coli* Repressor of Biotin Biosynthesis

*By* Dorothy Beckett and Brian W. Matthews

## *Escherichia coli* Biotin Regulatory System

The interactions outlined in Fig. 1 are responsible for control of biotin retention and biosynthesis in *Escherichia coli.*[1] Biotin is retained in two protein-bound forms in *E. coli;* covalently ligated to the biotin carboxyl carrier protein (BCCP) of the acetyl-CoA carboxylase or noncovalently bound in an adenylated form to the *E. coli* repressor of biotin biosynthesis (BirA). The latter protein, the key element of the biotin regulatory system, funnels biotin into metabolism by catalyzing its covalent linkage to BCCP and binds sequence specifically to the biotin operator sequence to repress initiation of transcription at the two promoters of the biotin biosynthetic operon (Fig. 2).[2,3] Transfer of biotin to BCCP is a two-step process involving activation of the biotin via formation of the acyl-adenylate, followed by transfer of the activated biotin to the $\varepsilon$-amino group of a unique lysine residue on the acceptor protein.[4] This two-step process is reminiscent of the tRNA synthetase-catalyzed linkage of amino acids to the 3′ termini of tRNAs.[5,6] A detailed discussion of BirA-catalyzed biotinyl-5′-adenylate (bio-5′-AMP) formation is contained in [43] in this volume.[7] Interestingly, the adenylate serves the additional function in this system of activating BirA for sequence-specific binding to the biotin operator sequence. This second role of the adenylate, based on the noted analogy between the biotin holoenzyme synthetases and the tRNA synthetases, was predicted by M. Eisenberg more than 20 years ago.[6]

[1] J. E. Cronan, Jr., *Cell* **58,** 527 (1989).
[2] D. F. Barker and A. M. Campbell, *J. Mol. Biol.* **146,** 451 (1981).
[3] D. F. Barker and A. M. Campbell, *J. Mol. Biol.* **146,** 469 (1981).
[4] M. D. Lane, K. L. Rominger, D. L. Young, and F. Lynen, *J. Biol. Chem.* **239,** 2865 (1964).
[5] A. Fersht, "Enzyme Structure and Mechanism," pp. 212–214. W.H. Freeman, New York, 1985.
[6] M. A. Eisenberg, *Adv. Enzymol.* **38,** 317 (1973).
[7] Y. Xu and D. Beckett, *Methods Enzymol.* **279** [43], 1997 (this volume).

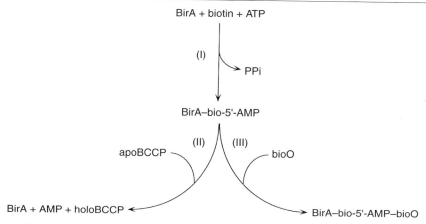

FIG. 1. Schematic representation of the biotin regulatory system. The repressor of biotin biosynthesis, BirA, (reaction I) catalyzes synthesis of biotinyl-5'-adenylate (bio-5'-AMP) from the two substrates biotin and ATP. The BirA–bio-5'-AMP complex can (reaction II) then bind to the unbiotinylated biotin carboxyl carrier protein (apoBCCP) subunit of acetyl-CoA carboxylase, which results in transfer of biotin to a lysine residue on the acceptor protein or (reaction III) bind to the biotin operator sequence, bioO, to repress transcription of the biotin biosynthetic operon.

The biotin regulatory system functions via feedback inhibition. In the presence of a low demand for biotin, signaled by low intracellular apoBCCP levels, holoBirA (BirA–bio-5'-AMP) occupies the biotin operator sequence and transcription of the biotin biosynthetic genes is repressed. As apoBCCP accumulates in the cell, it competes with the biotin operator (bioO) for

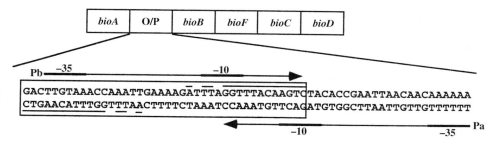

FIG. 2. Schematic representation of the biotin biosynthetic operon and the sequence of the biotin operon transcriptional control region, bioA through bioF are the genes that encode the biotin biosynthetic enzymes. O/P is the transcriptional control region of the operon and Pa and Pb are the promoters for leftward and rightward transcription, respectively. The boxed sequence is the biotin operator sequence. Note: The underlined sequences are identical in the two halves of the pseudopalindromic bioO sequence.

interaction with holoBirA and transcriptional initiation at the biotin biosynthetic operon becomes derepressed. On depletion of apoBCCP via its conversion to the biotinated form, BirA–bio-5′-AMP accumulates and binds to bioO to repress transcription. The functional state of BirA is thus modulated by fluctuations in the intracellular concentrations of apoBCCP and biotin. A unique feature of this system is that the same protein that is responsible for funneling biotin into metabolism, by catalyzing its covalent linkage to apoBCCP, also functions to repress synthesis of biotin at the transcriptional level.

The 321 amino acids of the BirA sequence encode a diverse array of small molecule and macromolecular ligand-binding sites, as well as two catalytic functions. The protein is, moreover, an allosteric site-specific DNA-binding protein that is activated via adenylate binding. A combination of structural and functional studies of the biotin repressor has provided some insight into the mechanism of action of this unique multifunctional protein.

## Structure of Biotin Repressor

### Three-Dimensional Structure of BirA

The structure of BirA has been determined by X-ray crystallography and refined to a crystallographic residual of 19.0% at 2.3-Å resolution.[8] In the crystals, adjacent molecules are related by a crystallographic twofold symmetry axis. The area of contact, however, is about 1100 Å$^2$, which is substantially less than the value of approximately 1600 Å$^2$ expected for dimeric proteins of comparable size.[9] The crystal structure is, therefore, consistent with the observation that BirA is predominantly monomeric in solution.[10,11]

The structure of BirA, shown in Fig. 3, consists of three domains. The amino-terminal domain is responsible for DNA binding. It contains a well-defined "helix–turn–helix" motif (residues 22–46) as was anticipated from sequence comparisons with other DNA-binding proteins as well as mutagenesis experiments.[12] The tenuous coupling of the DNA-binding domain to the remainder of the protein suggests that major changes in conformation could accompany DNA binding.

[8] K. P. Wilson, L. M. Shewchuk, R. G. Brennan, A. J. Otsuka, and B. W. Matthews, *Proc. Natl. Acad. Sci. U.S.A.* **89,** 9257 (1992).

[9] S. Miller, A. M. Lesk, J. Janin, and C. Chothia, *Nature (London)* **328,** 834 (1987).

[10] J. Abbott and D. Beckett, *Biochemistry* **32,** 9649 (1993).

[11] E. Eisenstein and D. Beckett, unpublished results (1996).

[12] M. R. Buoncristiani, P. K. Howard, and A. J. Otsuka, *Gene* **44,** 255 (1986).

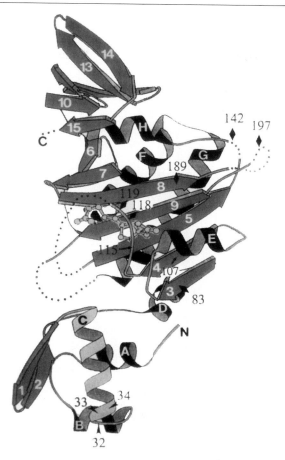

Fig. 3. Ribbon drawing showing the overall fold of the biotin repressor. The lower domain is the DNA-binding region of the molecule, the catalytic domain is above, and the carboxy-terminal domain is at the top of the molecule.[8] Dotted regions indicate parts of the backbone that cannot be seen in the electron density maps and are apparently disordered in the crystal structure.

The second domain encompasses the central part of the polypeptide chain and includes the active site for catalysis. It consists of five helices and a mixed β sheet (Fig. 4). The domain contains four loops that appear to be mobile or unstructured and cannot be visualized in electron density maps. These loops occur in pairs in the three-dimensional structure (Figs. 3 and 4). One disordered loop (residues 116–124) contains the sequence GRGRRG, previously identified as a consensus sequence (GXGXXG)

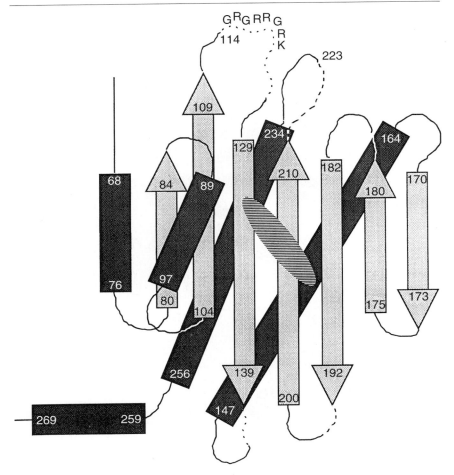

Fig. 4. Schematic illustration showing the topology of the central, catalytic domain of BirA. β Strands are drawn lightly shaded, with arrowheads. Helices are shown darker. The dashed lines show the parts of the backbone that are disordered in the crystal structure. The shaded oval shows the location of the biotin-binding site. [Original figure drawn by K. Wilson.]

associated with ATP binding.[13,14] In the nearby unstructured region (residues 212–223), limited proteolysis has revealed a subtilisin-sensitive bond that is partially protected from cleavage in the presence of the corepressor.[15] This also suggests that this region of the protein is at or close to the site of binding of corepressor.

To date, no high-resolution structure of the BirA–biotin complex is available. The biotin-binding site has, however, been inferred from a structure of BirA bound to biotinyllysine (biocytin),[8] an analog of the overall product of the biotin ligation reaction. The site includes parts of three $\beta$ strands (strands 5, 8, and 9), the N terminus of helix E, and main-chain atoms of residues 114–118 (Figs. 3 and 4). Residues 116 to 118 are disordered in the apoprotein. On biotinyllysine binding they become localized and partially bury the ligand. The binding pocket has hydrophobic sides with a more hydrophilic center. The biotin carbonyl and ureido nitrogens interact via hydrogen bonding with the hydrophilic interior of the cavity, while the hydrophobic tail and thiophene ring interact with the hydrophobic sides of the pocket. Specificity is derived from several hydrogen bonds between the biotin and the protein. Particularly striking are the hydrogen bonds from the amide nitrogen of Arg-116 to the carbonyl oxygen of the biotin molecule, and from the carbonyl oxygen of Arg-116 to one of the biotin ureido nitrogens. In this instance, the hydrogen bond donors and acceptors on biotin participate in hydrogen bonds with the protein that are characteristic of $\beta$-sheet structure. There is also a hydrogen bond between the biotin carbonyl oxygen and the side-chain oxygen of Ser-89. The remaining ureido nitrogen participates in a bidentate hydrogen bond with the side-chain oxygen atoms of Thr-90 and Gln-112.

No direct crystallographic evidence for the ATP-binding site has been obtained, although it is presumed that ATP binds adjacent to biotin.

The carboxy-terminal domain is composed primarily of $\beta$ sheet. Its contribution to BirA function is not known.

### Correlation of Genetic Studies of Mutants to Structure

Regions of BirA likely to interact with DNA are suggested by mutants that inactivate repressor functions but have little effect on enzymatic activity.[12] Such mutations are restricted to amino acids 32–34, which are located within the helix–turn–helix motif. The substitutions are approximately 40 Å away from the biotin-binding site, consistent with the observation that they have little effect on enzymatic activity. The remaining amino acid

[13] R. K. Wierenga and W. G. Hol, *Nature* (*London*) **302**, 842 (1982).
[14] G. E. Schulz, *Curr. Opin. Struct. Biol.* **2**, 61 (1992).
[15] Y. Xu, E. Nenortas, and D. Beckett, *Biochemistry* **34**, 16624 (1995).

A
DNA-binding fold of PheRS

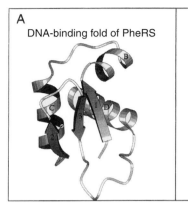

DNA-binding fold of BirA

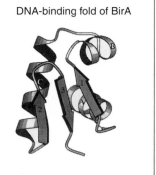

B
Catalytic-like module of PheRS

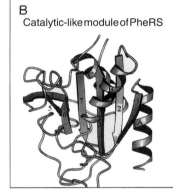

Biotin-binding domain of BirA

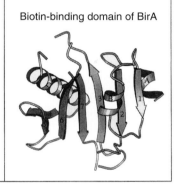

C
SH3-like fold of PheRS

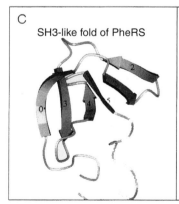

SH3-like fold of BirA

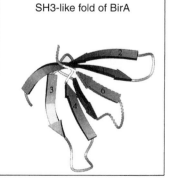

substitutions that are defective in repression (residues 118, 119, 142, and 197) are located at least 20 Å away from the presumed DNA recognition helix. These mutations could disrupt direct communication between the corepressor-binding site and DNA-binding domain, or they might have a more indirect effect such as interference with protein dimerization that accompanies DNA binding (see below).

Mutations that affect the enzymatic activity of BirA occur at residues 107, 115, and 189, each of which is within 10 Å of the biotin-binding site. These substitutions are, therefore, in a location where they could interfere directly or indirectly with the binding of biotin and/or ATP.[8]

Mutations in BirA that prevent growth are located at positions 83, 142, 207, and 235, and are distributed throughout the three-dimensional structure. They are believed to destabilize the folded structure.[8]

### Similarities between BirA and tRNA Synthetases

At the time when the structure of BirA was determined[8] a search of the Brookhaven Data Base suggested that the catalytic domain of the protein had a novel fold, not previously seen in any other protein. Subsequently, however, the coordinates have become available for two structures that are related. One of these is serine tRNA synthetase (SeRS)[16,17] and the other is phenylalanyl-tRNA synthetase.[18] In the latter case, the structural similarity encompasses not only the respective catalytic domains (Fig. 5B) but also includes the DNA-binding region (Fig. 5A) and a Src homology 3 (SH3)-like domain[19] that occurs in both enzymes (Fig. 5C). Notwithstanding the structural correspondence, there is no clear sequence homology.[18] Nevertheless, it seems likely that the tRNA synthetases and the biotin repressor evolved from a common precursor. More than 20 years ago, Eisenberg[6] pointed out functional similarities between the two classes of

[16] S. Cusack, G. Berthet-Colominas, M. Hartlein, N. Nassar, and R. Leberman, *Nature (London)* **347**, 249 (1990).

[17] P. J. Artymiuk, D. W. Rice, A. R. Poirrette, and P. Willet, *Struct. Biol.* **1**, 758 (1994).

[18] M. Safro and L. Mosyak, *Protein Sci.* **4**, 2429 (1995).

[19] M. E. M. Noble, A. Musacchio, M. Saraste, S. A. Courtneidge, and R. Wierenga, *EMBO J.* **12**, 2617 (1993).

FIG. 5. (A)–(C) show the structural and topological correspondence between parts of phenylalanyl-tRNA synthetase (left) and BirA (right). The common secondary structure elements are shown as sequentially labeled ribbons. (A) The DNA-binding regions. (B) The catalytic regions. (C) Regions of the two proteins that display SH3-like topologies. Dashed lines emphasize different exit pathways from the domains. [From M. Safro and L. Mosyak, *Protein Sci.* **4**, 2429 (1995). Reprinted with the permission of Cambridge University Press.]

enzymes. The synthetases activate amino acids by proceeding through an acyladenylate intermediate; BirA activates biotin by forming biotinyl-5'-adenylate. The finding that the two enzymes share structurally related catalytic domains is a striking vindication of Eisenberg's prescient observations.

## Solution Studies of Repressor of Biotin Biosynthesis

### Kinetic and Thermodynamic Studies of BirA–Small Ligand Interactions

The functional state of BirA is regulated by binding of small molecules. When biotin is bound to the protein it functions simply to catalyze synthesis of the adenylate. Adenylate binding activates the protein for site-specific DNA binding. Results of solution studies of the BirA–biotin and BirA–bio-5'-AMP interactions have begun to reveal how functional switching occurs in the protein. The equilibrium parameters governing interaction of BirA with the two small ligands have been determined from results of kinetic measurements.[15,20] Association of the ligands with BirA has been measured using stopped-flow fluorescence. This is made possible by the fact that binding of either ligand to the protein results in quenching of the intrinsic protein fluorescence signal. The kinetics of dissociation of the BirA–biotin complex have been determined using stopped-flow fluorescence. The rate of dissociation of the BirA–adenylate complex was determined from the time course of BirA-catalyzed synthesis of the adenylate from the two substrates biotin and ATP.

The kinetics of association of both biotin and bio-5'-AMP with BirA are characterized by two single exponential phases.[15] Results of measurement of the ligand concentration dependencies of the time constants of the two phases indicate that while the first displays a linear dependence on ligand concentration, the second is ligand concentration independent. These kinetics are consistent with the following two-step model:

$$\text{BirA} + \text{ligand} \underset{k_{-1}}{\overset{k_1}{\rightleftharpoons}} \text{BirA–ligand} \underset{k_{-2}}{\overset{k_2}{\rightleftharpoons}} {}^*\text{BirA–ligand} \qquad (1)$$

in which binding involves initial rapid formation of a collision complex, BirA–ligand, followed by a slow unimolecular conformational change in the complex to *BirA–ligand. The individual steps in the kinetic mechanism are governed by the four microscopic rate constants shown in the equation. Analysis of the kinetic data reveals that the bimolecular association constant

[20] Y. Xu and D. Beckett, *Biochemistry* **33**, 7354 (1994).

governing binding of either ligand to BirA is equivalent to $k_1$, the bimolecular rate constant governing formation of the collision complex.[15]

The unimolecular rate of dissociation of the BirA–biotin complex has been measured by stopped-flow fluorescence.[15] The rate constant for the process obtained from these measurements is 0.3 sec$^{-1}$. The unimolecular rate constant governing dissociation of the BirA–bio-5'-AMP complex has been obtained from measurements of the time course of BirA-catalyzed synthesis of bio-5'-AMP from the two substrates biotin and ATP.[20] Details of the methods utilized for these measurements are contained in [43] in this volume.[7] The time course of adenylate synthesis is characterized by an initial exponential burst followed by a slow linear phase. While the burst reflects synthesis of one molecule of bio-5'-AMP per molecule of enzyme, the linear phase is rate limited by the slow dissociation of the enzyme–adenylate complex.[20] Interestingly, the aminoacyl-tRNA synthetase-catalyzed synthesis of the aminoacyl adenylate is also characterized by burst kinetics.[21] The previously mentioned analogy between the chemistry of the overall reactions catalyzed by these two synthetases as well as the similarities in their three-dimensional structures thus extend to kinetic mechanism. Quantitation of the second slow phase in the time course of bio-5'-AMP synthesis allows for determination of the kinetic stability of the BirA–adenylate complex. The unimolecular rate constant governing dissociation of the complex obtained from the slow phase in the synthetic time course at pH 7.5, 200 m$M$ KCl, and 20° is 0.00027 sec$^{-1}$.[20] The 30-min half-life calculated from this rate constant indicates that the BirA–adenylate complex is kinetically stable.

Equilibrium parameters for interaction of BirA with biotin and the adenylate have been calculated from the kinetic data. Results of these calculations are shown in Table I. As indicated by the parameters, the thermodynamic stability of the BirA–adenylate complex is considerably greater than that of the BirA–biotin complex.

*Ligand-Linked Conformational Changes in BirA*

A number of experimental results support the occurrence of significant changes in the conformation of BirA concomitant with small ligand binding. First, diffusion of biotin or bio-5'-AMP into crystals of apoBirA is accompanied by cracking of the crystals.[22] Second, the biphasic kinetics of association of BirA with biotin and bio-5'-AMP are consistent with a model in which binding occurs via initial formation of a collision complex followed by a

[21] A. R. Fersht, J. S. Ashford, C. J. Bruton, R. Jakes, G. L. E. Koch, and B. S. Hartley, *Biochemistry* **14,** 1 (1975).
[22] K. Wilson, L. Shewchuk, and B. W. Matthews, unpublished observations (1992).

TABLE I
EQUILIBRIUM DISSOCIATION CONSTANTS AND GIBBS FREE ENERGIES
FOR BINDING OF BIOTIN AND Bio-5'-AMP TO BirA

| Complex | $K_D (M)^{a,b}$ | $\Delta G$ (kcal/mol) |
|---|---|---|
| BirA–bio-5'-AMP | $5 (\pm 2) \times 10^{-11}$ | $-13.9 \pm 0.3$ |
| BirA–biotin | $4.1 (\pm 0.5) \times 10^{-8}$ | $-9.9 \pm 0.3$ |

[a] Equilibrium dissociation constants were calculated from the measured bimolecular association rate constants for binding of the ligands to BirA, $k_1$, and unimolecular rate constants governing dissociation of the complexes, $k_{off}$.

[b] Measurements were performed in the following buffer: 10 m$M$ Tris-HCl (pH $7.50 \pm 0.02$ at $20.0 \pm 0.1°$), 200 m$M$ KCl, 2.5 m$M$ MgCl$_2$.

slow conformational change in the complex. Moreover, the kinetic results indicate that the conformational changes accompanying biotin binding are distinct from those accompanying adenylate binding, because both the amplitude of the fluorescence change that occurs in the second binding phase, as well as the rate constants governing the conformational change(s), are distinct for the two ligands.[15]

Results of other solution studies of biotin and bio-5'-AMP binding lend additional support to the occurrence of ligand-specific changes in the conformation of BirA. These results also provide information about one location of the ligand-induced changes in the protein structure. The kinetic techniques described previously have been applied to measure the temperature dependencies of biotin and bio-5'-AMP binding to BirA.[23] In addition, direct calorimetric measurements of the two binding processes have been performed.[23] Results of these combined measurements indicate that both ligand-binding reactions are, in the physiological range of temperature, accompanied by large, favorable enthalpic changes. The entropic contributions to the two binding processes are, however, opposite in sign for the two ligands. That is, an unfavorable entropic contribution to the energetics of the BirA–biotin interaction results in the observed lower overall stability of this complex relative to the BirA–bio-5'-AMP complex for which both enthalpic and entropic contributions to binding are energetically favorable. The results of thermodynamic measurements of the two binding processes have been interpreted using the structure-based thermodynamic analysis methods of Freire et al.[24] Results of the analysis indicate a loss of configurational entropy on binding of either ligand to BirA. The loss in configura-

[23] Y. Xu, C. R. Johnson, and D. Beckett, *Biochemistry* **34**, 5509 (1996).
[24] J. Gomez and E. Freire, *J. Mol. Biol.* **252**, 337 (1995).

tional entropy is significantly greater for biotin than for adenylate binding, which is consistent with the idea that ligand-induced structural changes in BirA are distinct for biotin and bio-5'-AMP binding.

Results of partial proteolytic digestion of BirA, as well as studies of an N-terminal deletion mutant of the protein, have allowed identification of one locus of ligand-linked structural change in BirA. Limited treatment of BirA with the endoprotease subtilisin results in initial cleavage of the protein at the peptide bond linking amino acid residues 217 and 218.[15] This cleavage site is located in the unstructured loop in the central domain of the protein that includes residues 212–223. Measurement of the rates of proteolytic digestion of BirA and its complexes with biotin and bio-5'-AMP indicates that binding of either ligand results in a significant decrease in the apparent first-order rate of subtilisin-catalyzed proteolysis. The magnitude of the decrease in rate is, however, significantly greater for adenylate than for biotin binding. An N-terminal domain deletion mutant of BirA, BirA65–321, has been constructed and purified. Biochemical characterization of the mutant protein revealed that its affinity for bio-5'-AMP is $10^3 - 10^4$ lower than the affinity of the intact repressor for the adenylate.[25] The DNA-binding domain of BirA is thus required for tight binding of the adenylate. As indicated in the previous discussion of the structural studies, the interaction between the N-terminal and central domains in the aporepressor structure is tenuous. The partial proteolysis studies in combination with the results of studies of the N-terminal domain deletion mutant support a model in which binding of the adenylate is accompanied by formation of additional contacts between the DNA-binding and catalytic domains in BirA. Residues in the unstructured loop 212–223 may lie in this interdomain interface in the holorepressor.

*Interaction of BirA with DNA*

The biotin operator sequence is a 40-base pair (bp) imperfect palindrome (Fig. 2). Stoichiometric studies of binding of holoBirA with bioO indicate that two protein monomers bind to the operator sequence.[10] As indicated in Fig. 3, apoBirA is monomeric in the crystal structure. Results of solution studies of the assembly properties of BirA indicate that while apoBirA and the BirA–biotin complexes are monomeric at concentrations as high as 200 $\mu M$, holoBirA does undergo limited assembly to dimers. The estimated $K_D$ for the dimerization process is 30 $\mu M$.[11]

Site-specific binding of holoBirA to bioO has been measured using the quantitative DNase footprint technique.[10] Results of these measurements

[25] Y. Xu and D. Beckett, *Biochemistry* **35**, 1783 (1996).

TABLE II
RATE PARAMETERS FOR BirA-CATALYZED TRANSFER OF BIOTIN FROM
Bio-5'-AMP TO ACCEPTOR PROTEIN[a,b]

| Protein | $k_{cat}/K_m$ ($M^{-1}$ sec$^{-1}$) | $k_{cat}$ (sec$^{-1}$) | $K_m$ ($\mu M$) |
|---------|-------------------------------------|------------------------|------------------|
| BCCP    | 10.4 ($\pm$0.5) $\times$ 10$^3$     | 8.6 $\pm$ 0.9          | 800 $\pm$ 100    |

[a] Adapted from Ref. 27 with permission.
[b] Buffer conditions are identical to those indicated in Table I.

indicate that the protein binds to the operator in the nanomolar range of repressor concentration. Analysis of footprinting data indicates that operator binding is characterized by cooperative association of two holoBirA monomers with the two operator half-sites. At 200 m$M$ KCl, 10 m$M$ Tris-HCl (pH 7.5 at 20°) the intrinsic free energies of binding are approximately −9.0 kcal/mol for each half-site and the cooperative free energy is between −2 and −3 kcal/mol. Additional DNase footprinting studies performed using mutants in which an entire operator half-site is replaced by nonspecific sequence indicate that DNA binding and dimerization are tightly coupled in the holoBirA–bioO binding interaction.[26] The cooperative free energy observed in binding of holoBirA to bioO may reflect, in part, formation of a protein–protein interface between the two monomers bound at the two operator half-sites. Given this apparent linkage between site-specific DNA binding and dimerization in the holoBirA–bioO interaction, the observed positive linkage of bio-5'-AMP binding and dimerization suggests that one mechanism of allosteric activation of BirA by the corepressor for binding to bioO is via activation of the protein dimerization function.

*Transfer of Biotin from Adenylate to Biotin Carboxyl Carrier Protein*

The kinetics of BirA-catalyzed transfer of biotin from the adenylate to apoBCCP have been measured by stopped-flow fluorescence.[27] Results of these measurements yielded the kinetic parameters in Table II. On the basis of these parameters and the kinetic parameters governing formation of the adenylate from biotin and ATP, a model for control of partitioning of BirA–bio-5'-AMP between ligase and DNA-binding functions has been proposed. This model is summarized in Fig. 6. Two kinetic processes govern the interconversion of BirA between its apo and holo forms: synthesis of bio-5'-AMP from biotin and ATP and binding of BirA–bio-5'-AMP to apoBCCP with concomitant conversion of apoBCCP to holoBCCP. While

[26] E. D. Streaker and D. Beckett, unpublished results (1997).
[27] E. Nenortas and D. Beckett, *J. Biol. Chem.* **271,** 7559 (1996).

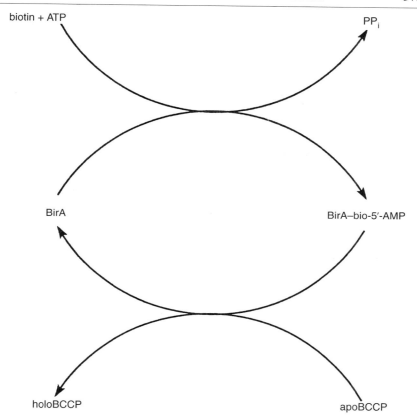

Fig. 6. Kinetic routes for interconversion of BirA between unliganded and active adenylate-bound forms. See text for detailed discussion. (Adapted from Ref. 27 with permission.)

the rate of the former process is dependent on the kinetic constants governing adenylate formation and the intracellular biotin concentration, the rate of the latter process depends on the kinetic parameters governing transfer of biotin to apoBCCP and the intracellular apoBCCP concentration. From this model it is evident that the half-life of BirA–bio-5′-AMP is directly related to the intracellular apoBCCP concentration; a low acceptor protein concentration results in a longer BirA–bio-5′-AMP half-life. The probability of formation of the transcriptional repression complex between holo-BirA and bioO is, in turn, related to the half-life of the BirA–bio-5′-AMP complex and the longer this half-life, the greater the probability of binding of holoBirA to bioO. Further quantitative analysis of the relationship of the BirA–bio-5′-AMP half-life to formation of the repression complexes awaits direct measurement of the kinetics of association of holoBirA with bioO.

## Concluding Remarks

The repressor of biotin biosynthesis is an allosteric site-specific DNA-binding protein that also catalyzes an essential posttranslational modification reaction. Results of the studies described in this chapter have begun to reveal structural and thermodynamic details of BirA function. These results, moreover, serve to direct the course of future studies of the protein. For example, elucidation of the structural details of regulation of BirA function via small ligand binding will require high-resolution structures of complexes of the protein bound to biotin and bio-5'-AMP. A high-resolution structure of the ternary holoBirA–bioO complex will reveal how a dimeric helix–turn–helix protein interacts with an unusually large 40-bp target site. Finally, further thermodynamic and kinetic studies of the multiple macromolecular interactions in which BirA participates will provide information critical to a complete understanding of the functional energetics of this unique regulatory protein.

## Acknowledgments

D.B. acknowledges the support of a Du Pont Young Professorship. Grants from the National Institutes of Health (GM46511 to D.B. and GM20066 to B.W.M.) are also gratefully acknowledged.

## [40] Structure of ATP-Dependent Carboxylase, Dethiobiotin Synthase

By Gunter Schneider and Ylva Lindqvist

## Introduction

In *Escherichia coli,* most of the enzymes involved in the biosynthetic pathway from pimeoyl-coenzyme A to biotin are encoded in a gene cluster, the *bio* operon.[1] This gene cluster consists of five genes and four of these have been shown to be linked to biotin synthesis. The gene *bioA* encodes a 7,8-diaminopelargonate synthase, *bioB* encodes a component of the biotin

---

[1] M. A. Eisenberg, *Ann. N.Y. Acad. Sci.* **447,** 335 (1985).

synthase, *bioD* encodes dethiobiotin synthetase, and *bioF* encodes 8-amino-7-oxopelargonate synthase.[1,2] There is evidence that similar enzymes are involved in the biosynthesis of biotin in other microorganisms[3] and plants.[4]

Dethiobiotin synthase (DTBS, EC 6.3.3.3) catalyzes the penultimate step in this pathway, the formation of the ureido ring of dethiobiotin from diaminopelargonic acid (DAPA), $CO_2$, and $Mg^{2+}$-ATP. More than two decades ago, Krell and Eisenberg[5] established that $CO_2$ and not $HCO_3^-$ is the substrate of the enzyme. Dethiobiotin synthase is structurally and mechanistically of considerable interest because it represents a unique class of carboxylases, different from the biotin-dependent carboxylase[5,6] and ribulose-1,5-bisphosphate carboxylase.[7]

The catalytic reaction of DTBS can be divided into a series of individual steps,[5] summarized in Fig. 1. In the first step of the reaction, one of the amino nitrogen atoms of the substrate DAPA will react with $CO_2$ and form the carbamylated DAPA, the first reaction intermediate. The carbamylation can proceed via two pathways, resulting in an $N^7$- or $N^8$-carbamylated DAPA intermediate; until recently, the regiospecificity of this reaction had not been resolved. The next step of the reaction consists of the formation of a carbamic-phosphate mixed anhydride as the second reaction intermediate. The phosphorylation of the carbamate oxygen activates the carbamyl carbon and facilitates nucleophilic attack by the second amino group. Ring closure and replacement of the phosphate then complete the reaction.

### Primary Structure

The amino acid sequences, derived from the DNA sequences for DTBS from five different species, are now known: *E. coli,*[2,8,9] *Serratia marcescens,*[10]

[2] A. J. Otsaku, M. R. Buoncristiani, P. K. Howard, J. Flamm, C. Johnson, R. Yamamoto, K. Uchida, C. Cook, J. Ruppert, and J. Matsuzaki, *J. Biol. Chem.* **263,** 19577 (1988).

[3] R. Glöckler, I. Ohsawa, D. Speck, C. Ledoux, S. Bernard, M. Zinsius, D. Villeval, T. Kisou, K. Kamogawa, and Y. Lemoine, *Gene* **87,** 63 (1990).

[4] P. Baldet, H. Gerbling, S. Axiotis, and R. Douce, *Eur. J. Biochem.* **217,** 479 (1993).

[5] K. Krell and M. A. Eisenberg, *J. Biol. Chem.* **245,** 6558 (1970).

[6] J. R. Knowles, *Annu. Rev. Biochem.* **49,** 877 (1989).

[7] G. Schneider, Y. Lindqvist, and C.-I. Brändén, *Annu. Rev. Biophys. Biomol. Struct.* **21,** 119 (1992).

[8] D. Alexeev, S. M. Bury, C. W. G. Boys, M. A. Turner, L. Sawyer, A. J. Ramsey, H. C. Baxter, and R. L. Baxter, *J. Mol. Biol.* **235,** 774 (1994).

[9] W. Huang, Y. Lindqvist, G. Schneider, K. J. Gibson, D. Flint, and G. Lorimer, *Structure* **2,** 407 (1994).

[10] SWISS-PROT Data Bank, accession code P36572.

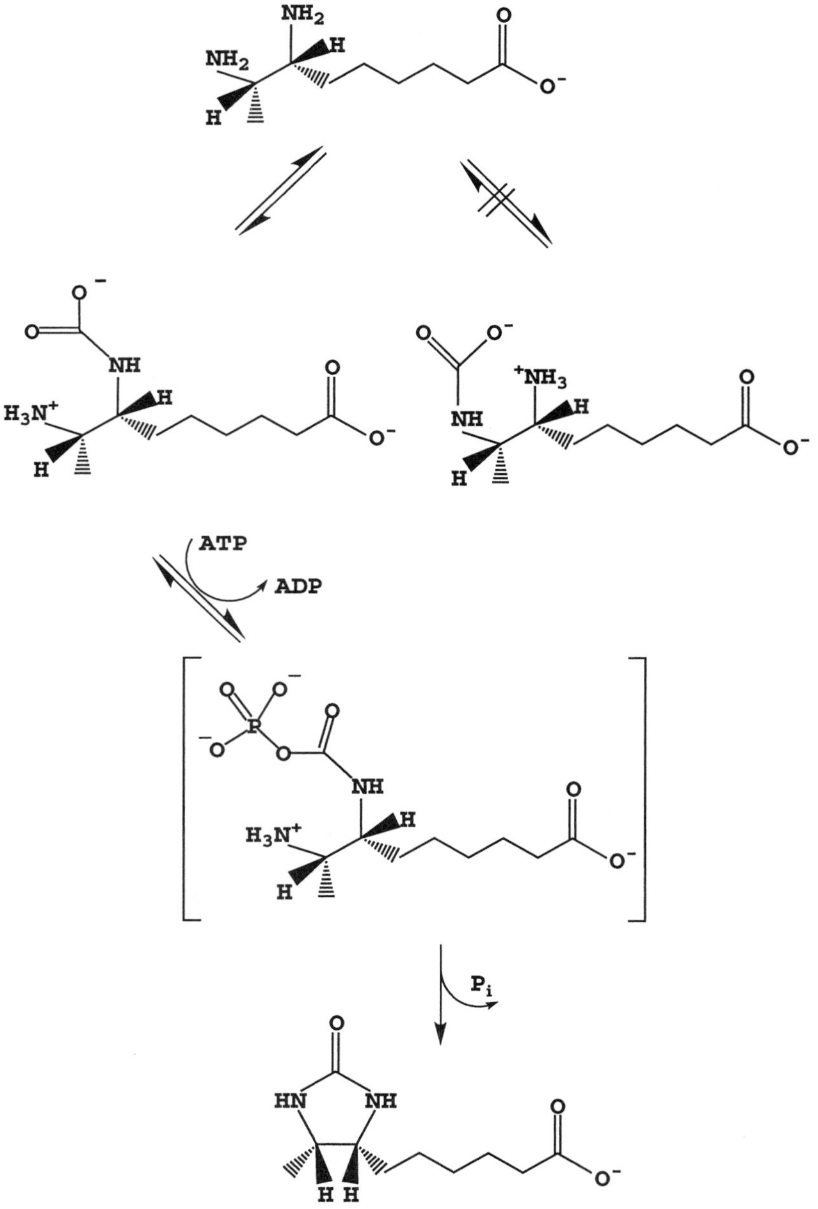

FIG. 1. Scheme of the reaction catalyzed by dethiobiotin synthase.

*Bacillus sphaericus,*[3] *Brevibacterium flavum,*[11] and *Mycobacterium leprae.*[12]
The polypeptide chains of these DTBS molecules vary in length from 223
to 234 amino acids. Alignment of these amino acid sequences reveals that
only 27 residues are invariant in all 5 species. The conserved stretches of
amino acid sequence are mostly involved in binding of the substrates and
the metal ion.


## Three-Dimensional Structure

### Crystallization

The three-dimensional structure of recombinant DTBS from *E. coli*
has been determined to high resolution (1.65-Å[9] and 1.8-Å[13] resolution,
respectively), using protein crystallographic methods. The crystals used in
the two independent studies were of the same space group, $C_2$, but showed
slightly different cell dimensions, $a = 73.2$ Å, $b = 49.2$ Å, $c = 61.8$ Å, $\beta =
107.1°$ and $a = 72.9$ Å, $b = 49.0$ Å, $c = 61.3$ Å, $\beta = 106.4°$. The crystals were
obtained using different precipitants (polyethylene glycol or ammonium
sulfate, respectively), which might have caused the observed variations
in the cell parameters. Nevertheless, the three-dimensional structures are
similar, with a root-mean-square (rms) deviation for 220 $C_\alpha$ atoms of 0.49 Å.

### Structure Determination

The structure of DTBS was solved by the multiple isomorphous replacement method, using six heavy metal derivatives.[9] The final protein model
for the unliganded enzyme subunit consists of residues 1–208, 213–224,
and 189 water molecules; residues 209–212 are localized in a disordered
loop. The crystallographic $R$ factor is 16.9% ($R$ free, 23.5%) at 1.65-Å
resolution, and the stereochemistry of the model is good: rms values for
bond lengths were 0.015 Å and for bond angles were 2.8°.

### Structure of Subunit

The structure of the subunit of DTBS from *E. coli* consists of a single
domain of the $\alpha/\beta$ type (Fig. 2). The core of the domain consists of a seven-
stranded parallel $\beta$ sheet surrounded by $\alpha$ helices. Strands 1, 4, 5, and 6

---

[11] K. Hatakayema, K. Hohama, A. A. Verttes, M. Kobayashi, Y. Kurusu, and H. Yukawa,
    *DNA Sequence* **4,** 177 (1994).
[12] SWISS-PROT Data Bank, accession code P45486.
[13] D. Alexeev, R. L. Baxter, and L. Sawyer, *Structure* **2,** 1061 (1994).

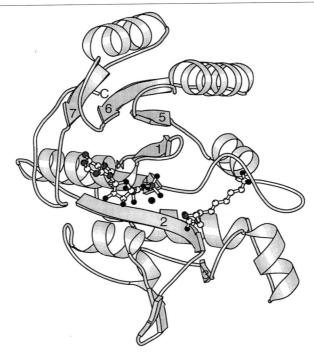

FIG. 2. Three-dimensional structure of the subunit of dethiobiotin synthase. The carbamylated DAPA substrate and ATP are included. The position of the metal ion is indicated by a black sphere. [Generated with MOLSCRIPT; P. Kraulis, *J. Appl. Crystallogr.* **24**, 946 (1991).]

form the classic mononucleotide-binding motif as found in other ATP- and GTP-binding proteins.[14]

The overall topology of the $\beta$ sheet in DTBS is similar to the central $\beta$ sheets in a number of ATP- and GTP-binding proteins (e.g., adenylate kinase[15] and p21[H-*ras*] protein).[16] The structural similarities indicate an evolutionary relationship and suggest that these proteins might have evolved from a common ancestral mononucleotide-binding fold.

The typical fingerprint motif for this class of mononucleotide-binding proteins consists of the sequence Gly-X-X-Gly-X-Gly-Lys-Thr/Ser. This stretch of the polypeptide chain is located in a loop between a $\beta$ strand and an $\alpha$ helix and interacts with the phosphates of the nucleotide.[15] In

[14] G. E. Schulz, *Curr. Opin. Struct. Biol.* **2**, 61 (1992).

[15] G. Dreusicke and G. E. Schulz, *FEBS Lett.* **208**, 301 (1986).

[16] E. F. Pai, U. Krengel, G. A. Petsko, R. S. Goody, W. Kabsch, and A. Wittinghofer, *EMBO J.* **9**, 2351 (1990).

TABLE I
CRYSTAL STRUCTURES OF DETHIOBIOTIN SYNTHASE

| Complex | Resolution (Å) | R factor (%) | Ref. |
|---|---|---|---|
| Native | 1.65 | 16.9 | 9 |
| | 1.80 | 17.2 | 13 |
| Native + DAPA + CO$_2$ | 1.70 | 17.1 | 17 |
| | 2.30 | 15.8 | 18 |
| Native + DAPA (CO$_2$ free) | 2.30 | 15.4 | 18 |
| Native + AEND[a] | 1.70 | 17.3 | 17 |
| Native + ADP | 1.60 | 18.3 | 17 |
| Native + ADP + DAPA | 1.70 | 17.2 | 17 |
| Native + AMPCP[b] + DAPA | 1.64 | 18.0 | 17 |
| Native + AMPCP[b] + Mn$^{2+}$ + DAPA | 1.64 | 17.4 | 17 |

[a] 3-(1-Aminoethyl)nonanedioic acid (AEND) is a chemical analog of the first reaction intermediate, DAPA-carbamylated at the N-7 nitrogen atom.
[b] Nonhydrolyzable ATP analog adenylyl($\beta,\gamma$-methylene) diphosphonate.

DTBS, a variant of this motif is found, with one glycine residue missing and one insertion, Gly-X-X-X-X-X-Gly-Lys-Thr, also involved in binding of the phosphate groups of ATP.

## Quaternary Structure

Dethiobiotin synthase from *E. coli* is a dimer built up of two identical subunits.[5,9,14] The dimer is formed through packing along a molecular two-fold axis, which in the crystals coincides with the crystallographic two-fold axis. The interface between the subunits is formed through tight interactions, involving helix $\alpha 5$, the loop between strand $\beta 4$ and helix $\alpha 4$, and the phosphate-binding loop between strand $\beta 1$ and helix $\alpha 1$. The dimer is the functional unit, because one of the substrates, DAPA, binds across the dimer interface and interacts with residues from both subunits.[17,18]

## Structure of Binary, Ternary, and Quaternary Complexes

A large variety of crystals of enzyme complexes (Table I), prepared by cocrystallization or soaking of native unliganded enzyme with substrates and substrate analogs,[17,18] have provided a detailed structural framework for the elucidation of the reaction mechanism.

[17] W. Huang, J. Jia, K. J. Gibson, W. S. Taylor, A. R. Rendina, G. Schneider, and Y. Lindqvist, *Biochemistry* **34**, 10985 (1995).
[18] D. Alexeev, R. L. Baxter, O. Smekal, and L. Sawyer, *Structure* **3**, 1207 (1995).

The DAPA substrate binds at the interface between the two subunits. The terminal carboxyl group of DAPA forms hydrogen bonds to the peptide nitrogens of residues 150 and 153, located at the N terminus of helix $\alpha5$. One of the carboxyl oxygens forms an additional hydrogen bond to the side chain of Tyr-187. The hydrophobic tail of the molecule spans the space between the two subunits. The diamino end of the substrate reaches into the active site of the second monomer and is bound at the switch point between the ends of the $\beta$ strands $\beta4$ and $\beta2$. The N-7 nitrogen atom forms a hydrogen bond to the side chain of Ser-41, as does the N-8 nitrogen atom to a solvent molecule.

In enzyme–DAPA complexes prepared in an atmosphere containing $CO_2$, the DAPA substrate is bound in its carbamylated form, suggesting that these structures represent the complex of the enzyme with the first reaction intermediate. The high resolution of the X-ray data allows an unambiguous assignment of the N-7 nitrogen atom of DAPA as the position of carbamate formation.[17,18] In binary complexes prepared in an inert nitrogen atmosphere devoid of $CO_2$ the noncarbamylated DAPA species is found at the enzyme active site.[18]

Binding of ADP is similar in all complexes studied,[17] and its interactions with the protein are identical irrespective of the presence of the second substrate, DAPA. The adenine ring binds in a pocket formed by the ends of strands $\beta6$ and $\beta7$. The ribose moiety of the nucleotide forms only one weak hydrogen bond with its O2R atom to the side chain of Glu-211. The diphosphate points to the active site and is bound to the phosphate-binding loop formed by residues $Gly^8$-$Thr^{16}$. The $\alpha$-phosphate points to the N terminal of helix $\alpha1$ and forms hydrogen bonds to main-chain nitrogen atoms of residues 16 and 17 and two solvent molecules. The $\beta$-phosphate interacts with main chain nitrogen atoms of residues 15 and 16 and also forms hydrogen bonds to the $O_\gamma$ atom of Thr-16 and a solvent molecule.

In complexes of DTBS with the ATP analog adenylyl($\beta,\gamma$-methylene) diphosphonate, the $\gamma$-phosphate is bound close to the phosphate-binding loop. It bridges the space between the $\beta$-phosphate of ATP and one of the oxygens of the carboxyl group of the carbamylated substrate, DAPA. This phosphate group interacts with the $O_\gamma$ atom of Thr-11, the N-8 nitrogen atom, and one of the oxygens of the carbamate and two solvent molecules.

Dethiobiotin synthase uses divalent metal ions as cofactors and the enzyme–substrate complexes have been prepared in the presence of $Mg^{2+}$, $Ca^{2+}$, or $Mn^{2+}$, respectively. In complexes with the ATP analog, the metal ion is bound to two oxygen atoms from the $\beta$- and $\gamma$-phosphate group of the ATP analog and to the side chains of the conserved residues Asp-54, Thr-16, and Glu-115. This metal-binding site is similar to the metal-binding site found in other GTP-binding proteins such as p21$^{ras}$.[16] In binary complexes of the enzyme with ADP, a second metal-binding site was found, about 4 Å away from the first metal site. This site is formed by an oxygen atom of the $\alpha$-phosphate, the side chain of Asp-54, and a number of solvent molecules. The side chain of Asp-54 must change its position to be able to coordinate to the metal ion in site 2 and both sites are therefore not likely to be occupied simultaneously. The mechanistic significance of this metal site is not quite clear, but it seems as if the presence of the $\gamma$-phosphate group is required for the formation of the first metal-binding site.

The structural information from these various enzyme complexes can be assembled into a schematic view of the enzyme with bound substrates, $N^7$-carbamylated DAPA, and divalent metal ($Me^{2+}$)-ATP, which describes the situation before the phosphoryl transfer step is occurring (Fig. 3).

## Conformational Changes on Substrate Binding

All conformational changes that occur on substrate binding can be accommodated in the crystal lattice, because binding of substrates does not destroy crystal packing.[17,18]

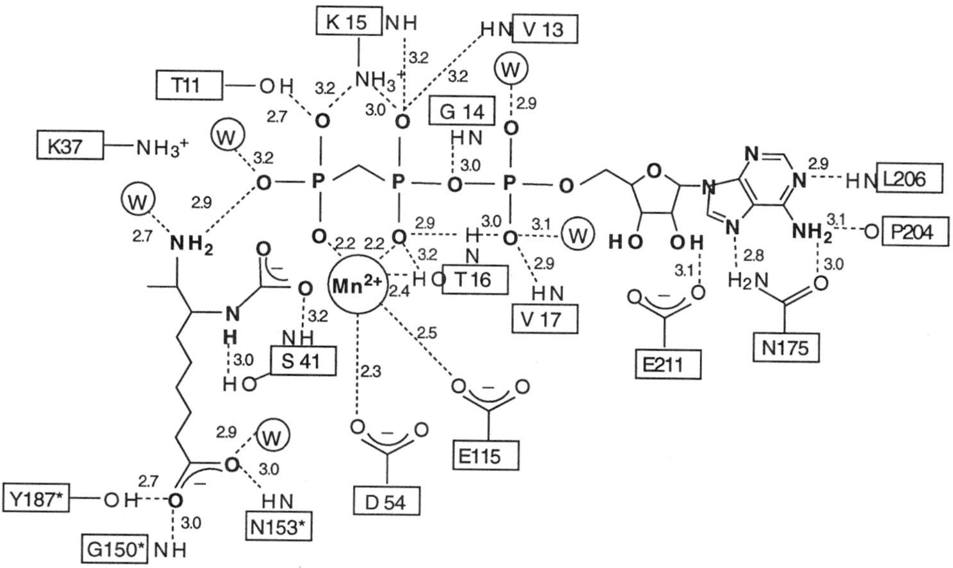

FIG. 3. Schematic view of the active site of dethiobiotin synthase with a bound ATP analog, adenylyl($\beta,\gamma$-methylene) diphosphonate, the carbamylated DAPA substrate, and metal ion. [Reprinted with permission from W. Huang, J. Jia, K. J. Gibson, W. S. Taylor, A. R. Rendina, G. Schneider, and Y. Lindqvist, Biochemistry **34,** 10985 (1995). Copyright 1995 American Chemical Society.]

Binding of the substrates induces local conformational changes that are mostly confined to loop regions. The binding of the mononucleotides ADP and ATP results in residue displacements at several areas in the structure. One such structural change involves the disordered region, residues 209–212. This part of the polypeptide chain folds over the adenine part of the nucleotide and has a more ordered conformation in the complexes with bound nucleotides. Other significant residue displacements involve residues 9–14 of the phosphate-binding loop, which change conformation on binding of the nucleotide, the largest atomic shifts in this area being around 2.5 Å. This loop also changes its conformation on binding of DAPA, albeit to a lesser extent. Another significant area of conformational change involves residues 117–119 in the loop after strand $\beta 4$, which participate in binding of DAPA and interact with the phosphate-binding loop.

The region around residues 146–152 displays significant atomic displacements on binding of DAPA. This loop is directly involved in subunit–subunit interactions of the homodimer and binds to the carboxyl group of the DAPA substrate in the dimer.

In summary, most of these structural changes on binding of the nucleotide substrate and DAPA can be described as a tightening of the active site to maximize enzyme–substrate interactions.

## Mechanistic Conclusions

### Regiospecificity of Carbamate Formation

The regiospecificity of carbamate formation had been addressed by a number of different techniques, trapping experiments,[19,20] nuclear magnetic resonance (NMR),[20] and protein crystallography.[17,18] While initial diazomethane trapping experiments[19] implicated the N-8 nitrogen atom as the site of carbamylation, compelling evidence for the N-7 nitrogen atom as the site of carbamylation was obtained by NMR measurements[20] and protein crystallography.[17,18] High-resolution studies of the complex of enzyme with the substrate DAPA, obtained in the presence of atmospheric $CO_2$, reveal that the enzyme-bound species is the $N^7$-DAPA carbamate.[17,18] Binding of the noncarbamylated DAPA to the enzyme can be achieved when crystals of this binary complex are prepared in a $CO_2$-free nitrogen atmosphere.[18]

[19] R. L. Baxter, A. J. Ramsey, L. A. McIver, and H. C. Baxter, *J. Chem. Soc. Commun.* 559 (1994).
[20] K. J. Gibson, G. H. Lorimer, A. R. Rendina, W. S. Taylor, G. Cohen, A. A. Gatenby, W. G. Payne, D. C. Roe, B. A. Lockett, A. Nudelman, D. Marcovici, A. Nachum, B. A. Wexler, E. L. Marsili, I. M. Turner, L. D. Howe, C. E. Kalbach, and H. Chi, *Biochemistry* **34**, 10976 (1995).

## Phosphoryl Transfer

Biochemical experiments have suggested a carbamic-phosphoric mixed anhydride as a second intermediate.[21] Direct crystallographic evidence for this proposed intermediate of the catalytic reaction is as yet lacking. However, studies of ternary and quaternary complexes with bound DAPA and the nonhydrolyzable ATP analog adenylyl($\beta,\gamma$-methylene) diphosphonate[17] have given insights into the structure of the active site just before phosphoryl transfer is about to occur (Fig. 3). In this structure, one of the carbamate oxygens is close to the $\gamma$-phosphate of the ATP analog, almost perfectly aligned for nucleophilic attack on the phosphorus atom of the $\gamma$-phosphate.

## Role of Metal Ion

As in other ATP- or GTP-dependent enzymes, the metal ion is bound by oxygen atoms of the $\beta$- and $\gamma$-phosphate groups of the nucleotide and groups on the enzyme, thus providing additional interactions between the nucleotide and the protein. It has been suggested that in DTBS the metal ion helps to neutralize the negative charges of the $\gamma$-phosphate and to position this group such that nucleophilic attack by the carbamate oxygen can occur.[17] A role for $Mg^{2+}$ in binding of the DAPA substrate has also been suggested.[18]

## Acknowledgments

We thank the many colleagues, in particular, Weijun Huang, Jia Jia, Katherine Gibson, George Lorimer, Helena Käck, Dennis Flint, Alan Rendina, and Wendy Taylor, who have been involved at various stages of this work. Support from the Swedish Agricultural Research Council is gratefully acknowledged.

[21] R. L. Baxter and H. C. Baxter, *J. Chem. Soc. Commun.* 759 (1994).

## [41] Purification and Properties of Bovine and Human Holocarboxylase Synthetases

*By* YOICHI SUZUKI and KUNIAKI NARISAWA

### Introduction

Holocarboxylase synthetase (HCS; EC 6.3.4.10, biotin–[propionyl-CoA-carboxylase] ligase) catalyzes the incorporation of biotin into carboxylases. Biotinylation of the apocarboxylases includes two partial reactions as follows:

$$\text{Biotin} + \text{ATP} \rightleftharpoons \text{biotinyl-5}'\text{-AMP} + \text{PP}_i$$
$$\text{Biotinyl-5}'\text{-AMP} + \text{apocarboxylase} \rightarrow \text{holocarboxylase} + \text{AMP}$$

Biotin is first converted to biotinyl-5'-AMP. The biotinyl group then becomes covalently bonded to the ε-amino group of a lysine residue of the apo-enzyme.

In mammalian cells, HCS is distributed intracellularly in both the cytosol and mitochondria, and more than 70% of HCS activity is in the cytosolic fraction.[1]

### Assay Methods

#### Principle

Holocarboxylase synthetase activity is assessed by measuring holopropionyl-CoA carboxylase (holo-PCC) formed by the HCS reaction. Holo-PCC is formed from the apo-PCC (the liver apo-PCC or the lymphoblast apo-PCC) in the presence of *d*-biotin and ATP. Holo-PCC activity is determined using a modification of the procedure of Weyler *et al.,*[2] by fixation of $NaH^{14}CO_3$ into acid nonvolatile products using propionyl-CoA and ATP.

#### Preparation of Apopropionyl-CoA Carboxylase

Apopropionyl-CoA carboxylase is prepared from biotin-deficient rat livers.[3] The washed mitochondria-containing pellet is suspended in 20–30 ml of buffer P [20 m$M$ potassium phosphate (pH 7.0), 1 m$M$ EDTA, 1 m$M$

[1] N. D. Cohen, M. Thomas, and M. Stock, *Ann. N.Y. Acad. Sci.* **447,** 393 (1985).
[2] W. Weyler, L. Sweetman, D. C. Maggio, and W. L. Nyhan, *Clin. Chim. Acta* **76,** 321 (1977).
[3] B. J. Burri, L. Sweetman, and W. L. Nyhan, *J. Clin. Invest.* **68,** 1491 (1981).

reduced glutathione (GSH)] and sonicated. The sonicate is centrifuged at 106,000 $g$ for 1 hr at 4°. The resulting supernatant fluid is applied to a DEAE-Sepharose 6B column (2.5 × 9 cm) previously equilibrated with buffer P and eluted at 2 ml/min using a linear gradient of 300 ml of buffer P and 250 ml of 0.3 $M$ KCl in buffer P. Apo-PCC coelutes with holo-PCC between 220 and 350 ml. Fractions enriched in PCC activity are pooled and concentrated by precipitation with ammonium sulfate at a final concentration of 55% saturation. The precipitate is redissolved in about 3 ml of buffer P, applied to a column of Sepharose 6B (1.5 × 75 cm), and eluted with additional buffer P. The HCS is separated from the apo-PCC and holo-PCC on this column. Fractions containing apo-PCC and holo-PCC are pooled and passed through an avidin–Sepharose column (0.5 × 15 cm) equilibrated, and eluted, with buffer P. Holo-PCC is retained on the column while apo-PCC is eluted with the first 10 ml of buffer P. Apo-PCC is stored in 3.2 $M$ ammonium sulfate and is stable for about 3 weeks at 4°, and is much less stable when frozen. The final preparation contains negligible amounts of holo-PCC or other holocarboxylases.

Lymphoblast apo-PCC is obtained by the following procedure.[4] Lymphoblasts immortalized by infection with Epstein–Barr virus from a patient with HCS deficiency are stored at −80° (the unpurified apoenzyme is stable for at least 1 year). Immediately before the HCS assay, the frozen cells are thawed quickly and suspended in 50 m$M$ Tris-HCl buffer (pH 8.0), 3 m$M$ EDTA, and 2.5 m$M$ GSH (lysate buffer). After freezing and thawing the cell lysate is centrifuged at 10,000 $g$ at 4° for 10 min to remove cell debris. Two volumes of saturated ammonium sulfate is added to the supernatant, and the precipitate is collected by centrifugation (10,000 $g$, 10 min, 4°). The pellet is dissolved in the lysate buffer at a concentration of 10–13 mg/ml (1 × 10$^8$ cells/ml).

## Procedure

### Holocarboxylase Synthetase Reaction (Reaction 1)

Holopropionyl-CoA carboxylase is formed by mixing 10 $\mu$l of enzyme solution with 20 $\mu$l of the apo-PCC (20 $\mu$g of the liver apo-PCC or 200 $\mu$g of the lymphoblast apo-PCC) as a substrate and 10 $\mu$l of reaction buffer [400 m$M$ Tris-HCl (pH 8.0), 32 m$M$ MgCl$_2$, 200 m$M$ KCl, 0.4 m$M$ EDTA, 10 m$M$ GSH, 10 m$M$ ATP, and biotin (400 ng/ml)]. This mixture is incubated at 30° for 60 min and the reaction is stopped on ice.

[4] Y. Chiba, Y. Suzuki, Y. Aoki, Y. Ishida, and K. Narisawa, *Arch. Biochem. Biophys.* **313,** 8 (1994).

*Propionyl-CoA Carboxylase Reaction (Reaction 2)*

Propionyl-CoA carboxylase activity is assayed by mixing 10 $\mu$l of reaction buffer [200 m$M$ Tris-HCl (pH 8.0), 16 m$M$ MgCl$_2$, 100 m$M$ KCl, 1.5 m$M$ EDTA, 5 m$M$ GSH, and 6 m$M$ ATP], 5 $\mu$l of 28 m$M$ n-propionyl-CoA, and 5 $\mu$l of 100 m$M$ NaH$^{14}$CO$_3$ (9.2 mCi/mmol) with reaction 1 mixture. The mixture is incubated for 15 min at 30°. The reaction is stopped by adding 30 $\mu$l of 1 $N$ H$_2$SO$_4$ and 60 $\mu$l of the mixture is then transferred to a scintillation vial. It is dried on a hot plate to remove unreacted bicarbonate. Five milliliters of ACS-II scintillation cocktail (Amersham, Arlington Heights, IL) is added to the vial and nonvolatile radioactivity is measured in a Beckman (Palo Alto, CA) LS-1000 liquid scintillation counter. Assay blanks are prepared by omitting biotin and ATP from reaction 1 and these values are then subtracted from those of the complete system.

One unit of holo-PCC activity is defined as the amount of enzyme that catalyzes fixation of 1 pmol of $^{14}$CO$_2$ per minute. One unit of HCS activity is defined as the amount of enzyme that can synthesize 1 unit of holo-PCC activity per hour.

*Comments*

In the application of the assay method to crude tissue preparations, holo-PCC needs to be previously eliminated from the enzyme solution because it elevates the background values. For this purpose, an avidin–Sepharose column,[3] polyethylene glycol,[4] and anti-PCC antibody[5] have been used.

Holocarboxylase synthetase activity is much higher for apo-PCC from biotin-deficient rats than for lymphoblast apo-PCC. The HCS assay method using rat liver apo-PCC as a substrate is useful for enzymatic diagnosis of HCS deficiency and characterization of HCS. However, rat liver apo-PCC is unstable[3] and the procedure for its preparation is labor intensive and time consuming. However, the method using lymphoblast apo-PCC is suitable for assay of multiple fractions on column chromatography but not sufficiently sensitive to detect lower HCS activity in cultured cells.

*Alternative Assay Method*

A new method for the assay of HCS that is based on measuring the incorporation of [$^3$H]biotin into apocarboxylase has been developed.[6] In

[5] M. E. Saunders, W. G. Sherwood, M. Duthie, L. Surh, and R. A. Gravel, *Am. J. Hum. Genet.* **34,** 590 (1982).
[6] Y. Suzuki, Y. Aoki, O. Sakamoto, X. Li, S. Miyabayashi, Y. Kazuta, H. Kondo, and K. Narisawa, *Clin. Chim. Acta* **251,** 41 (1996).

this procedure, the substrate for the reaction, recombinant apocarboxyl carrier protein (apo-CCP), is produced according to the method of Kondo and Kazuta.[7]

*Preparation of Apocarboxyl Carrier Protein*

The gene for *Escherichia coli* carboxyl carrier protein (CCP)[8] has been amplified by the polymerase chain reaction and cloned into the pTrc99A expression vector. This construct encodes CCP with three amino acids (Met-Glu-Phe) added at the amino terminus. *Escherichia coli* JM105 is transformed with this recombinant plasmid. Transformants are cultured in LB broth for 2 hr in the presence of 1 m$M$ isopropylthioglucoside. Bacteria are then suspended in potassium phosphate buffer and disrupted by sonication. The solution is centrifuged, and the supernatant is purified by fractionation with ammonium sulfate and calcium phosphate gel. The resultant fraction is purified further with column chromatographies on butyl-Toyopearl and Sephadex G-100. Analysis of the final preparation on a sodium dodecyl sulfate (SDS)–polyacrylamide gel reveals a single band. Apo-CCP can be used as a substrate for at least 6 months when stored at −80°.

*Procedure*

To determine HCS activity, incorporation of [³H]biotin into acid-insoluble material per unit time is monitored.[6] In a standard assay, holo-CCP formation is performed in 20 $\mu$l of mixture that contains HCS preparation (5 $\mu$l, 10–20 $\mu$g of protein), 180 $\mu M$ apo-CCP, 120 m$M$ Tris-HCl (pH 8.0), 8 m$M$ MgCl$_2$, 50 m$M$ KCl, 1.4 m$M$ EDTA, 3 m$M$ ATP, 2.5 m$M$ mercaptoethanol, and 1.5 $\mu M$ [³H]biotin (specific activity, 14 Ci/mmol). The mixture is incubated at 30° for 40 min. The reaction is stopped by adding 250 $\mu$l of 10% (w/v) TCA. The mixture is kept on ice for 15 min, then centrifuged at 18,000 g for 15 min at 4°. The resultant pellet is washed two times with 250 $\mu$l of 10% (w/v) TCA. Finally, the pellet is dissolved in 200 $\mu$l of 0.1 $N$ NaOH and transferred to a scintillation vial. After adding 5 ml of scintillation cocktail (ACS-II; Amersham), radioactivity is measured in a Beckman LS-6500 scintillation counter. Blanks for HCS determination are prepared by omitting ATP from the reaction buffer.

[7] H. Kondo and Y. Kazuta, *Vitamins* **69,** 249 (1995).
[8] H. Kondo, K. Shiratsuchi, T. Yoshimoto, T. Masuda, A. Kitazono, D. Tsuru, M. Anai, M. Sekiguchi, and T. Tanabe, *Proc. Natl. Acad. Sci. U.S.A.* **88,** 9730 (1991).

## Purification of Holocarboxylase Synthetase

The enzyme from bovine liver has been purified by the procedure described below. All operations are performed at 0–4°. The activity of HCS is assayed using the lymphoblast apo-PCC as a substrate.[4]

### Step 1: Preparation of Cytosolic Fraction

Bovine livers (1.5 kg) are passed through a meat grinder and homogenized in a Waring blender at a high setting for 2 min with 2 vol of buffer P [20 m$M$ potassium phosphate buffer (pH 7.0), 1 m$M$ EDTA, 1 m$M$ GSH] containing 0.25 $M$ sucrose. The homogenate is centrifuged at 600 $g$ for 10 min at 4° to remove nuclei and cell debris. The supernatant is further centrifuged at 100,000 $g$ (maximum) for 90 min at 4° in a RPZ-35T rotor (Hitachi, Tokyo, Japan).

### Step 2: Ammonium Sulfate Fractionation

The supernatant is brought to 45% saturation with solid ammonium sulfate, and the mixture is stirred for 30 min. The resultant precipitate is collected by centrifugation and dissolved in 100 ml of 5 m$M$ potassium phosphate buffer (pH 7.0), 0.25 m$M$ EDTA, and 0.25 m$M$ GSH and desalted in a Sephadex G-25 column (5 × 60 cm) equilibrated with the same buffer. The eluate from this column is diluted to a protein concentration of 40 mg/ml with the same buffer.

### Step 3: Alumina C$\gamma$ Fractionation

Alumina C$\gamma$ (aged) is prepared to 40 mg dry weight per milliliter. This gel suspension is added slowly by stirring to the eluate until a gel-to-protein ratio (weight to weight) of 0.3 is obtained. The mixture is stirred slowly for 60 min and centrifuged at 600 $g$ for 10 min at 4°. The supernatant is discarded and the gel is resuspended in 10% saturated ammonium sulfate (0.1 ml/mg of alumina C$\gamma$ originally added). After 60 min of stirring, the solution is centrifuged and the supernatant is recovered. The gel eluate is brought to 55% saturation with solid ammonium sulfate followed by centrifugation. The precipitate is then dissolved in buffer P and applied to a Sephadex G-25 column (5 × 40 cm) equilibrated with buffer P containing 150 m$M$ KCl to remove ammonium sulfate.

### Step 4: DEAE-Sepharose CL-6B Chromatography

The enzyme solution is applied to a DEAE-Sepharose CL-6B column (5 × 18 cm) previously equilibrated with buffer P containing 150 m$M$ KCl.

The column is then washed with the same buffer until the absorbance of the eluate at 280 nm returns to the base line. The column is eluted with buffer P containing 400 mM KCl.

## Step 5: EAH-Sepharose 4B Chromatography

The eluate from the previous column (140 ml) is diluted fivefold with buffer P and applied to an EAH-Sepharose 4B column (2.5 × 9 cm) equilibrated with buffer P containing 80 mM KCl. The absorbed proteins are eluted with a linear 80 to 500 mM KCl gradient in 420 ml of buffer P. The flow rate is 24 ml/hr and the fraction sizes are 5 ml. Fractions containing HCS activity are pooled.

## Step 6: Sephacryl S-200 HR Chromatography

The pooled fractions are concentrated to 10 ml using an Amicon (Danvers, MA) PM10 membrane. The sample is then applied to a Sephacryl S-200 HR column (2.6 × 96 cm) equilibrated with buffer P containing 150 mM KCl and eluted at a flow rate of 60 ml/hr. Each 5-ml fraction is collected.

## Step 7: Bio-Gel HTP Chromatography

The active fractions from step 6 are collected. The buffer of the eluate is changed to 10 mM potassium phosphate buffer (pH 7.0), 1 mM GSH (buffer P′), using a Sephadex G-25 column (2.5 × 35 cm). The eluate is then applied to a Bio-Gel HTP column (1.5 × 13 cm) equilibrated with buffer P′. After washing the column with the same buffer, the column is eluted with 210 ml of a linear gradient from 10 to 100 mM potassium phosphate buffer (pH 7.0) and 1 mM GSH. The flow rate is 24 ml/hr and the fraction sizes are 5 ml.

## Step 8: Phenyl-Superose HR 5/5 Chromatography

Fractions that contain activity are mixed and solid ammonium sulfate is added to a final concentration of 400 mM. The solution is then applied to a phenyl-Superose HR 5/5 column equilibrated with buffer P containing 400 mM ammonium sulfate. After washing the column with the same buffer, the absorbed enzyme is eluted with 25 ml of a linear gradient of 400 to 0 mM ammonium sulfate in buffer P. The flow rate is 0.4 ml/min. Fractions (1.0 ml) are collected and monitored for enzyme activity and protein con-

TABLE I

PURIFICATION OF HOLOCARBOXYLASE SYNTHETASE FROM BOVINE LIVER[a]

| Step | Total protein[b] (mg) | Total activity[c] (units) | Specific activity (units/mg) | Yield (%) |
|---|---|---|---|---|
| Supernatant (100,000 g) | 140,000 | 4,000,000 | 28.5 | 100.0 |
| Ammonium sulfate precipitate | 50,300 | 2,310,000 | 46 | 57.8 |
| Alumina Cγ elute | 10,200 | — | — | — |
| DEAE-Sepharose CL-6B | 1,390 | 636,000 | 459 | 15.9 |
| EAH-Sepharose 4B | 123 | 425,000 | 3,460 | 10.6 |
| Sephacryl S-200 HR | 40.3 | 331,000 | 8,200 | 8.26 |
| Bio-Gel HTP | 3.17 | 117,000 | 36,800 | 2.91 |
| Phenyl-Superose HR 5/5 | 0.181 | 93,000 | 514,000 | 2.32 |

[a] Results of a purification from 1.5 kg of bovine liver.
[b] Determined by the method of Bradford.
[c] Assayed with apopropionyl-CoA carboxylase from human lymphoblasts as substrate.

centration. A summary of the purification of bovine liver HCS is presented in Table I.

*Polyacrylamide Gel Electrophoresis*

The protein (64,000 Da) with HCS activity purified through phenyl-Superose HR 5/5 (step 8) is clearly separated from an inert protein of 34,000 Da on a nondenaturing polyacrylamide gel at pH 9.5 or 8.0. After separation from this inert protein, HCS is found to be extremely labile and its enzymatic activity is totally lost within a few hours.

Properties of Holocarboxylase Synthetase

*Molecular Properties*

It is likely that human and bovine liver HCS are monometoric enzymes[4] with a similar molecular mass of 81,000–85,000 Da.[9] Human HCS cDNA clones have been isolated by Suzuki *et al.*[9] The amino acid sequence deduced from this cDNA predicts a molecular mass of 80,759 Da composed of 726 amino acid residues. The 64,000-Da protein found at step 8 of purification of bovine HCS[4] seems to be the protein partially degraded during the purification.

[9] Y. Suzuki, Y. Aoki, Y. Ichida, Y. Chiba, A. Iwamatsu, T. Kishino, N. Niikawa, Y. Matsubara, and K. Narisawa, *Nature Genet.* **8,** 122 (1994).

*Catalytic Properties*

The catalytic properties of bovine liver HCS were determined using liver apo-PCC as a substrate. The Michaelis constant ($K_m$) for biotin of the purified HCS from bovine liver cytosol was 13 n$M$.[4] This value is nearly equal to the $K_m$ of human fibroblast HCS (15 n$M$) assayed on the same substrate,[3,10] whereas it is appreciably different from that (260 n$M$) assayed using *E. coli* CCP as a substrate.[6] The $K_m$ for ATP was likewise estimated to be 20 m$M$.[4] Concerning the nucleoside triphosphate specificity, the bovine liver HCS was not strictly specific to ATP, that is, CTP could totally replace ATP, and approximately half the activity was detected with ITP and GTP.[4] These results are different from those obtained with rat liver[11] and rabbit liver HCS,[12] both of which were specific for ATP, and from chicken HCS, which could utilize UTP as well as ATP.

[10] B. J. Burri, L. Sweetman, and W. L. Nyhan, *Am. J. Hum. Genet.* **37,** 326 (1985).
[11] D. P. Kosow, S. C. Huang, and M. D. Lane, *J. Biol. Chem.* **237,** 3633 (1962).
[12] L. Siegel, J. L. Foote, and M. J. Coon, *J. Biol. Chem.* **240,** 1025 (1965).

# [42] Biotin Uptake in Cultured Cell Lines

*By* DAVID L. DYER and HAMID M. SAID

## Introduction

In humans, the water-soluble vitamin biotin serves as a carboxyl carrier in the ATP-dependent reactions catalyzed by acetyl-CoA carboxylase, pyruvate carboxylase, propionyl-CoA carboxylase, and $\beta$-methylcrotonyl-CoA carboxylase. Because these enzymes are essential to the processes of lipogenesis, gluconeogenesis, amino acid metabolism, and odd-chain fatty acid catabolism, biotin is essential for the maintenance of normal cell functions and overall physiological homeostasis.[1] Owing to its central role in these important cellular functions, severe biotin deficiency can be manifested as a wide range of clinical abnormalities, including seborrheic dermatitis, alopecia, abnormal development of reproductive organs, decreased immune function, and neurodegeneration.[2-5]

[1] K. Dakshinamurti and J. Chauan, *Annu. Rev. Nutr.* **8,** 211 (1988).
[2] J. P. Bonjour, *Int. J. Vitam. Nutr. Res.* **47,** 107 (1977).
[3] L. Sweetman, L. Surh, H. Baker, R. M. Peterson, and W. L. Nylander, *Pediatrics* **48,** 553 (1981).

Biotin is not synthesized *de novo* by mammalian cells, but must be obtained from dietary sources through absorption in the small intestine. Other important sites that play a role in normal biotin physiology and homeostasis include the liver (the principal site of biotin utilization), the kidney (the site of reabsorption of filtered biotin), and the placenta (the site of transport of biotin from the maternal circulation to the developing embryo). Biotin is present in the diet both as free and protein-bound forms. The protein-bound form is first processed by gastrointestinal proteases and peptidases, to produce biocytin and biotin-containing short peptides. These compounds are then further hydrolyzed to free biotin by the action of biotinidase. Because of its importance to cellular functions and overall health, and the fact that mammalian cells cannot synthesize biotin, it is important to understand the mechanisms of cellular uptake of biotin in the above tissues, and the means by which such uptake mechanisms are regulated.

Many model systems have been developed with which to study the phenomenon of biotin uptake into mammalian cells. Naturally, such models have varied depending on the tissue and cell types to be studied. In the case of the intestine, for example, commonly used techniques include *in vivo* models such as whole-animal studies and *in vivo* perfusion, while *in vitro* models include everted sacs, everted rings, isolated cells, and brush border membrane and basolateral membrane vesicles.

To study the fundamental intracellular mechanisms governing regulation of biotin transport, however, cultured cells are the preferred model. Such a system presents several advantages over other models, including the availability of homogeneous cell populations, the ease with which certain manipulations may be done that are difficult or impossible to carry out in living animals, and the lack of systemic factors, (e.g., growth factors and hormones) that may interfere with the transport process under investigation. Primary cultures of isolated cells obtained from specific tissues have been used to study biotin uptake. However, such a system may have several complications, including limited cell viability and a loss of functional polarity (as in the case of primary cultures of intestinal epithelial cells) or a loss of certain functional characteristics (as in the case of hepatocytes). Furthermore, primary cultures of cells from organs such as the intestine, for example, consist of mixed cell populations, making the interpretation of results quite difficult. These and other problems associated with primary

[4] C. S. Paulose, J. A. Thliveris, M. Viswanathan, and K. Dakshinamurti, *Horm. Metabol. Res.* **21,** 661 (1989).

[5] D. P. Bousanis, P. R. Camfield, and B. Wolf, *J. Nutr.* **123,** 2101 (1993).

cell culture can generally be overcome by using continuous cell lines derived from the tissue of interest. In this way, issues of cell viability, heterogeneity, and the possible influences of extraneous systemic factors are circumvented. Given the wide variety of continuous cell lines available, and the different physiological characteristics of each one of these, the choice of the proper cell line for modeling any particular phenomenon is not a trivial issue. In general, one should attempt to utilize a well-studied, accepted cell line model that retains a majority, if not all, of the normal functions of the native cells.

This chapter examines some of the published information concerning biotin uptake in different tissues (intestine, liver, kidney, placenta), and focuses on some of the studies that have examined the biotin uptake process in cell culture systems.

## Uptake by Intestine

As mentioned in the previous section, biotin is not synthesized by mammalian cells but must be obtained from exogenous sources by absorption in the small intestine. For this reason, understanding the mechanism and regulation of the intestinal biotin uptake process is important. Carefully designed experiments have improved our understanding of the mechanism of biotin uptake in the small intestine. In contrast to the original notion that uptake of this vitamin occurred in this tissue through a passive, nonspecific, diffusional process, it is presently recognized that biotin uptake occurs through a specialized, carrier-mediated system present in the absorptive epithelial cells of the small intestinal mucosa.

The general characteristics of biotin uptake have been defined by different preparations derived from native human and animal intestinal tissue. Brush border membrane vesicles prepared from human and animal small intestine have been used to show that uptake of physiological concentrations of biotin occurs through a saturable, carrier-mediated process (apparent $K_m \sim 5$–$15\ \mu M$), which is $Na^+$ gradient dependent, specific (being inhibited by structural analogs of biotin but not by structually unrelated compounds), and electroneutral in nature; moreover, the cotransport stoichiometry of biotin to $Na^+$ is $1:1$.[6,7] Studies using intestinal basolateral membrane vesicles have shown that biotin uptake at this membrane domain also occurs by a saturable, carrier-mediated (apparent $K_m \sim 1$–$3\ \mu M$), and specific process.[8] Unlike the uptake process at the brush border, biotin uptake at the basolat-

[6] H. M. Said, R. Redha, and W. Nylander, *Am. J. Physiol.* **253**, G631 (1987).
[7] H. M. Said and I. Derweesh, *Am. J. Physiol.* **261**, R94 (1991).
[8] H. M. Said, R. Redha, and W. Nylander, *Gastroenterology* **94**, 1157 (1988).

eral membrane domain was found to be $Na^+$ independent, and electrogenic in nature. Because the transport process of biotin at the basolateral membrane domain is facilitative, while that at the brush border membrane domain is secondary active, driven by the $Na^+$ gradient (i.e., which is maintained by the activity of the ATP-driven $Na^+$–$K^+$ pump), it appears that the uptake process at the brush border membrane is the rate-limiting step in the overall absorption of the vitamin in these cells. Certain aspects of the molecular mechanisms of biotin uptake have also been addressed with the use of group-specific reagents. These studies have demonstrated the involvement of essential sulfhydryl groups and histidine residues in the normal functioning of the biotin uptake carrier.[9]

Within the small intestine, regional differences in biotin uptake have also been seen. It has been demonstrated that the uptake of biotin is higher in the duodenum than jejunum, which is in turn higher than uptake in the ileum. This difference is due to differences in the $V_{max}$, but not the apparent $K_m$, of the biotin uptake process.[10,11] The expression of biotin uptake is also developmentally regulated. In suckling animals, biotin uptake is greater than in weanling animals, and uptake in weanling animals is in turn higher than in adult animals.[12,13] Furthermore, in the course of such overall developmental changes, the site of maximum uptake of biotin in the small intestine also changes, with biotin uptake being higher in the ileum than the jejunum in suckling rats, similar in jejunum and ileum in weanling rats, and higher in the jejunum than the ileum in adult rats.

The uptake of biotin is also regulated by dietary substrate level. Rats fed a biotin-deficient diet have significantly upregulated biotin uptake compared to pair-fed controls, whereas rats fed a diet oversupplemented with biotin have significantly downregulated uptake compared to controls. These changes are mediated through changes in the $V_{max}$, but not the apparent $K_m$, of the biotin uptake process.[14] Biotin uptake is also affected by other factors, such as chronic ethanol exposure, renal failure, and the use of anticonvulsant drugs (carbamazepine and primidone).[15–17]

With the use of the standard techniques mentioned earlier, a good understanding of the mechanism and driving force of intestinal biotin up-

[9] H. M. Said and R. Mohammadkhani, *Biochim. Biophys. Acta* **1107**, 238 (1992).

[10] H. M. Said and R. Redha, *Am. J. Physiol.* **252**, G52 (1987).

[11] H. M. Said, R. Redha, and W. Nylander, *Gastroenterology* **95**, 1312 (1988).

[12] H. M. Said and R. Redha, *Gastroenterology* **94**, 68 (1988).

[13] H. M. Said, A. Sharifan, and A. Bagherzadeh, *Pediatr. Res.* **28**, 266 (1990).

[14] H. M. Said, D. M. Mock, and J. C. Collins, *Am. J. Physiol.* **256**, G306 (1989).

[15] H. M. Said, A. Sharifian, A. Bagerzadeh, and D. Mock, *Am. J. Clin. Nutr.* **52**, 1083 (1990).

[16] H. M. Said, N. D. Vaziri, F. Oveisi, and S. Husseinzadah, *J. Lab. Clin. Invest.* **120**, 471 (1992).

[17] H. M. Said, R. Redha, and W. Nylander, *Am. J. Clin. Nutr.* **49**, 127 (1989).

take has been obtained. Nothing, however, is known about the intracellular regulation of the intestinal biotin uptake process. As mentioned earlier, such an issue is best addressed by using a cell culture model that possesses the same biotin uptake characteristics as that of the native intestine. Primary cultures of intestinal epithelium cells cannot be used in such studies owing to their limited viability,[18] and the loss of their functional polarization *in vitro*.[19] Furthermore, membrane vesicles lack the intracellular biochemical pathways believed to be involved in such regulation. These problems are circumvented by the use of cultured intestinal epithelial cell line models. Examples of such models include IEC-6 and IEC-18 cells[20] (both derived from rat small intestine), NCM 460 cells[21] (derived from normal human colon), and T84 cells[22] and Caco-2 cells[23] (both being derived from human colonic carcinomas).

In our laboratory we have used the human-derived intestinal epithelial cell line Caco-2 as a model. Although these cells are derived from a colonic adenocarcinoma, they differentiate spontaneously in culture on reaching confluence to become mature, enterocyte-like cells.[23] Because of this, these cells have been widely used as a model to study different aspects of intestinal functions, including the transport of a variety of nutrients.[24]

Our work has shown that biotin uptake in these cells is Na$^+$, temperature, and energy dependent, saturable as a function of concentration (apparent $K_m \sim 9.5 \ \mu M$), specific, and regulated by substrate level in the culture medium.[25] Studies by Ng and Borchardt[26] have further demonstrated that the apical-to-basolateral transport of biotin in Caco-2 cells grown on Transwell polycarbonate membrane (Cambridge, MA) is greater than the basolateral-to-apical transport, demonstrating the functional polarization of biotin transport in these cells. These properties are similar to, and further validate some of the findings previously described in native human and animal small intestinal preparations, and demonstrate the suitability of Caco-2 cells as an *in vitro* model system with which to study the details of intracellular regulation of the biotin intestinal uptake process.

[18] J. Kumaji and L. R. Johnson, *Am. J. Physiol.* **254**, G81 (1990).
[19] C. A. Ziomek, S. Schulman, and M. Edidin, *J. Cell. Biol.* **86**, 849 (1980).
[20] A. Quaroni, J. Wands, R. L. Trelstad, and K. J. Isselbacher, *J. Cell. Biol.* **80**, 248 (1979).
[21] J. S. Stauffer, L. A. Manzano, G. C. Balch, R. L. Merriman, L. L. Tanzer, and M. P. Moyer, *Gastroenterology* **106**, A273 (1994).
[22] K. Dharmsathaphorn, J. A. McRoberts, K. G. Mandel, L. D. Tisdale, and H. Masui, *Am. J. Physiol.* **246**, 6204 (1984).
[23] S. von Kleist, E. Chany, P. Burton, M. King, and J. Fogh, *J. Natl. Cancer Inst.* **55**, 555 (1975).
[24] J. Fogh, J. M. Fogh, and T. Orfeo, *J. Natl. Cancer Inst.* **59**, 221 (1977).
[25] T. Y. Ma, D. L. Dyer, and H. M. Said, *Biochim. Biophys. Acta* **1189**, 81 (1994).
[26] K.-Y. Ng and R. T. Borchardt, *Life Sci.* **53**, 1121 (1993).

Uptake of Biotin by Liver

The liver is the primary tissue of biotin utilization and metabolism. In addition to its role in the activation of the apocarboxylases described in the previous section, biotin is also involved in several other aspects of hepatocyte function, including the regulation of guanylate cyclase activity,[27] and of the expression of asialoglycoprotein receptor, which is involved in asialoglycoprotein degradation.[28] The liver has the greatest concentration of biotin compared to other organs. However, it has a relatively limited capacity to store the vitamin, compared to its capacity for the storage of other water-soluble vitamins, such as folate, riboflavin, and cyanocobalamin.[29] Also, it has been demonstrated that the site of biotin recycling is outside of the cell.[30] These properties make the hepatocyte dependent on extracellular biotin, which is imported primarily across the basolateral membrane. Therefore, characterization of the mechanism of biotin uptake at this membrane domain is important for overall understanding of biotin physiological homeostasis.

Information about transport mechanisms of biotin in this organ has come from studies using freshly isolated hepatocytes in suspension, basolateral membrane vesicles, and cultured cell lines. Collectively, these studies have demonstrated that biotin uptake in the liver is carrier mediated in nature (apparent $K_m \sim 1.2$–$20 \, \mu M$). The work with basolateral membrane vesicles prepared from rat has also shown the process to be $Na^+$ gradient dependent and electrogenic; the stoichiometric cotransport coupling ratio of biotin to $Na^+$ has been found to be at least $1:2$.[31] In human liver, biotin was also found to be transported by a carrier-mediated and $Na^+$ gradient-dependent process.[32] As with the intestine, the use of these liver preparations has provided insight into the mechanism and driving force of the biotin uptake process. Nothing, however, is known about the intracellular regulation of this process. These studies are best carried out in a viable, homogeneous cell population that possesses transport characteristics similar to those of the native hepatocyte.

One of the problems with using primary hepatocyte cultures is that various culture conditions can greatly influence long-term cell viability. For

[27] J. T. Spence and A. P. Koudelka, *J. Biol. Chem.* **259,** 6393 (1984).

[28] J. C. Collins, E. Paietta, R. Green, A. G. Morell, and R. J. Stockert, *J. Biol. Chem.* **263,** 11820 (1988).

[29] D. E. Danford and H. N. Munro, *in* "The Liver: Biology and Pathobiology" (J. Arias, H. Pepper, D. Schlachter, and D. A. Schifritz, eds.), p. 367. Raven, New York, 1982.

[30] S. O. Freytag and M. F. Utter, *J. Biol. Chem.* **258,** 6307 (1983).

[31] H. M. Said, S. Korchid, D. W. Horne, and M. Howard, *Am. J. Physiol.* **259,** G865 (1990).

[32] H. M. Said, J. Hoefs, R. Mohammadkhani, and D. W. Horne, *Gastroenterology* **102,** 2120 (1992).

example, when cultured on plastic and grown in the presence of dexamethasone, isolated hepatocytes remain viable for several days,[33] and when they are plated onto a biomatrix, such as collagen, they remain viable for several weeks. With extended culture, however, hepatocytes rapidly lose certain functional characteristics of the native tissue. Notable among these is a loss of the ability to transport biotin by a carrier-mediated pathway.[34] Thus, a continuous cell line is highly desirable for studies of intracellular regulation of the liver biotin uptake process. Examples of available liver cell lines include the Hep G2 and Hep-derivative C3A and Hep 3B cells.

In our laboratory, we have used the cell line Hep G2 as a model of biotin uptake in the hepatocyte. Hep G2 is a human-derived hepatoma cell line that possesses many of the normal hepatocyte functions, including similar membrane transport processes, and more general aspects of liver physiology.[35,36] Biotin transport in Hep G2 cells has been shown to be similar to biotin transport in the native hepatocyte, being $Na^+$, temperature, and energy dependent, and saturable as a function of concentration (apparent $K_m \sim 19.2\ \mu M$).[37] As in the intestine, the uptake process in the liver is also specific for biotin; moreover, biotin analogs with a free carboxyl group on the valeric acid side-chain moiety, such as thioctic acid and dethiobiotin, inhibit biotin uptake, while analogs with a blocked carboxyl group, such as biocytin and biotin methyl ester, do not. The information obtained about biotin uptake in Hep G2 so far is quite similar to that obtained from preparations of normal human and rat liver hepatocytes and membrane vesicles, and demonstrates the suitability of this cell line for detailed studies about the cellular and molecular regulation of biotin transport in the liver.

## Uptake of Biotin in Kidney

The reabsorption of biotin in the renal proximal tubule is necessary to prevent excessive biotin loss in the urine. Because of this, the kidney plays an important role in regulating biotin homeostasis. Understanding the mechanisms and regulation of biotin reabsorption process in this tissue is therefore of great importance.

[33] R. I. Freshney, in "Culture of Animal Cells: A Manual of Basic Technique," 3rd Ed., p. 309. Wiley-Liss, New York, 1994.
[34] D. Weiner and B. Wolf, Biochem. Med. Metabol. Biol. 44, 271 (1990).
[35] C. M. Divino and G. C. Schussler, J. Biol. Chem. 265, 1425 (1990).
[36] W. Stremmel and H. E. Diede, Biochim. Biophys. Acta 1013, 218 (1989).
[37] H. M. Said, T. Y. Ma, and V. S. Kamanna, J. Cell. Phys. 161, 483 (1994).

The process of biotin transport in the kidney has been studied using purified brush border and basolateral membrane vesicle preparations.[38–40] From these studies, it was determined that biotin uptake by the brush border membrane domain of mammalian renal cortex cells is carrier mediated and saturable as a function of substrate concentration (with an apparent $K_m$ ~ 28–50 $\mu M$), $Na^+$ dependent, and electroneutral, with a biotin-to-$Na^+$ cotransport coupling ratio of 1:1. Biotin uptake at the brush border is specific, being inhibited only by analogs with a free carboxyl group. Furthermore, analogs with shortened side chains have a reduced ability to inhibit biotin uptake.[40] In contrast, the uptake process at the basolateral membrane domain has also been shown to be carrier mediated and specific in nature, but is $Na^+$ independent.[38]

Presently, there are no published studies dealing with the uptake of biotin in cultured renal proximal tubule cells. However, several renal-proximal tubule cell lines are now available that could serve as useful models to study intracellular regulation of the biotin uptake process. Examples of these include vEPT, a rabbit proximal tubule-derived epithelial cell line,[41] OK, an opossum-derived normal kidney cell line,[42] and HK, which is a human-derived proximal tubule epithelial cell line.[43]

Uptake of Biotin by Placenta

Biotin is important for the proper development of the fetus, as it is involved in key metabolic pathways, and its deficiency in animals has been associated with fetal malformation.[44,45] Biotin present in the maternal circulation is delivered to the developing fetus through the placental epithelium. The syncytiotrophoblast is the cellular element responsible for the net transfer of metabolites and utilizes a polarized distribution of transport processes to accomplish this. Therefore, understanding the biotin uptake mechanism present in the placental trophoblast is important.

The kinetic aspects of biotin transport by the human placenta have been defined by the use of membrane vesicles, cultured trophoblasts and isolated

[38] R. A. Podevin and B. Barbarat, *Biochim. Biophys. Acta* **856,** 481 (1986).

[39] B. Baur and E. R. Baumgartner, *Pflugers Arch.* **422,** 499 (1993).

[40] B. Bauer, H. Wick, and E. R. Baumgartner, *Am. J. Physiol.* **258,** F840 (1990).

[41] M. F. Romero, J. G. Douglas, R. L. Eckert, U. Hopfer, and J. W. Jacobberger, *Kidney Int.* **42,** 1130 (1992).

[42] M. J. Ryan, G. Johnson, J. Kirk, S. M. Feurstenberg, R. A. Zager, and B. Torok-Storb, *Kidney Int.* **45,** 48 (1994).

[43] H. Kozmann, C. Goodpasture, M. M. Miller, R. L. Teplitz, and A. D. Riggs, *In Vitro* **14,** 239 (1978).

[44] T. Watanabe and A. Endo, *J. Nutr.* **119,** 255 (1989).

[45] T. Watanabe, *J. Nutr.* **113,** 574 (1983).

perfused human placenta. In maternal-facing membrane vesicles, biotin uptake was found to be carrier mediated (apparent $K_m \sim 21-26 \ \mu M$), Na$^+$ dependent, and electrogenic, with a cotransport coupling ratio of biotin to Na$^+$ of 1:2. The biotin uptake process is also specific in nature,[46] with the range of this specificity apparently limited to biotin analogs with a free carboxyl group at the valeric acid side chain, as is the case with the intestine and liver tissues. However, it has been postulated that a second anionic nucleus of charge, possibly the keto-oxygen group at the 2-position of the imadazole ring, may also be needed to confer substrate recognition for transport.[46] Finally, even though isolated plasma membrane vesicles offer a number of advantages for the study of transport functions, such as ion requirements, kinetics, and drug interactions, they are limited in studies of regulation of transport function owing to their lack of intact intracellular second-messenger pathways.

As with the intestine and liver, the preferred system with which to study intracellular regulation of biotin transport is cultured cells. *In vitro* cultured trophoblasts have been shown to possess a biotin uptake process that is similar to that of the native tissue, being Na$^+$ and temperature dependent, and saturable as a function of concentration (apparent $K_m \sim 7.0 \ \mu M$).[47] Although the method of Kliman *et al.*[48] can provide purified populations of cytotrophoblasts, such cultures may be maintained for only a relatively short period of time (up to 5 days) compared to continuous cell lines.

To date there has been no published study examining the suitability of continuous cell lines derived from the placenta as models of biotin uptake. However, the JAR-1 cell line[49] (derived from a human choriocarcinoma) has been used as a model with which to study the transport regulation of various other solutes,[49,50] and might be useful in studies of intracellular regulation of biotin transport in the trophoblast.

### Uptake of Biotin by Other Cells

*Astrocytes*

It has been postulated that a mechanism exists in the central nervous system for the homeostatic regulation of the total biotin concentration.[51,52]

[46] S. M. Grassl, *J. Biol. Chem.* **267,** 17760 (1992).
[47] P. I. Karl and S. E. Fisher, *Am. J. Physiol.* **262,** C302 (1992).
[48] H. J. Kliman, J. E. Nestler, E. Semasi, J. M. Sanger, and J. F. Strauss III, *Endocrinology* **118,** 1567 (1986).
[49] P. D. Prased, F. H. Leibach, V. B. Mahesh, and V. Ganapathy, *Endocrinology* **134,** 547 (1994).
[50] D. R. Cool, F. H. Lebach, V. K. Bhalla, V. B. Mahesh, and V. Ganapathy, *J. Biol. Chem.* **266,** 15750 (1991).
[51] H. N. Bhagavan and D. B. Coursin, *J. Neurochem.* **17,** 289 (1970).
[52] P. N. S. Murthy and S. P. Mistry, *Prog. Food Nutr. Sci.* **2,** 405 (1977).

Because astrocytes act as support cells for neurons in the central nervous system and are responsible for the absorption from cerebrospinal fluid of some metabolites and nutrients required for neuronal survival and maintenance of central nervous system homeostasis, they are a candidate for such a mechanism. Studies on biotin transport in the intact central nervous system are quite difficult to evaluate because of cellular heterogeneity, and therefore *in vitro* studies using pure cultures of astrocytes have been quite useful and are extensively employed.

In contrast to primary preparations of isolated cells from absorptive epithelium (e.g., intestine), virtually pure populations of astrocytes are easily prepared and have a high viability.[53] To determine whether a carrier-mediated biotin uptake process exists in the central nervous system, Rodriguez-Pombo and Ugarte[54] examined the transport of biotin in primary cultures of astrocytes. As in the other tissues previously mentioned, biotin uptake in these cells was found to occur by a carrier-mediated and temperature-dependent mechanism that is specific and saturable as a function of concentration (apparent $K_m = 18$ n$M$). The process, however, is Na$^+$ independent in nature. Overall, it appears that the biotin uptake process in astrocytes is compatible with a facilitative, carrier-mediated process, somewhat similar to that detected at the basolateral membrane domains of the enterocyte and the renal proximal tubule cells. Although the reason for this large difference in apparent $K_m$ between the biotin uptake mechanism in the astrocyte and those present in other tissues is unclear, it is possible that the glial biotin uptake mechanism is different from those demonstrated in other tissues. Interestingly, as in other tissues, biotin uptake was also found to be regulated by substrate available from the medium. Prolonged culture of astrocytes in medium restricted in biotin induced a 10-fold increase in total uptake of biotin. This difference was attributed to changes in the $V_{max}$ of the carrier-mediated process, with little change in the apparent $K_m$ of the process.

## Xenopus Oocytes

The *Xenopus* oocyte has proven to be an extremely useful system with which to examine a wide range of biological processes, including membrane transport and its regulation. For understanding the intracellular regulatory mechanisms of biotin uptake, the *Xenopus* oocyte presents several unique advantages over cultured mammalian cells. Its large size allows easy delivery

[53] M. Sensenbrenner, *in* "Cell, Tissue and Organ Cultures in Neurobiology" (S. Federoff and L. Hertz, eds.), p. 191. Academic Press, New York, 1977.
[54] P. Rodriguez-Pombo and M. Ugarte, *J. Neurochem.* **58,** 1460 (1992).

by microinjection of drugs and other compounds into the cell interior. Also, the cell is arrested in metaphase, and therefore changes associated with active cell division do not influence the uptake process. Furthermore, regulation of biotin uptake can be studied at the single-cell level.

In the oocyte, the biotin uptake mechanism is similar in nature to that found in mammalian tissues.[55] The uptake of biotin is saturable as a function of biotin concentration (apparent $K_m \sim 3.9 \ \mu M$), and is $Na^+$, temperature, and energy dependent. Furthermore, the specificity of the transport process is identical to the uptake systems expressed in mammalian cells, with uptake being inhibited by biotin analogs with a free carboxyl group at the valeric acid side chain, but unaffected by compounds such as biocytin and biotin methyl ester, in which this carboxyl group is blocked. A possible involvement of a sulfhydryl group or groups present at the exofacial domain of the carrier has been postulated owing to the fact that the membrane-impermeant sulfhydryl group blocker $p$-(chloromercuri)phenyl sulfonate strongly inhibits biotin transport. These findings indicate that the characteristics of biotin uptake in the oocyte are similar to those described in mammalian intestinal, liver, and kidney cells. Therefore, the oocyte is a suitable model for studies of the regulation of the biotin transport process across biological membranes in higher organisms.

The oocyte has also proven to be extremely useful in the isolation, by expression cloning, of cDNAs encoding certain transmembrane nutrient transporter proteins. Among the numerous transport proteins isolated using this system are the sodium-dependent (SGLT-1 and SGLT-2)[56] and -independent (GLUT family)[57,58] glucose transporters, and the high-affinity $Na^+$-dependent glutamate transporter (EAAC-1).[59] Furthermore, studies have demonstrated that the oocyte can express cDNA encoding the folate transporter of the intestinal epithelial cell from several animal species, including rat, mouse, and human.[60,61] The ability of the oocyte to functionally express exogenous genes encoding transmembrane solute transporters in general raises the possibility of its potential usefulness in the expression cloning of the $Na^+$-dependent biotin transport protein from any of the previously

[55] H. M. Said, L. Polenzani, S. Khorchid, D. Hollander, and R. Miledi, *Am. J. Physiol.* **259,** C397 (1990).

[56] M. A. Hediger, M. J. Coady, T. S. Ikeda, and E. M. Wright, *Nature (London)* **330,** 379 (1987).

[57] G. I. Bell, T. Kayans, J. B. Buse, C. F. Burant, J. Takeda, D. Lin, H. Fukumoto, and S. Seino, *Diabet. Care* **13,** 198 (1990).

[58] M. Mueckler, *Diabetes* **9,** 6 (1990).

[59] Y. Kanai and M. A. Hediger, *Science* **360,** 467 (1992).

[60] H. M. Said, T. T. Nguyen, D. L. Dyer, and S. A. Rubin, *Biochim. Biophys. Acta* **1281,** 164 (1996).

[61] D. L. Dyer, unpublished observations, (1996).

mentioned tissues. Furthermore, it is important to point out that the identification of the cDNA that encodes the biotin transporter(s) of mammalian cells will facilitate a detailed understanding of the expression and regulation of the uptake process at the molecular level.

Techniques

*Assay Protocols of Biotin Uptake by Cultured Cells*

In our laboratory, we have used a number of cell lines to model biotin uptake in normal epithelial cells. Although the growth conditions for each of these cell lines can vary widely, the assay protocol for radiolabeled biotin uptake is generally the same. With this in mind, we have chosen to present the growth and assay protocols used in our laboratory for studies involving the colonic adenocarcinoma Caco-2 cell line, which is fairly standard in its culture requirements compared to other cell lines.

In our laboratory, Caco-2 are started at passage 17, and used between passage 19 and 24 for uptake studies. These cells are routinely grown in 75-cm$^2$ flasks in Dulbecco's modified Eagle's medium supplemented with D-glucose (4.5 g/liter), 1% (v/v) nonessential amino acids, 10% (v/v) fetal bovine serum, and antibiotics in a humidified 37° incubator, with medium changes every 2–3 days. Near-confluent cell cultures (70–90% confluent) are trypsinized following standard procedures. Briefly, cells are washed with 5 ml of prewarmed Ca$^{2+}$- and Mg$^{2+}$-free Hanks' balanced salt solution, and then incubated in 0.25% (w/v) trypsin containing 0.9 $\mu M$ ethylenediaminetetraacetic acid (EDTA). Cells are plated onto 12-well cluster plates (Costar, Cambridge, MA) at a concentration of 500,000 cells/well, and all uptake studies are done at 7–10 days after these cultures reach confluence. The state of cell confluence is monitored by visual inspection using an inverted light microscope, and culture viability is determined by trypan blue dye exclusion.

Uptake studies are generally performed at 37° in Krebs–Ringer phosphate buffer [in m$M$: NaCl, 123; KCl, 4.93; MgSO$_4$, 1.23; CaCl$_2$, 0.85; glucose, 5; glutamine, 5; NaH$_2$PO$_4$, 20 (pH 6.5)] containing $^3$H-labeled biotin (NEN-Du Pont, Boston, MA) with or without unlabeled biotin. A trace amount of [$^{14}$C]inulin is added to measure the contribution of adhering extracellular medium to total nonspecific binding. The prewarmed incubation solution is added to the cultures at the onset of the experiment, and then removed by aspiration at termination of the uptake. Following this, monolayers are washed twice with ice-cold buffer, and then digested by adding 0.5 ml of 1 $N$ NaOH and incubating at 70° for 60 min. Digests are neutralized with 10 $N$ HCl, scintillation fluid is added and radioactivity contained in the samples is determined by liquid scintillation counting. Uptake values are expressed in terms of moles of substrate uptake per milligram of protein; moreover, protein can be determined using the standard Lowry protein assay procedure.[62] The determination of various kinetic parameters of the uptake process can be facilitated through the use of various data analysis software, including the excellent data analysis paradigm developed by Wilkinson.[63]

*Further Techniques*

The biotin uptake process at the brush border membrane and the basolateral membrane domains of cultured epithelial cells may be further examined using purified membrane vesicles isolated from these cells. The procedure for preparation of brush border membrane vesicles involves the standard $Ca^{2+}$ or $Mg^{2+}$ precipitation method, while the procedure for preparation of basolateral membrane vesicles involves the Percoll gradient centrifugation method. Both of these procedures are described in detail elsewhere.[6–8,64]

Acknowledgments

This chapter was supported by a grant from the Departments of Veterans Affairs (H.M.S.), and by grants from the National Institutes of Health (DK47203-01, H.M.S.; and DK09040-02, D.L.D.).

[62] O. H. Lowry, N. J. Rosebrough, A. L. Farr, and R. J. Randall, *J. Biol. Chem.* **193,** 269 (1951).
[63] G. N. Wilkinson, *Biochem. J.* **80,** 324 (1961).
[64] S. Ramamoorthy, D. R. Cool, V. B. Mahesh, F. H. Leibach, H. E. Melikian, R. D. Blakely, and V. Ganapathy, *J. Biol. Chem.* **268,** 21626 (1993).

# [43] Biotinyl-5'-Adenylate Synthesis Catalyzed by *Escherichia coli* Repressor of Biotin Biosynthesis

*By* YAN XU and DOROTHY BECKETT

## Biotin Holoenzyme Synthetases

In biotin-dependent carboxylases from all organisms the biotin moiety is covalently linked to a unique lysine residue on the enzyme via an amide bond. Formation of the linkage is catalyzed by a class of enzymes termed the biotin holoenzyme synthetases or biotin ligases. The ligation reaction

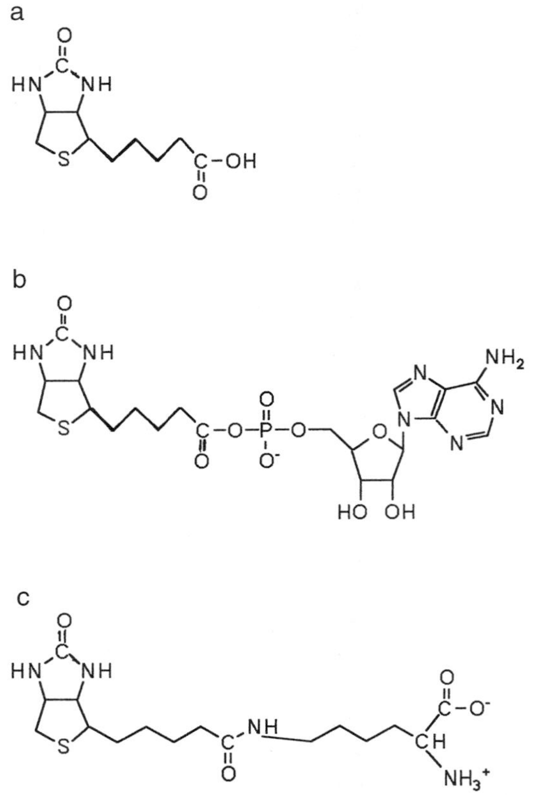

a

b

c

FIG. 1. Chemical structures of (a) biotin, (b) biotinyl-5'-AMP, and (c) biotinyllysine or biocytin. In the holocarboxylases the linkage shown in (c) exists between the biotin moiety and the nitrogen of the ε-amino group of an enzyme-based lysine residue.

is a two-step process involving initial synthesis of an adenylate of biotin from the substrates biotin and ATP followed by transfer of the activated biotin to the ε-amino group of the lysine residue.[1] These two steps are summarized in Eqs. (1) and (2).

$$\text{Biotin} + \text{ATP} \rightarrow \text{biotinyl-5'-AMP} + \text{PP}_i \qquad (1)$$
$$\text{Biotinyl-5'-AMP} + \text{apocarboxylase} \rightarrow \text{holocarboxylase} + \text{AMP} \qquad (2)$$

The structures of biotin and its derivatives relevant to the biotin ligase reaction are shown in Fig. 1. Adenylation of biotin in the first step of the reaction serves to activate the carboxylate group on biotin for subsequent

[1] M. D. Lane, K. L. Rominger, D. L. Young, and F. Lynen, *J. Biol. Chem.* **239**, 2865 (1964).

nucleophilic attack by the nitrogen of the unique lysine residue on the carboxylase that is biotinated in the second step. This two-step chemical reaction is conserved across evolution.

The subunit organizations of the biotin-dependent carboxylase substrates for the biotin ligase reaction fall into two classes. All of these enzymes contain biotin carboxylase (BC), biotin carboxyl carrier (BCC), and transcarboxylase (TC) functionalities. Biotin is linked to a unique lysine residue on the BCC functional unit of the enzyme. In prokaryotic biotin-dependent carboxylases, the three functionalities are located on separate polypeptide chains. Most eukaryotic biotin-dependent carboxylases contain the BC and BCC functionalities on the same polypeptide chain and the TC functionality may or may not reside on the same polypeptide chain. These evolutionary differences in quaternary structure are evidenced in results of analyses of the acetyl-CoA carboxylases from *Escherichia coli*,[2] the nematode *Tubatrix aceti*,[3] and the yeast *Saccharomyces cerevisiae*,[4] which possess the first, second, and third organizations, respectively.

Holocarboxylase synthetases from diverse sources including rat liver, yeast, and bacteria have been shown to catalyze biotination of apocarboxylases from other species. Cronan (1990) demonstrated that the *E. coli* enzyme, BirA, in addition to catalyzing linkage of biotin to its homologous substrate, the BCCP subunit of the acetyl-CoA carboxylase, catalyzes covalent ligation of biotin to fusion proteins containing the 1.3 S subunit of *Propionibacterium shermanii* transcarboxylase, the α subunit of *Klebsiella pneumoniae* oxaloacetate decarboxylase, and a protein sequence encoded by a cDNA from tomato.[5] Gravel and co-workers showed that BirA also biotinates the α subunit of human propionyl-CoA carboxylase and that the human holocarboxylase synthetase when expressed in *E. coli* biotinates BCCP.[6] This cross-species reactivity has been ascribed to the high degree of sequence conservation noted among biotin-accepting segments (Fig. 2).[7–11]

[2] R. B. Guchait, S. E. Polakis, P. Dimroth, E. Stoll, J. Moss, and M. D. Lane, *J. Biol. Chem.* **249,** 6636 (1974).

[3] H. Meyer, B. Nevaldine, and F. Meyer, *Biochemistry* **17,** 1822 (1978).

[4] A.-F. Walid, S. S. Chirala, and S. J. Wakil, *Proc. Natl. Acad. Sci. U.S.A.* **89,** 4534 (1992).

[5] J. E. Cronan, Jr. *J. Biol. Chem.* **265,** 10327 (1990).

[6] A. Leon-Del-Rio, D. Leclerc, B. Akerman, N. Wakamatsu, and R. A. Gravel, *Proc. Natl. Acad. Sci. U.S.A.* **92,** 4626 (1995).

[7] S.-J. Li and J. E. Cronan, Jr. *J. Biol. Chem.* **267,** 855 (1992).

[8] P. Marini, S. Li, D. Gardiol, J. E. Cronan, Jr., and D. de Mendoza, *J. Bacteriol.* **177,** 7003 (1995).

[9] W. Al-Feel, S. S. Chirala, and S. J. Wakil, *Proc. Natl. Acad. Sci. U.S.A.* **89,** 4534 (1992).

[10] L. Abu-Elheiga, A. Jayakumar, A. Baldini, S. S. Chirala, and S. J. Wakil, *Proc. Natl. Acad. Sci. U.S.A.* **92,** 4100 (1995).

[11] A.-M. Lamhonwah, T. J. Barankiewicz, H. F. Willard, D. J. Mahuran, F. Quan, and R. A. Gravel, *Proc. Natl. Acad. Sci. U.S.A.* **83,** 4864 (1986).

| Ec-ACC | HIVRSPMVGT | FYRTPSPDAK | AFIEVGQKVN | VGDTLCIV**EA** | (120) |
|---|---|---|---|---|---|
| Bs-ACC | HKITSPMVGT | FYASSSPEAG | PYVTAGSKVN | ENTVVCIV**EA** | (121) |
| yt-ACC | LEVENDPTQL | RTPSPGKLVK | FLVENGEHII | KGQPYAEI**EV** | (733) |
| hum-ACC | FEKENDPSVM | RSPSAGKLIQ | YIVEDGGHVF | AGQCYAEI**EV** | (784) |
| hum-PCC | KVTEDTSSVL | RSPMPGVVVA | VSVKPGDAVA | EGQEICVI**EA** | (667) |

| Ec-ACC | **MKM**MNQIEAD | K**SG**TVKAILV | ESGQPVEFDE | PLVVIE | (156) |
|---|---|---|---|---|---|
| Bs-ACC | **MKL**FIEIEAE | V**KGE**IVEVLV | ENGQLVEYGQ | PLFLVKAE | (159) |
| yt-ACC | **MKM**QMPLVSQ | ENGIVQLLKQ | PGSTIVAGDI | MAIMTL... | (769) |
| hum-ACC | **MKM**VMTLTAV | ESGCIHYVKR | PGAALDPGCV | LAKMQL... | (820) |
| hum-PCC | **MKM**QNSMTAG | K**TG**TVKSVHC | QAGDTVGEGD | LLVELE | (703) |

FIG. 2. Sequence comparison of biotin-accepting domains of carboxylases. Sequences shown are *E. coli* (Ec) acetyl-CoA carboxylase,[7] *Bacillus subtilis* (Bs) acetyl-CoA carboxylase,[8] yeast (yt) acetyl-CoA carboxylase,[9] human liver acetyl-CoA carboxylase (hum-ACC),[10] and human mitochondrial propionyl-CoA carboxylase α chain (hum-PCC).[11] The numbers in parentheses indicate the termini of the segments of primary sequences from the five proteins that are being compared. The conserved sequence Glu/Ser-Ala/Val-Met-Lys-Met/Leu, which contains the site of biotination in each domain, is in boldface. Four additional residues in boldface are conserved in all sequenced biotin-dependent carboxylases.

## Escherichia coli Biotin Holoenzyme Synthetase

The biotin holoenzyme synthetase (biotin repressor) from *E. coli*, BirA, is a 35.3-kilodalton (kDa) protein that catalyzes covalent linkage of biotin to a single protein substrate in the organism, the biotin carboxyl carrier protein (BCCP) subunit of the acetyl-CoA carboxylase.[12] As noted previously, biotin-accepting sequences from a variety of organisms will also serve as substrates for the enzyme. BirA catalyzes biotin ligation via the two-step process shown in Eqs. (1) and (2). In addition to its function as a biotin holoenzyme synthetase, BirA functions as a transcriptional repressor by binding to the transcriptional control region of the biotin biosynthetic operon.[13] The reader is referred to Beckett and Matthews[13a] in this volume for additional details of the transcriptional repression function of BirA.

[12] J. E. Cronan, Jr. *Cell* **58**, 427 (1989).
[13] D. F. Barker and A. M. Campbell, *J. Mol. Biol.* **146**, 451 (1981).
[13a] D. Beckett and B. W. Matthews, *Methods Enzymol.* **279** [39], 1997 (this volume).

Several features of the *E. coli* biotin holoenzyme synthetase render it an excellent model for kinetic studies of all enzymes in this class. The first, which has already been discussed, is the cross-reactivity of the enzyme in catalyzing linkage of biotin to protein segments isolated from organisms ranging from other bacteria to humans. This evolutionary conservation ensures that results of studies of BirA will be generally useful for understanding the properties of all biotin holoenzyme synthetases. A second reason is the existence of an extensive body of research on BirA. Barker and Campbell isolated a number of mutations in the gene that encodes the enzyme.[13] These mutants have been characterized *in vivo* and were subsequently subcloned and sequenced.[14] The gene has also been cloned into vectors that allow for overexpression of the protein in *E. coli*.[15] This, of course, enables purification of the enzyme in the large quantities required for detailed kinetic studies. Finally, a high-resolution structure of the enzyme has been determined by X-ray crystallographic techniques.[16] The availability of structural data allows for correlation of results of functional studies with structural detail.

## Assays of Biotin Holoenzyme Synthetase Activity

Assays previously developed for biotin holoenzyme synthetase activity have been reviewed in an earlier contribution to this series.[17] Two general methods have been used to determine the enzyme activity. The first involves direct measurement of the incorporation of radioactive biotin into the appropriate apocarboxylase substrate. In the second assay, the appearance of the active holotranscarboxylase is monitored. Although it has been demonstrated that biotinyl-5'-AMP is an intermediate in the biotin transfer reaction catalyzed by all biotin holoenzyme synthetases, no direct measurement of the kinetics of synthesis of the adenylate was published until 1994.[18] The availability of large quantities of homogeneous BirA preparations has been critical to performing these measurements. A method for purification of the protein is described as follows. In addition, details of the methods utilized for measurement of the time course of BirA-catalyzed synthesis of biotinyl-5'-AMP are provided. Finally, methods for determination of the thermodynamic stability of the BirA–adenylate complex are presented.

---

[14] M. R. Buoncristiani, P. K. Howard, and A. J. Otsuka, *Gene* **44,** 255 (1986).

[15] M. R. Buoncristiani and A. J. Otsuka, *J. Biol. Chem.* **263,** 1013 (1988).

[16] K. P. Wilson, L. M. Shewchuk, R. G. Brennan, A. J. Otsuka, and B. W. Matthews, *Proc. Natl. Acad. Sci. U.S.A.* **89,** 9257 (1992).

[17] N. H. Goss and H. G. Wood, *Methods Enzymol.* **107,** 261 (1984).

[18] Y. Xu and D. Beckett, *Biochemistry* **33,** 7354 (1994).

Methods

*Purification of Escherichia coli Biotin Holoenzyme Synthetase*

Construction of plasmids that allow for high levels of expression of the *E. coli* biotin holoenzyme synthetase (BirA) has enabled purification of the protein in quantities previously unobtainable. Preparations of the enzyme obtained utilizing the procedure described as follows have been shown to be 100% active in catalyzing synthesis of the adenylate from biotin and ATP.[19] These preparations have, moreover, been shown to retain this same level of activity for several years, provided that proper storage and handling procedures are followed.

*Buffers*

Buffer 1: 100 m$M$ Sodium phosphate (pH 6.5), 200 m$M$ NaCl, 5% (v/v) glycerol, 0.1 m$M$ dithiothreitol (DTT)

Buffer 2: 100 m$M$ Sodium phosphate (pH 6.5), 100 m$M$ NaCl, 5% (v/v) glycerol, 0.1 m$M$ dithiothreitol (DTT)

Buffer 3: 100 m$M$ Sodium phosphate (pH 6.5), 5% (v/v) glycerol, 0.1 m$M$ dithiothreitol (DTT)

Buffer 4: 50 m$M$ Potassium phosphate (pH 6.5), 5% (v/v) glycerol, 0.1 m$M$ dithiothreitol (DTT)

Buffer 5: 50 m$M$ Tris-HCl (pH 7.5 at 4°), 50 m$M$ KCl, 5% (v/v) glycerol, 0.1 m$M$ dithiothreitol (DTT)

Buffer 6: 50 m$M$ Tris-HCl (pH 7.5 at 4°), 200 m$M$ KCl, 5% (v/v) glycerol, 0.1 m$M$ dithiothreitol (DTT)

*Cell Growth.* The plasmid utilized for overproduction of BirA, pJMR1, encodes a fusion of the *birA* gene to the *tac* promoter (A. J. Otsuka and M. Junichi, unpublished results, 1990). The protein is purified from *E. coli* strain X90 [*ara* Δ(*lac-pro*) *nalA argE*am *rif thi⁻*/F′ *¹lacIqlac⁺, pro⁺*].[20] Cells are grown at 37° in 2X YT medium containing ampicillin at 100 $\mu$g/ml. A 75-ml aliquot of a late log culture is utilized as an inoculum for each 2 liters of medium. Cultures are allowed to grow until the measured optical density at 600 nm is 1.0 (approximately 3 hr). At this point a fresh solution of IPTG (isopropyl-$\beta$-D-thiogalactoside; Sigma, St. Louis, MO) is added to a final concentration of 75 $\mu$g/ml and induction is allowed to proceed for 5 hr. Cells are harvested by centrifugation at 3500 $g$ for 20 min at 4° and the resulting cell pellet is resuspended in twice its weight of buffer 1. From

[19] J. Abbott and D. Beckett, *Biochemistry* **32**, 9649 (1993).
[20] E. Amman, J. Brosius, and M. Ptashne, *Gene* **25**, 167 (1983).

this point on in the procedure all manipulations are conducted at 4°. Cells are disrupted by sonication and lysis is considered complete when the measured optical density at 600 nm is decreased by 90% from its initial value. The lysate is diluted fivefold into buffer 1 and cleared by centrifugation at 13,000 $g$ for 45 min. Nucleic acids are removed by addition of a 10% (v/v in $H_2O$, pH 7.0) solution of polyethyleneimine to a final concentration of 0.2%. After gentle stirring for 20 min, the resulting precipitate is removed by centrifugation at 13,000 $g$ for 30 min. A saturated solution of ammonium sulfate in buffer 1 is added to a final concentration of 60% and the mixture is allowed to stir overnight. The precipitate, which is recovered by centrifugation at 13,700 $g$ for 45 min, is then resuspended in buffer 2 and exchanged into buffer 3 by dialysis.

*Chromatography.* Three chromatographic steps are required to obtain homogeneous preparations of the BirA. These include cation-exchange chromatography on Whatman (Clifton, NJ) P11 resin, chromatography on hydroxyapatite, and, finally, a second cation-exchange step on SP-Sepharose Fast Flow resin.

*Whatman P11 Chromatography.* The sample obtained from 10 liters of culture is chromatographed on P11 cellulose phosphate resin (Whatman) (column dimensions = 5.5 × 35 cm) equilibrated in buffer 3. After sample loading, the column is washed extensively with buffer 3 to remove unbound proteins. The bound proteins are eluted in a linear NaCl gradient (0–0.8 $M$) in buffer 3, in which the repressor is found to elute at approximately 0.6 $M$ NaCl. Fractions are assayed for the presence of BirA by sodium dodecyl sulfate-polyacrylamide gel electrophoresis [SDS–PAGE; 12.5% (w/v) separating] and the appropriate fractions are pooled. The resulting sample is exchanged into buffer 4 by dialysis.

*Hydroxyapatite Ultrol Chromatography.* The sample from the first chromatography step is loaded onto Hydroxyapatite Ultrol resin (BioSepra, Marborough, MA) equilibrated in buffer 4 (column dimensions = 2.5 × 35 cm) and unbound proteins are removed by extensive washing with the same buffer. Bound proteins are eluted in a linear potassium phosphate gradient (50–500 m$M$) in which BirA elutes at a concentration of approximately 300 m$M$. Fractions containing BirA are pooled and the sample is exchanged into buffer 5 by dialysis.

*SP-Sepharose Fast Flow Chromatography.* The sample obtained from chromatography on hydroxyapatite is loaded onto SP-Sepharose Fast Flow resin (Pharmacia, Piscataway, NJ) equilibrated in buffer 5 (column dimensions = 2.3 × 13 cm). Unbound proteins are removed by washing with the same buffer and BirA is eluted in a linear KCl gradient (0.05–0.8 $M$). The protein elutes at approximately 0.3 $M$ KCl in this gradient. Fractions containing BirA are pooled and exchanged into buffer 6 by dialysis. The protein is stored in 1-ml aliquots at −70°. The final protein sample is found to be >98% pure as judged by results of Coomassie Brilliant Blue staining of samples subjected to SDS–PAGE. An extinction coefficient of 1.3 mg$^{-1}$ ml at 280 nm is utilized to determine the protein concentration. The yield of pure protein obtained from 1 g of wet cell paste is typically 1.4 mg.

## Measurement of Fractional Activity of Biotin Repressor Preparations

The fractional activity of a ligase preparation is determined by measuring binding of biotinyl-5′-adenylate to the protein under stoichiometric conditions. The stoichiometry of the enzyme–intermediate complex has been independently shown to be 1:1 and the fractional activity of enzyme preparations can be ascertained by performing a simple breakpoint titration of a known amount of protein with the ligand. The binding is conveniently monitored by steady-state fluorescence because the intensity of intrinsic protein fluorescence signal is quenched by 40% on saturation of BirA with the ligand. The biotinyl-5′-AMP is readily synthesized in the laboratory using a modification of the method originally published by Lane et al.[1] These modifications are described in Ref. 14. The adenylate is stored desiccated at −20° and solutions are prepared in water. Fluorescence titrations are performed by preparing a 0.8-ml solution of BirA at 1 $\mu M$ final concentration in standard buffer [10 m$M$ Tris-HCl (pH 7.50 ± 0.01) at 20.0 ± 0.1°, 200 m$M$ KCl, 2.5 m$M$ MgCl$_2$] and transferring this to a fluorescence cuvette. Small aliquots of concentrated biotinyl-5′-AMP (freshly prepared in standard buffer) are added to the protein and the mixture is allowed to equilibrate for at least 2 min. The intrinsic protein fluorescence spectrum is then recorded. Fluorescence spectra are measured using an excitation wavelength of 295 nm and emission is monitored from 310 to 450 nm. All spectra are integrated and corrected for both dilution and the buffer background. Plotting of the corrected integrated spectra as a function of the biotinyl-5′-AMP concentration yields a simple breakpoint curve. Nonlinear least-squares fitting of the data provides an estimate of the activity of the preparations. In our hands, the BirA preparations are, within error, 100% active.

## Measurement of Time Course of BirA-Catalyzed Synthesis of Biotinyl-5′-AMP

The time course of BirA-catalyzed synthesis of the adenylate from the substrates biotin and ATP can be montiored by direct quantitation of the amount of product formed as a function of time.[18] The radioactively labeled substrate [$\alpha$-$^{32}$P]ATP is utilized in these measurements so that product formation can be quantitated by monitoring incorporation of the label into the adenylate. All measurements are performed in standard buffer.

## Preparation of Reagents for Measurements

*Biotin Stocks.* Unfortunately, biotin concentrations cannot be determined by a convenient spectrophotometric technique. Solutions of a given concentration are, therefore, prepared by careful weighing of the dry biotin powder (Sigma) using a high-precision analytical balance and adjustment of solutions to their final desired volumes using a volumetric flask. We typically prepare 10 ml of a concentrated solution (800–850 $\mu M$) directly in the buffer utilized for measurements. The dry biotin is diluted into standard buffer that has not yet been titrated with acid. Once the biotin is dissolved, HCl is added to achieve the final desired pH at the temperature at which experiments are conducted and the volume is adjusted to its final value in a 10-ml volumetric flask.

### ATP Stocks

Concentrated ATP (Sigma) stocks are prepared in distilled water and the pH is adjusted to pH 7.0 by addition of KOH. The concentration of the stock is determined by ultraviolet (UV) absorbance.

Both substrates are stored as 1-ml aliquots at $-20°$.

*Reaction Mixtures.* Time courses of adenylate synthesis are typically carried out in a total volume of 50 $\mu$l. Mixtures containing the $[\alpha\text{-}^{32}P]ATP$ (5,000,000 cpm) (6000 Ci/mmol; Amersham, Arlington Heights, IL), ATP at the desired concentration, inorganic pyrophosphatase (2 U/ml), and BirA at the desired concentration are first prepared and incubated at the assay temperature. The inorganic pyrophosphatase (Boehringer-Mannheim, Indianapolis, IN), which is utilized in the reactions to drive the equilibrium to completion, is exchanged into standard buffer by dialysis prior to use. Reactions are initiated by rapid addition of biotin to the final desired concentration and the incubation is continued. Time points are obtained by removing small aliquots (3.0 $\mu$l) of the reaction mixture and diluting them into 1 $\mu$l of a 4 m$M$ solution of unlabeled chemically synthesized biotinyl-5'-AMP to quench the reaction. The resulting mixtures are spotted onto a cellulose thin-layer chromatography (TLC) plate (Kodak, Rochester, NY) and the chromatograph is developed using a mobile phase containing $H_2O$–formic acid–*tert*-amyl alcohol (1/1/1, v/v/v). The $R_f$ values for biotinyl-5'-AMP, biotin, and ATP in this TLC system are 0.14, 0.65, and 0.9, respectively.

*Quantitation.* Quantitation of the time course of biotinyl-5'-AMP synthesis is achieved using a Molecular Dynamics (Menlo Park, CA) phosphorimaging system. Typically the phosphor screen is exposed to the TLC plate for 5 hr and then scanned. The optical density in the spots on the image corresponding to biotinyl-5'-AMP and ATP is integrated at each time point and the biotinyl-5'-AMP concentration is calculated by dividing the integrated optical density in the spot corresponding to the product, biotinyl-5'-AMP, by the sum of the integrated optical densities from both spots and multiplying the result by the initial ATP concentration.

### Measurement of Initial Rates of Biotinyl-5'-AMP Synthesis

The initial rates of BirA-catalyzed synthesis of biotinyl-5'-AMP are monitored by stopped-flow fluorescence.[18] As indicated previously, binding of biotinyl-5'-AMP to BirA results in quenching of 40% of the intrinsic protein fluorescence signal. Binding of biotin to the protein leads to quenching of only 15% of the intrinsic protein fluorescence signal. These spectroscopic changes can be exploited to monitor the kinetics of synthesis of biotinyl-5'-AMP from biotin and ATP. Measurements are made using a dual-syringe system. A complex of BirA with either biotin or ATP is present in one syringe and the other substrate is present in the second syringe. The contents of the two syringes are rapidly mixed and the resulting time-dependent decrease in the intrinsic fluorescence spectrum is monitored.

*Reagents.* Solutions of biotin and ATP are prepared as described in the previous section. Measurements were made in standard buffer.

*Data Analysis.* The raw data from the stopped-flow is analyzed using the software provided with the KinTek (State College, PA) stopped-flow instrument. The time courses are in all cases well described by a single exponential model. At least six traces are collected for each set of conditions of ATP or biotin concentrations.

### Kinetics of Association of BirA with Biotinyl-5'-AMP

As with determination of the initial rates of biotinyl-5'-AMP synthesis, the kinetics of the bimolecular association of BirA with biotinyl-5'-AMP can be monitored by observing the time-dependent change in the intrinsic protein fluorescence that occurs following rapid mixing of the protein with ligand.

### Analysis of Results of Kinetic Measurements

### Time Course of Adenylate Synthesis Characterized by Burst Kinetics

The time course of BirA-catalyzed synthesis of biotinyl-5'-AMP is shown in Fig. 3. As indicated in Fig. 3, the time course is characterized by an initial rapid exponential burst of synthesis followed by a slow linear phase. The following chemical equation [Eq. (3)] can adequately describe this process:

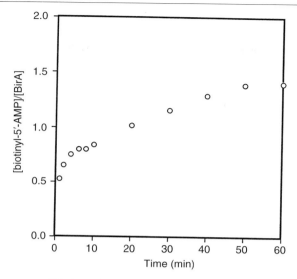

FIG. 3. Time course of BirA-catalyzed synthesis of biotinyl-5′-AMP from biotin and ATP. Concentrations were as follows: BirA, 5 $\mu M$; biotin, 106 $\mu M$; ATP, 50 $\mu M$; pyrophosphatase, 2 U/ml. The time course was measured in standard buffer at 20°.

$$\text{BirA} + \text{biotin} + \text{ATP} \xrightarrow{k_{\text{syn}}}$$

$$\text{BirA–biotinyl-5′-AMP} \xrightarrow{k_{\text{off}}} \text{BirA} + \text{biotinyl-5′-AMP} \quad (3)$$

where the first step in the reaction, involving synthesis of biotinyl-5′-AMP, corresponds to the rapid burst in the time course and the second step, which is rate limited by release of product from BirA, corresponds to the slow linear phase. The apparent rate constant for the enzyme-catalyzed synthesis of biotinyl-5′-AMP from biotin and ATP is $k_{\text{syn}}$ and $k_{\text{off}}$ is the rate constant governing release of the adenylate from the BirA–adenylate complex. Oh the basis of this reaction scheme, the concentration of biotinyl-5′-AMP synthesized at any time in the course of the reaction is expressed as follows[21]:

$$[\text{Biotinyl-5′-AMP}]_t$$
$$= n[\text{BirA}]_0 \left( \frac{k_{\text{syn}}}{k_{\text{syn}} + k_{\text{off}}} \right) \left[ \frac{k_{\text{syn}}}{k_{\text{syn}} + k_{\text{off}}} - \frac{k_{\text{syn}} e^{-(k_{\text{syn}} + k_{\text{off}})t}}{k_{\text{syn}} + k_{\text{off}}} + k_{\text{off}} t \right] \quad (4)$$

[21] A. Fersht, "Enzyme Structure and Mechanism." W. H. Freeman, New York, 1985.

where $k_{syn}$ and $k_{off}$ are the same as described previously, $[BirA]_0$ is the total concentration of the enzyme, and $n$ is the stoichiometry of the burst complex. If the reaction conditions are such that [biotin] and [ATP] are much greater than $[BirA]_0$ and if $k_{syn} \gg k_{off}$ and $t \gg (k_{syn} + k_{off})^{-1}$, Eq. (2) reduces to the following linear form:

$$[Biotinyl-5'-AMP]_t = n[BirA]_0 + n[BirA]_0 k_{off} t \qquad (5)$$

where $n$ corresponds to the stoichiometry of the burst complex. The slope of the line, $n[BirA]_0 k_{off}$, provides information about the rate of release of biotinyl-5'-AMP from the BirA–biotinyl-5'-AMP complex. Provided that the enzymatic reactions are carried out according to the stipulations of the simplification of Eq. (3), the slopes and intercept of the linear phase in the time course are directly proportional to the total enzyme concentration utilized for measurement of the time course. Experiments performed at multiple total enzyme concentrations confirm the prediction of Eq. (3). The stoichiometry of the burst complex obtained from such an analysis was found to be 1:1 and the unimolecular rate constant for dissociation of this complex obtained from the slope of the linear portion of the curve is 0.00027 $sec^{-1}$. The half-life of the complex calculated from this rate constant is approximately 30 min.

### Specificity Constants for Biotin and ATP in BirA-Catalyzed Synthesis of Biotinyl-5'-AMP

The initial burst phase in BirA-catalyzed synthesis of biotinyl-5'-AMP can be measured using stopped-flow fluorescence.[18] Mixing of the BirA–biotin complex with the second substrate, ATP, results in a single exponential decrease in the intrinsic protein fluorescence owing to the synthesis and tight binding of the adenylate product (Fig. 4). As previously indicated, the half-life of the enzyme-biotinyl-5'-AMP complex is approximately 30 min and, therefore, in the time scale of the initial rate measurements (40 sec), only the production of 1 mole of adenylate per mole of enzyme is observed.

The dependence of the pseudo first-order rate of adenylate synthesis on the concentration of substrate has been measured. Technically these experiments are limited with respect to the lower limit of enzyme concentration that can be reliably used because the observable is the intrinsic protein fluorescence. This lower limit is approximately 0.5 $\mu M$. All measurements are made using a great excess of each substrate over enzyme (pseudo first-order conditions). Because these conditions are saturating in biotin concentration, no information about the dependence of the initial rate on biotin concentration has been obtained. The dependence of the initial rate

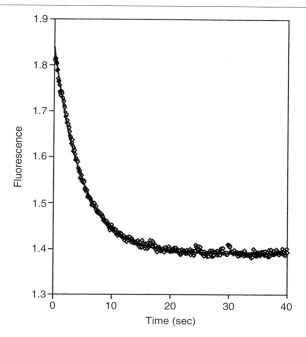

FIG. 4. Stopped-flow fluorescence trace obtained from measurement of the "burst" phase in BirA-catalyzed synthesis of biotinyl-5'-AMP. The measurement was performed in standard buffer at 20° using a KinTek 2001 instrument. The excitation wavelength was set at 295 nm and emission was monitored about 340 nm, using a cutoff filter. Concentrations were as follows: BirA, 1 $\mu M$; biotin, 100 $\mu M$; ATP, 1.5 m$M$. The solid line represents results of a simulation using parameters obtained from nonlinear least-squares analysis of the data, using a single exponential model.

on ATP concentration at saturating biotin concentration has, however, been determined. Analysis of the kinetic data yields the results shown in Table I. The $K_m$ for ATP determined from this analysis is high.

Values of $k_{cat}/K_m$ for both substrates, biotin and ATP, can be estimated from the results of the kinetic analysis as well as independent measurements of binding of biotin to BirA. The equilibrium dissociation constant for the BirA–biotin binding interaction has been calculated from the measured values of the bimolecular association rate constant for complex formation and the unimolecular dissociation rate constant for decay of the complex.[22] The resulting equilibrium dissociation constant is $4 \times 10^{-8}$ $M$. If we assume that the value of the $K_m$ for biotin in adenylate synthesis is identical to this

[22] Y. Xu, E. Nenortas, and D. Beckett, *Biochemistry* **34,** 16624 (1995).

TABLE I

KINETIC PARAMETERS GOVERNING BirA-CATALYZED FORMATION OF BIOTINYL-5'-AMP[a]

| Substrate | $k_{cat}$ (sec$^{-1}$)[b] | $K_m$ ($M$)[b] | $k_{cat}/K_m$ ($M^{-1}$ sec$^{-1}$) |
|---|---|---|---|
| ATP | $0.6 \pm 0.3$ | $3 (\pm 1) \times 10^{-3}$ | $2 (\pm 1) \times 10^2$ |
| Biotin | $0.6 \pm 0.3$ | $4.1 (\pm 0.5) \times 10^{-8}$ [c] | $2 (\pm 1) \times 10^7$ |

[a] Values of $k_{cat}$ for both substrates and the value of $K_m$ for ATP were estimated from analysis of the dependence of the pseudo first-order rate of adenylate synthesis on ATP concentration at saturating biotin concentration. Buffer conditions: 10 m$M$ Tris-HCl (pH 7.50 $\pm$ 0.02 at 20°), 200 m$M$ KCl, 2.5 m$M$ KCl.

[b] The large errors associated with the parameters result from the fact that measurements of the dependence of the pseudo first-order rate of adenylate synthesis on ATP concentration were limited to the $k_{cat}/K_m$ region of the curve.

[c] The $K_m$ for biotin is based on the assumption that $K_m = K_D$ for this substrate. The $K_D$ was estimated from stopped-flow fluorescence measurements of the binding kinetics.[22]

value of $K_D$ we can calculate the $k_{cat}/K_m$ for biotin at saturating ATP concentration. The resulting value shown in Table I is close to the diffusion-controlled rate. By contrast, the value of $k_{cat}/K_m$ for ATP at saturating biotin concentration is small. The high specificity of BirA for biotin and its low specificity for ATP are consistent with a system in which synthesis of the adenylate is highly sensitive to biotin, and not ATP, concentration.

*Thermodynamics of Interaction of BirA with Biotinyl-5'-AMP*

The equilibrium dissociation constant for interaction of BirA with biotinyl-5'-AMP has been obtained from kinetic measurements of the constants governing the binding reaction. The unimolecular dissociation rate constant was obtained from the measured time course of BirA-catalyzed synthesis of the adenylate from biotin and ATP previously described. Association kinetics were measured using stopped-flow fluorescence.[22] BirA was rapidly mixed with the adenylate and the resulting time-dependent decrease in the intrinsic protein fluorescence was monitored. A representative stopped-flow trace is shown in Fig. 5. As indicated in the trace, the association kinetics are biphasic. The time constant for the first phase has been shown to depend on the ligand concentration while the second is ligand concentration independent. This is consistent with the following general model for a two-step reaction:

$$\text{BirA} + \text{ligand} \underset{k_{-1}}{\overset{k_1}{\rightleftharpoons}} \text{BirA--ligand} \underset{k_{-2}}{\overset{k_2}{\rightleftharpoons}} {}^*\text{BirA--ligand} \qquad (6)$$

where the first step corresponds to the formation of a collision complex and the second corresponds to a unimolecular conformational change in

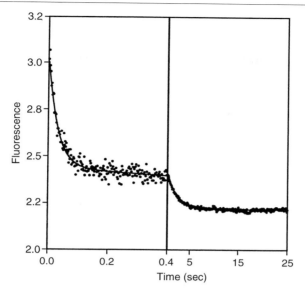

FIG. 5. Stopped-flow fluorescence trace obtained from measurement of the bimolecular association of BirA with biotinyl-5′-AMP. Concentrations: BirA, 0.5 $\mu M$; biotinyl-5′-AMP, 12.5 $\mu M$. The measurement was performed in standard buffer at 20° using a KinTek 2001 stopped-flow instrument. The settings are the same as those indicated in Fig. 4.

the complex. Analysis of the dependence of the magnitudes of individual time constants on ligand concentration using Eqs. (7) and (8)[21]:

$$\frac{1}{\tau_1} + \frac{1}{\tau_2} = k_1[\text{ligand}] + k_{-1} + k_2 + k_{-2} \tag{7}$$

$$\frac{1}{\tau_1} \times \frac{1}{\tau_2} = k_1(k_2 + k_{-2})[\text{ligand}] + k_{-1}k_{-2} \tag{8}$$

yields the microscopic kinetic constants governing the protein–ligand interaction (Table II). However, the value of $k_{-1}$ determined from this analysis has a large error associated with it. A more accurate estimate of $k_{-1}$ has been determined by direct measurement of partitioning of the intermediate complex. This can be readily accomplished by double mixing stopped-flow measurements. The protein and biotinyl-5′-AMP are initially allowed to mix for a sufficiently long time to maximally populate the intermediate [see Eq. (6)]. The contents of the third syringe, which consists of a large molar excess of biotin, a competing ligand, are then rapidly mixed with the intermediate. Any protein that partitions toward dissociation is trapped by

TABLE II
Microscopic Rate Constants Governing Binding of BirA to Biotinyl-5'-AMP[a]

| $k_1$ ($M^{-1}$ sec$^{-1}$) | $k_{-1}$ (sec$^{-1}$) | ($k_2 + k_2$) (sec$^{-1}$) | $k_{off}$ (or $k_{-2}$) (sec$^{-1}$) |
|---|---|---|---|
| 5.9 ($\pm$0.8) $\times$ 10$^6$ | 1 $\pm$ 3 (0.2)[b] | 0.32 $\pm$ 0.09 | 2.7 ($\pm$0.3) $\times$ 10$^{-4}$ |

[a] Values of $k_1$, $k_{-1}$, and ($k_2 + k_{-2}$) were obtained from analysis of data obtained from stopped-flow measurements of the bimolecular association of BirA with biotinyl-5'-AMP using Eqs. (7) and (8).[22]

[b] The value of $k_{-1}$ in parentheses was estimated from simulations of the data obtained from measurements of the partitioning of the collision complex formed between BirA and biotinyl-5'-AMP.[22]

the excess competing ligand. While conversion of the intermediate to the final complex is accompanied by a decrease in intrinsic protein fluorescence, dissociation of the intermediate results in an increase in fluorescence. Simulation of the partitioning data using a range of assumed values of $k_{-1}$ indicates that it is approximately equal to 0.2 sec$^{-1}$.[22]

The equilibrium dissociation constant for the BirA–biotinyl-5'-AMP-binding reaction is related to the bimolecular rate constant for association of the protein and ligand and the unimolecular rate constant for dissociation of the complex by the familiar expression $K_D = k_{off}/k_{on}$. For a two-step binding reaction of the type shown in Eq. (6), the microscopic kinetic constants obtained from the analysis of the stopped-flow kinetic data can be related to $k_{off}$ and $k_{on}$ using the following expressions.

$$k_{on} = \frac{k_1 k_2}{k_{-1} + k_2} \tag{9}$$

$$k_{off} = \frac{k_{-2} k_{-1}}{k_{-1} + k_2} \tag{10}$$

Given that $k_2$ and $k_{-1}$ are approximately equivalent in magnitude for this system, $K_D = k_{-2}/k_1$. The value of $k_1$ obtained from measurements of the bimolecular association of BirA with biotinyl-5'-AMP is shown in Table II and $k_{-2}$ is equivalent to the overall unimolecular rate constant governing dissociation of the BirA–biotinyl-5'-AMP complex obtained from measurements of the linear phase in the time course of BirA-catalyzed synthesis of the adenylate (see Analysis of Results, above). The value of $K_D$ calculated using these values of the microscopic rate constants is 5 $\times$ 10$^{-11}$ $M$ and the calculated Gibbs free energy based on this constant at 20° is $-14$ kcal/mol.[22] This large favorable Gibbs free energy indicates that the biotin repressor–adenylate complex is thermodynamically stable.

Additional Remarks

The methods described in the previous section have allowed the first characterization of the kinetics of biotinyl-5'-adenylate synthesis catalyzed by a biotin holoenzyme synthase. The results obtained with the *E. coli* enzyme indicate that the reaction is characterized by burst kinetics, reflecting initial rapid formation of 1 mol of adenylate per mole of enzyme followed by slow unimolecular dissociation of the enzyme–adenylate complex. This phenomenology has allowed for determination of the dissociation rate of the BirA–adenylate complex from the time course of adenylate synthesis. On the basis of the value of the unimolecular rate constant governing dissociation of the complex, we conclude that the enzyme–adenylate complex is kinetically stable. This kinetic stability may be significant for ensuring that BirA is in the "activated" adenylate-bound state required for both binding to the acetyl-CoA carboxylase and concomitant biotin transfer, as well as for site-specific binding to the biotin operator sequence.

The methods for measurement of BirA-catalyzed synthesis of biotinyl-5'-AMP described in this chapter are generally useful for characterizing this reaction in other enzymes in this class. Cloning of the biotin holoenzyme synthetase genes from *Bacillus subtilis*,[23] *Paracoccus denitrificans*,[24] and *Saccharomyces cerevisiae*,[25] as well as the cDNA encoding the human protein,[6] should render all of these enzymes available in quantities required for detailed kinetic analysis. Comparison of the BirA sequence to those of these other biotin holoenzyme synthetases indicates, not surprisingly, the existence of considerable homology. Functionally, although all four proteins catalyze the same two-step reaction, only one (*B. subtilis*) of the four enzymes appears also to serve as a transcriptional regulatory protein. In light of this functional distinction, comparison of results of kinetic studies of BirA-catalyzed biotinyl-5'-AMP synthesis with the reactions catalyzed by these other enzymes will further clarify the physiological significance of the unusual kinetic stability of the BirA–adenylate complex.

Acknowledgment

Funding for the work described in this chapter was provided by NIH Grant GM46511 and a Du Pont Young Professorship.

[23] S. Bower, J. Perkins, R. R. Yocum, P. Serror, A. Sorokin, P. Rahaim, C. L. Howitt, N. Prasad, S. D. Ehrlich, and J. Pero, *J. Bacteriol.* **177**, 2572 (1995).
[24] X. Xu, A. Matsuno-Yagi, and T. Yagi, *Biochemistry* **32**, 968 (1993).
[25] J. E. Cronan, Jr. and J. C. Wallace, *FEMS Microbiol. Lett.* **130**, 221 (1995).

# [44] Biotinidase in Serum and Tissues

By Jeanne Hymes, Kristin Fleischhauer, and Barry Wolf

## Introduction

Biotin, a water-soluble vitamin, is the coenzyme for four carboxylases in humans: acetyl-CoA carboxylase, pyruvate carboxylase, propionyl-CoA carboxylase, and $\beta$-methylcrotonyl-CoA carboxylase.[1] The carboxyl group of biotin is covalently attached through an amide bond to the $\varepsilon$-amino group of specific lysyl residues of these carboxylases by holocarboxylase synthetase.[2] The carboxylases are subsequently degraded proteolytically to biocytin (biotinyllysine) or small biotinyl peptides. Biotinidase then cleaves these molecules, releasing biotin, and thereby recycling the vitamin.[3]

There are two major types of inherited biotin-responsive disorders: early-onset multiple carboxylase deficiency, which is due to a deficiency of holocarboxylase synthetase, and late-onset multiple carboxylase deficiency, which is due to a deficiency of biotinidase in serum and other tissues.[4]

## Functions

### Hydrolytic Activity

Biotinidase (biotin-amide amidohydrolase, EC 3.5.1.12) hydrolyzes biocytin to biotin and lysine. Biotinidase is necessary for the endogenous recycling of the vitamin and probably for the liberation of the vitamin from dietary protein-bound sources.[3] Biotinidase specifically cleaves *d*-biotinylamides and esters,[5,6] but does not cleave biotin from intact holocarboxylases.[7,8] The rate of hydrolysis of biotin from natural and synthetic biotinylated peptides decreases with increasing size of the peptide.[7] The specific requirements for the hydrolysis of biotinyl substrates by biotinidase

---

[1] J. Moss and M. D. Lane, *Adv. Enzymol.* **35,** 321 (1971).

[2] M. D. Lane, D. L. Young, and F. Lynen, *J. Biol. Chem.* **239,** 2858 (1964).

[3] B. Wolf, R. E. Grier, J. R. Secor McVoy, and G. S. Heard, *J. Inher. Metab. Dis.* **8,**(1), 53 (1985).

[4] L. Sweetman, *J. Inher. Metab. Dis.* **4,** 53 (1981).

[5] J. Pispa, *Ann. Med. Exp. Biol. Fenn.* **43**(5), 1 (1965).

[6] J. Knappe, W. Brommer, and K. Biederbick, *Biochem. Z.* **228,** 599 (1963).

[7] D. V. Craft, N. H. Goss, N. Chandramouli, and H. G. Wood, *Biochemistry* **24,** 2471 (1985).

[8] J. Chauhan and K. Dakshinamurti, *J. Biol. Chem.* **261,** 4268 (1986).

have been reviewed previously.[9] Because hydrolysis of biocytin by biotinidase requires acid pH and nonphysiological (micromolar) concentrations of biocytin, it has been suggested that hydrolysis is probably not the primary role of biotinidase unless the $K_m$ is decreased by a modifier.[10]

When biotinylated derivatives are used either *in vivo* or *in vitro*, and biotinidase is present, it is important to determine if these derivatives are substrates for biotinidase. For example, deferodesaminolysylbiotin used for imaging and radiotherapy is converted to biotin and desaminolysyldeferoxamine by biotinidase.[11] In addition, biotinidase cleaves biotin that is attached through a spacer arm to erythrocytes.[12]

## Biotinyltransferase Activity

Biotinidase is biotinylated when incubated with biocytin, but not biotin, at pH 7 to 9.[13] Biotinidase can then transfer the biotin from the biotinyl acyl-enzyme to nonspecific nucleophiles such as hydroxylamine[5] or specific acceptors including histones.[14] Biotinidase biotinyltransferase activity (i.e., biotinylation of histones) occurs at physiological pH and at physiological (nano- to picomolar) concentrations of biocytin.

## Lipoamidase Activity

Human serum biotinidase hydrolyzes lipoyl-*p*-aminobenzoate (PABA), but the $K_m$ value is high (0.6 m$M$) relative to that for the hydrolysis of biocytin.[15] Lipoamidase activity in serum is due to biotinidase.[15,16] Purified biotinidase from rat serum hydrolyzes both lipoyl-PABA and lipoyllysine at a concentration of about 1 m$M$.[17] Therefore, it is unlikely that serum biotinidase acts as a lipoamidase *in vivo*.

## Biotin-Binding Protein

Chauhan and Dakshinamurti[10] reported that biotinidase is the only protein in serum that binds tritiated biotin and that the enzyme contains two biotin-binding sites ($K_d = 3$ n$M$ and $K_d = 59$ n$M$). Mock and Malik,[18]

[9] B. Wolf, J. Hymes, and G. S. Heard, *Methods Enzymol.* **184**, 103 (1990).
[10] J. Chauhan and K. Dakshinamurti, *Biochem. J.* **256**, 265 (1988).
[11] S. F. Rosebrough, *J. Pharmacol. Exp. Ther.* **265**, 408 (1993).
[12] M. Magnani, L. Chiarantini, and U. Mancini, *Biotechnol. Appl. Biochem.* **20**, 335 (1994).
[13] J. Hymes, K. Fleischhauer, and B. Wolf, *Clin. Chim. Acta* **233**, 39 (1995).
[14] J. Hymes, K. Fleischhauer, and B. Wolf, *Biochem. Mol. Med.* **56**, 76 (1995).
[15] C. L. Garganta and B. Wolf, *Clin. Chim. Acta* **189**, 313 (1990).
[16] L. Nilsson and B. Kagedal, *Biochem. J.* **291**, 545 (1993).
[17] L. Nilsson and B. Kagedal, *Eur. J. Clin. Chem. Clin. Biochem.* **32**, 501 (1994).
[18] D. Mock and M. I. Malik, *Am. J. Clin. Nutr.* **56**, 427 (1992).

also using tritiated biotin, found that most of the biotin in serum is not bound to protein. Hymes *et al.*[13] have demonstrated that peroxidase-conjugated avidin reacts with biotinidase on transblot (after sodium dodecyl sulfate-polyacrylamide gel electrophoresis) when the enzyme is incubated at pH 7 to 9 in the presence of biocytin, but not biotin.

## Molecular Biology

The cDNA that encodes normal human serum biotinidase has been cloned from a human liver cDNA library and has been sequenced.[19] The gene is located on human chromosome 3p25.[20] The cDNA sequence contains two possible ATG initiator codons (numbered as bases 1 and 60) and an open reading frame of 1629 bp, relative to the first ATG codon and the termination codon at base 1629. This cDNA encodes a protein of 543 amino acids, including 41 amino acids of a putative signal peptide sequence. Starting at nucleotide 123, the sequence that encodes the N terminus of the mature serum biotinidase is in the same reading frame with both of the ATG codons. Either sequence (the entire 41-amino acid sequence or the shorter 21-amino acid sequence) is consistent with the motif for a secretory signal peptide. Biotinidase mRNA is a low-abundance message in all tissues examined.[19]

## Physical Properties

### Isoforms

Human serum biotinidase migrates as an $\alpha_1$ protein on serum electrophoresis.[21-23] Isoelectric focusing (IEF) of serum from normal individuals reveals at least nine isoforms, four major and five minor, between pH 4.15 and 4.35.[24] Human serum biotinidase migrates to the $\beta$ region on electrophoresis after neuraminidase treatment, which indicates that the serum enzyme is extensively sialylated.[22] Most of the microheterogeneity found on IEF is due to differences in the degree of sialylation of the enzyme, because neuraminidase treatment reduces the number of isoforms to one

[19] H. Cole, T. R. Reynolds, G. B. Buck, J. M. Lockyer, T. Denson, J. E. Spence, J. Hymes, and B. Wolf, *J. Biol. Chem.* **269**, 6566 (1994).

[20] H. Cole, S. Weremowicz, C. C. Morton, and B. Wolf, *Genomics* **22**, 662 (1994).

[21] M. Koivusalo and J. Pispa, *Acta Physiol. Scand.* **58**, 13 (1963).

[22] D. A. Weiner and B. Wolf, *Ann. N.Y. Acad. Sci.* **447**, 435 (1985).

[23] B. Wolf, G. S. Heard, L. G. Jefferson, V. K. Proud, W. E. Nance, and K. A. Weissbecker, *New Engl. J. Med.* **313**, 16 (1985).

[24] P. S. Hart, J. Hymes, and B. Wolf, *Clin. Chim. Acta* **197**, 257 (1991).

major species (p$I$ 5.27) and three minor species (p$I$ 5.01, 5.14, and 5.30).[24] On preparative IEF, serum biotinidase was detected at p$I$ 4.0 and 4.4.[16] The isoform at p$I$ 4.4 may be due to a cysteine adduct of biotinidase because it occurs when cysteine, but not mercaptoethanol, is used as a stabilizer.

## Molecular Mass

Human serum biotinidase is a single glycosylated polypeptide with a molecular mass of about 66 to 76 kDa.[7,8,25] Treatment of biotinidase with $N$-glycanase results in a lower molecular mass species of 60 kDa, indicating that biotinidase contains about 20% carbohydrate.[24] The cDNA for mature human serum biotinidase contains 502 amino acid residues with a molecular mass of 56,771 Da.[19] It contains six potential glycosylation sites (Asn-X-Thr/Ser). Assuming varying degrees of glycosylation of all six sites, the molecular mass of the glycosylated enzyme is predicted to be between 70 and 80 kDa.

Biotinidase from human serum is purified to homogeneity at around 20,000-fold purification.[8,17,25] One study identified two proteins in serum with biotinidase activity, a 76-kDa and a 110-kDa species, and suggested that the larger species was specific for biocytin.[26] These proteins were purified 349- and 11.2-fold, respectively. Moreover, the N-terminal amino acid sequence of both of these proteins did not correspond to the N terminus of biotinidase ascertained by others or to that predicted from the cDNA nucleotide sequence of biotinidase.[19] The same investigators have reported that biotinidase cleaves enkephalins[27] and that biotinidase in human milk is capable of synthesizing biocytin in the presence of biotin and ATP.[28]

## Heat Stability

Biotinidase in human serum[29] and partially purified from hog serum and hog liver is stable both during the enzyme assay and during preincubation for up to 30 min at 37°.[5] Denaturation of the porcine enzyme occurs after incubation for 30 min at 60°. Human serum enzyme is relatively heat stable and is denatured 60% after incubation for 15 min at 60° and 100% after incubation for 15 min at 70°.[8,30]

[25] B. Wolf, J. B. Miller, J. Hymes, J. Secor McVoy, Y. Ishikana, and E. Shapira, *Clin. Chim. Acta* **164,** 27 (1987).

[26] J. Oizumi and K. Hayakawa, *Clin. Chim. Acta* **215,** 63 (1993).

[27] J. Oizumi and K. Hayakawa, *Biochim. Biophys. Acta* **1074,** 433 (1991).

[28] J. Oizumi and K. Hayakawa, *J. Chromatogr.* **612,** 156 (1993).

[29] B. Wolf, unpublished results (1986).

[30] K. Hayakawa and J. Oizumi, *Clin. Chim. Acta* **178,** 95 (1988).

Catalytic Properties

*pH Dependence*

Hydrolysis of biocytin by human serum is maximal at pH 4.5 to 6.[8] Hydrolysis of biotinyl-*p*-aminobenzoate (BPABA) has a broader pH profile with considerable activity from pH 5 to 8 and maximal activity from pH 5.5 to 6.5. Biotinyltransferase activity of human serum biotinidase occurs at pH 7 to 9 and is undetectable below pH 7.[14]

A broad pH range has also been reported for biotinidase from rat plasma and liver using BPABA as the substrate.[17] BPABA is hydrolyzed by biotinidase from hog liver and serum at pH 6.5 and from hog kidney at pH 6.0 at the same rate as is biocytin.[5,6]

Dialysis of hog serum biotinidase in acetate buffer below pH 5.0 and in potassium phosphate or Tris–malate buffers above pH 6.0 results in loss of enzyme activity.[5] Purified human serum biotinidase can be stored at 4° in 0.1 $M$ phosphate buffer, pH 6.0, containing 1 m$M$ 2-mercaptoethanol and 1 m$M$ EDTA for 1 month without loss of activity.[8]

*Acyl-Enzyme Formation*

The following is evidence for biotinylacylbiotinidase formation during catalysis: (1) biotinylhydroxamate is formed when biotinidase is incubated in the presence of high concentrations of hydroxylamine with either BPABA or biocytin as substrate[5,6]; (2) biotinidase exhibits a considerably higher $K_i$ for the same inhibitor at pH 7 than at pH 5.5 (i.e., the lack of inhibition by biotin at pH 7 may be due to slower dissociation of biotin from the acyl-enzyme, resulting in less available, inhibitable enzyme[5,8]); (3) biotinylated biotinidase can be detected by avidin reactivity on transblot after incubation at pH 7 to 9 in the presence of biocytin, but not biotin[13]; (4) biotinylated biotinidase is not detected in the presence of high concentration of mercaptoethanol or hydroxylamine (reagents known to cleave thioesters), suggesting that when biocytin is cleaved, biotin forms an acyl-enzyme (thioester) with the cysteine in the active site of the enzyme; and (5) biotinidase transfers biotin to histones at pH 7 to 9 in the presence of biocytin, but not biotin.[14]

Mercaptoethanol (0.1–0.2 m$M$) or dithiothreitol (1–2 m$M$) protects or stabilizes biotinidase.[5,7,8,31] Hydrolytic activity of biotinidase in human serum is enhanced by addition of mercaptoethanol.[32] This enhanced activity may be due to the protection of the cysteine in the active site or to the

[31] K. Hayakawa and J. Oizumi, *Clin. Chim. Acta* **168,** 109 (1987).
[32] K. Hayakawa and J. Oizumi, *J. Biochem.* **103,** 773 (1988).

increased turnover of biotinylated biotinidase that results from removal of biotin from the acyl-enzyme by mercaptoethanol, thereby allowing more substrate to be cleaved.[14]

*Kinetics*

Kinetics for hydrolysis of various related substrates by biotinidase and for its inhibition has been reviewed previously.[9] The following are the most important findings. The $K_m$ value for BPABA hydrolysis by human serum biotinidase is 10 to 34 $\mu M$[8,33] and that for biocytin is 5 to 7.8 $\mu M$.[8,21] The $K_m$ values of BPABA and biocytin for human plasma biotinidase are 10 and 6.2 $\mu M$, respectively.[7] The $K_m$ value for BPABA hydrolysis by biotinidase in crude rat liver homogenate is 15 $\mu M$.[17]

The $K_m$ of biocytin for biotinyltransferase activity has not been determined precisely, but the transfer of biotin to histones can be detected at picomolar concentrations of biocytin.[14] The degree of biotinylation of histones is greater when biocytin is the substrate than when BPABA is used.

*Inhibition*

Biotin is a competitive inhibitor of biotinidase at acidic pH when either biocytin or BPABA is the substrate. At pH 7.4, biotin is not inhibitory when BPABA is the substrate.[5] Biotinidase activity is not competitively inhibited by *dl*-dethiobiotin, *l*-biotin, *l*-lysine, $\varepsilon$-aminocaproic acid, fatty acids, *d*-norbiotin, *d*-biotin sulfone, or PABA.[5,8]

Sulfhydryl inhibitors inactivate biotinidase. Biotinidase from hog liver and serum is inactivated completely by 0.1 m$M$ *p*-chloromercuribenzoate[5] and the enzymes in human plasma[7] and hog kidney[6] are inactivated completely by 10 $\mu M$ *p*-chloromercuribenzoate.[5] Monoiodoacetate also inhibits enzyme activity, whereas *N*-ethylmaleimide or arsenite does not.[5] Partially purified biotinidases from *Lactobacillus casei* and *Streptococcus faecalis* are not inhibited by 0.1 m$M$ *p*-chloromercuribenzoate.[6,34]

Evidence for the presence of a serine-hydroxyl group in or near the active site of biotinidase is equivocal.[9]

## Tissue Distribution and Cellular Localization

In animals, including humans, the highest specific activity of biotinidase is in serum.[5,21] Activities are high in liver, kidney, and adrenal glands, and

[33] B. Wolf, R. E. Grier, R. J. Allen, S. I. Goodman, and C. L. Kien, *Clin. Chim. Acta* **131**, 273 (1983).
[34] M. Koivusalo, C. Elorriaga, Y. Kaziro, and S. Ochoa, *J. Biol. Chem.* **238**, 1038 (1963).

low in brain.[5,17,35] Biotinidase is also present in secretory cells, including fibroblasts and leukocytes, and in pancreatic juice and zymogen granules.[36,37] Biotinidase activity has been detected in the urine of humans.[38]

Because biotinidase activity in blood correlates positively with the concentration of albumin, biotinidase probably originates from liver.[39] Subcellular fractionation studies of rat hepatocytes suggest that the enzyme is enriched in the microsomal fraction and, to a lesser extent, in the lysosomal fraction.[5,37] Studies of subcellular fractions of rat liver confirm these results.[17] Microsomal biotinidase is localized in the rough endoplasmic reticulum.

Biotinidase Deficiency

Children with biotinidase deficiency exhibit neurologic and cutaneous abnormalities in association with ketolactic acidosis, organic aciduria, and mild hyperammonemia.[40] There is considerable variability in the clinical features of affected children. Symptomatic children have improved markedly after treatment with 5–20 mg of oral biotin per day. Because symptoms can be prevented by biotin treatment, many states have included screening for biotinidase deficiency in their newborn screening programs.

Biotinidase deficiency has been divided into profound enzyme deficiency (less than 10% of mean normal serum biotinidase activity) and partial enzyme deficiency (10–30% of mean normal serum activity).[41] Profound deficiency includes children who are untreated and become symptomatic and those who are detected by newborn screening. Partial biotinidase deficiency includes children who are detected by newborn screening. If stressed with an infection or starvation a small number of partial biotinidase-deficient children may become symptomatic.

Individuals with profound biotinidase deficiency can be classified into nine distinct biochemical phenotypes on the basis of the presence of cross-reacting material (CRM) to antibody prepared against human serum bio-

[35] S. F. Suchy, J. R. McVoy Secor, and B. Wolf, Neurology 35, 1510 (1985).

[36] B. Wolf, G. S. Heard, J. S. McVoy, and H. M. Raetz, J. Inher. Metab. Dis. 7(2), 121 (1984).

[37] G. S. Heard, R. E. Grier, D. L. Weiner, J. R. Secor McVoy, and B. Wolf, Ann. N.Y. Acad. Sci. 447, 259 (1985).

[38] C. De Felice, K. Hayakawa, T. Tanaka, and E. Terentieva, Nephron 70, 115 (1995).

[39] D. L. Weiner, R. E. Grier, P. Watkins, G. S. Heard, and B. Wolf, Am. J. Hum. Genet. 34, 56A (1983).

[40] B. Wolf, in "The Metabolic and Molecular Bases of Inherited Disease" (C. R. Scriver, A. L. Beaudet, W. S. Sly, and D. Valle, eds.), p. 3151. McGraw-Hill, New York, 1995.

[41] J. R. McVoy, H. L. Levy, M. Lawler, M. S. Schmidt, D. D. Eberts, P. S. Hart, D. D. Pettit, M. G. Blitzer, and B. Wolf, J. Pediatr. 116, 78 (1990).

tinidase. Phenotypes differ in number of isoforms and distribution and frequency of the isoforms.[42] Individuals with partial biotinidase deficiency are similarly classified into six distinct biochemical phenotypes.[43]

Biotinyltransferase activity has been determined in sera from patients with profound deficiency of biotinyl hydrolytic activity.[14] Cross-reacting material to anti-biotinidase was also determined in this group. None of the sera from symptomatic children have biotinyltransferase activity, whereas 67% of the sera with CRM and none of the sera without CRM from children detected by newborn screening have transferase activity.

A common 7-bp deletion and 3-bp insertion in the putative leader sequence of the biotinidase gene were demonstrated in at least 1 allele of 10 of 20 symptomatic children with profound biotinidase deficiency.[44] This mutation results in a frameshift and the premature termination of the translated enzyme, resulting in profound biotinidase deficiency.

### Assays of Biotinidase Hydrolytic Activity

Natural and artificial substrates have been used in various assays of biotinidase activity; these include secondary enzymatic,[45] radiometric,[46,47] colorimetric,[6,33] and fluorometric[48–50] measurements. Some methods require derivatization[48] or chromatographic separation of the reaction products.[46,49]

Serum and tissues, such as liver with high biotinidase activity, can be measured using the colorimetric assay described below.[33] Biotinidase activity can also be measured in cultured cells, such as hepatocytes and fibroblasts, by this method.

Biotinidase activity in tissues, such as peripheral blood leukocytes and brain, requires a more sensitive method, such as the radioassay described below.[47] A high-performance liquid chromatography (HPLC) method for measuring biotinidase activity using biotinyl-6-aminoquinoline as substrate has been developed.[51] An HPLC radioassay uses biotinylmono[$^{125}$I]iodotyramine and biotinyldi[$^{125}$I]iodotyramine as substrates.[52]

[42] P. S. Hart, J. Hymes, and B. Wolf, *Am. J. Hum. Genet.* **50,** 126 (1992).

[43] P. S. Hart, J. Hymes, and B. Wolf, *Pediatr. Res.* **31,** 261 (1992).

[44] R. J. Pomponio, T. R. Reynolds, H. Cole, G. A. Buck, and B. Wolf, *Nature Genet.* **11,** 96 (1995).

[45] D. L. Weiner, R. E. Grier, and B. Wolf, *J. Inher. Metab. Dis.* **8,** 101 (1985).

[46] L. E. Thuy, B. Zielinska, L. Sweetman, and W. L. Nyhan, *Ann. N.Y. Acad. Sci.* **447,** 434 (1985).

[47] B. Wolf and J. R. Secor McVoy, *Clin. Chim. Acta* **135,** 275 (1984).

[48] H. Ebrahim and K. Dakshinamurti, *Anal. Biochem.* **154,** 282 (1986).

[49] K. Hayakawa and J. Oizumi, *J. Chromatogr.* **383,** 148 (1986).

[50] H. Wastell, G. Dale, and K. Bartlett, *Anal. Biochem.* **140,** 69 (1984).

[51] K. Hayakawa, K. Yoshikawa, J. Oizumi, and K. Yamauchi, *J. Chromatogr.* **617,** 29 (1993).

[52] E. Livaniou, A. K. Roboti, S. E. Kakabakos, J. Nyalala, G. P. Evangelatos, and D. S. Ithakissio, *J. Chromatogr. B Biomed. Appl.* **656,** 215 (1994).

The simplest and most commonly used method for measuring biotinidase activity uses BPABA as substrate. The two methods described as follows are based on measurement of the PABA released from BPABA by the hydrolytic action of biotinidase. The first method is a modification of the method of Knappe *et al.* and Wolf *et al.*[33] and is usually used for the diagnosis of biotinidase deficiency. A procedure based on this method, for semiquantitatively measuring biotinidase activity in blood-soaked filter paper disks, is used to screen newborn infants for biotinidase deficiency.[23,53] The second method, that of Wolf and Secor McVoy,[47] is 100-fold more sensitive than the colorimetric assay and can measure biotinidase activity in tissues, such as leukocytes and amniotic cells, that have low enzyme activities.

### Colorimetric Assay of Biotinidase Activity in Serum

*Principle*

The hydrolysis of BPABA by biotinidase results in production of biotin and PABA. The primary aromatic amine, PABA, is released, diazotized, and coupled to a naphthol reagent producing purple color, which is quantitated spectrophotometrically at 546 nm.

*Specimen Requirements*

This method is applicable to specimens that contain relatively high enzyme activity, including serum or plasma, and homogenates of liver, kidney, and adrenal glands. Serum should be separated from whole blood within 1 hr after collection and should be stored at $-70°$. Long-term storage of serum samples at $-20°$ is not recommended because biotinidase activity decreases with time. One to 2 ml of serum or plasma is required for testing.

Heavy metals, such as copper and zinc, inhibit biotinidase activity.[5] Therefore, 1–10 m$M$ EDTA is included in storage buffers to prevent this inhibition. Anionic detergents, such as sodium dodecyl sulfate (SDS), denature biotinidase, whereas low concentrations (0.1 to 0.5%) of nonionic detergents, such as Triton X-100 or Nonidet P-40, can be used without loss of activity.[31] Triton X-100 was included in the incubation buffer when biotinidase activity was determined by autoanalyzer.[54]

[53] G. S. Heard, B. Wolf, L. G. Jefferson, K. A. Weissbecker, W. E. Nance, Jr., J. R. Secor McVoy, A. Napolitano, P. L. Mitchell, F. W. Lambert, and A. S. Linyear, *J. Pediatr.* **108,** 40 (1986).
[54] K. A. Weissbecker, H. D. Gruemer, G. S. Heard, and W. G. Miller, *Clin. Chem.* **35,** 831 (1989).

*Reagents*

Potassium phosphate buffer (KPB; 50 m$M$, pH 6.0): Combine 0.70 g of $K_2HPO_4$ and 2.98 g of $KH_2PO_4$, bring to a final volume of 500 ml with HPLC water, and store at 0–5°

Substrate buffer (0.05 m$M$ biotinyl-PABA): Add 9.64 mg of biotin-4-amidobenzoic acid (Sigma, St. Louis, MO) dissolved in 0.1–0.2 ml of 1 $M$ sodium bicarbonate and 8.41 mg of disodium EDTA to 500 ml of HPLC water and stir until the EDTA is in solution. Store at 0–5°

Standard solution (0.2 m$M$ PABA standard): Dilute a 2 m$M$ PABA stock solution with KPB; dissolve 13.71 mg of analytically pure PABA (potassium salt) in 50 ml of 50 m$M$ KPB. Aliquots of the stock can be stored indefinitely at −70°. The standard solution is stable at 0–5° for 1–2 months

Trichloroacetic acid solution (TCA; 30%, w/v): Add 30 ml of TCA solution (6.1 $N$) to 70 ml of HPLC water and store at room temperature

Sodium nitrite solution (0.1%, w/v): Prepare fresh daily; bring 0.1 g of sodium nitrite (crystalline) to a final volume of 100 ml with HPLC water

Ammonium sulfamate solution (0.5%, w/v): Bring 0.5 g of ammonium sulfamate (ACS reagent) to a final volume of 100 ml with HPLC water and store at room temperature

$N$-1-Naphthylethylenediamine dihydrochloride solution (NEDD; 0.1%, w/v): This solution (0.1 g of NEDD per 100 ml of HPLC water) is light sensitive and should be stored in a dark bottle at room temperature

*Calibration*

The standard must be included in each assay performed. Enzyme activity is calculated on the basis of the absorbance of this standard.

*Quality Control*

Controls for biotinidase testing are not commercially available. A normal serum control is obtained from an individual ascertained to have normal biotinidase activity. A control with low biotinidase activity can be prepared by incubating serum from an individual with normal enzyme activity at 37° for 72 hr. This partially inactivated enzyme can then be stored at −70°.

*Procedure*

Serum samples should be assayed for enzyme activity in duplicate and an additional sample, which does not contain substrate buffer, should be included to determine background absorbance and to detect the presence of substances that can interfere with color development.

1. Pipette 1.9 ml of substrate buffer into subject and control sample tubes and pipette 1.9 ml of KPB into subject and control background tubes. Pipette 1.9 ml of KPB into the tube used for the PABA standard. Pipette 2 ml of KPB into a tube that is designated as the standard blank. The tubes are incubated at 37° for 10 min.

2. Pipette 0.1 ml of the standard, and 0.1 ml of subject and control serum into appropriate sample and background tubes and gently vortex each tube.

3. Incubate the tubes in a water bath at 37° for 30 min and then pipette 0.2 ml of 30% TCA into each tube. Vortex and centrifuge the tubes for 10 min at room temperature at 700 g.

4. Pipette 1.5 ml of the supernatant from each tube into clean test tubes.

5. Pipette 0.5 ml of water into each tube and vortex.

6. Pipette the following color-developing reagents: 0.2 ml of 0.1% sodium nitrite, 0.2 ml of 0.5% ammonium sulfamate, and 0.2 ml of 0.1% NEDD, consecutively into each tube at 3-min intervals and vortex after each addition.

7. Allow color development to proceed for 10 min and then measure the absorbance of the solution in each tube at 546 nm within 30 min.

*Calculation of Activity*

Biotinidase activity is expressed as nanomoles of PABA liberated per minute per milliliter of serum. The standard contains 20 nmol of PABA and the activity of the subject sample is calculated using the following equation:

$$\text{Activity} = \frac{(\text{Abs}_{\text{sample}} - \text{Abs}_{\text{background}}) \times 20 \text{ nmol} \times 10 \text{ (conversion factor)}}{(\text{Abs}_{\text{std}} - \text{Abs}_{\text{background}}) \times 30 \text{ min}}$$

The conversion factor expresses the final activity per milliliter of serum. The incubation time is 30 min. The mean biotinidase activity in normal human serum is 7.1 nmol/min/ml serum with a range from 4.4 to 12.0 nmol/min/ml serum. Individuals with profound biotinidase deficiency have activities that are less than 10% of mean normal activity ($<0.7$ nmol of PABA/min/ml serum); individuals with partial biotinidase deficiency have activities that are 10–30% of mean normal activity (0.7–2.1 nmol of PABA/min/ml serum); and obligate heterozygotes for biotinidase deficiency have activities that are 30–50% of mean normal activity (2.2–5.2 nmol of PABA/min/ml serum).

Radioassay of Biotinidase Activity in Serum, Cells, and Tissues

*Principle*

Biotinidase is assayed by measuring the hydrolysis of $N$-($d$-biotinyl)-[$^{14}$C-carboxyl]PABA. The [$^{14}$C]PABA released is separated from unhydrolyzed radioactive substrate by reacting it with avidin and precipitating the biotin–avidin and BPABA–avidin complexes with bentonite. The [$^{14}$C]PABA remaining in the supernatant is determined by a liquid scintillation counter.

*Specimen Requirement*

Serum is diluted to 1% (v/v) with KPB. Extracts of leukocytes,[55] fibroblasts, and other tissues are prepared by homogenizing in KPB in the presence of 0.1% (v/v) Triton X-100.[47] Extracts should contain 2–3 mg of protein per milliliter.

*Reagents*

Potassium phosphate buffer (KPB; 200 m$M$, pH 6.0): Bring 2.39 g of $KH_2PO_4$ and 0.56 g of $K_2HPO_4$ to 100 ml with HPLC water

$N$-($d$-Biotinyl)-[$^{14}$C-carboxyl]PABA solution (2.3 mCi/mmol): Synthesize as previously described[33] and make 3.6 m$M$ by adding cold biotinyl-PABA

$p$-hydroxymercuribenzoate (PHMB; 2 m$M$): Dissolve 14.41 mg of PHMB in 0.1 ml of 1 $N$ NaOH and dilute with HPLC water in 100 ml. PHMB inhibits enzyme activity

Avidin (1%, w/v): Prepare fresh daily 10 mg of avidin per milliliter of KPB

Bentonite (6.67%, w/v): Suspend 0.67 g of bentonite in 10 ml of KPB. Make fresh daily and vortex thoroughly before use

*Procedure*

Each sample is assayed in duplicate and an identical sample with inhibitor is prepared as a background sample. One standard with avidin and one that does not contain avidin are run per assay. Tubes are prepared as follows: Sample: Add 0.018 ml of KPB and 0.1 ml of cell supernatant (or diluted serum) to duplicate sample tubes. Background sample: Add 0.018 ml of PMBA and 0.1 ml of cell supernatant (or diluted serum) to each back-

[55] B. Wolf and L. E. Rosenberg, *J. Clin. Invest.* **62**, 931 (1978).

ground tube. Standard with avidin: Add 0.118 ml of KPB to a tube. Standard without avidin: Add 0.118 ml of KPB to a tube.

1. After all of the tubes have been prepared, place the tubes in a water bath at 37° for 10 min.

2. Add 0.015 ml of [$^{14}$C]biotinyl-PABA to each tube.

3. Incubate the tubes for 4 hr at 37° (2 hr for serum).

4. After the incubation, add 0.018 ml of 2 m$M$ HMBA to all sample tubes and 0.018 ml of KPB to all background tubes. Add 0.018 ml of 2 m$M$ HMBA to both standard tubes. Gently vortex all tubes.

5. Add 0.025 ml of 1% avidin to all tubes, except for the standard tube without avidin. To this tube, add 0.025 ml of KPB.

6. Vortex all tubes and incubate at room temperature for 30 min.

7. Add 0.075 ml of 6.67% bentonite to all tubes. Vortex and incubate at room temperature for 15 min.

8. Centrifuge the tubes in a microcentrifuge at approximately 2000 $g$ for 4 min.

9. Transfer 0.15 ml of the supernatant from each tube into individual scintillation vials. Add 8 ml of scintillant to each vial.

10. Determine the radioactivity in each vial in a scintillation counter.

*Calculation of Enzyme Activity*

The activity in each sample is calculated according to the following formula:

$$\text{Activity} = \frac{(\text{cpm}_{\text{sample}} - \text{cpm}_{\text{sample blank}}) \times \text{assay volume}}{\text{cpm}_{\text{substrate}} \times \text{time (in min)} \times \text{volume counted} \times \text{mg protein}}$$

in which cpm$_{\text{sample}}$ and cpm$_{\text{sample blank}}$ are the counts per minute (cpm) associated with the sample and sample blank tubes, respectively, and cpm$_{\text{substrate}}$ is the counts per minute per picomole of radioactive substrate (typically about 1500 cpm). Volumes (in milliliters) are of the total stopped-reaction mixture in the numerator and of the aliquot counted in the denominator. Incubation time is in minutes.

# [45] Determination of Biotinidase Activity with Biotinyl-6-aminoquinoline as Substrate

*By* Kou Hayakawa, Kazuyuki Yoshikawa, Jun Oizumi, and Kunio Yamauchi

## Introduction

Biotinidase (EC 3.5.1.12) is a hydrolytic enzyme (amidase), which hydrolyzes biocytin to produce biotin and lysine. The enzyme also hydrolyzes some artificially synthesized biotin derivatives such as biotin 4-amidobenzoate (BPAB; **1**)[1] and biotin 6-amidoquinoline (BAQ; **2**).[2] Furthermore,

biotinidase from human serum has been purified as a single $M_r$ 76,000 polypeptide,[3] and found to be a sialoglycoprotein.[4] Human serum biotinidase is a typical thiol-type enzyme.[5] This enzyme also recognizes lipoyl-*p*-aminobenzoic acid (LPAB)[6–9] and opioid neuropeptide of enkephalin (aminoexopeptidase activity).[10]

A sensitive high-performance liquid chromatographic (HPLC)–fluorimetric biotinidase assay method, which uses BPAB as substrate, has been developed.[11] The method is relatively free from interference owing to prior separation by HPLC, and enables detection of biotinidase activity

---

[1] J. Knappe, W. Brümmer, and K. Biederbick, *Biochem. Z.* **338**, 599 (1963).
[2] H. Wastell, G. Dale, and K. Bartlett, *Anal. Biochem.* **140**, 69 (1984).
[3] D. V. Craft, N. H. Goss, N. Chandramouli, and H. G. Wood, *Biochemistry* **24**, 2471 (1985).
[4] J. Chauhan and K. Dakshinamurti, *J. Biol. Chem.* **261**, 4268 (1986).
[5] K. Hayakawa and J. Oizumi, *J. Biochem. (Tokyo)* **103**, 773 (1988).
[6] J. Oizumi and K. Hayakawa, *Biochem. Biophys. Res. Commun.* **162**, 658 (1989).
[7] C. L. Garganta and B. Wolf, *Clin. Chim. Acta* **189**, 313 (1990).
[8] L. Nilsson and E. Ronge, *Eur. J. Clin. Chem. Clin. Biochem.* **30**, 119 (1992).
[9] L. Nilsson and B. Kagedahl, *Biochem. J.* **291**, 545 (1993).
[10] J. Oizumi and K. Hayakawa, *Biochim. Biophys. Acta* **1074**, 433 (1991).
[11] K. Hayakawa and J. Oizumi, *J. Chromatogr.* **383**, 148 (1986).

in turbid specimens, such as human breast milk[12,13] and porcine cerebrum.[14] Biotinidases from these two specimens have already been isolated, and these biotinidases show a distinctly different molecular weight ($M_r$ 68,000) from that of human serum ($M_r$ 76,000).[13,14] The method measures the intrinsic fluorescent activity of one of the PAB products at a relatively short ultraviolet (UV) wavelength (excitation, 276 nm; emission, 340 nm).

Another fluorescent synthetic substrate using BAQ for the biotinidase assay was previously described by Wastell et al.[2] In their method the intensity of the fluorescent signal is measured at a pair of wavelengths (excitation, 350 nm; emission, 550 nm) longer than those of 4-aminobenzoate (PAB) (excitation, 276 nm; emission, 340 nm).

This fluorescent biotinidase substrate of BAQ was combined with an HPLC–fluorimetric biotinidase assay method.[11,12] The changes in the detection wavelength and in the gradient program for HPLC analysis were sufficient for the application. The analysis time for each sample was as short as 12 min. This method, using BAQ instead of BPAB, resulted in a remarkable improvement in accuracy, because detection can be performed in the visible region. This analytical method proved to be an accurate way to determine biotinidase (BAQ hydrolase) activity even at low levels.[15] Applications to human cerebrospinal fluid[16] and human urine[17,18] have been achieved.

## Assay Methods

### Specimens

Human cerebrum (autopsy; 1 year 2 months, male, died of an illness; from the Showa University School of Medicine, Tokyo, Japan)
Human livers (autopsies; died from liver diseases; from the Gunma University School of Medicine, Maebashi, Gunma, Japan)
Human intestine (autopsy; died from liver cirrhosis; from the Gunma University School of Medicine, Maebashi, Gunma, Japan)
Human serum (from healthy volunteers)

[12] J. Oizumi and K. Hayakawa, Am. J. Clin. Nutr. 48, 295 (1988).
[13] J. Oizumi, K. Hayakawa, and M. Hosoya, Biochimie 71, 1163 (1989).
[14] J. Oizumi and K. Hayakawa, Arch. Biochem. Biophys. 278, 381 (1990).
[15] K. Hayakawa, K. Yoshikawa, J. Oizumi, and K. Yamauchi, J. Chromatogr. 617, 29 (1993).
[16] C. DeFelice, K. Hayakawa, K. Nihei, S. Higuchi, T. Tanaka, T. Watanabe, and I. Hibi, Brain Dev. (Tokyo) 16, 156 (1994).
[17] C. DeFelice, K. Hayakawa, T. Tanaka, T. Watanabe, I. Hibi, T. Kohsaka, and K. Itoh, J. Liquid Chromatogr. 17, 2641 (1994).
[18] C. DeFelice, K. Hayakawa, T. Tanaka, and E. Terentyeva, Nephron 70, 115 (1995).

Human milk (from the Tokyo Boshi-hoken-in Hospital, Tokyo, Japan)
Human bile (from the Saiseikai Maebashi Hospital, Maebashi, Gunma, Japan)
Human cerebrospinal fluid (from the National Children's Hospital, Tokyo, Japan)
Human urine (from a healthy volunteer)
Porcine liver and serum (from Pel-Freez Biologicals, Rogers, AR)
Cockroach (caught from this laboratory)
The specimens were stored at −80°.

*Instruments.* A Waters (Milford, MA) model 600 HPLC with a gradient elution unit is used. The column is a 50 × 4.0 mm i.d. stainless steel tube packed manually with spherical, 10-$\mu$m silica gel particles chemically bonded with octadecylsilane (ODS) (Develosil ODS-10; Nomura Chemical Co., Seto, Aichi, Japan). The sample injection unit used is a diaphragm-type injector (model U6K; Waters) with a 2-ml sample loading loop. Detection is carried out with a fluorimeter (F-3000; Hitachi Co., Tokyo, Japan) using a flow-through cell (cell volume, 18 $\mu$l).

*6-Aminoquinoline and p-Aminobenzoic Acid Assay.* Stepwise elutions are carried out as shown in Table I. Solvent A is 0.1 $M$ sodium phosphate buffer (pH 2.1).[19] Solvent B (washing solvent) is methanol. Both 6-amino-quinoline (AQ) and *p*-aminobenzoic acid (PAB) assays require 12 min/ sample.

*Biotinidase Assay*

*Reagents*

BPAB containing substrate buffer: BPAB is dissolved at 0.275 m$M$ (106 mg/liter) in 0.1 $M$ sodium phosphate buffer (pH 7.0) containing 1 m$M$ (452 mg/liter) EDTA-4Na salt and 10 m$M$ (781 mg/liter) 2-mercaptoethanol

BAQ containing substrate buffer: BAQ is dissolved at 0.044 m$M$ (16.3 mg/liter) in the previously described phosphate buffer

*Procedure.* Substrate containing the reaction buffer of 0.09 ml is mixed with 0.01 ml of enzyme solution (10 × 75 mm Pyrex test tube). Thus, the reaction mixture (0.10 ml) contains 0.00147 mg (3.96 nmol) of BAQ, 0.040 mg of EDTA, and 0.070 mg of 2-mercaptoethanol, respectively. The reaction proceeds for an appropriate time interval at 37 or 22°, and is stopped by adding 0.20 ml of acetonitrile–methanol (1:1, v/v) (i.e., the reaction mixture is diluted threefold with acetonitrile–methanol solution to precipi-

[19] W. Mönch and W. Dehnen, *J. Chromatogr.* **147,** 415 (1978).

TABLE I
ELUTION PROGRAMS ROUTINELY USED FOR BAQ AND BPAB HYDROLASE ACTIVITIES[a,b]

| Time (min) | Flow rate (ml/min) | Solvent A (%) | Solvent B (%) | Curve type |
|---|---|---|---|---|
| **Program 1: For BAQ hydrolase activity (AQ assay)** | | | | |
| 0.0 | 3.00 | 100 | 0 | b |
| 4.0 | 3.00 | 0 | 100 | 11 |
| 5.00 | 3.00 | 0 | 100 | 11 |
| 5.01 | 3.00 | 100 | 0 | 1 |
| 10.0 | 3.00 | 100 | 0 | 1 |
| **Program 2: For BPAB hydrolase activity (PAB assay)** | | | | |
| 0.0 | 1.50 | 100 | 0 | b |
| 2.5 | 1.50 | 100 | 0 | 11 |
| 2.51 | 3.00 | 0 | 100 | 6 |
| 5.50 | 3.00 | 0 | 100 | 1 |
| 5.51 | 3.00 | 100 | 0 | 1 |
| 9.50 | 3.00 | 100 | 0 | 1 |
| 9.51 | 1.50 | 100 | 0 | 1 |

[a] Amended version for acidic phosphate buffer eluent. Proportioning valves were used; solvent A, 0.1 $M$ sodium phosphate buffer (pH 2.1); solvent B, methanol. Curve types in numbers were according to the manufacturer (i.e., 1 and 11 were step gradient and 6 was linear gradient).

[b] Amended and reprinted from *J. Chromatogr.* **617,** K. Hayakawa, K. Yoshikawa, J. Oizumi, and K. Yamauchi, p. 32. Copyright 1993 with kind permission from Elsevier Science–NL, Sara Burgerhartstraat 25, 1055 KV Amsterdam, The Netherlands.

tate the enzyme proteins). After centrifugation and deproteinization at 1500 $g$ for 15 min at 4°, a portion (0.005 ml) of the clear supernatant (0.30 ml) is injected into the HPLC system under Program 1 (Table I). The supernatant is stored either at room temperature or at 4°. The product of AQ is measured at an excitation wavelength of 350 nm and an emission wavelength of 550 nm.[2,20] External AQ standard solution (0.005 ml containing 20 pmol of AQ) is also injected into the HPLC system. One cycle of AQ determination by HPLC requires 12 min.

BPAB hydrolase activity is also assayed by using BPAB containing substrate buffer as shown previously. A portion of the sample solution (0.005 ml of the supernatant after centrifugation) is also injected onto the HPLC system under gradient Program 2 (Table I). One cycle of PAB determination also requires 12 min.

[20] P. J. Brynes, P. Bevilacqua, and A. Green, *Anal. Biochem.* **116,** 408 (1981).

*Purification of Human Serum Biotinidase.* Human serum biotinidase is purified from 5 liters of human serum by the procedure described in Ref. 21; 1.56 mg of protein of human serum biotinidase is obtained. Serum is sequentially purified in five steps: ammonium sulfate precipitation; DEAE-Cellulofine (Chisso, Tokyo, Japan) ion-exchange chromatography; Sephadex G-200 (Pharmacia, Piscataway, NJ), gel permeation chromatography (GPC), and hydroxyapatite HPLC (Koken, Tokyo, Japan). Biotinidases are purified without using EDTA and 2-mercaptoethanol in order to maintain the enkephalin hydrolysis activity.[10]

*Kinetic Study.* Michaelis constants are determined by the calculation method of Lineweaver and Burk.[22] The substrate concentrations are 40, 20, 15, 12.5, and 10 $\mu M$.

*Protein Content.* Protein content is determined by the bicinchoninic acid (BCA) protein assay kit (Pierce Chemical Co., Rockford, IL), using bovine serum albumin as a standard.

*Definition of Unit and Specific Activity.* One unit is defined as the amount of enzyme that hydrolyzes 1 $\mu$mol of substrate (BAQ or BPAB) per minute. Therefore, specific activity (picomoles per minute per milligram of protein) is expressed as microunits per milligram of protein.

### Results of Application of Biotin 6-Aminoquinoline Assay Method to Various Specimens

The HPLC biotinidase (BPAB hydrolase) assay has been used in our laboratory since 1986.[11] The method prevents interference due to turbidity, and has been successfully applied to human milk[12,13] and various tissues, such as homogenates of guinea pig livers[23] and pig cerebrum.[14] However, because the wavelengths that we used for the detection of PAB were similar to those of various nucleotides, the possibility of such interference remained. Indeed, the retention times of nucleotides by reversed-phase HPLC separation are close to those of PAB. Therefore, we used a more suitable fluorimetric substrate, BAQ,[2] for the biotinidase HPLC assay.

As shown in Table I, the same HPLC system can be used for both PAB and AQ assays, simply by changing the elution program and the wavelengths of fluorimetric detection. The detection limit for AQ was 2 pmol [signal-to-noise ($S/N$) of 3] in this system using fluorimetric detection at an excitation wavelength of 350 nm and an emission wavelength of 550 nm. The $S/N$ for AQ was 6 with 4 pmol of the compound, compared with an $S/N$ of 3 with PAB under the same analytical conditions. The detection threshold could

[21] K. Hayakawa, C. DeFelice, T. Watanabe, and T. Tanaka, *J. Chromatogr.* **616,** 327 (1993).
[22] H. Lineweaver and D. Burk, *J. Am. Chem. Soc.* **56,** 658 (1934).
[23] J. Oizumi and K. Hayakawa, *Biochim. Biophys. Acta* **991,** 410 (1989).

thus be halved. The intraassay coefficient of variation (CV) was 0.8% ($n = 6$) when 50 pmol of AQ was used.

The biotinidase reaction was studied using BAQ as substrate. The biotinidase reaction was found to proceed stoichiometrically.[15] The results of mechanistic studies using BAQ as substrate with purified human serum biotinidase are summarized in Table II. As a reference for affinity, kinetic results on another three biotinidase substrates (biocytin, BPAB, and enkephalin) are also included in Table II. These data indicate that there are differences in $K_m$ values among biotinidase substrates. There are also kinetic differences that are mainly due to the thiol nature of human serum biotinidase[5]; however, the $K_m$ for BAQ is most stably measurable. The effect of storage of enzyme on $K_m$ measurement for BAQ is also shown in Table II (i.e., the effect of the storage of the enzyme occurs mainly on $V_{max}$ values). Thus, BAQ as substrate is expected to be a useful tool with which

TABLE II

DETERMINATION OF MECHANISTIC PARAMETERS FOR BAQ AND SUBSTRATES USING PURIFIED HUMAN SERUM BIOTINIDASE[a]

| Substrate | $K_m$ ($\mu M$) | $V_{max}$ (microunits/mg) | $k_{cat}/K_m$ (sec$^{-1}$ $M^{-1}$) |
|---|---|---|---|
| BAQ | 12[b] | 2,468[b] | 260.5[b] |
|  | 10[c] | 6,560[c] | 798[c] |
|  | 10[d] | 3,220[d] | 401[d] |
|  | 10[e] | 1,350[e] | 171[e] |
| BPAB | 30[f] | 38,000[f] | 1,599[f] |
|  | 13[d] | 8,520[d] | 831[d] |
| Biocytin | 5.1[f] | 238[f] | 59.4[f] |
|  | 7.3[c] | 11,100[c] | 1,936[c] |
|  | 13.8[d] | 5,320[d] | 490[d] |
| Methionine-enkephalin | 385[f] | 2,170[f] | 71.2[f] |
|  | 241[g] | 11,400[g] | 60.2[g] |
|  | 392[h] | 878[h] | 2.24[h] |

[a] The enzyme was purified without 2-mercaptoethanol and EDTA, although these chemicals were included in the reaction mixture.
[b] Data published in Ref. 15. Reaction was at 37°.
[c] At 22°, 0.1 $M$ phosphate buffer (pH 7.0).
[d] At 22°, 0.1 $M$ Tris-HCl buffer (pH 7.0), and the enzymes used were fresh.
[e] As for d, but the enzymes used were stored ($-80°$, 2 years).
[f] Data published in Ref. 10. The enzyme was purified with 2-mercaptoethanol and EDTA. Reaction was at 37°.
[g] At 22°, 0.1 $M$ phosphate buffer (pH 7.0), without 2-mercaptoethanol and EDTA in the reaction buffer.
[h] As for g, but with 2-mercaptoethanol and EDTA in the reaction buffer.

TABLE III
BAQ Activity in Various Specimens

| Specimens | Specific activity (microunits/mg of protein) | |
|---|---|---|
| | BAQ hydrolase | BPAB hydrolase |
| Human cerebrum ($n = 1$; 1 year 2 months, male) | | |
|   Gray matter | 3.0[a] | 9.7[b] |
|   White matter | 2.0[a] | 6.5[b] |
| Human liver ($n = 4$) (liver cirrhosis) | 34.1[a] | ND[c] |
| Human intestine ($n = 1$) | 48.6[a] | ND |
| Human serum ($n = 3$) | 75[d] | 117[d] |
| Human milk ($n = 4$) | 2.8[a] | 17[b] |
| Human bile ($n = 6$) | 1.2[b] | 8.2[b] |
| Human urine ($n = 1$) | 82[b] | 290[b] |
| Human cerebrospinal fluid ($n = 1$) | 19[b] | 8[b] |
| Human fibroblast ($n = 3$)[e] | 533 | —[f] |
| Porcine liver ($n = 3$) | 19[b] | 41[b] |
| Porcine serum ($n = 1$) | 193[b] | 194[b] |
| Cockroach ($n = 1$) | 24[b] | 62[b] |

[a] Data published in Ref. 24.
[b] Our unpublished observation.
[c] ND, Not determined.
[d] Activity expressed by volume; $3.36 \times 10^3$ microunits/ml of serum ($n = 3$). Wastell et al.[2] reported $3.34 \times 10^3$ microunits/ml of serum ($n = 15$), determined by assay with a fluorimeter.
[e] Data published in Ref. 2.
[f] —, Not described.

to study the tissue- and species-related enzyme differences in the active center of biotinidase molecule(s).

The new method is directly applicable to the analysis of various biological fluids (human and porcine serum, human and bovine milk) without pretreatment.[15] The biotinidase (BAQ hydrolase) assays have also been applied to human body fluids, such as cerebrospinal fluid[16] and human urine.[17,18] A summary of the specific activities is presented in Table III.[24] As a reference, BPAB hydrolase activity is also indicated. BAQ and BPAB hydrolase of human and porcine cerebrum are present in the soluble cytoplasmic fraction,[14] and that of porcine liver is also present in the soluble fraction (our unpublished observation, 1996). On the other hand, the enzyme of guinea pig livers is present in the membrane subfractions.[14,23]

[24] K. Yoshikawa, K. Hayakawa, N. Katsumata, T. Tanaka, T. Kimura, and K. Yamauchi, J. Chromatogr. B **679**, 41 (1996).

Therefore, biotinidases of pig and guinea pig seem to show different subcellular distributions. Although species-specific and tissue-specific differences seem to present, the BAQ hydrolase assay is also expected to be a helpful tool in basic research on various diseases in humans.

Acknowledgment

This work was supported by the Ministry of Health and Welfare, Japan.

## [46] Determination of Serum Biotinidase Activity with Radioiodinated Biotinylamide Analogs

*By* Evangelia Livaniou, Sotiris E. Kakabakos,
Stavros A. Evangelatos, Gregory P. Evangelatos,
and Dionyssis S. Ithakissios

Biotinidase (EC 3.5.1.12) frees biotin from biocytin (*N*-biotinyl-L-lysine) or other biotinyl peptides, liberating this vitamin from dietary sources as well as during its endogenous recycling.[1,2] Biotinidase deficiency may lead to death due to irreversible neurological and other damages.[3,4] Early diagnosis of the disease can be life saving because treatment with biotin can reverse most symptoms and prevent damage.[3-5] Biotinidase activity in biological fluids has been determined by measuring the hydrolysis of either natural or artificial substrates in various assay systems.[6] The most important radioassays reported so far are those using *N*-(*d*-biotinyl)[*carboxyl*-¹⁴C]PABA[7] (PABA, *p*-aminobenzoic acid), [¹⁴C]biocytin,[8] and biotinyl-

[1] D. V. Craft, N. H. Goss, N. Chandramouli, and H. G. Wood, *Biochemistry* **24**, 2471 (1985).
[2] B. Wolf, G. S. Heard, J. R. Secor McVoy, and H. M. Raetz, *J. Inher. Metab. Dis.* **7**(2), 121 (1984).
[3] B. Wolf, R. E. Grier, W. D. Parker, S. I. Goodman, and R. J. Allen, *N. Engl. J. Med.* **308**, 161 (1983).
[4] B. Wolf, R. E. Grier, R. J. Allen, S. I. Goodman, and C. L. Kien, *Clin. Chim. Acta* **131**, 272 (1983).
[5] B. Wolf, R. E. Grier, R. J. Allen, S. I. Goodman, C. L. Kien, W. D. Parker, D. M. Howell, and D. L. Hurst, *J. Pediatr. (St. Louis)* **103**, 233 (1983).
[6] B. Wolf, J. Hymes, and G. S. Heard, *Methods Enzymol.* **184**, 103 (1990).
[7] B. Wolf and J. R. Secor McVoy, *Clin. Chim. Acta* **135**, 275 (1983).
[8] L. P. Thuy, B. Zielinska, L. Sweetman, and W. L. Nyhan, *Ann. N.Y. Acad. Sci.* **447**, 434 (1985).

mono[125I]iodotyramine[9,10] radiotracers in their substrate mixture.[10a] The methods based on [14]C-labeled substrates have been reviewed in this series previously.[6] In this chapter, we describe radioassays for the determination of human serum biotinidase activity by following the hydrolysis of the radiotracer biotinylmono[125I]iodotyramine, which has been properly diluted with a "cold" carrier molecule. Two different substances have been used as carrier molecules. The first, biotinylmonoiodotyramine, is the non-radiolabeled counterpart moiety of the radioiodinated substrate (homogeneous radioassay),[9] whereas the second is considered the natural enzyme substrate biocytin (heterogeneous radioassay).[10]

## Methods

### Materials

All reagents are of analytical grade.

The water is doubly distilled. Carrier-free Na[125]I (17 kCi/g; radiochemical purity, 99.9%; iodates, <2%) is purchased from Nordion Europe (Fleurus, Belgium). Avidin, biocytin, $p$-hydroxymercuribenzoate ($p$-HMB), $N$-hydroxysuccinimidobiotin, and tyramine are products of Sigma Chemical Co. (St. Louis, MO). The thin-layer chromatography (TLC) plates (silica gel 60 without fluorescent indicator), the trifluoroacetic acid (TFA) (amino acid sequencing grade), as well as the acetonitrile, water [high-performance liquid chromatography (HPLC) grade], and all other reagents are obtained from Merck-Schuchardt (Darmstadt, Germany), except as otherwise indicated.

### Reagents

#### Synthesis of Radiotracer Biotinylmono[125I]iodotyramine

The assay radiotracer may be prepared (Fig. 1) using either a two-step or a one-step synthetic protocol. According to the two-step protocol, tyramine is radioiodinated and the mono[125I]iodotyramine produced

[9] S. A. Evangelatos, S. E. Kakabakos, G. P. Evangelatos, and D. S. Ithakissios, *J. Pharm. Sci.* **82,** 1228 (1993).

[10] S. A. Evangelatos, E. Livaniou, S. E. Kakabakos, G. P. Evangelatos, and D. S. Ithakissios, *Anal. Biochem.* **196,** 385 (1991).

[10a] Biotinylmono[125I]iodotyramine and biotinylmonoiodotyramine for $N$-[$\beta$-(4-hydroxy-3-[125I]iodophenyl)ethyl]biotinamide and $N$-$\beta$-(4-hydroxy-3-iodophenyl)ethyl]biotinamide, respectively. Biotinyldi[125I]iodotyramine and biotinyldiiodotyramine for $N$-[($\beta$-(4-hydroxy-3,5-di[125I]iodophenyl)ethyl]biotinamide and $N$-($\beta$-(4-hydroxy-3,5-diiodophenyl)ethyl]biotinamide, respectively.

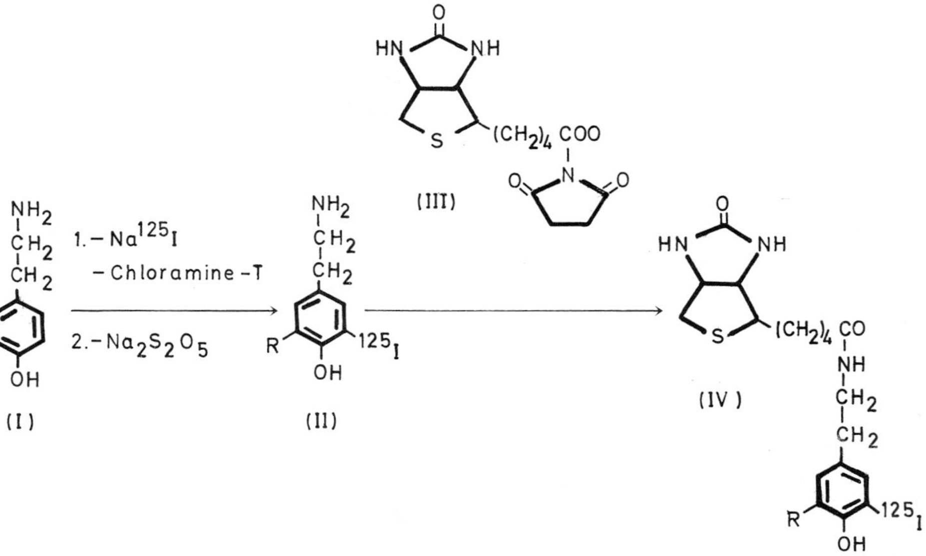

Fig. 1. Synthesis of the radiotracer biotinylmono[$^{125}$I]iodotyramine (**IV**, R = H). Tyramine (**I**); mono[$^{125}$I]iodotyramine (**II**, R = H); N-hydroxysuccinimidobiotin (**III**). Adapted by permission of the Society of Nuclear Medicine from: Livaniou E, Evangelatos GP, Ithakissios DS. Radioiodinated biotin derivatives for in vitro radioassays. *Journal of Nuclear Medicine* 1987; 28:1430–1434.)

is isolated from the reaction mixture using paper electrophoresis (first step); the mono[$^{125}$I]iodotyramine is then biotinylated, and the biotinyl-mono[$^{125}$I]iodotyramine synthesized is isolated by analytical TLC (second step). According to the one-step protocol, biotinylation proceeds immediately after radioiodination, without purifying the intermediate product mono[$^{125}$I]iodotyramine, and the biotinylmono[$^{125}$I]iodotyramine finally synthesized is isolated by analytical HPLC.

*Two-Step Protocol.* Ten microliters of Na$^{125}$I solution (250 Ci/liter in 0.25 $M$ phosphate buffer, pH 7.5) and 5 $\mu$l of chloramine-T solution (3.55 m$M$ in 0.25 $M$ phosphate buffer, pH 7.5) are incubated, in a 3-ml test tube, with 10 $\mu$l of aqueous tyramine solution (0.73 m$M$) for 70 sec at 22°.[11] The reaction is terminated by adding 5 $\mu$l of sodium metabisulfite solution (5.26 m$M$ in 0.25 $M$ phosphate buffer, pH 7.5).

[11] E. Livaniou, G. P. Evangelatos, and D. S. Ithakissios, *J. Nucl. Med.* **28**, 1430 (1987).

The mono[$^{125}$I]iodotyramine synthesized is isolated from the reaction mixture by paper electrophoresis as follows.[12]

The radioiodination mixture (in two 15-$\mu$l aliquots) is put in the middle of an electrophoresis paper strip (3MM, 34 × 3.5 cm; Whatman, Clifton, NJ) and undergoes electrophoresis (500 V, 0.2 $M$ barbital buffer, pH 8.4) for 120 min at 4°. The radioactive band of interest (moved 5 cm from the origin) is cut off, eluted (three 0.5-ml washes) with ethanol–water (2:1, v/v), dried under a stream of nitrogen (providing carrier-free mono[$^{125}$I]iodotyramine with an estimated specific radioactivity of 2500 Ci/mmol), and redissolved in 3 ml of borate buffer (0.2 $M$, pH 8.5) to obtain a 2.33 $\mu M$ solution.

Thirty microliters of the above mono[$^{125}$I]iodotyramine solution is added to 90 $\mu$l of N-hydroxysuccinimidobiotin solution (7 $\mu M$ in anhydrous dioxane), and the mixture is incubated for 90 min at 22°.

The biotinylmono[$^{125}$I]iodotyramine synthesized is isolated from the reaction mixture by analytical TLC[9,11] [0.25-mm thick silica gel plates; solvent system: $n$-butanol–NH$_4$OH (2 $M$)–ethanol, 3:1:1, v/v/v]. The radioactive spots are detected by exposing the plate for 2 min to X-ray film (industrial D-7; Agfa-Gevaert, Leverkusen, Germany). The radiotracer biotinylmono[$^{125}$I]iodotyramine is obtained by scraping off the spot area around $R_f$ 0.67 and eluting with 1 ml of an equivolume ethanol–water solution.

*One-Step Protocol.* Tyramine is radioiodinated as described in the two-step method.[11] Without any previous purification, the pH of the radioiodination mixture is adjusted to pH 8.5 with 30 $\mu$l of borate buffer (0.2 $M$, pH 9.0), 90 $\mu$l of N-hydroxysuccinimidobiotin solution (0.98 m$M$ in anhydrous dioxane) is added, and the mixture is incubated for 90 min at 22°.

The biotinylmono[$^{125}$I]iodotyramine synthesized is isolated from the reaction mixture by analytical HPLC as follows (Fig. 2).[13]

Twenty-microliter aliquots of the reaction mixture are injected into a Waters (Milford, MA) model 600E HPLC system, connected with a Merck (Darmstadt, Germany) Li-Chrospher RP-18 column (250 × 4.6 mm i.d., 5-$\mu$m particle size), which is protected by a Waters (Milford, MA) $\mu$Bondapak C$_{18}$ HPLC precolumn insert, and equipped with a Beckman (Palo Alto, CA) model 171 $\gamma$-radioactivity detector. Solvent A is 0.05% (v/v) TFA in water and solvent B is 60% (v/v) acetonitrile in solvent A. A linear gradient from 10 to 53% solvent B within 15 min is first applied, and then an isocratic

[12] S. E. Kakabakos, E. Livaniou, S. A. Evangelatos, G. P. Evangelatos, and D. S. Ithakissios, *Eur. J. Nucl. Med.* **18,** 952 (1991).

[13] E. Livaniou, A. Roboti, S. E. Kakabakos, J. Nyalala, G. P. Evangelatos, and D. S. Ithakissios, *J. Chromatogr. B* **656,** 215 (1994).

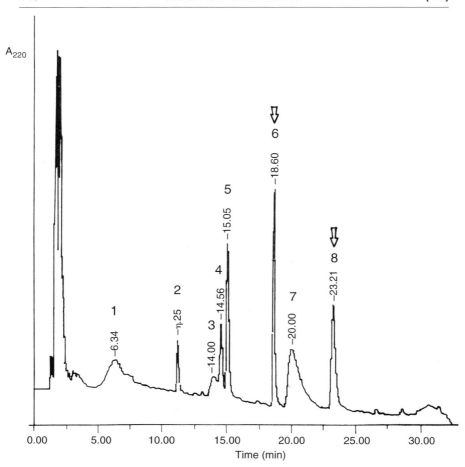

$A_{220}$

FIG. 2. HPLC chromatogram obtained for a mixture consisting of tyramine (1), biotin (2), monoiodotyramine (3), biotinyltyramine (4), $N$-hydroxysuccinimidobiotin (5), biotinylmono-iodotyramine (6, open arrow), diiodotyramine (7), and biotinyldiiodotyramine (8, open arrow). The chromatographic conditions are described in text. (Reprinted with permission from *J. Chromatogr. B* **656,** E. Livaniou, A. Roboti, S. E. Kakabakos, J. Nyalala, G. P. Evangelatos, and D. S. Ithakissios, 215, Copyright 1994 with kind permission of Elsevier Science–NL, Sara Burgerhartstraat 25, 1055 KV Amsterdam, The Netherlands.)

one, at 53% solvent B, is applied for 5 min, followed by a linear gradient from 53 to 100% solvent B within 10 min; the flow rate is 1.6 ml/min. The radiotracer biotinylmono[$^{125}$I]iodotyramine is obtained by collecting the radioactive peak area eluted at 18.6 min retention time and concentrated to 1 ml.

*Substrate Working Solutions*

*Homogeneous Radioassay.* To prepare the substrate working solution, the radiotracer biotinylmono[$^{125}$I]iodotyramine is diluted with 0.5 $M$ phosphate buffer, pH 6.5, containing 0.18 m$M$ biotinylmonoiodotyramine, to give a final specific radioactivity of 3.7 mCi/mmol.

*Preparation of biotinylmonoiodotyramine:* Monoiodotyramine is synthesized by adding slowly, at 22°, 2 ml of 1 $N$ iodine solution into 20 ml of a tyramine solution [50 m$M$ in 30% (v/v) ammonia]. The solution is condensed under reduced pressure to almost 5 ml and left overnight at 0°. The precipitate formed is filtered, washed with ice-cold water, and recrystallized twice from water. Monoiodotyramine is further purified using preparative TLC [2.0-mm thick silica gel plates; solvent system: $n$-butanol–NH$_4$OH (2 $M$)–ethanol, 3:1:1, v/v/v]. Compounds on the TLC plate are detected by spraying a part of the plate with Folin–Ciocalteu reagent. Monoiodotyramine is obtained by scraping off the spot area at $R_f$ = 0.46, eluting with an equivolume solution of ethanol and water, and drying under reduced pressure. The purified monoiodotyramine (almost colorless needles) is dissolved in 0.2 $M$ borate buffer, pH 8.5, to give a 0.178 $M$ solution, 500 $\mu$l of which is added to 2.3 ml of an $N$-hydroxysuccinimidobiotin solution (0.35 $M$ in anhydrous dioxane). The mixture is incubated for 90 min at 22°. The biotinylmonoiodotyramine synthesized is isolated by preparative TLC [2.0-mm thick silica plate; solvent system: $n$-butanol–NH$_4$OH (2 $M$)–ethanol, 3:1:1, v/v/v]. Compounds on the TLC plate are detected as previously described, by using the Folin–Ciocalteu reagent. The biotinylmonoiodotyramine product is obtained by scraping off the spot area around $R_f$ = 0.67 and eluting with an equivolume solution of ethanol and water. After evaporating the solvent and recrystallizing twice from water the cream-colored crystals obtained are characterized by infrared (IR) and nuclear magnetic resonance (NMR) spectroscopy.[9]

*Heterogeneous Radioassay.* To prepare the substrate working solution, the radiotracer biotinylmono[$^{125}$I]iodotyramine is diluted with 0.5 $M$ phosphate buffer, pH 6.5, containing a 0.18 m$M$ concentration of the commercially available biotinidase substrate biocytin to give a final radiotracer concentration of 332 p$M$.

Avidin solution: Avidin, with a specific activity of 13,000 units (U)/g (1 unit binds 1 $\mu$g of biotin) is dissolved in water to give a 10% (w/v) solution

Biotinidase inhibitor suspension: $p$-Hydroxymercuribenzoate is added in water to give a 2 m$M$ suspension

Polyethylene glycol (PEG) solution: Prepare by dissolving 200 g of PEG 6000 in 1 liter of 0.05 $M$ phosphate buffer, pH 7.4, containing 0.1% (w/v) NaN$_3$

*Specimens*

Human sera should be assayed immediately after collection or stored immediately at $-20°$ for up to 2 months.

*Biotinidase Radioassay Protocol*

One hundred microliters of unknown serum sample (diluted 1:25, v/v, in 0.1 $M$ phosphate buffer, pH 6.5), or 100 $\mu$l of reference pooled serum with completely inactivated biotinidase at zero time (diluted 1:25, v/v, in 0.1 $M$ phosphate buffer, pH 6.5), and 50 $\mu$l of substrate working solution are pipetted into 3-ml polystyrene tubes and incubated for 30 min at 37°. After terminating the enzymatic reaction with 25 $\mu$l of biotinidase inhibitor suspension, 25 $\mu$l of avidin solution is added and incubated for 30 min at 37°. Fifty microliters of protein source (rabbit or human serum) and 2 ml of PEG solution are added, the mixture is incubated for 15 min at 22°, centrifuged (10 min, 3000 $g$, 22°), and the radioactivity of the precipitate is measured.

*Data Analysis and Interpretation*

The enzymatic activity ($B_\alpha$) is determined according to the equation

$$B_\alpha = A(\mathrm{Pr} - P)/T$$

and expressed as nanomoles of substrate cleaved per minute per milliliter (homogeneous radioassay), or as femtomoles of radiotracer cleaved per minute per milliliter (heterogeneous radioassay), where $A$ is the product of the substrate concentration (9 nmol/assay tube, homogeneous radioassay) or the radiotracer concentration (16.6 fmol/assay tube, heterogeneous radioassay), the incubation time (30 min), and the volume of undiluted serum used (4 $\mu$l); Pr is the mean radioactivity of the precipitate of reference pooled serum samples (cpm); $P$ is the mean radioactivity of the precipitate of unknown samples (cpm); and $T$ is the total radioactivity added per assay tube (cpm).

Thus, in the homogeneous radioassay, $A$ is equal to 75 nmol min$^{-1}$ ml$^{-1}$; in the heterogeneous radioassay, $A$ is equal to 138.3 fmol min$^{-1}$ ml$^{-1}$.

Comments

Various assays for determining biotinidase activity, using either natural or artificial substrates, have been reported in the literature. Among them, sensitive radioassays using $^{14}$C-labeled substrates have been described. Despite their high sensitivity, however, these radioassays have not been widely applied, owing to the difficulties in synthesizing [14]C-labeled compounds as well as to the type of [14]C-isotype radioactive characteristics. The radioassay described here uses the [125]I-labeled substrate biotinylmono[[125]I]iodotyramine as radiotracer. This compound, apart from offering high analytical sensitivity and optimal radioisotope characteristics, can also be easily prepared (Fig. 1). The previous radiosubstrate has been selected between biotinylmono- and biotinyldi-[[125]I]iodotyramine derivatives, which were previously synthesized and evaluated as biotinidase substrates.[9]

Radioiodination of tyramine is performed with Na[125]I, by the well-established chloramine-T method, in which minute amounts of reagents are used. Because this method is not applicable to the preparation of high quantities of iodinated products, nonradioactive iodination of tyramine is achieved with electrophilic substitution of the phenolic ring using 1 $N$ iodine solution.

The ratio of mono- to diiodinated tyramines produced (radioactive or nonradioactive) depends on the molar ratio of tyramine to Na[125]I or nonradioactive iodine used,[12] respectively. Thus, a molar ratio of tyramine to Na[125]I of 7:1 provided 90% mono[[125]I]iodotyramine, whereas a molar ratio of 2:1 provided 80% di[[125]I]iodotyramine. Likewise, in the nonradioactive iodination of tyramine, a molar ratio of tyramine to iodine of 1:1 provided higher than 90% monoiodotyramine, whereas a molar ratio of 1:2.5 led to >95% diiodotyramine. Under the conditions described, the monoiodotyramine derivatives, radioactive or nonradioactive, are mainly obtained. However, small amounts of the diiodotyramine derivatives, which can lead to the corresponding biotinyldiiodotyramine derivatives on biotinylation, are present in the iodination mixture. It is therefore necessary to isolate the final product of interest, because the extent of aromatic substitution in the substrate molecule affects the enzymatic activity, and a slight contamination of the substrate may alter the $K_m$ and $V_{max}$ values estimated. For biotinylmonoiodotyramine, this isolation is achieved by purifying the monoiodotyramine derivative, prior to the biotinylation reaction, by preparative TLC. Concerning the radiotracer biotinylmono[[125]I]iodotyramine, two strategies can be followed depending on the facilities available: One can use the one-step synthetic protocol and purify the radiotracer by analytical HPLC, which is capable of completely isolating biotinylmono[[125]I]iodotyramine from every other substance that may be present in the reaction mixture (Fig. 2). Otherwise, one can use the two-step protocol, at first purifying the mono[[125]I]iodotyramine by paper electrophoresis and then proceeding to the biotinylation reaction.

The $K_m$ value of human serum biotinidase determined for the biotinylmono[[125]I]iodotyramine radiosubstrate is 15.8 $\mu M$ and the corresponding $V_{max}$ value is 27 nmol min$^{-1}$ ml$^{-1}$. However, the values determined for the biotinyldi[[125]I]iodotyramine substrate are 27 $\mu M$ and 8.7 nmol min$^{-1}$ ml$^{-1}$, respectively. Thus, biotinylmono[[125]I]iodotyramine was considered a better biotinidase substrate. The optimum pH of the enzymatic reaction with both the mono- and diiodinated substrates is in the range pH 6.0–6.8.

Biocytin can be used as a "cold" carrier molecule, although being structurally different from the radiosubstrate, because it was shown to inhibit the enzymatic cleavage of the latter ($K_i$ 7.3 $\mu M$), competing with it for the same active site of human serum biotinidase.[9] Thus, in the working substrate solution, biotinylmono[[125]I]iodotyramine can be diluted with either its cold counterpart molecule (homogeneous radioassay), or with biocytin (heterogeneous radioassay).

The homogeneous biotinidase radioassay described, even using 100 $\mu$l of a 1:400 diluted serum,[9] is considerably more sensitive than the widely used colorimetric assay. This may prove to be important when neonate

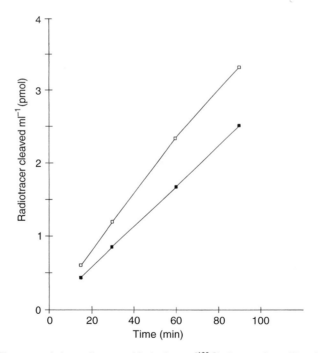

FIG. 3. Cleavage of the radiotracer biotinylmono[[125]I]iodotyramine, diluted with either cold monoiodotyramine (homogeneous substrate, □) or biocytin (heterogeneous substrate, ■), that is caused by serum biotinidase in a period of up to 90 min. (Adapted with permission of the American Pharmaceutical Association, from Ref. 9.)

sera, or other samples of limited quantity, are to be assayed. The enzyme activities determined by the two assays are well correlated,[9] although the values estimated with our homogeneous substrate are almost twofold higher than that determined with the biotinyl-PABA.[4] This difference may be related to the fact that the amino group in the *p*-aminobenzoate molecule is directly connected to the aromatic ring, whereas in iodotyramine a two-carbon atom chain is interposed.[9]

The heterogeneous radioassay was also found to be superior, in terms of sensitivity, compared with the colorimetric method. The results provided by the two assays are also well correlated.[10]

The homogeneous and heterogeneous radioassays developed have excellent correlation,[9] and in some cases may be used alternatively. Nevertheless, the homogeneous substrate provided higher values of reaction velocity, owing to the difference in the enzyme kinetics in each radioassay system. The reaction velocity obtained with human serum biotinidase is constant up to 90 min at 37° for both the homogeneous and the heterogeneous substrates (Fig. 3). The use of PEG solutions in both radioassays as separation reagent provides high reproducibility and improves the assay performance.

In conclusion, the radioassays described are sensitive, fast, and easy to perform, and are proposed as the methods of choice when samples of limited amount and/or low enzyme activity are to be assayed. The homogeneous assay may be helpful for gaining information providing basic insight on the enzyme behavior and characteristics in different biological specimens. However, the heterogeneous radioassay, which uses a substrate that is nonhomogeneous in nature but more easily obtained, may be the assay of choice for routine clinical use.

## [47] Preparation and Properties of Anti-Biotin Antibodies

*By* Fortüne Kohen, Hasan Bagci, Geoff Barnard, Edward A. Bayer, Batya Gayer, Daniel G. Schindler, Elena Ainbinder, and Meir Wilchek

Introduction

The affinity of avidin or streptavidin for biotin is one of the strongest interactions ($K_d$ $10^{-15}$ $M$) known in biology. This strong interaction and the tetrameric structure of avidin, which can accommodate four residues of biotin, has led to the widespread exploitation of this system for a variety of purposes, including affinity chromatography, affinity cytochemistry, histochemistry, hybridoma technology, as well as immunoassay procedures.[1,2] Use of the avidin–biotin system, however, may be restricted owing to the high basicity of avidin (p$I$ 10.5) and the presence of carbohydrate moieties on the avidin molecule that cause high levels of nonspecific binding. However, use of streptavidin may reduce some of the nonspecific binding because streptavidin is not glycosylated and exhibits a neutral p$I$.[2]

During the course of our studies on the use of avidin–biotin technology for immunoassay procedures, we were interested in reducing the nonspecific binding properties. As one of the possible approaches, we produced a high-affinity monoclonal antibody to biotin.[3] We then compared the binding properties of the monoclonal anti-biotin with that of streptavidin in the development of noncompetitive immunoassays for polypeptide hormones and haptens. We sequenced and cloned the $V_H$ and the $V_L$ cDNA of this antibody. Comparison of the $V_H$ sequence of the anti-biotin antibody with those of avidin and streptavidin revealed a similarity in the CDR2 and CDR3 regions of the antibody with known biotin-binding motifs in two of the homologous stretches of avidin and streptavidin.[3] The $V_L$ sequence showed no similarity to such stretches of avidin or streptavidin.

## Reagents

Tris(hydroxymethyl)aminomethane(Tris), bovine serum albumin [BSA, radioimmunoassay (RIA) grade], Tween 20, diethylenetriaminepentaacetic acid (DTPA), bovine gamma globulin, Freund's incomplete and complete adjuvant, 1-ethyl-3-(3-dimethylaminopropyl)carbodiimide (WSC, water-soluble carbodiimide), ovalbumin, and biotin-$N$-hydroxysuccinimide are purchased from Sigma (St. Louis, MO). Affinity-purified rabbit anti-mouse immunoglobulin is obtained from Dakopatts (Glostrup, Denmark). Sepharose–protein A and Sephadex G-25 are products of Pharmacia (Uppsala, Sweden). Anti-mouse immunoglobins (used for Ouchterlony immunodiffusion) are from Serotec (Oxford, UK). The labeling reagent [$N$-1-($p$-isothiocyanatophenyl)diethylenetriamine-$N^1,N^2,N^3$-tetraacetic acid (DTTA)], chelated with Eu$^{3+}$ was kindly provided by I. Hemmila (Wallac Oy, Turku, Finland).

[1] M. Wilchek and E. A. Bayer, *Methods Enzymol.* **184,** 746 (1990).
[2] C. J. Strasburger, Y. Amir-Zaltsman, and F. Kohen, in "Non-Radiometric Assays: Technology and Application in Polypeptide and Steroid Hormone Detection" (B. Albertson and F. Hazeltine, eds.), p. 87. Alan R. Liss, New York, 1988.
[3] H. Bagci, F. Kohen, U. Kuşçuoglu, E. A. Bayer, and M. Wilchek, *FEBS Lett.* **322,** 47 (1993).

## Preparation of Europium-Labeled Reagents

### Europium-Labeled Biotinylated Ovalbumin

Dissolve biotin-$N$-hydroxysuccinimide (1.1 mg) in dimethylformamide (50 $\mu$l) and add slowly to a solution of ovalbumin (15 mg in 1 ml of 50 m$M$ sodium phosphate, pH 8.0). Stir the mixture overnight at 4°, and dialyze against 0.15 $M$ phosphate-buffered saline (PBS), pH 8.0. Dialyze a portion (0.130 ml) of the solution against 50 m$M$ carbonate buffer, pH 9.5, and bring to 1 ml with the same buffer. Dissolve 1 mg of the europium-chelated labeling reagent (DTTA) into 500 $\mu$l of double-distilled water, and add 176 $\mu$l of the solution to the biotinylated ovalbumin in carbonate buffer. Stir the reaction mixture overnight at 4°, and purify the labeled protein by gel filtration on Sephadex G-25M (eluent: 50 m$M$ of Tris-HCl buffer, pH 7.75).

### Europium-Labeled Proteins

Add an 80-fold excess of europium labeling reagent (140 $\mu$l, 4 mmol/liter) to a solution of anti-estradiol immunoglobulin G (IgG) [1 mg, clone 2F$_9$[4])] or anti-biotin IgG[3] in 1 ml of carbonate buffer (pH 9.5). Allow the reaction mixture to stir overnight at 4° and purify on Sephadex G-25M as described previously for the europium-labeled, biotinylated ovalbumin. The europium-labeled proteins are stored at 4° in 5 $\mu M$ of Tris buffer, pH 7.7, containing 0.1% (w/v) BSA and 0.1% (w/v) sodium azide.

## Preparation of Monoclonal Anti-Biotin Antibodies

Bovine serum albumin (BSA) is biotinylated with biotin $N$-hydroxysuccinimide ester as described previously,[5] and the biotinylated BSA is used as an immunogen (50 $\mu$g/mouse) to immunize female CB6/black mice (age, 2 months). Subsequently, two booster injections are given using biotin conjugate in Freund's incomplete adjuvant. After 2 months of immunization, the antibody titer is checked using rabbit anti-mouse IgG-coated plates (prepared as described previously[6]), europium-labeled biotinylovalbumin as a label, and time-resolved fluorescence as an end point. Three months after the initial immunization, the spleen cells that show the highest titer of antibodies are fused with a mouse myeloma cell line (NSO, kindly provided by C. Milstein, Cambridge) using the hybridoma technique of

---

[4] G. Barnard and F. Kohen, *Clin. Chem.* **36,** 1945 (1990).
[5] E. A. Bayer and M. Wilchek, *Methods Enzymol.* **184,** 138 (1990).
[6] A. Altamirano-Bustamante, G. Barnard, and F. Kohen, *J. Immunol. Methods* **138,** 95 (1991).

Köhler and Milstein.[7] The culture supernatants of growing hybridomas are screened for antibody activity using rabbit anti-mouse IgG-coated plates and europium-labeled biotinylovalbumin. Four hybridomas, secreting antibodies against biotin, are selected, and the one (clone $F_1$) exhibiting the highest affinity are propagated *in vitro,* and used for RNA preparation. For other purposes, clone $F_1$ is propagated as ascites in pristane-primed mice. Clone $F_1$ belongs to the $IgG_1$ class, and the ascites derived from this clone are purified by affinity chromatography on Sepharose–protein A as described previously.[8]

Immunoglobulin Sequencing

The $F_1$ hybridoma cell line, which secretes anti-biotin antibodies ($IgG_1$), is grown in Dulbecco's modified Eagle's medium (DMEM), supplemented with 10% (v/v) horse serum. Total cellular RNA is prepared from $F_1$ cells ($5 \times 10^6$) using the Nonidet P-40 (NP-40)/sodium dodecyl sulfate (SDS) technique as described by Gough.[9] Poly(A)-containing RNA is purified by affinity chromatography over magnetic oligo(dT).[10] First-strand DNA synthesis is performed by incubating poly(A) mRNA ($\sim$0.1 $\mu$g) at 37° for 90 min with a mixture of 50 m$M$ Tris-HCl (pH 8.3), 50 m$M$ KCl, 10 m$M$ $MgCl_2$, 10 m$M$ dithiothreitol (DTT), RNase inhibitor, 1 m$M$ deoxynucleotides, 2.5 ng of oligo(dT)$_{15}$ as primer, and avian myeloblastosis virus (AMV) reverse transcriptase in a reaction volume of 60 $\mu$l. Ten microliters of the cDNA reaction mixture is used directly in a PCF (polymerase chain reaction) format, using 2.5 units of *Taq* DNA polymerase with 25 pmol of the appropriate primers and deoxynucleotides in Tag reaction buffer[11] from Promege (Madison, WI).

For the heavy chain, the following primers are used:

Mouse $V_H$ forward primer: 5' TGA GGA GAC **GGT GAC** CGT GGT CCC TTG GCC CC

Mouse $V_H$ backward primer: 5' AGG T(C/G)(A/C) A(A/G) **CTG CAG** (C/G)AG TC(A/T) GG

For the light chain, the 5' primer consists of the following sequence:

5' CCC **AAG CTT** GAC ATT GTG GTG ACC CAG TCT CCA 3'

and the 3' primer consists of either:

[7] G. Köhler and C. Milstein, *Eur. J. Immunol.* **6,** 511 (1976).

[8] C. J. Strasburger and F. Kohen, *Methods Enzymol.* **184,** 481 (1990).

[9] N. M. Gough, *Anal. Biochem.* **173,** 93 (1988).

[10] K. S. Jakobsen, E. Breivold, and E. Hornes, *Nucleic Acids Res.* **18,** 12 (1990).

[11] R. Orlandi, D. H. Güssow, P. T. Jones, and G. Winter, *Proc. Natl. Acad. Sci. U.S.A.* **86,** 3833 (1989).

5' CCC **GAA TTC** TTA GAT CTC CAG CTT GGT CCC 3' from the J region, or

5' GCG CCG **TCT AGA** ATT AAC ACT CAT TCC TGT TGA A
3' from the constant region to give the entire $\kappa$ chain

For the heavy and light chain, the cDNA reaction mixture is amplified for 30 cycles, and an ethidium bromide-stained 2% (w/v) agarose gel is used to visualize the PCR fragments. The amplified product is purified by electroelution. For the heavy chain cDNA the product is blunt ended with Klenow enzyme and $T_4$ polynucleotide kinase and is inserted into pBluescriptKS$^-$ vector (Stratagene, La Jolla, CA), which is digested with SmaI and dephosphorylated. The ligation mixture is used to transform *Escherichia coli* TG1 competent cells, and transformants are selected on LB plates, containing ampicillin, isopropylthiogalactoside (IPTG), and 5-bromo-4-chloro-3-indolyl-$\beta$-D-galactopyranoside CX-Gal). From the resulting colonies, DNA minipreparations are performed; the size of the inserts is analyzed by an XbaI/SalI digestion. Clones containing inserts of the correct size (approximately 350 bp) are grown in 5 ml of LB medium. Following growth, cells are collected by centrifugation and lysed. The supernatant fluids are then decanted into a Magic minicolumn (from Promega, Madison, WI), and plasmid DNA is eluted according to the procedure recommended by the manufacturer. The purified DNA (approximately 0.5 $\mu$g) is sequenced on an ABI 373 DNA sequencer (Applied Biosystems, Foster City, CA), using SK or KS primer and cycle-sequencing reactions.[12] For the light chain the PCR products are digested either with HindIII and EcoRI for the variable-region PCR or with HindIII and XbaI for the whole $\kappa$ chain PCR. They are cloned into similarly cut Bluescript. The plasmid DNA is then purified and sequenced using the same conditions as for the heavy chain cDNA. Sequences of the products from the PCR of the variable region and of the whole $\kappa$ chain have been found to be identical.

## $V_H$ and $V_L$ Sequences of Anti-Biotin Clone $F_1$

The sequence of the end of anti-biotin clone $F_1$ is shown in Fig. 1.[12a] A notable feature of the sequence in the CDR2 and CDR3 regions is the similarity of two homologous stretches in avidin and streptavidin, which are known to interact with the bicyclic ring system of biotin (see Fig. 2). Specifically, the second homologous stretch of avidin and streptavidin

---

[12] W. R. McCombie, C. Heiner, J. M. Kelley, M. G. Fitzgerald, and J. D. Gocayne, *J. DNA Sequencing Mapping* **2**, 289 (1992).

[12a] E. A. Kabat, T. T. Wu, M. Reid-Miller, H. M. Perry, and K. S. Gottsman, "Sequences of Proteins of Immunological Interest." U.S. Public Health Service, Bethesda, Maryland, 1987.

Light Chain

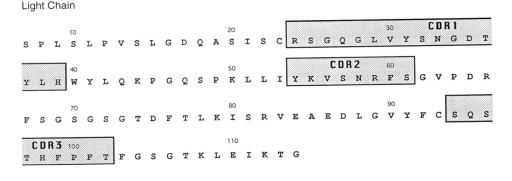

Heavy Chain

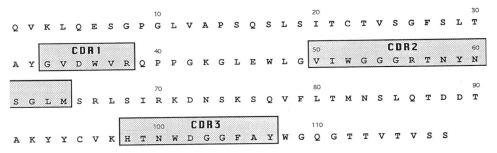

Fig. 1. Deduced amino acid sequence of the $V_H$ and $V_L$ gene of the anti-biotin antibody. Single-letter amino acid code is used. The complementarity-determining regions (CDRs) are indicated by shaded areas. The numbering is according to Kabat et al.[11a]

(which encompasses the β3 strand of both proteins) contains the sequence **XXXYXT(S)XX**, and the CDR2 region of the heavy chain of the antibody contains **RTNYNSGL**. Similarly, the fourth homologous stretch of avidin and streptavidin (which encompasses the β6 strand) contains the sequence **XXTXF(W)XG**, and the CDR3 region of the antibody heavy chain contains **HTNWDG**. These results seem to indicate a similar pattern in the sequences that may dictate the quality of biotin binding.

A comparison of the sequences of the complementarity-determining regions of different antibodies, deposited in the SWISS-PROT Data Bank (Release 23), revealed that the sequence YNS is common in the CDR2 region of the heavy chain. However, we could not find the sequence **TXWXG** in any other CDR3 region of known antibody sequences.

In contrast to the $V_H$ sequence, we could not detect distinct similarity

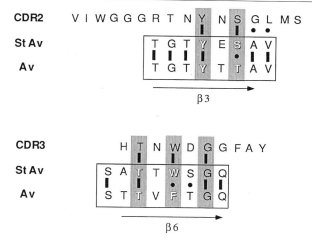

Fig. 2. Alignment of the amino acid sequences of complementarity-determining regions 2 and 3 of the anti-biotin antibody with relevant homologous stretches of avidin (Av) and streptavidin (StAv). Identical residues are denoted by vertical bars and similar conserved residues are indicated by dots. Identity in the antibody is emphasized by shading. Amino acid residues in the binding sites of avidin and streptavidin, which are known to interact with the bicycling ring of the biotin molecule, are shown in white lettering. The second and fourth homologous stretches of avidin and streptavidin encompass the $\beta 3$ and $\beta 6$ strands of the two proteins. [Adapted from *FEBS Lett.* **322**, 49 (1993).]

between any of the CDR regions of the $V_L$ gene and the conserved stretches in either avidin or streptavidin.

## Use of Anti-Biotin Antibodies in Immunoassay Procedures

### Two-Site Immunofluorometric Assay for Human Growth Hormone

#### Reagents

Purified immunoglobulin fractions of monoclonal antibodies against human growth hormone (hGH), which bind to different epitopes of the hGH molecule[13]:

Clone 2, affinity constant, $K_A$ $9 \times 10^9$ $M^{-1}$
Clone 7, affinity constant, $K_A$ $6 \times 10^8$ $M^{-1}$. Clone 7 is biotinylated using the conditions previously described[13]

---

[13] C. J. Strasburger, J. Kostyo, T. Vogel, G. J. Barnard, and F. Kohen, *Endocrinology* **124**, 1548 (1989).

Recombinant hGH: Biotechnology General (Rehovot, Israel)

Europium-labeled anti-biotin and europium-labeled streptavidin: Wallac (Turku, Finland)

Coating buffer: Carbonate buffer, 50 mmol/liter, pH 9.6 containing 2.93 g of anhydrous sodium hydrogen carbonate, 1.59 g of anhydrous sodium carbonate, and 0.2 g of sodium azide, dissolved in 1 liter of double-distilled water

Assay buffer: 50 m$M$ Tris-HCl buffer, pH 7.75, containing BSA (5 g/liter), bovine gamma globulin (0.5 g/liter), 20 $\mu$mol of DTPA, sodium chloride (9 g/liter), Tween 20 (0.5 ml/liter), and sodium azide (0.5 g/liter)

Wash solution: 50 m$M$ Tris-HCl buffer, pH 7.75, containing sodium chloride (9 g/liter), Tween 20 (0.5 ml/liter), and sodium azide (0.5 g/liter)

Enhancement solution: 1 g of Triton X-100, 6.8 mmol of potassium hydrogen phthalate, 100 mmol of acetic acid, 50 $\mu$mol of tri-$n$-octyl-phosphine oxide, and 15 $\mu$mol of 2-naphthoyltrifluoroacetone in 1 liter of double-distilled water

Antibody-coated microtiter strips: Dilute the purified anti-hGH monoclonal antibodies (clone 2) to 5 $\mu$g/ml in coating buffer. Add 200 $\mu$l (1 $\mu$g) of anti-hGH IgG, clone 2, to the wells of polystyrene microtiter strips (Labsystems, Helsinki, Finland). After overnight incubation at 4°, aspirate and discard the coating buffer; wash the strips three times with wash solution. Cover the strips and store them dry at 4° until use

*Procedure for Two-Site Time-Resolved Immunofluorometric Assay.* Prepare hGH standards in concentrations ranging from 0 to 40 ng/ml in assay buffer. Add 200 $\mu$l of standard, in duplicate, to the coated microtiter wells. Incubate the strips for 1 hr at room temperature. Wash the strips three times and add biotinylated anti-hGH monoclonal antibodies [100 ng (clone 7) in 200 $\mu$l of assay buffer]. After a 1-hr incubation at room temperature, aspirate the reaction mixture and wash the strips three times with wash solution. Add europium-labeled streptavidin (20 ng in 200 $\mu$l of assay buffer) or europium-labeled anti-biotin antibody (34 ng in 200 $\mu$l of assay buffer) to the microtiter wells. After a further incubation at room temperature for 30 min in an automatic plate shaker, aspirate the reaction mixture and wash the strips six times with wash solution. Then add 200 $\mu$l of enhancement solution, and agitate the strips on the shaker for 10 min. Finally, measure the fluorescence (we use an Arcus time-resolved fluorometer; Wallac Oy, Turku, Finland).

*Evaluation.* Typical dose–response curves for the two-site immuno-fluorometric assay for hGH using the different europium-labeled biotin-

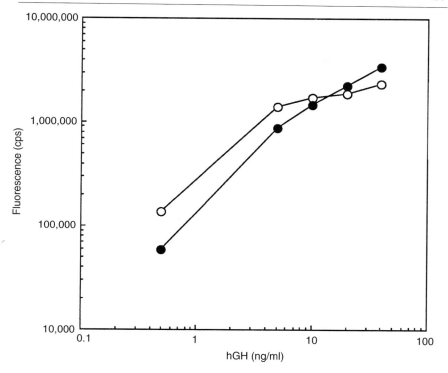

FIG. 3. Dose–response curve for human growth hormone using europium-labeled strepta-vidin (●) or anti-biotin (○) as the labeled protein and fluorescence as an end point.

binding proteins are shown in Fig. 3. The sensitivity of the method (de-fined as the zero-standard signal plus 2 standard deviations) is 0.1 ng/ml. This technique provides a working range of 0.25–40 ng of hGH/ml. When the two labeling reagents were compared, the performance of the two reagents was similar (see Fig. 3).

*Noncompetitive Idiometric Immunoassay for Estradiol Using Streptavidin or Anti-Biotin as Captured Proteins*

*Principle.* In previous studies we have reported the development of a noncompetitive assay, which we termed an "idiometric assay," for the direct measurement of estradiol[4,14] and progesterone[15] in serum, and of the estrone

[14] A. Mares, J. De Boever, J. Osher, S. Quiroga, G. Barnard, and F. Kohen, *J. Immunol. Methods* **181,** 83 (1995).
[15] G. Barnard, J. Osher, S. Lichter, B. Gayer, J. De Boever, R. Limor, D. Ayalon, and F. Kohen, *Steroids* **60,** 824 (1995).

3-glucuronide in diluted urine.[16] This method utilized two types of anti-idiotypic antibody that recognize different epitopes within the hypervariable region of the specific primary anti-estradiol antibody. The first anti-idiotype ($\beta$ type) competes with the analyte for an epitope at the binding site (paratope) of the primary antibody. The second anti-idiotypic antibody ($\alpha$ type) recognizes an epitope within the variable region of the primary antibody ($Ab_1$). The $\alpha$ type is not sensitive to the presence or absence of the analyte, but is sterically hindered from binding to $Ab_1$ in the presence of the $\beta$ type owing to epitope proximity. The availability of these matched antibodies ($\beta$ type, $\alpha$ type, and $Ab_1$) permitted the development of a noncompetitive immunoassay for small molecules.[4,17] Independently, Self[18,19] has described conceptual approaches of noncompetitive assay formats for the measurement of small molecules.

In the present format (see Fig. 4), the europium-labeled $Ab_1$ is incubated with the ligand (estradiol) in streptavidin- or anti-biotin antibody-coated microtiter strips, resulting in a population of both occupied and unoccupied sites. Subsequently, the $\beta$-type anti-idiotypic antibody is added to this reaction mixture to block the unoccupied sites. After a short incubation period, the biotinylated $\alpha$-type anti-idiotypic antibody is added to the microtiter strips. At this stage of the method, the biotinylated $\alpha$-type antibody immobilizes the sites occupied by the analyte. After a short incubation and a wash step, the light yield of the bound fraction is measured by time-resolved fluorescence.

### Reagents

Europium-labeled primary anti-estradiol antibody, clone 2F$_9$: A hybrid monoclonal antibody produced by a rat mouse heterohybridoma[4]

$\beta$-Type, analyte-sensitive, anti-idiotypic antibody clone 1D$_5$[4]: The $\alpha$-type, analyte-insensitive, biotinylated anti-idiotypic antibody, clone 1D$_7$[4]

Streptavidin-coated microtiter strips: Wallac (Turku, Finland)

The coating buffer, assay buffer, wash solution, and enhancement solution are of the same composition as those used in the hGH assay described in the previous section.

[16] G. Barnard, S. Lichter, B. Gayer, and F. Kohen, *J. Steroid Mol. Biol.* **55,** 107 (1995).
[17] F. Kohen, J. DeBoever, and G. Barnard, in "Textbook of Immunological Assays" (E. P. Diamandis and T. K. Christopoulos, eds.), pp. 405–421. Academic Press, San Diego, California, 1996.
[18] C. H. Self, World Intellectual Property Organisation International Publication Number 85/04422 (1985).
[19] C. H. Self, World Intellectual Property Organisation International Publication Number 89/05453 (1989).

A. Capture analyte with the labeled primary antibody

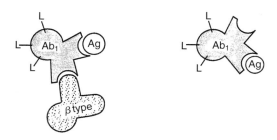

Occupied site          Unoccupied site

B. Add β type to block unoccupied sites

C. Add biotinylated α type

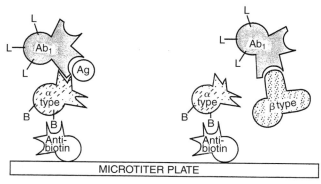

Note: β type sterically
hinders the binding of
the biotinylated α type

MICROTITER PLATE

D. Measure fluorescence of the bound fraction

FIG. 4. Schematic diagram of the noncompetitive immunoassay procedure for haptens, which we have termed an "idiometric assay." L, Europium label.

*Anti-Biotin-Coated Microtiter Strips.* Monoclonal anti-biotin IgG (clone $F_1$, $IgG_1$ class) is purified by affinity chromatography of ascites fluid (containing anti-biotin antibodies) on Sepharose–protein A, using previously described conditions.[8] Dilute the purified IgG to 1.5 μg/ml in phosphate-buffered saline, pH 7.4 (PBS), containing 0.01% (w/v) sodium azide, and add 200 μl of the IgG solution to each well of polystyrene microtiter strips (Maxisorb plates; Nunc, Roskilde, Denmark). After overnight incubation at 4°, decant the coating buffer, and wash the strips twice. Block the strips for 1 hr at room temperature with 2% (w/v) BSA in PBS. Aspirate the blocking solution and wash twice. Store the strips at 4° pending further use.

## Idiometric Assay

Prepare estradiol standards ranging from 10 to 5000 pg/ml in assay buffer. Add europium-labeled anti-estradiol antibody (100 μl, 1 : 1000 dilution) and estradiol standards (50 μl) to each well of anti-biotin- or

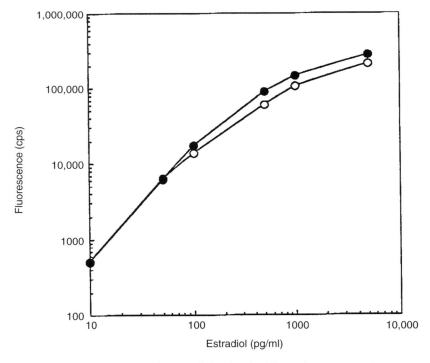

FIG. 5. Dose-response curves for estradiol, using the idiometric assay procedure, streptavidin (●) or anti-biotin (○) as capture proteins, and fluorescence as an end point.

streptavidin-coated microtiter strips. Incubate the reaction mixture with shaking at room temperature for 1 hr. Add the $\beta$-type anti-idiotypic antibody (clone $1D_5$; 50 $\mu$l, 4 : 500, v/v), and incubate for another hour. In the third step of the reaction, add the biotinylated $\alpha$-type antibody (clone $1D_7$; 50 $\mu$l, 1 : 500, v/v) to each well. Incubate the microtiter strips for 1 hr at room temperature on a plate shaker and wash six times. Finally, add enhancement solution (200 $\mu$l) to each well, agitate the strips on the shaker for 10 min, and measure the fluorescence.

*Evaluation.* Typical dose–response curves, using anti-biotin IgG or streptavidin as capture proteins in the idiometric assay for estradiol, are shown in Fig. 5. The sensitivity of the method is 10 pg/ml. The light yield of the bound fraction increased with increasing estradiol concentrations over the range 0.01–2.5 ng/ml. When anti-biotin IgG and streptavidin were compared as capture proteins, the performance of the two reagents was similar (see Fig. 5). This approach allows an efficient and universal capture system, in which the selectivity of the biotinylated capture antibody can be changed according to the desired target ligand. In addition, the idiometric assay offers an extended working range and precision; the method is suitable for dipstick technology and home use. Using such an idiometric assay, we have measured estradiol directly from serum.[14]

## Conclusions

In this chapter we have described the production, molecular characteristics, and performance of a monoclonal anti-biotin antibody preparation. We compared the suitability of the anti-biotin antibody versus those of streptavidin as probes in immunoassay procedures. In one case, the respective biotin-binding proteins served as a europium-labeled detection system, and in the other, as capture proteins for immobilizing the desired biotinylated antibody. The results demonstrate that the monoclonal antibody can be an excellent substitute for streptavidin-based immunoassay systems and provides an alternative probe for avidin–biotin technology.

## Acknowledgments

We thank Dr. Orit Leitner for the generation of monoclonal antibodies against biotin, Malka Kopelowitz for secretarial assistance, and Prof. Israel Hanukoglu and Dr. Moshe Raikhinstein for technical advice on molecular cloning. Part of this work was supported by a grant to M.W. and to E.A.B. from the United States–Israel Binational Science Foundation (BSF), Jerusalem, Israel, and to F.K. from Gesellschaft für Biotechnologische Forschung mBH, Germany.

## [48] Competitive Binding Assays for Biotin-Binding Proteins

*By* Harold B. White III

Radioligand binding assays of egg white avidin and egg yolk biotin-binding protein (BBP) are complicated by the presence of indeterminate amounts of biotin in samples.[1] In each case, the exceptionally tight binding of biotin ($K_d$ values of ~1 f$M$ and ~1 p$M$, respectively) and its slow rate of dissociation mask occupied binding sites from exchange with added radiolabeled biotin. Heat-accelerated exchange to isotopic equilibrium between free and bound biotin at several sample concentrations followed by separation and measurement of protein-bound radioactivity enables the quantitation of total biotin-binding sites and the proportion of those sites occupied by endogenous biotin.

## Method

### Stock Reagents

Sodium acetate (50 m$M$, pH 5.5), containing 50 m$M$ sodium chloride and 0.02% (w/v) sodium azide: Make fresh weekly in deionized water and store at room temperature. Except for [$^3$H]biotin, the following reagents are made up in this buffer and all dilutions used this buffer

$d$-[8,9-$^3$H(N)]Biotin (30 to 50 Ci/mmol), 5 $\mu$Ci/ml in ethanol containing 0.3% by volume of 2-mercaptoethanol and stored at $-20°$: Dilute fivefold into the assay buffer before use

$d$-Biotin, 50 $\mu M$

Avidin, 0.2 mg/ml (~13 U/mg): Store frozen. Dilute to 0.5 $\mu$g/ml before use

Sodium chloride, 1.0 and 2.0 $M$

*Preparation of Samples Containing Avidin or Biotin-Binding Protein.* To be within the range of the assay, dilute egg white samples 200- to 400-fold and egg yolks samples 40- to 200-fold in assay buffer. These ranges may be somewhat different for sources other than chicken.

*Preparation of Phosphocellulose Minicolumns for Separation of Free and Protein-Bound Biotin.* Both avidin and BBP bind to phosphocellulose while unbound biotin does not. (Note that purified chicken BBP does not

---

[1] R. W. Schreiber, Jr., M. A. Letavic, T. J. McGahan, and H. B. White III, *Anal. Biochem.* **192**, 392 (1991).

bind to phosphocellulose and DEAE-cellulose must be used with a different buffer.[2]) Small, reusable ion-exchange minicolumns are prepared from Macro/Micro pipette tips (Sarstedt, Princeton, NJ).[3] The minicolumns are hung vertically in a test tube rack and a small plug of polyester or glass wool is seated in the tip of each. A slurry of phosphocellulose (P-11; Whatman, Clifton, NJ) is equilibrated with assay buffer (settled volume occupies about one-third of the total volume). Add 0.5 ml of the slurry to each minicolumn. The columns are ready to use as soon as the excess buffer has drained. They can be reused almost indefinitely provided they are recycled with 1.0 ml of 2 $M$ sodium chloride followed by 3.0 ml of assay buffer prior to each use.

*Incubation Conditions.* For each sample to be assayed, prepare seven incubation mixtures in $10 \times 75$ mm polypropylene tubes (Sarstedt). Add increasing amounts of diluted sample to successive tubes (e.g., 10, 20, 30, 40, 60, 80, 120 $\mu$l) and assay buffer to a total volume of 900 $\mu$l. Initiate the exchange reaction with 100 $\mu$l of 1.0 $\mu$Ci/ml [3H]biotin, mix, and transfer to a water bath at 85° for 30 min (avidin assay) or 45° for 40 min (BBP assay).

*Blanks and Controls.* Several tubes to measure nonspecific "binding" should be incubated with each set of samples. These include duplicate tubes for binding in the absence of protein, binding in the presence of excess unlabeled biotin (500 $\mu$l of 50 $\mu M$ $d$-biotin), and binding in the presence of protein (120 $\mu$l of sample) and excess biotin (500 $\mu$l of 50 $\mu M$ $d$-biotin). In addition, because not all of the radioisotope is in a bindable form, the total amount of bindable radioactive biotin must be determined in each experiment. This is done by preparing reaction mixtures containing increasing amounts of avidin (40 to 200 $\mu$l of a 0.5-$\mu$g/ml solution of avidin), over a range where [3H]biotin becomes the limiting reagent. A stock solution of egg white, appropriately diluted, can be used in place of purified avidin.

*Separation and Measurement of Protein-Bound [3H]Biotin.* At the end of the incubation, cool the reaction mixtures to room temperature and quantitatively transfer the contents to corresponding minicolumns with two 1.0-ml buffer rinses. Elute and discard into radioisotopic waste any remaining unbound biotin with two successive 1.0-ml buffer washes, making sure that each wash has completely drained before the next is added to the column. Position the rack of minicolumns such that each minicolumn drains directly into a scintillation vial. Elute the protein-bound biotin from the phosphocellulose with two successive 0.5-ml washes of 1.0 $M$ sodium chloride. Add 10 ml of a water-compatible scintillation fluid (e.g., Liquiscint; National Diagnostics, Atlanta, GA) and measure the amount of radioactivity in a liquid scintillation counter.

[2] L. Bush and H. B. White III, *Biochem. J.* **256,** 797 (1988).
[3] H. B. White III, *Methods Enzymol.* **122,** 221 (1986).

Analysis of Data

A hyperbolic equation that describes the binding of radiolabeled biotin ($*B_B$) as a function of sample volume ($V$) at equilibrium can be transformed into the following linear equation [Eq. (1)],

$$*B_T/*B_B = (1/[P_T])(*B_T/V) + B_T/P_T \qquad (1)$$

where $*B_T$ is the total bindable radioactive biotin, $B_T$ is the total unlabeled biotin in the sample, and $[P_T]$ is the concentration of biotin-binding protein. If $*B_T/*B_B$ is plotted as a function of $*B_T/V$, the slope of the graph is the reciprocal of the binding protein concentration, $1/[P_T]$, and the $y$ intercept, $B_T/P_T$, is the fractional saturation of the biotin-binding protein in the original sample.

This analysis applies only for incubation conditions of biotin excess (saturation of biotin-binding sites). Thus any values of $*B_B$ that are equal to $*B_T$ should be omitted from the analysis. The analysis also makes the reasonable assumption that negligible amounts of protein-bound biotin dissociate during the nonequilibrium separation of free and bound biotin.

*Limitations of Assay.* Here, as with other linearly transformed equations such as the Lineweaver–Burk equation for enzyme assays, it is convenient to apply linear regression analysis to the data to obtain values for the slopes and intercepts. However, this is statistically inappropriate because of the unequal weighting of data in double-reciprocal plots.[4] Nonlinear regression analysis of the hyperbolic form of Eq. (2) yields greater precision and accuracy, particularly when there is scatter in the data.

$$*B_B/*B_T = ([P_T]V/*B_T)/(1 + [B_T]V/*B_T) \qquad (2)$$

The high specific radioactivity of commercially available [³H]biotin provides sensitivity in the subpicomolar range, a considerable improvement over [¹⁴C]biotin.[5] However, at the high dilutions, several technical problems can intervene. One of the two forms of chicken BBP becomes unstable at 65° at high dilution and thus the incubation temperature for its assay is 45°.[6] This thermal sensitivity has not been observed for BBP from reptilian egg yolk, for which incubation at 65° is appropriate. Avidin, which has a high isoelectric point, adsorbs to many surfaces, a problem accentuated at high dilution. This problem can be reduced by including another basic protein such as lysozyme (10 $\mu$g/ml) in the assay in addition to the 50 m$M$ sodium chloride.[1]

[4] W. W. Cleland, *Methods Enzymol.* **63,** 103 (1979).
[5] H. Meslar and H. B. White III, *Methods Enzymol.* **62,** 316 (1979).
[6] H. B. White III and C. C. Whitehead, *Biochem. J.* **241,** 667 (1987).

# Author Index

Numbers in parentheses are footnote reference numbers and indicate that an author's work is referred to although the name is not cited in the text.

Ryan, M. J., 400
Rydon, H. N., 135

# S

Sable, H. Z., 135, 149
Sabolic, I., 124
Sacquet, E., 213
Safro, M., 369
Sahashi, Y., 349
Said, H. M., 393, 395–399, 403, 405(6–8)
Saito, H., 343
Saito, J., 203
Sakaitani, M., 255
Sakamoto, O., 388, 393(6)
Sakata, J., 36, 42(12)
Sakurai, N., 344, 346(32)
Sambrook, J., 186, 334, 342, 350
Sanemori, H., 68–69, 75, 80(7)
Sanger, J. M., 401
Santhosh-Kumar, K. L., 13–14, 16, 18(8), 19(8), 21–22, 22(8), 23, 23(5)
Santucci, A., 31, 33(12), 34(12)
Sanyal, I., 341, 344, 349–350, 350(7), 351, 351(7, 8), 352(7, 8), 353(7), 354(7, 8)
Sapper, H., 3
Sarai, K., 68, 75, 80(7)
Saraste, M., 369
Satoh, Y., 348
Satow, Y., 316, 319(13)
Sauberlich, H. E., 52
Saunders, M. E., 388
Sawyer, L., 377, 379, 381, 381(13), 382(18), 383(18), 384(18), 385(18)
Sax, M., 135(15), 140, 143(15), 144
Scarpa, M., 30
Schagger, H., 257
Schellenberger, A., 131–132, 135, 135(17), 136–137, 140, 140(2), 141(5), 143, 143(5), 144, 144(5, 17)
Schenker, S., 119(1), 121
Scheutjens, J. M., 314
Schindler, D., 451
Schmidt, M. S., 428
Schmidt, U., 166
Schnackerz, K. D., 141
Schneider, G., 135(17), 140, 144, 144(17), 376–377, 381, 381(9), 382(17), 383, 383(17), 384(17), 385(17)

Schoffeniels, E., 69, 74–75, 78(8), 79(8), 127, 130
Schray, K. J., 321, 326(4, 5)
Schreiber, R. W., Jr., 464, 466(1)
Schreiner, J. M., 295
Schreurs, W. H. P., 75
Schrijver, J., 69, 75
Schron, C. M., 124
Schruers, W. H. P., 69
Schuhmacher, G., 138
Schulman, S., 397
Schultz, G. A., 144
Schulz, A., 346
Schulz, G. E., 367, 380, 381(14)
Schumacher, S., 135
Schussler, G. C., 399
Schwaiberger, R., 135
Schweingruber, A.-M., 112
Schweingruber, M. E., 112
Schwoch, G., 126
Scott, B. C., 167
Secor McVoy, J. R., 422, 425, 428–430, 430(47), 433(47), 442
Seino, S., 403
Sekiguchi, M., 389
Selavka, C. M., 277
Self, C. H., 460
Seliger, B., 135, 137, 141(5), 143(5), 144(5)
Semasi, E., 401
Semenza, G., 124, 129(12)
Sempuku, K., 116–117
Sensenbrenner, M., 402
Serbinova, E., 167
Serror, P., 421
Severin, S. E., 135
Shapira, E., 425
Sharifan, A., 396
Shaw, N. M., 341, 349, 354(9), 355(9)
Shaw, P. E., 8
Shaw-Goldstein, L. A., 135
Sheehan, J. J., 119(6), 121
Sheehan, T., 4
Sheiner, L. B., 47, 49(13)
Shellhammer, J., 327
Shemin, D., 220
Shen, B., 255
Shen, Z., 182, 183(18)
Sherlock, S., 167
Sherman, D. H., 255
Sherratt, H. S. A., 240, 245, 246(13)

# Subject Index

## A

Acetolactate synthase (ALS)
  catalytic mechanism, general, 131–132
  thiamin diphosphate analogs
    activity assay, 137–138
    aminopyrimidine modifications, 141,
      143–144
    binding constant determination by spectroscopy, 138–139
    mechanistic insights, 144, 146
    synthesis, 136–137
    thiazolium modification and isotope effects, 140–141
    types and inhibition properties,
      134–135
ACP, *see* Acyl-carrier-protein
Acyl-carrier-protein (ACP) synthase
  biological functions, 254–255
  holoenzyme conversion assays
    high-performance liquid chromatography
      chromatography, 260
      materials, 260
      principle, 259
    radioassay
      autoradiography, 259
      materials, 258
      principle, 257–258
      reaction conditions, 258–259
  purification of recombinant *Escherichia coli* apoprotein
    chromatography, 256–257
    holoenzyme preparation, 261–262
    overproduction in *Escherichia coli*,
      255–256
    physical properties, 262
  specificity, 255
Acyl-coenzyme A, *see* Acyl-carrier-protein
  synthase; Acyl-coenzyme A oxidase;

Acyl-coenzyme A synthase; Coenzyme
  A
Acyl-coenzyme A oxidase, synthesis of dicarboxyl-2-enoyl-coenzyme A esters,
  242
Acyl-coenzyme A synthase, synthesis of dicarboxylylmono-coenzyme A esters,
  241–242
Acyldethia(carba)pantetheine 4-phosphate,
  synthesis, 229–232
Adenosine 2′,3′-cyclophosphate 5′-phosphomorpholidate, synthesis, 232–233
Aequorin
  bioluminescence
    avidin effects, 299–300
    biotinylated protein, 299
    calcium triggering, 296
  biotin competitive binding assays with
    avidin–aequorin
    avidin effects on bioluminescence,
      299–300
    emission, 299
    heterogeneous assay, 296–297, 302–303
    homogeneous assay, 300–302
    instrumentation, 297–298
    sensitivity, 303
  commercial sources, 298–299
ALS, *see* Acetolactate synthase
4-Amino-2-nitrobenzoylthiamin
  photoaffinity labeling of thiamin transporter, 116–117
  synthesis, 115–116
L-Ascorbic acid
  antioxidant activity, 30
  ascorbate cascade, 10, 12
  bioavailability determination, 49, 51, 52
  biosynthesis, *see* Glutathione-dependent
    dehydroascorbate reductase;
    L-Gulono-γ-lactone oxidase

ISBN 0-12-182180-3

90038

9 780121 821807

Yu-Hui Rogers